Natural Products and Medicinal Properties of Carpathian (Romanian) Plants

Romanian ethnomedicinal knowledge extends as far back as the 16th century to the Geto-Dacian priests who used medicinal plants and practiced integrated holistic medicine. The ethnomedicine continued in monasteries by monks who used cultivated medicinal plants and wild harvested plants. There are now over 800 species of medicinal plants in Romania. An earlier work last century entitled "Pharmaceutical Botany: the Culture and Harvest of Pharmaceutical Plants" by Grinţescu refers to approximately 500 Romanian healing plants, although most of them are not recognized in modern medicine. There is clear evidence of ethnomedicine in this important region, particularly those that are endangered.

Features:

- Provides an understanding of indigenous plant-derived natural medicines of Romania.

- Discusses selected plant families that are representative members of the most important medicinal plants in the region.

- Includes discussions and critical views on the potential and challenges for further development of the selected plants in a modern setting.

- Details the important plants and organizes the chapters based on either taxonomy or medical use.

- Covers traditional and folk medicine of Romania.

Natural Products Chemistry of Global Plants

Series Editor: Clara Bik-san Lau
Founding Editor: Raymond Cooper

This unique book series focuses on the natural products chemistry of botanical medicines from different countries such as Bangladesh, Borneo, Brazil, Cameroon, China, Ecuador, India, Iran, Laos, Romania, Sri Lanka, Turkey, etc. These fascinating volumes are written by experts from their respective countries. The series will focus on the pharmacognosy, covering recognized areas rich in folklore as well as botanical medicinal uses as a platform to present the natural products and organic chemistry. Where possible, the authors will link these molecules to pharmacological modes of action, reflecting the ethnopharmacological uses. The series intends to trace a route through history from ancient civilizations to the modern day showing the importance to man of natural products in medicines, in foods and a variety of other ways. With special emphasis on plant parts for medicinal uses, phytochemistry and biological activities, this book series will be of useful reference to scientists/ pharmacognosists/ pharmacists/ chemists/ graduates/ undergraduates/ researchers in the fields of natural products, herbal medicines, ethnobotany, pharmacology, chemistry and biology, Furthermore, pharmaceutical companies may also found valuable information on potential herbs and lead compounds for the future development of health supplements and western medicines.

Recent Titles in this Series:

Natural Products Chemistry of Botanical Medicines from Cameroonian Plants, *Xavier Siwe Noundou*

Medicinal Plants and Mushrooms of Yunnan Province of China, *Clara Lau and Chun-Lin Long*

Medicinal Plants of Borneo, *Simon Gibbons and Stephen P. Teo*

Natural Products and Botanical Medicines of Iran, *Reza Eddin Owfi*

Natural Products of Silk Road Plants, *Raymond Cooper and Jeffrey John Deakin*

Brazilian Medicinal Plants, *Luzia Modolo and Mary Ann Foglio*

Medicinal Plants of Bangladesh and West Bengal: Botany, Natural Products, and Ethnopharmacology, *Christophe Wiart*

Traditional Herbal Remedies of Sri Lanka, *Viduranga Y. Waisundara*

Medicinal Plants of Ecuador, *Pablo A. Chong Aguirre, Patricia Manzano Santana, Migdalia Miranda Martínez (Eds)*

Medicinal Plants of Laos, *Djaja Djendoel Soejarto, Bethany Gwen Elkington and Kongmany Sydara*

Edible and Medicinal Mushrooms of the Himalayas: Climate Change, Critically Endangered Species and the Call for Sustainable Development, *Ajay Sharma, Garima Bhardwaj and Gulzar Ahmad Nayik*

Medicinal Plants of Turkey, *Ufuk Koca-Caliskan and Esra Akkol*

Natural Products and Medicinal Properties of Carpathian (Romanian) Plants, *Adina-Elena Segneanu*

Natural Products and Medicinal Properties of Carpathian (Romanian) Plants

Edited By

Adina-Elena Segneanu

Leading Researcher, Department of Biology-Chemistry,
Institute for Advanced Environmental Research-West
University of Timisoara, Romania

CRC Press
Taylor & Francis Group
Boca Raton London New York

CRC Press is an imprint of the
Taylor & Francis Group, an **informa** business

First edition published 2024
by CRC Press
2385 Executive Center Drive, Suite 320, Boca Raton, FL 33431

and by CRC Press
4 Park Square, Milton Park, Abingdon, Oxon, OX14 4RN

Library of Congress Cataloging-in-Publication Data
Names: Segneanu, Adina-Elena, editor.
Title: Natural products and medicinal properties of Carpathian (Romanian)
plants / edited by Adina-Elena Segneanu, Professor, Department of
Biochemical-Pharmaceutical Technology, School of Pharmaceutical
Sciences, University of São Paulo (USP), Brazil.
Description: First edition. | Boca Raton : CRC Press, 2024. |
Series: Natural products chemistry of global plants |
Includes bibliographical references and index. |
Summary: "Romanian ethnomedicinal knowledge extends as far back as the
16th century to the Geto-Dacian priests who used medicinal plants and practiced
integrated holistic medicine. The ethnomedicine continued in monasteries by monks
who used cultivated medicinal plants and wild harvested plants. There are now
over 800 species of medicinal plants in Romania. An earlier work last century
entitled "Pharmaceutical Botany: the Culture and Harvest of Pharmaceutical Plants"
by Grințescu refers to approximately 500 Romanian healing plants, although most of
them are not recognized in modern medicine. There is clear evidence of ethnomedicine in this
important region, particularly those that are endangered"– Provided by publisher.
Identifiers: LCCN 2023051041 (print) | LCCN 2023051042 (ebook) |
ISBN 9781032219080 (hbk) | ISBN 9781032219035 (pbk) |
ISBN 9781003270515 (ebk)
Subjects: LCSH: Medicinal plants–Romania. | Materia medica, Vegetable–Romania.
Classification: LCC QK99.R6 N38 2024 (print) | LCC QK99.R6 (ebook) |
DDC 581.6/3409498–dc23/eng/20240130
LC record available at https://lccn.loc.gov/2023051041
LC ebook record available at https://lccn.loc.gov/2023051042

ISBN: 9781032219080 (hbk)
ISBN: 9781032219035 (pbk)
ISBN: 9781003270515 (ebk)

DOI: 10.1201/9781003270515

Typeset in Palatino
by Newgen Publishing UK

Contents

CONTENTS

CONTENTS

Preface

This book, *Natural Products and Medicinal Properties of Carpathian (Romanian) Plants*, aims to explore one of the most significant parts of Romanian cultural heritage: ethnomedicine. Romania possesses a unique and highly complex biodiversity resulting from Romania's particular geographical location. With over 4000 identified species, the flora is exceptionally diverse, including alpine, forest, and steppe.

The book delves into the roots of Romanian medicine, tracing it back to Dacian medicine, which scholars of the ancient world regarded as highly advanced for its time. Moreover, the Dacian psychosomatic medical concept has been only recognized many centuries later in medical science, further emphasizing its importance.

From ancient times, Romanian popular tradition has emphasized the importance of a perfect balance between man and nature, which is why the gathering of therapeutic plants takes place only after certain rituals and only during particular periods.

Modern studies have confirmed the therapeutic properties of the medicinal plants used by the Dacians preserved and transmitted over time in Romanian ethnomedicine.

This book aims to bring knowledge of the most representative endemic medicinal plants from the Romanian Carpathian to the international community. It provides information on their metabolic profile and biological activity based on the latest studies, along with data related to their use in folk medicine and other lesser-known uses such as culinary preparations, cosmetics, and others.

Overall, the book provides valuable insights into the roots of Romanian ethnomedicine and the therapeutic properties of medicinal plants.

Acknowledgments

This book would have remained a distant aspiration without the unwavering support, encouragement, and sustained efforts of the exceptional people who made it possible.

Grateful thanks to Ray Cooper for the invitation and for believing in and encouraging me to embark on this endeavor.

I express my profound appreciation to the invaluable team of contributors for their admirable work and commendable efficiency, which allowed for the successful completion of this book.

Special recognition to Ms. Clara Bin-san Lau for her precious support and advice offered throughout the book creation process.

I wish to express my gratitude to the team of CRC Press, Ms. Hilary Lafoe, Ms. Sukirti Singh, and Ms. Christina Edwards for their exceptional assistance and for allowing me to bring this project to fruition.

Finally, yet importantly, thanks to Thara Kanaga from Newgen Knowledge Works for ensuring the book production.

Acknowledgements

Contributors

Cornelia Bejenaru
Department of Pharmaceutical Botany, Faculty of Pharmacy, University of Medicine and Pharmacy of Craiova, 2 Petru Rareș Street, Craiova, Dolj County, Romania

Ludovic Everard Bejenaru
Department of Pharmacognosy & Phytotherapy, Faculty of Pharmacy, University of Medicine and Pharmacy of Craiova, 2 Petru Rareș Street, Craiova, Dolj County, Romania

Andrei Biță
Department of Pharmacognosy & Phytotherapy, Faculty of Pharmacy, University of Medicine and Pharmacy of Craiova, 2 Petru Rareș Street, Craiova, Dolj County, Romania

George Dan Mogoșanu
Department of Pharmacognosy & Phytotherapy, Faculty of Pharmacy, University of Medicine and Pharmacy of Craiova, 2 Petru Rareș Street, Craiova, Dolj County, Romania

Diana Obistioiu
Banat's University of Agriculture and Veterinary Medicine, Faculty of Veterinary Medicine Timisoara, 119 Calea Aradului Street, Timisoara, Romania

Antonia Radu
Department of Pharmaceutical Botany, Faculty of Pharmacy, University of Medicine and Pharmacy of Craiova, 2 Petru Rareș Street, Craiova, Dolj County, Romania

Adina-Elena Segneanu
Institute for Advanced Environmental Research–West University of Timisoara (ICAM–WUT), 4 Oituz Street, Timisoara, Romania

Editor Biography

Adina-Elena Segneanu has almost 20 years of experience in bioactive compounds with an emphasis on the extraction, isolation, and chemical screening of metabolite profiling from different plant species. Her current research is primarily in the development of innovative phyto-carrier systems with predetermined biological properties.

She has a PhD in Chemistry (Organic Chemistry) at the University Politehnica Timisoara, Romania and an MSc. in Physics (Physics and Technology of Advanced Materials) at West University of Timisoara. She holds the highest category of researcher granted by Romanian Ministry of Research, Innovation and Digitalization.

Medicinal Plants of the Romanian Carpathian Mountains

Adina-Elena Segneanu
Institute for Advanced Environmental Research–West University of
Timisoara (ICAM–WUT), 4 Oituz Street, Timisoara, Romania

TRADITION AND HISTORY OF ETHNOMEDICINE

Much of ethnomedicine has been lost in the mists of the history of humankind. Maintaining health and identifying effective ways to heal or treat various ailments have always been constant concerns of humanity. Worldwide, medicinal plants have always played a key role in traditional medicine. A complete separation between humanity and nature cannot be made. Man was not designed to exist in its absence.

Medicinal plants can be considered as an integral part of the historical heritage of the Romanian people. In Romania, the ancient folk medicine based on the healing properties of plants has been preserved and pass over the years the unwritten rituals of harvesting plants and their various uses for various ailment.[1–50]

Among the Romanian people, the popular therapy has a millennial duration, its origin being attributed to the Geto-Dacian priests. Their traditional ethnomedicinal knowledge on plant use was attested in the manuscripts of the greatest personalities of ancient culture: Herodot, Homer, Dioscorides, Vegilius, Iordanes, Ovidius, and Pseudo-Apuleius.[1–55]

The Geto-Dacians practiced integrated holistic medicine, in which the body was considered in its entirety: astral, mental, spiritual, energetic, and physical. In their view, the body is a temporal temple of the soul that was considered immortal. For this reason, it is necessary to take care of inseparable components that form the "*whole*" to maintain the health of the human body. If this whole is sick, none of its components can be healthy.[1–2,7–57]

Later, this concept was taken over by Hippocrates and then by Paracelsus. However, for a long time, holistic medicine has been overtaken by allopathic medicine, which focuses exclusively on physical symptoms.[1–2,7–57]

Nevertheless, in recent decades, holistic medicine has been rediscovered: it is more and more appreciated today. [1–2,7–57]

In the Geto-Dacians concept, the state of health is, in fact, a state of harmonization between the human body, spirit, and the balance of the universe. The holistic therapeutic approach restored the health of the body (material component) and the rebalancing of the soul (energy component). Indeed, those initiated in healing techniques used a complex ritual consisting of a remedy for the body and another for the soul (so-called *enchantment*). The *enchantment* is basically a *sine qua non* in Romanian ethnomedicine. It is also described in the practice of folk medicine in the Middle Ages and later included in the modern era in rural communities. However, for the profane who witnessed these ancient medical rituals performed on sick people, it was considered magic.[1–2,7–57] Depending on the type of disease, the Geto-Dacians used phytotherapeutic remedies (medicinal plants), balneotherapy, mud therapy, heliotherapy, apitherapy, and gemotherapy, which were usually accompanied by extrasensory therapies including biotherapy, psychotherapy, melotherapy, and so on. These holistic extrasensory therapies have long been considered a magical practice and were meant to heal the soul.

"*Magic was born of medicine,*" Pliny wrote in Naturalis Historia.[39] Later, specialists appreciated *enchantment* as a primary form of primitive man or empirical psychotherapy.[58]

In fact, *the enchantment* is only an energy transfer by word of mouth, through which the Geto-Dacian nobles and priests, and later so-called healers sought to restore balance in the weakened body of a patient. The healing herbs were used to regulate a local imbalance, due to a disease, without intervening in the unitary energy balance of the human body.

Geto-Dacian priests and nobles conducted studies on the therapeutic properties of medicinal plants, preparation of tinctures, decocts, extracts, infusions, and ointments. In Dacian medicine, in addition to botanicals, various products of animal origin or animal minerals were also used.

Archaeological evidence discovered in the Thracian-Geto-Dacian territories such as that of Cucuteni (5000 years BC), Mangalia, Tartaria, Gumelnita, Turdas, Vinca, and Hamangia showed that the Geto-Dacian civilization was highly advanced.[1–2,11–12, 26,32,36–40,42,46,53–55,59] Ceramic objects were found with motifs and symbolism similar to ancient Chinese pottery: vessels that were used for various medicinal preparations, clay anthropomorphic figurines (dating from the ninth millennium

DOI: 10.1201/9781003270515-1

BC) on which acupuncture meridians were drawn (a therapeutic technique later found in Chinese and (tantra-yoga) Indian traditional medicine).[59]

Dacian medicine is recognized as being highly advanced, and therapeutic properties of medicinal plants was taken over by great scholars of the time.[1–2,4,8,11–12,21–22,26,32,36–41,44,46,53–55] More recently these principles have been found in modern phytotherapy[5–6,10,14,18,20,25,28–29,31,33,35,41,45,47,60–64].

Two examples of Dacian medicinal plants are mentioned here: salia (*Datura stramonium L., Solanaceae*) and dielleina (*Hyoscyamus niger L., Solanaceae*). The Dacian names of these plants are *salia* and *dielleina* and their curative properties were recorded by Discorides and Pseudo-Apuleios.[11,47,49]

Dacians used salia (*Datura stramonium L.*) for poultices, antiasthmic fumigations, psychomotor relaxant, anesthetic, analgesic, and insomnia. Dielleina was used for analgesic, anti-inflammatory, antifebrile, antispasmodic, antiasthmic, sedative and hypnotic effects.[11,47,49]

Scientific studies show these plants contain scopolamine and hyoscyamine, compounds with high biological activity from the class of tropane alkaloids.[49,65]

A selection of the most appreciated plants with therapeutic properties[66] is given in Table 1.

Table 1: Selection of Dacian Healing Herbs

Romanian plant name	Scientific name	Diseases
angelică	*Angelica archangelica*	injury, heart disease
rostopască	*Chelidonium majus*	jaundice
fierea pământului	*Centaurium umbellatum*	fever
alior	*Euphorbia cyparissias*	rheumatism, ringworm, warts, corns
captalan	*Petasites hybridus*	consumption, diuretic, analgesic, asthma, expectorant, antispasmodic
brusture	*Arctium lappa*	blood cleansing
brustan	*Telekia speciosa*	stomach ache
scrântitoare	*Potentilla anserina*	antihemorrhagic
coada-calului	*Equisetum arvense*	diuretic, urinary disorders
coada-șoricelului	*Achillea millefolium*	wound healing
pojarniță, sunătoare	*Hypericum perforatum*	digestive disorders, analgesic, liver detoxification, insomnia
mușețel	*Matricaria recutita*	digestive and rheumatic disorders, wound healing
zămoșiță	*Hibiscus trionum*	stomach aches, anti-inflammatory, antimicrobial
boz	*Sambucus ebulus*	detoxifying, purgative
busuioc	*Ocimum basilicum*	tonic, energizing, regenerating
pelin	*Artemisia absinthium*	vomiting, nausea, stomach pains
brânca ursului	*Heracleum sphondylium*	amenorrhea (absence of menstruation), menopausal disorders, general and sexual asthenia, urinary infections, respiratory diseases (laryngitis, bronchitis, asthma), tonic, antiaging, aphrodisiac, infertility
cătină	*Hippophaë rhamnoides*	respiratory infections, gastrointestinal disorders, wound healing, anticholesteremia, expectorant
păducel	*Crataegus monogyna*	angina pectoris, arrhythmia, heart failure, arterial hypertension, atherosclerosis
gălbenele	*Calendula officinalis*	gastritis, indigestion, ulcer, constipation, liver disease, sore throat, menstrual pain, hemorrhoids, wound healing (eczema, acne, burns, burns, etc.) osteoporosis, immunostimulant
ciuboțica cucului	*Primula veris*	wound healing, antispastic, emollient, diaphoretic, sedative, sudorific
mac	*Papaver somniferum*	insomnia

Table 1: (Continued)

Romanian plant name	Scientific name	Diseases
grașiță, iarbă grasă	*Portulaca oleracea*	antiscorbutic
limba boului	*Anchusa officinalis*	respiratory diseases, urinary disorder
valeriană	*Valeriana officinalis*	sedative
tătăneasă	*Symphytum officinale*	analgesic, hernia, diarrhea
măselariță	*Hyoscyamus niger*	toothaches
măceș	*Rosa canina*	vitaminized, eye disinfectant
drăgaică, sânziene	*Galium verum*	diuretic, depurative, sudorific, astringent, hemostatic, sedative, antispastic, wound healing
urzică	*Urtica dioica*	astringent, antiallergic, hemostatic, wound healing, anti-inflammatory, antidiarrheal, antianemic, remineralizing
traista ciobanului	*Capsella bursa-pastoris*	uterotonic, hemostatic, chills, epilepsy, antitumor
stregoaie	*Veratrum album*	rheumatism, gout, pneumonia, dermatological diseases, mental disorders (eczema, itching, scabies, etc.), stomach cramps, dyspepsia
cimbrișor	*Thymus serpyllum*	urinary infections, carminative, diuretic, expectorant, antiseptic, antispasmodic
omag	*Aconitum napellus*	neuralgia, rheumatism, ascites, gout, fever, respiratory diseases (whooping cough, bronchitis, laryngitis, premonitis) liver diseases, angina, coryza, acute congestive inflammations
tei	*Tilia cordata*	sedative, diaphoretic, antiseptic, emollient, insomnia
toporaș, viorea	*Viola odorata*	female infertility, expectorant, anti-inflammatory, cicatrizing, laxative, diuretic, antiseptic, disinfectant, sedative, antirheumatic
cinci degete	*Potentilla reptans*	antidiarrheal, anti-inflammatory, antihemorrhagic, antipyretic, antirheumatic
spânz	*Helleborus purpurascens*	rheumatism, spondylosis, arthrosis, myalgia, neuralgia, purgative, thrombosis, varicose veins, sprains, cancer
salvie	*Salvia officinalis*	wound healing, digestive tonic, infections, regulates menstruation, antispasmodics, sedative
mărul lupului	*Aristolochia clematitis*	infections, wound healing, abscesses, dermatological disorders (eczema, dermatosis, cuts, burns, scars, gangrene, skin cancer), anal fissures, hemorrhoids, varicose veins, varicose ulcers, gynecological conditions (vaginitis, candidiasis, uterine fibroids, genital cancer), prostate adenoma, colon cancer
iederă	*Hedera helix*	expectorant, asthma, bronchitis
stânjenel	*Iris germanica*	expectorant, diuretic, whooping cough, pneumonia, bronchial asthma, headache, tuberculosis, biliary dyskinesia, constipation, menstrual symptoms, dental neuralgia, hemorrhoids, dermatological disorders (burns, mycosis, skin ulcers)
ghimpe pădureț	*Ruscus aculeatus*	detoxification, heavy legs, hemorrhoids, bone fractures
știr	*Amaranthus retroflexus*	cholesterol reduction, enuresis, anemia, detoxification
păpălău	*Physalis alkekengi*	urinary disorders (kidney stones, urinary retention, nephritis), diuretic, laxative, gout, antipyretic, rheumatism, enuresis, facial paralysis, jaundice, diabetes
tămâiță de câmp	*Ajuga chamaepitys*	diuretic, jaundice

(continued)

Table 1: (Continued)

Romanian plant name	Scientific name	Diseases
vindecea	Betonica officinalis	digestive disorders, toothache, anuria, hemoptysis, dropsy, gout, migraines, ulcers, apoplexy, circulatory disorders, liver diseases, hypertension
părul fetei	Adiantum capillus veneris	diuretics, detoxification, stimulation of menstruation, wound healing, decrease in blood pressure and cholesterol, hepatoprotective, cough, increase in appetite
scoruş de munte	Sorbus aucuparia	depurative, diuretic, hemostatic, antitussive, antirheumatic, astringent
pătlagină	Plantago major	wound healing, antihemorrhagic, dermatological disorders (stings, burns, inflammation, dermatitis), constipation, cough
sulfină	Melilotus officinalis	emollient, antiedematous, sedative, antispasmodic, diuretic, gastrointestinal disorder
tulichină, lemn câinesc	Daphne mezereum	dental neuralgia
ceapă	Allium cepa	vermifuge, diuretic, cough, diabetes, hypertension, stomachic, carminative, diuretic, asthenia, expectorant, vasodilator, immunostimulant, stimulates, cold, rickets
mătăciune, răchitan	Dracocephalum moldavica	astringent, antidiarrheal, fermentation colitis, diarrhea, dysentery, epistaxis, irregular menstrual flow, hemoptysis, gastrointestinal bleeding, metrorrhagia
limba câinelui	Cynoglossum officinale	pulmonary catarrh, dysentery, intestinal pain, hemorrhages, hemorrhoids, thrombophlebitis, wound healing
cucută	Conium maculatum	asthma, analgesic, cough, intestinal spasms, strychnine antidote, cancer
iarbă neagră, buberic	Scrophularia nodosa	astringent, hemolyzing, anthelmintic, antifebrile, cicatrizing, hemorrhoids, scrofulous
vârnanț	Ruta graveolens	antispasmodic, sedative, laxative, antiulcer, carminative, vermifuge, diuretic, sudorific, arthrosis, rheumatism, gout, cicatrizing, antispastic, menstrual disorders, eye secretions, insect bites, snake bites
brusture	Lappa major	laxative, depurative, choleretic, cholagogue, hypoglycemic, diuretic, sudorific, antirheumatic, astringent, cicatrizing, disinfectant, capillary tonic, antivenomous
iarba vântului	Aster linosyris	digestive disorder
mutătoare, brâu	Bryonia alba	digestive disorders, diuretic, constipation, rheumatoid arthritis, headaches
pin	Pinus sylvestris	disinfectant, decongestant, antiseptic, diuretic, antirheumatic, anti-inflammatory, antimicrobial, expectorant, sedative, antipyretic, diaphoretic
mojdrean	Fraxinus ornus	gout, rheumatism, laxative, diuretic
curpen	Clematis vitalba	antimalarial, immune tonic, phytoestrogenic, diuretic, stimulation of hair growth
mei păsăresc, mărgeluțe	Lithospermum arvense	diuretic, urinary stones
dracilă	Berberis vulgaris	constipation, kidney disease, stomach pain, hepato-biliary disorders, anorexia, gout, colic cystitis, metrorrhagia, nephritis, febrifuge, hypertension
lumânărică	Verbascum phlomoides	cough, expectorant, emollient, diuretic
sparcetă, iarbă săracă	Onobrychis viciifolia	diuretic, sudorific

Table 1: (Continued)

Romanian plant name	Scientific name	Diseases
varga ciobanului, scăiuș	*Dipsacus pilosus*	diuretic, sudorific, astringent, depurative, dermatological conditions (boils, wounds), warts, freckles, indigestion, rheumatism, gout, edema
țintaură	*Erythraea centaurium*	indigestion, bloating, heartburn, detoxification, liver protection, diabetes, hypertension, spots and freckles, jaundice, antipyretic
gențiană	*Gentiana lutea*	lack of appetite, cold, chronic fatigue, neurasthenia, anorexia, digestive disorders
mentă	*Mentha × piperita*	digestive disorders, decongestant in inflammatory conditions of the nose, sinuses, and respiratory tract
albăstrele	*Centaurea cyanus*	eye pains, colds, respiratory disorders, colds, cuts, diuretic
bobornic	*Veronica beccabunga*	calming, expectorant, depurative, asthma, scrofulous, liver diseases
avrămească	*Gratiola officinalis*	cough, stomach pain, colds, infertility
podbal	*Tussilago farfara*	cough, expectorant, respiratory diseases (tracheitis, bronchitis, laryngitis, bronchial asthma, pulmonary emphysema) convalescence, biliary dyskinesia, dyspepsia, hepatobiliary diseases, scurvy

Only at the end of the 16th century did the first manuscripts in Romanian appear, which attest to the extensive development of ethnomedicine based on the use of the therapeutic properties of medicinal plants in the country. There can be noted only a few examples such as Scheian Psalter, Herbarium from Cluj, *Sanitatis studium* or *Codex of Matei Basarab*. In fact, Paul Kyr's *Sanitatis studium*, published in 1551 at Brasov, Romania is considered to be the first medical treatise printed in the Romanian countries. And, the manuscript entitled *"The use of medicinal plants"* dated in the 16th century is considered to be one of the oldest texts in the Romanian language in the field of phytotherapy.[16,37] This document includes presentations of the therapeutic uses of medicinal plants in common use during that period: burdock, gentian (*Gentiana lutea*), cut grass (*Stachys officinalis*), plantain (*Plantago major*), pimpernel (*Pimpinella saxifraga*), schinel (*Cnicus benedictus*), ventrilica (*Veronica officinalis*), shepherd's purse (*Capsella bursa-pastoris*), and nettle (*Urtica dioica*).[16,37]

According to current estimates, there are over 800 species of medicinal plants in Romania. It is worth mentioning that, in the middle of the last century, the work entitled *"Pharmaceutical Botany. The culture and harvest of pharmaceutical plants"* written by Dr. Grințescu Gh.P. includes about 500 Romanian healing plants, although most of them are not recognized in modern medicine, even though there is clear evidence of knowledge of ethnomedicine (the popular name of these healing herbs and their therapeutic effects are recognized throughout the country). [16,37,42,48]

Some of these curative plants are presented in Table 2.

During the second half of the 19th century, after the appearance of synthetic drugs, phytotherapy experienced a significant decline. Subsequently, the majority of phytotherapeutic preparations were replaced with synthetic chemicals. However, at the same time there was a steady increase in adverse side effects. Towards the end of the twentieth century, more attention was paid to natural products in the study of pharmaceutical research.

Plants contain a complex mixture of secondary metabolites, including stereo-structures with high biological activity, due to several centers of chirality. The chemical synthesis of these biologically active compounds is often difficult, and large amounts are required directly from plant extracts.

SUPERNATURAL DIMENSION IN ROMANIAN ETHNOMEDICINE

Herbs have played an undeniably crucial role in folk traditions throughout history, extending beyond their nutritional and healing properties. In Romanian folk tradition, numerous symbols and rituals have been associated with a diversity of herbs. These herbs were often believed to possess mystical properties and were used in different contexts, from healing to divination. The

Table 2: Selection of Healing Herbs from Romanian Ethnomedicine

Romanian plant common name	Scientific name	Curative properties
tiliscă. iarba vrăiitoarei	*Circaea lutetiana*	sedative. diuretic. would healing
scrântitoare	*Potentilla anserina*	pneumonia, rheumatic pains
rocoţea	*Stellaria graminea*	bone dislocation, chills
sică	*Statice gmelinii*	madness
drob	*Cytisus heufelianus*	typhoid fever
negruşcă	*Nigella arvensis*	gonorrhea
ciucuşoară, albiţă	*Berteroa incana*	gonorrhea
turiţă mare	*Agrimonia eupatoria*	respiratory disorders
cinci degete	*Potentilla reptans*	respiratory disorders
pidosnic	*Cerinthe minor*	rheumatism
cervană, iarba lui ceas rău	*Lycopus europaeus*	rheumatism
răchită	*Salix fragilis*	rheumatism
iarba voinicului	*Nasturtium sylvestre*	anemia
solovârfiţă	*Phlomis tuberosa*	anemia
răchitan, sburătoare	*Lytrum salicaria*	anemia, paralysis
bobornic	*Veronica beccabunga*	diuretic, detoxifying
slăbănog, buruiana celor slabi	*Impatiens noli-tangere*	diuretic
iarba gălbenelii	*Potentilla recta*	hepatitis
volbură	*Convolvulus arvensis*	stomach pain
studeniţă	*Arenaria serpyllifolia*	hernia
sânziene	*Galium mollugo*	hernia
rostogol, măciuca ciobanului	*Echinops sphaerocephalum*	antrax
izma broaştei	*Mentha aquatica*	tonic
silnic, rotundioară	*Glechoma hederacea*	epilepsy, mumps
pufuliţă	*Epilobium hirsutum*	epilepsy
unghia găii	*Astragalus glycyphyllos*	epilepsy, immunostimulant
colţul lupului	*Cirsium erisithales*	cancer
crăstăval	*Cirsium oleraceum*	cancer
captalan	*Cirsium rivulare*	cancer
cătuşnică, iarba vântului	*Nepeta cataria*	astringent, antirheumatic, sedative, antispasmodic, anesthetic
priboi, pălăria cucului	*Geranium phaeum*	astringent, diuretic, hypotensive, disinfectant
cioroi	*Inula salicina*	dysentery
zârnă, umbra nopţii	*Solanum nigrum*	mumps
crăiţe, vâzdoage	*Tagetes erecta*	digestive disorders
talpa gâştei	*Leonurus cardiaca*	prostate
rocoină	*Stellaria media*	prostate
bănuţei	*Bellis perennis*	chronic bronchitis
saschiu	*Vinca minor*	hypotensive, antianemia, diuretic, antidiabetic, depurative, astringent

significance of these herbs in Romanian culture has been the subject of extensive research, and their use continues to be an important aspect of traditional practices in the region.

Thus, the healing properties of butcher's broom (*Ruscus aculeatus*), in Romanian ghimpele paduret, have been recognized from ancient times for kidney disorders, according to Plini and Discorides. In the Middle Ages, it was used as a diuretic and then as poultices for bone fractures. Modern research reported the vasoconstrictor properties. This plant is associated with good luck, warding off evil spirits and offering hope, fertilization, and abundance. However, due to excessive harvesting, it is now considered a species protected by law.[11,13,67–69]

Another well-known herb is wormwood (*Artemisia absinthium*), known in Romanian as "pelin" or "iarba fecioarelor." It has been used since ancient times for digestive disorders such as vomiting, nausea, and stomach pains. It has also been used for insomnia and fumigations to disinfect homes, among other purposes. On a spiritual level, the plant is connected to Whitsuntide, excitation, and protection.[70]

Mandrake (*Atropa belladonna*) in Romanian matraguna is connected with a series of magical practices related to human or natural forces, love, material prosperity, fertility, marriage, curses, and more. In the popular tradition, special rituals are dedicated to this medicinal plant harvesting, including two women and a payment (bread, salt, and money) in exchange for the collected plant.[11,13,67,71]

Henbane bell (*Scopolia carniolica*), also known as matraguna mica or iarba codrului, holds significant cultural value in Romania for its association with luck, love, and death. The plant is picked only by older women or children to prevent its energy from being loaded with worldly sin, which could potentially worsen the patient's health. For magical purposes, such as predicting the future, life span, and serious illnesses, the gathering ritual involves lighting, incense, and offerings.[11,13,67,71]

Fern (*Dryopteris filix-mas*), or snake grass, is believed to be associated with wealth, abundance, and marriage in Romanian folklore. The diverse traditional uses and beliefs surrounding these herbs and plants provide fascinating insight into the cultural practices and beliefs of the region.[11,13,67,71]

Cynanchum vincetoxicum, in Romanian Iarba fiarelor (whose root is used in traditional medicine as an expectorant, depurative, diuretic, vermifuge, or as emetic in large doses), is considered a plant with outstanding magical properties, facilitating access to any "lock," preventing the chaining of the man who possesses it, protecting against iron weapons, and attracting material prosperity. Moreover, the plant offers the ability to master the language of living things and plants and access to the land of the fairies, located in another spiritual dimension.[11,13,67,71]

Angelica archangelica, angelica in Romanian, is an elixir plant related to health and youth and attracts the protection of angels. The plant's name is related to the feast of Saint Michael the Archangel, the period when its blooms. In the Middle Ages, it was cultivated even in monasteries, being used as a plant that can fight the plague.[11,13,67,71]

Hawthorn (*Crataegus monogyna*), also known as paducel in Romanian, is unequivocally recognized for its healing properties in Dacian medicine. The plant was appreciated by famous doctors, including Discorides, Galen, Paracelsus, and later Carl Linnaeus, for its indisputable benefits on heart diseases and gout. In Romanian ethnomedicine, it is recommended as a first-line treatment for heart disorders and metabolic diseases associated with puberty or menopause. Hawthorn is also related to numerous mystical properties, such as protection from evil spirits, which have been validated by the Romanian popular tradition. It is widely accepted that fumigating hawthorn leaves and flowers cleanses negative energies and brings good luck.[11,13,67,72–73]

Hedge hyssop (*Gratiola officinalis*), avrameasca in Romanian, has the property of repelling evil spirits and the undead. It is used in fumigations for nightmares in children, depressive psychoses, epilepsy, and diseases associated with the presence of dark entities.[11,13,67,71]

Sambucus ebulus, Boz in Romanian beyond, its therapeutic properties, is used for rain invocation for crops.[11,13,67,71]

Galium verum, Dragaica sau sanziene in Romanian, for example, is believed to possess the ability to repel dark entities, and its name is linked to June 24, an important feast in Romanian culture. This day marks the harvesting of medicinal plants, as their therapeutic properties are believed to be at their maximum. It is also believed that during the night of June 24, the material world and the spiritual realm come into contact, and the heavens open up.[11,13,67,71]

Cannabis sativa, canepa in Romanian, protects against undead or moroi (children born dead or killed at birth).[11,13,67,71]

Birch (*Betula pendula*), mesteacan in Romanian, represents rebirth and regeneration, and is related to fertility, creativity, health, and protection.[11,13,67,71]

Viscum album, mistletoe in Romanian, is placed above the entrance door to the house during the winter holidays to attract luck, prosperity, love, health, and angels and ward off diseases.[11,13,67,71]

It might be interesting to note that in the Romanian folk tradition nature has a sacred role and a spiritual significance. Fir is a symbol of the link between earth and heaven. This tree marks the three major events in human life: birth (planting a Christmas tree), marriage, or death. In some regions of the country, when a new house is built, a small fir tree is placed on the roof. Other examples are represented by the hazelnut that symbolizes fertility, and the apple is the tree of wisdom and even life itself.

In Romanian ethnomedicine there are different plants whose popular name indicates their high toxicity; for example: devil's mushroom (*Boletus satanas*), devil's grass (*Datura stramonium*), wolf's cherry or wolf's grass (*Atropa belladonna*), unded (*Veratrum album*), dracila (*Berberis vulgaris*), old lady fangs (*Tribulus terrestris*), etc.

Even today, many Romanians use various ethnomedical preparations for colds, increased immunity (extracts from fir buds, plantain, elder), tonics (raspberries, blackcurrant), weight loss (sea buckthorn vinegar), digestive disorders (basil, St. John's wort, propolis), dermatological diseases (propolis, marigold), rheumatological (comfrey), natural dermato-cosmetic products (lavender, mint, wormwood, yarrow), etc.[5–6,8,14,20,28,31,48,61–62,67]

THERAPEUTIC PROPERTIES OF ROMANIAN MEDICINAL PLANTS – FROM ETHNOMEDICINE TO MODERN SCIENCE

Chemical screening of medicinal plants provides numerous bioactive compounds with complex chemical structures. The approach indicates there are often the multiple therapeutic properties of one single medicinal plant.[5–6,8,14,20,28,31,48,61–62,67]

Medicinal plants contain many phytoconstituents with therapeutic properties: flavonoids, phenolic compounds, glycosides, carbohydrates, alkaloids, volatile oils, sterols, anthocyanins, derivatives of polyphenolic carboxylic acids, coumarins, vitamins, terpenes, and so on.[62,74–75]

In addition, medicinal plants demonstrate many other advantages:

- favorable benefit/risk ratio;

- allows the transition from short-term to long-term treatment;

- synergistic and complex action;

- high tolerance of the human body;

- very low incidence of side effects;

- known toxicity;

- allow long-term administration;

- low costs.

A very important role in phytotherapy is the identification of phytoconstitutions with curative properties and the determination of the optimal isolation/extraction method.

Research in the field of medicinal plants has confirmed that different groups of phytoconstituents possesses certain therapeutic properties. Table 3 shows the curative actions of the different secondary metabolites of vegetal origin.[18,20,24,28–30,33–35,41,43,47,61–64,67]

It is particularly interesting that the harvesting of medicinal plants according to these ancient rituals corresponds to the maximum concentration of active compounds. Moreover, through the very old ethnomedical techniques, preserved until today, the active compounds from the healing plants are highlighted.[1–2,5,11–12,14,16–17,20,25,27–29,32,36,41,46–47,49,51]

Thus, a very brief presentation of the harvesting schedule of some of the most well-known medicinal plants corresponds to a certain cosmic calendar. Only very few herbs can be harvested at any one time and maintain their therapeutic benefits (e.g., garlic, wormwood). In the spring cowslip primrose, nettles, ramsons, and hawthorn are collected. In summer yarrow, common horsetail, plantain, dandelion, thyme, sandalwood, marigold, St. John's wort, hollyhock, shepherd's purse, etc., are harvested. And in autumn blueberries, sea buckthorn, rosehips, and blackthorn.[1–2,5,11–12,14,16–17,20,25,27–29,32,36,41,46–47,49,51]

At present, it is considered that the optimal harvesting period of medicinal plants must correspond to the moment when the content of bioactive compounds is maximum.

Table 4 summarizes the period suitable for harvesting each organ from medicinal plants.

Drying of medicinal plants is recommended to be done by natural processes, and preservation is done in well-ventilated areas, at constant temperatures, away from exposure to light sources, in the absence of moisture. The duration of drying varies depending on the season and the organ of the plant harvested. Thus, summer can last between three and seven days for leaves, flowers, or stems and up to a month for root or bark. For medicinal plants, only paper or cardboard packaging is used.[1–2,5,11–12,14,16–17,20,25,27–29,32,36,41,46–47,49,51]

Table 3: Therapeutic Properties of Main Secondary Metabolites from Different Healing Plant Species

Secondary metabolites group	Exemples of plant species	Therapeutic properties	Ref
glycosides	*Apocynaceae, Liliaceae, Scrophulariaceae*	cardiotonics	18,24,35,61,76,77
alkaloids	*Lamiaceae, Liliaceae, Papaveraceae, Ranunculaceae, Solanaceae*	antitumoral, antiviral, antibacterial, antispasmodic, anestesic, expectorant, antitussive	78
terpenoids	*Apiaceae, Asteraceae, Oleaceae, Ranunculaceae*	antitumoral, anti-inflammatory	79,80
flavonoids	*Asteraceae, Ericaceae, Fabaceae, Rosaceae, Scrophulariaceae*	anti-inflammatory, diuretic, antibacterial, antiviral, antifungal, antioxidant	18,24,35,61,76,77
coumarins and furanocoumarins	*Asteraceae, Apiaceae, Fabaceae, Lamiaceae, Rubiaceae, Solanaceae*	vasoprotective, antibiotic, diuretic, analgesic, anti-inflammatory, antispasmodic, anticoagulant, thrombolytic, estrogen, photosensitizer, anthelmintic	61,62,81
phenolic acids	*Amaryllidaceae, Apiaceae, Asteraceae, Ericaceae, Fabaceae, Lamiaceae, Liliaceae, Rosaceae, Rubiaceae, Scrophulariaceae, Solanaceae*	antimutagenic, antidiabetic, anticarcinogenic, hypolipidemic, immunostimulatory, hypoglycemic, antimicrobial	62,82,83
tannins	*Aceraceae, Actinidiaceae, Ericaceae, Fabaceae, Solanaceae, Vitaceae*	astringent, hemostatic, antifungal, antiviral, antiseptic	18,24,35,61,76,77,84
essential oils	*Apiaceae, Asteraceae, Cupressaceae, Ericaceae, Fabaceae, Lamiaceae*	sedative, decongestant, anti-inflammatory, carminative, antiseptic, antiviral, antifungal, antiparasitic, hormonal, hepatoprotective, cholecystokinetic	18,24,35,61,76,77
saponins	*Amaryllidaceae, Asteraceae, Liliaceae, Scrophulariaceae*	expectorant, depurative, wound healing (stimulating collagen synthesis), antimicrobial, antifungal, antiviral, vasoprotector, anti-inflammatory, antiedematous, analgesic, anthelmintic, cytostatic, immunomodulatory	18,24,35,61,76,77
carbohydrates (mucilages, pectins, gums, etc.)	*Lamiaceae, Liliaceae, Fabaceae, Oleaceae, Plantaginaceae*	antioxidant, emollient, laxative/antidiarrheic, antiparasitic, antitumor, anti-inflammatory, immunostimulant, antiarthritic, gastroprotective, decrease cholesterol, hemostatic	18,24,35,61,76,85–88

Table 4: Harvesting Period of Different Medicinal Plant Organs

Plant organ	Optimal harvesting period
Roots or other underground parts	Early spring or late autumn (during the rest period)
Leaf buds	Early spring, before it unfolds (manual harvesting), sunny weather, no humidity
Leaves, barks	Spring (manual harvesting), sunny weather, no humidity
Flowers	Before or in flowering time (manual harvesting), sunny weather, no humidity
Fruits, seeds	Ripening time, cool weather (morning or evening)

In phytotherapy it is essential to know how to prepare and administer medicinal plants.[5,10,19–20,25,33,35,52,89]

ROMANIAN CARPATHIAN MEDICINAL FLORA

The term "Carpathians" has Geto-Dacian origin, and comes from the original word "Karpate" signifying rock, and from which derives the Dacian name "*Karpathos-Horos*" meaning Carpathian Dacians of Carpi or Carpiani, an Dacian tribe, known in history, after the Roman conquest, as free Dacians.[4,37,44,53–55,90–92]

Over 50% of the entire Carpathian Mountains are found in Romania. The Romanian Carpathian Mountains form a natural arched barrier, whose older name "*Corona Montium*," denotes their physical and geographical form.[22–23,27,46,56,92–94]

There is a conventional delimitation of the Romanian Carpathians as Eastern Carpathians, Southern Carpathians, and western Carpathians.

Each of these mountain regions have had other names over time. For example, Eastern Carpathians were also called the "Carpathians of Dacia" or Easterners; Southern Carpathians or Southerners, "*Getic Carpathians*," the South or the Transylvanian Alps; and Western Carpathians were the Carpathians of Romania or the West.[17,22–24,27,46,56,93–94]

The Carpathians form a natural bridge between Western and Eastern Europe for species migration and genetic exchange. The Carpathian Mountains are recognized as one of the main European areas in terms of rich biodiversity (ecosystems, species, and genetic diversity). More than a third of the entire European flora is found here.

Romania, through its geographical position, benefits from absolutely unique characteristics including diversity, complexity, variety, proportionality, symmetry, and complementarity. In this sense, only the existence of the relatively balanced ratio between the types of relief must be mentioned: plains (approx. 30%), hills and plateaus (approx. 35%), mountains (approx. 35% – related to the total surface of the country).[95] In addition, the existence of biogeographical regions such as the continental, Pannonian, steppe, alpine, and Pontic influenced the development of Romania's flora and its division into its main categories: alpine, forest, and steppe.[95]

From the point of view of the flora, in Romania approximately 4000 species and subspecies of plants have been identified. The endemic and subendemic plants are in a percentage of about 4% and they are part of national natural heritage.[96–99]

Romanian plant species can be cataloged according to their main use:

✓ food use (including aromatic or oilseeds) represents approximately 180 species;

✓ dyes (from which food or textile dyes are extracted) approximately 50 plants;

✓ tannins (extracted tannins are used in the leather industry, especially in the tannery) about 30 plants; and

✓ medicinal plants: over 800 plant species.[84–87]

The great majority of these endemic and subendemic plants can be found only in the Carpathian Mountains.[96–100]

A summary of some of these endemic and subendemic plants[96–100] is given in Table 5.

The category of endemic plants includes a series of plants considered monuments of nature, that is, vulnerable species, endangered species, and rare species. But what is remarkable about them, beyond the extremely limited geographical region, is their extremely high economic potential.[96–100] Some of these are listed in Table 6.

PHYTOTHERAPY IN MODERN TIMES

Like any other living organism, medicinal plants have a vital energy, an energy field that intertwines with a complex macromolecular architecture composed of a mixture of metabolites with high biological activity causing a synergistic mechanism. Recent research in the field of drug development based on bioactive compounds identified from medicinal plants confirms the impossibility of elucidating the synergistic mechanism upon which secondary metabolites of plant compounds act.[101] And this can be justified by the fact that the interaction of energy field-plant matter with curative effect was not taken into account.

In Romania there is an ascending trend regarding the re-evaluation of the importance of phytotherapy and the complementary implementation with allopathic medicine, in the therapy of various diseases.

Table 5: Selection of Endemic and Subendemic Plants from Romanian Carpathian Mountains

Carpathian endemic plants	Carpathian subendemic plants
Achillea coziana	*Achillea schurii*
Aconitum toxicum	*Aconitum moldavicum*
Androsace chamaejasme	*Aconitum tauricum* subsp. *hunyadense*
Andryala levitomentosa	*Allium fuscum* subsp. *fussii*
Anthemis carpatica subsp. *pyrethriformis*	*Alopecurus laguriformis*
Aquilegia nigricans subsp. *subcaposa*	*Androsace villosa* subsp. *arachnoidea*
Alyssum petraeum	*Aquilegia transsilvanica*
Arenaria tenella	*Armeria pocutica*
Aster alpinus	*Asperula carpatica*
Astragalus roemeri	*Aubrieta intermedia* subsp. *falcata*
Asperula capitata	*Betula pubescens* subsp. *carpatica*
Barbarea lepuznica	*Campanula rotundifolia* subsp. *polymorpha*
Bruckenthalia spiculifolia	*Campanula serrata*
Campanula carpatica	*Cardamine glanduligera*
Campanula transsilvanica	*Cardaminopsis neglecta*
Campanula patula	*Chrysosplenium alpinum*
Carduus kerneri subsp. *lobulatiformis*	*Dactylorhiza cordigera* subsp. *siculorum*
Carex parviflora	*Dactylorhiza maculata* subsp. *schurii*
Comastoma tenellum	*Erigeron acer polymorpha*
Crepis conyzifolia	*Erigeron nanus*
Cerastium transsilvanicum	*Eritrichium nanum* subsp. *jankae*
Delphinium simonkaianum	*Erysimum witmannii*
Dianthus callizonus	*Festuca carpatica*
Dianthus glacialis subsp. *gelidus*	*Festuca porcii*
Dianthus henteri	*Festuca rupicola* subsp. *saxatilis*
Dianthus spiculifolius	*Festuca tatrae*
Digitalis grandiflora	*Gentiana phlogifolia*
Dianthus kitaibelii	*Heracleum carpaticum*
Draba haynaldii	*Heracleum palmatum*
Draba dorneri	*Hesperis nivea*
Festuca bucegiensis	*Hieracium fagarasense*
Festuca nitida subsp. *flaccida*	*Hieracium magocsyanum*
Festuca pachyphylla	*Hieracium pojoritense*
Festuca versicolor subsp. *dominii*	*Hieracium porphiriticum*
Gentiana lutea	*Koeleria macrantha* subsp. *transsilvanica*
Gentiana verna	*Larix decidua* subsp. *polonica*
Helictotrichon decorum	*Leontodon montanus* subsp. *pseudotaraxaci*
Hepatica transsilvanica	*Leontodon repens*
Hesperis matronalis subsp. *moniliformis*	*Melampyrum herbichii*
Hesperis oblongifolia	*Melampyrum saxosum*
Hieracium transsilvanicum	*Minuartia verna* subsp. *oxypetala*
Leontopodium alpinum	*Oxytropis carpatica*
Moehringia pendula	*Phyteuma tetramerum*
Nigritella carpatica	*Phyteuma wagneri*
Onobrychis montana subsp. *transsilvanica*	*Pinus nigra* subsp. *banatica*
Pedicularis baumgarteni	*Poa rehmannii*
Pedicularis verticillata	*Primula elatior* subsp. *leucophylla*

(continued)

Table 5: (Continued)

Carpathian endemic plants

Poa granitica subsp. *disparilis*
Potentilla haynaldiana
Potentilla ternata
Pulmonaria rubra
Primula baumgarteniana
Primula minima
Rhododendron kotschyi
Rosa coziae
Saxifraga cymosa
Saxifraga demissa
Saxifraga mutata subsp. *demisa*
Sempervivum heuffelii
Senecio integrifolius
Seseli rigidum
Symphytum cordatum
Thesium kernerianum
Thelapsi dacicum subsp. *banaticum*
Thymus bihariensis
Trollius europaeus
Tulipa hungarica
Viola alpina
Viola dacica
Viola jooi

Carpathian subendemic plants

Pulmonaria filarszkyana
Pulsatilla halleri subsp. *slavica*
Pyrola carpatica
Ranunculus carpaticus
Saussurea porcii
Scabiosa lucida subsp. *barbata*
Sesleria heufleriana
Silene zawadskii
Silene nutans
Syringa josikaea
Thlaspi dacicum
Thlaspi dacicum subsp. *dacicum*
Thymus comosus
Thymus pulcherrimus
Trisetum fuscum
Trisetum macrotrichum

Table 6: Main Carpathian Endemic Plants Classified as Vulnerable, Endangered, and Rare Species

Carpathian vulnerable, endangered and rare species endemic plants

Aquilegia nigricans
Atocion rupestre
Andryala levitomentosa
Athamanta turbith subsp. *hungarica*
Cardamine glauca
Carex bigelowii
Centaurea phrygia subsp. *rarauensis*
Centaurea phrygia subsp. *retezatensis*
Cochlearia borzaeana
Cypripedium calceolus
Draba dorneri
Draba simonkaiana
Fumaria jankae
Gentiana lutea
Galium baillonii
Hieracium praebiharicum
Linaria alpina
Lilium martagon
Lychnis nivalis
Nigritella carpatica

Table 6: (Continued)

Carpathian vulnerable, endangered and rare species endemic plants

Nigritella nigra
Nigritella rubra
Papaver corona-sancti-stephani
Potentilla haynaldiana
Primula auricula subsp. *serratifolia*
Primula wulfeniana subsp. *baumgarteniana*
Silene dinarica
Trollius europaeus

The extremely vast tradition of ethnomedicine and the richness and great diversity of Romanian medicinal flora (over 800 species) are some of the reasons the majority of the population (over 50%) prefers phytotherapy.

The quality of Romanian medicinal plants has also been recognized by the top position it occupies on the European and American markets.[102]

REFERENCES

1. Petrescu-Dîmboviţa M., Vulpe A. (eds). 2001. *Istoria Românilor*, vol. I. Ed. Enciclopedică, Bucureşti, 1020 pp.

2. Iliescu V., Popescu V.I., Ştefan Gh. (eds). 1964. *Izvoare privind Istoria României (Fontes ad Historiam Dacoromaniae pertinentes). Vol. I: Ab Hesiodo usque Itinerarium Antonini.* Academia Reipublicae Popularis Dacoromanae, Institutum Archaeologicum, Bucureşti.

3. Ardelean A., Mohan G. 2008. *Flora medicinală a României.* Ed. All, Bucureşti.

4. Bîrsan C. 2003. *Revanşa Daciei.* Ed. Obiectiv, Craiova.

5. Bojor O., Popescu O. 1998. *Fitoterapie tradiţională şi modernă.* Ed. Fiat Lux, Bucureşti.

6. Butură V. 1979. *Enciclopedie de etnobotanică românească.* Ed. Ştiinţifică şi Enciclopedică, Bucureşti, 281 pp.

7. Candrea A. 1999. *Folclorul medical român comparat. Privire generală. Medicina magică (reprint).* Ed. Polirom, Iaşi.

8. Carciumaru M. 1987. Plants used by traco-geto-dacians (Attempt of synthesis) (V). *Thraco-Dacica* VIII(1–2):171–176.

9. Claudian I. 1939. *Food of the Romanian people in anthropogeography and history.* King Carol II Foundation for Literature and Art, Bucureşti.

10. Constantinescu C., Agopian A. 1962. *Plante medicinale din flora spontană.* Ed. Centrocoop, Bucureşti.

11. Crişan H.I. 2013. *Medicina în Dacia.* Ed. Dacica, Hunedoara.

12. Desunsianu N. 2002. *Dacia preistorică.* Ed. Arhetip, Bucureşti.

13. Diaconu T. 2008. *Medicina sacră – miracolul medicinei geto-dace (leacuri pentru boli incurabile).* Ed. Obiectiv, Craiova.

14. Mihăescu E. 2008. *Dicţionarul plantelor de leac.* 2nd edition, Ed. Călin, Bucureşti.

15. Dioscorides P. 2000. *De Materia Medica*. Ibidis Press, Johannesburg, South Africa.

16. Drăgulescu C., Drăgulescu R. 2000. *Contribuții la cunoașterea limbii geto-dacilor. Denumiri dacice de plante*. Ed. Universității "Lucian Blaga" Sibiu.

17. Dulgheru V. 2010. *Din istoria Neamului*. Ed. Tehnica Info, Chișinău, Republic of Moldova.

18. Dobrescu D., Baloescu C. 1993. *Farmacopeea Română ediția a X-a*. Ed. Medicală, București.

19. Geiculescu V.T. 1986. *Bioterapie. Rețete medicale fără medicamente*. Ed. Științifică și Enciclopedică, București.

20. Grigorescu E. 1987. Din ierburi s-au născut medicamentele. Ed. Albatros, București.

21. Herodot. 1961. *The Histories*, vol. 1–2. Ed. Științifică, București.

22. Iordanes. 1939. *De origine Actibusque getarum (Getica)*. Ed. George Papu-Lisseanu, București.

23. Iordanes. 2014. *Getica*. Ed. Uranus, București.

24. Istudor V. 2001. *Farmacognozie. Fitochimie. Fitoterapie. Vol. II: Aetherolea, rezine, iridoide, principii amare, vitamine*. Ed. Medicală, București.

25. Ivan S. 1990. *Să ne tratăm fără medicamente*. Ed. Științifică și Enciclopedică, București.

26. Iliescu V. (ed). 1964. *Izvoare privind Istoria României (Fontes ad Historiam Dacoromaniae pertinentes). Vol. I: De la Hesiod la Itinerarul lui Antoninus*. Institutul de Arheologie, București.

27. Lăzărescu G. 1992. *Dicționar de mitologie*. Casa Editorială Odeon, București.

28. Mazăre G.C. 2019. Plantele medicinale în spațiul carpato-danubiano pontic. *Bucovina Forestieră* 19(1):46–50.

29. Milică C., Roman C.N., Troia D. 2012. *Flora medicinală a României*. Ed. Doxologia, Iași.

30. Muntean L.S., Tămaș M., Muntean S., Muntean L., Duda M., Vârban D., Florian S. 2016. *Tratat de plante medicinale cultivate și spontane*, ediția a II-a. Ed. Risoprint, Cluj-Napoca.

31. Muntean S., Muntean L. 2011. *Curs de plante medicinale*. Ed. Risoprint, Cluj-Napoca.

32. Bejinariu C. (ed). 2020. *Sănătate, că-i mai bună decât toate! Oameni, epoci, tratamente medicale – expoziție online*. Muzeul Județean de Istorie și Artă, Zalău.

33. Nădășan V. 2003. *Incursiune în Fitoterapie*. Ed. Viață și Sănătate, București.

34. Nistreanu A. 2020. Contributions to the study of medicinal plants. *Buletinul Academiei de Științe a Moldovei. Științe Medicale* 66(2):50–54.

35. Pârvu C. 2002. *Universul plantelor. Plante din flora României*. Vol. I, ediția a II-a, Ed. Tehnică, București.

36. Pârvan V. 1928. *Dacia: An outline of the early civilizations of the Carpatho-Danubian Countries*. Cambridge University Press, Cambridge, UK, 214 pp.

37. Pârvan V. 1992. *O protoistorie a Daciei*. Ed. Universitas, Petroșani.

38. Plato. 2008. *Charmides*. Translated by Jowett B. (1998), The Project Gutenberg EBook of Charmides [EBook #1580].

39. Plinius. 2001. *Naturalis Historia*. Ed. Polirom, Iaşi.

40. Pogăciaş A. 2017. The Dacian Society – Fierce warriors and their women: sources and representations. *Hiperboreea Journal* 4(1):5–22.

41. Popescu H. 1984. *Resurse medicinale în flora României*. Ed. Dacia, Cluj-Napoca.

42. Popovici R. 2011. *At the table with the ancestors*. Ed. Dacia, Cluj-Napoca.

43. Bone K., Mills S. (eds). 2013. Principles of herbal treatment, in principles and practice of phytotherapy. 2nd edition, Churchill Livingstone, London, UK, pp. 83–88.

44. Ptolemy C. 2011. *Geography of Paperback*. Cosimo Classics, New York, USA.

45. Săvulescu T. (ed). 1976. *Flora RSR*, vol. XII. Ed. Academiei Române, Bucureşti.

46. Scorobete M. 2006. *Dacia edenică*. Ed. Renaşterea, Cluj-Napoca, pp. 11–32.

47. Segneanu A.E., Cepan C., Grozescu I., Cziple F., Olariu S., Ratiu S., Lazar V., Murariu S.M., Velciov S.M. 2019. Therapeutic use of some Romanian medicinal plants. In: Perveen S., Al-Taweel A. (eds). *Pharmacognosy – Medicinal Plants*. IntechOpen, London, UK.

48. Senciuc A. 2012. *Scientia: Istoria ideilor şi a descoperirilor ştiinţifice*. www.scientia.ro/stiinta-la-minut/48-scurta-istorie-descoperiri-stiintifice/4043-medicina-in-dacia.html

49. Stănescu U. 2007. Farmacia verde – de la traco-daci la epoca modernă. *Proceeding Simpozionului Internaţional Cucuteni 5000 REDIVIVUS*. Ştiinţe exacte şi mai puţin exacte. Din istoria neamului. Ediţia a 2-a, Chişinău, Republic of Moldova, pp. 74–82. www.researchgate.net/publication/318723180_Din_istoria_Neamului;

50. Vulcănescu R. 1985. *Mitologia română*. Ed. Academiei Republicii Socialiste România, Bucureşti, 712 pp.

51. Ábrán Á., Cosma V.S. 2017. Poveşti din hotare. Lumea rurală postsocialistă. *GAP – Gazeta de Artă Politică* 5(18):8–9.

52. Winston J.C. 1996. *The use and safety of common herbs and herbal teas*. Golden Harvest Books, Eau Claire, Michigan, USA.

53. Gimbutas M. 1989. *Civilizaţie şi cultură*. Ed. Meridiane, Bucureşti.

54. Gimbutas M. 1965. *Bronze Age cultures in Central and Eastern Europe*. Mouton & Co., Paris–Haga–London.

55. Gimbutas M. 1991. *The Civilization of the Goddess: The World of old Europe*. Harper, San Francisco, USA.

56. Cioacă A., Dinu M.S. 2010. Romanian Carpathian landscapes and cultures. In: Martini I.P., Chesworth W. (eds). *Landscapes and Societies*. Springer Science + Business Media B.V., Berlin–Heidelberg, Germany, pp. 257–269.

57. Fernando W.G.D. 2012. Plants: an international scientific open access journal to publish all facets of plants, their functions and interactions with the environment and other living organisms. *Plants (Basel)* 1(1):1–5.

58. Gligor M. 2012. Între filosofie şi medicină – Folclorul medical în viziunea lui Mircea Eliade şi Valeriu Bologa. Ed. Presa Universitară Clujeană, Cluj-Napoca.

59. Merlini M. 2014. The Sacred Cryptograms from Tărtăria: Unique or widespread signs? Putting the asserted literate content of the tablets under scrutiny. In: Marler J. (ed). *Fifty Years of Tărtăria Excavations. Festschrift in Honor of Gheorghe Lazarovici on the Occasion of His 73rd Birthday*. Institute of Archaeomythology, Sebastopol, California, USA, Ed. Lidana, Suceava, Romania, pp. 73–119.

60. Bojor O. 2003. *The Guide of Medicinal and Aromatic Plants from A to Z*. Ed. Fiat Lux, Bucureşti.

61. Cercasov C., Oprea E., Popa C.V., Fărcăşanu I.C. 2009. *Compuşi naturali cu acţiune terapeutică*. Ed. Universităţii din Bucureşti.

62. Moţa C., Roşu A., Câmpeanu G. 2016. *Compuşi bioactivi de origine vegetală. Abordări biotehnologice*. Ed. Universităţii din Bucureşti.

63. Stănescu U., Hăncianu M., Miron A., Aprostoaie C. 2004. *Plante medicinale de la A la Z: monografii ale produselor de interes terapeutic*. Ed. Grigore T. Popa University of Medicine and Pharmacy, Iaşi, pp. 280–282.

64. Stănescu U., Hăncianu M., Cioancă O., Aprotosoaie A., Miron A. 2014. *Medicinal plants from A to Z*. Ed. Polirom, Iaşi.

65. Kukula-Koch W.A., Widelski J. 2017. Alkaloids. In: Badal S., Delgoda R. (eds). *Pharmacognosy: Fundamentals, Applications and Strategies*. Academic Press–Elsevier, London, UK.

66. Crişan I.H. 1986. *Spiritualitatea geto-dacilor. Repere istorice*. Ed Albatros, Bucureşti.

67. Fierascu R.C., Fierascu I., Ortan A., Avramescu S.M., Dinu-Pirvu C.E., Ionescu D. 2017. Romanian aromatic and medicinal plants: from tradition to science. In: El Shemy H.A. (ed). *Aromatic and Medicinal Plants*. IntechOpen, London, UK.

68. https://fares.ro/dictionar-plante/ghimpele/

69. https://antropocene.it/en/2017/05/20/ruscus-aculeatus/

70. Gorovei A. 2003. *Credinţi şi superstiţii ale poporului român*. Ed. Grai şi Suflet – Cultura Naţională, Bucureşti.

71. Ruff A. 2023. *Leacuri magice din Carpaţi. Despre locul etnomedicinei în spaţiul carpatic*. www.academia.edu/7749146/DESPRE_LOCUL_ETN0MEDICINEI_%C3%8 EN_ SPA%C5%A2IUL_CARPATIC?auto=download

72. Holubarsch C.J.F., Colucci W.S., Eha J. 2018. Benefit-risk assessment of *Crataegus* extract WS 1442: An evidence-based review. *American Journal of Cardiovascular Drugs* 18(1):25–36.

73. https://aquiahora.ro/cum-alunga-paducelul-vindecatorul-inimii-spiritele-rele-si-atrage-protectia-divina/

74. Balandrin M.F., Klocke J.A. 1988. Medicinal, aromatic, and industrial materials from plants. In: Bajaj Y.P.S. (ed). *Biotechnology in Agriculture and Forestry*. Vol. 4, Medicinal and Aromatic Plants I, Springer-Verlag, Berlin–Heidelberg, Germany, pp.191–199.

75. Schmauder H.P., Doebel P. 1990. Plant cell cultivation as a biotechnological method. *Engineering in Life Sciences* 10(6):501–516.

76. Chomel M., Guittonny-Larchevêque M., Fernandez C., Gallet C., DesRochers A., Paré D., Jackson B.G., Baldy V. 2016. Plant secondary metabolites: a key driver of litter decomposition and soil nutrient cycling. *Journal of Ecology* 104(6):1527–1541.

77. Pagare S., Bhatia M., Tripathi N., Pagare S., Bansal Y.K. 2015. Secondary metabolites of plants and their role: overview. *Current Trends in Biotechnology and Pharmacy* 9(3):293–304.

78. Heinrich M., Mah J., Amirkia V. 2021. Alkaloids used as medicines: structural phytochemistry meets biodiversity – An update and forward look. *Molecules* 26(7):1836.

79. Yang W., Chen X., Li Y., Guo S., Wang Z., Yu X. 2020. Advances in pharmacological activities of terpenoids. *Natural Product Communications* 15(3):1–13.

80. Masyita A., Mustika Sari R., Dwi Astuti A., Yasir B., Rahma Rumata N., et al. 2022. Terpenes and terpenoids as main bioactive compounds of essential oils, their roles in human health and potential application as natural food preservatives. *Food Chemistry X* 13:100217.

81. Hosseinzadeh Z., Ramazani A., Razzaghi-Asl N. 2019. Plants of the genus *Heracleum* as a source of coumarin and furanocoumarin. *Journal of Chemical Reviews* 1(2):78–98.

82. Goleniowski M., Bonfill M., Cusido R., Palazón J. 2013. Phenolic Acids. In: Ramawat K., Mérillon J.M. (eds). *Natural Products*. Springer-Verlag, Berlin–Heidelberg, Germany.

83. Kiokias S., Proestos C., Oreopoulou V. 2020. Phenolic acids of plant origin – A review on their antioxidant activity *in vitro* (O/W emulsion systems) along with their *in vivo* health biochemical properties. *Foods* 9(4):534.

84. Melo L.F.M.D., Aquino-Martins V.G.D.Q., Silva A.P.D., Oliveira Rocha H.A., Scortecci K.C. 2023. Biological and pharmacological aspects of tannins and potential biotechnological applications. *Food Chemistry* 414:135645.

85. Camarda L., Di Stefano V., Pitonzo R. 2011. Natural resins: Chemical constituents and medicinal uses. In: Song D.B. (ed). *Resin Composites: Properties, Production and Applications*. Materials Science and Technologies Series, Nova Science Publishers, New York, NY, USA, pp. 353–374.

86. Hamdani A.M., Wani I.A., Bhat N.A. 2019. Sources, structure, properties and health benefits of plant gums: A review. *International Journal of Biological Macromolecules* 135:46–61.

87. Srivastava P., Malviya R. 2011. Sources of pectin, extraction and its applications in pharmaceutical industry – An overview. *Indian Journal of Natural Products and Resources* 2(1):10–18.

88. Upadhyay R. 2017. Nutritional, therapeutic, and pharmaceutical potential of plant gums: A review. *International Journal of Green Pharmacy* 11(1):S30–S48.

89. Simionescu I. 1939. *Flora României*. Fundaţia pentru Literatură şi artă Regele Carol II, Bucureşti.

90. Bichir G. 1976. *The archaeology and history of the Carpi from the second to the fourth century A.D. Parts I and II*. British Archaeological Reports, Oxford, UK.

91. Burian J. 2006. Carpi. In: Cancik H., Schneider H., Salazar C.F., Landfester M., Gentry F.G. (eds). *Brill's New Pauly*. Antiquity Volumes, Classical Tradition, Brill–De Gruyter, Berlin, Germany.

92. von Bredow I. 2006. Carpathians. In: Cancik H., Schneider H., Salazar C.F., Landfester M., Gentry F.G. (eds). *Brill's New Pauly*. Antiquity Volumes, Classical Tradition, Brill–De Gruyter, Berlin, Germany.

93. Brătianu G.I., Manuilă S., Vulcănescu M., Conea I., Golopenţia A. (eds). 1941. *Geopolitică şi geoistorie*. Ed. Societăţii Române de Statistică, Bucureşti.

94. Mehedinți S. 1938. *Geografie și geografi la începutul sec. XX – Însemnări cu privire la desvoltarea științelor și a învățământului în România.* Ed. Librăriei Socec & Co., București.

95. Pătru I. 2006. *Geografia fizică a României: climă, ape, vegetație, soluri.* Ed. Universitară, București, pp. 35–111.

96. Coldea G., Stoica I.A., Pușcaș M., Ursu T., Oprea A., The IntraBioDiv Consortium. 2009. Alpine–subalpine species richness of the Romanian Carpathians and the current conservation status of rare species. *Biodiversity and Conservation* 18(6):1441–1458.

97. Hurdu B.I., Pușcaș M., Turtureanu P.D., Niketić M., Vonica G., Coldea G. 2012. A critical evaluation of the Carpathian endemic plant taxa list from the Romanian Carpathians. *Contribuții Botanice – Grădina Botanică "Alexandru Borza" Cluj-Napoca* XLVII:39–47.

98. Muica E.C., Popova-Cucu A. 1993. The composition and conservation of Romania's plant cover. *GeoJournal* 29(1):9–18.

99. Romanian Ministry of Waters, Forests and Environment Protection. 1998. *Romania – Convention on Biological Diversity.* Bucharest.

100. Oprea A., Sîrbu C. 2013. The vascular flora of Rarău Massif (Eastern Carpathians, Romania). Note II. *Memoirs of the Scientific Sections of the Romanian Academy* XXXVI:17–52.

101. Yuan H., Ma Q., Ye L., Piao G. 2016. The traditional medicine and modern medicine from natural products. *Molecules* 21(5):559.

102. Hurdu B., Barina Z., Mráz P., Novikov A., Pușcaș M., Ronikier M., Šibík J. 2018. Endemic flora of the Carpathians: The importance of digitally integrating scientific information of major Carpathian region herbaria. *Visnyk of the Lviv University Biological Series* 78:56–59.

1   *Alliaceae* J.G. Agardh Family

Cornelia Bejenaru[1], Antonia Radu[1], George Dan Mogoşanu[2],
Ludovic Everard Bejenaru[2], Andrei Biţă[2], and Adina-Elena Segneanu[3*]

[1] Department of Pharmaceutical Botany, Faculty of Pharmacy, University of Medicine
and Pharmacy of Craiova, 2 Petru Rareş Street, Craiova, Dolj County, Romania
[2] Department of Pharmacognosy & Phytotherapy, Faculty of Pharmacy, University of
Medicine and Pharmacy of Craiova, 2 Petru Rareş Street, Craiova, Dolj County, Romania
[3] Institute for Advanced Environmental Research–West University of
Timisoara (ICAM–WUT), 4 Oituz Street, Timisoara, Romania

BOTANICAL DESCRIPTION

Family with perennial herbaceous plants, with tunicate bulbs attached to a short rhizome. The
leaves are basal, and the stem is scapiform. The flowers are hermaphroditic, actinomorphic,
trimerous, arranged in uniparous cymes inflorescences, in umbeliform or capituliform groups.
The inflorescences are protected by protective bracteate leaves (spathes), and the flowers have
septate nectaries. The fruit is a loculicidal capsule, and the seeds are black. The family includes 2
genera: *Nectaroscordum* Lindl. with a single species and *Allium* L. with 26 species in the spontaneous
flora of Romania.[1–3]

Allium L.

Genus with perennial herbaceous plants, having a tunicate bulb attached to an oblique or
horizontal rhizome. The stem is fistulous or solid, cylindrical or compressed-angular, smooth
or striated, and the leaves are flat or fistulous, with a characteristic odor. The flowers are
hermaphroditic, small, campanulate, hexamerous, with petaloid tepals, single-nerved, non-
unguiculate, arranged in two whorls of three. The stamens have widened filaments, often with
appendages, and a superior ovary that is trilocular with 2 ovules in each locule. The flowers
are melliferous, forming umbeliform, globose inflorescences, protected by fused bracts (fused
spathes). The capsule contains numerous seeds and dehisces into three valves. The seeds are black,
spherical, or compressed-angular, typically 1–2 (6) in each locule. This genus comprises 400 species
distributed in Europe, Asia, North America, and North Africa. The genus name derives from the
Latin name for garlic, given by the Romans.[1–4]

Allium ursinum L. (Wild garlic)

Scientific name: *Allium ursinum* L.
Common name: Wild garlic, Ramsons
Romanian name: Leurdă
Taxonomic classification
Kingdom: *Plantae*
Subkingdom: *Cormobionta*
Phylum: *Spermatophyta*
Subphylum: *Magnoliophytina (Angiospermae)*
Class: *Liliatae (Liliopsida, Monocotyledonatae)*
Subclass: *Liliidae*
Order: *Liliales*
Family: *Alliaceae*
Genus: *Allium*
Species: *ursinum* L.[1–3]
Subspecies: – *ursinum* is rarely distributed in our country, found in the Caraş–Severin and Sibiu
counties. It has scabrous pedicels with papillae (Figure 1.1).[2–3]

   – *ucrainicum* Kleopow et Oxner is frequently distributed and has smooth pedicels.[2–3]

SPREAD

It grows in deciduous forests on humus-rich, relatively moist soils, ranging from oak forests to the
beech forest zone.[1–4]

DOI: 10.1201/9781003270515-2

Figure 1.1 *Allium ursinum* L. (Wild garlic).

Source: 5.

DESCRIPTION

It is a perennial herbaceous species with a height of 20–50 cm and a strong garlic-like odor. The bulb is solitary, occasionally 2–3, lacks a rhizome, and is attached to the tip of an elongated axis. It has few parallel fibers, and the basal part has several rigid, sclerified, brown scales. The basal leaves (2, occasionally 3) are long-petiolate, with petioles longer than 5 cm, elliptic in shape, and emit a distinct garlic smell. They are glabrous, with a sharp tip and an attenduated base, featuring 15–20 anastomosed veins. The stem is straight, arranged laterally in relation to the leaves, and is triangular and solid. The fragrant flowers have tepals measuring 7–12 mm and form umbeliform inflorescences with a white membranous spathe that is deciduous and equal in length to the inflorescence. and one seed, occasionally 2, in each locule. The flowering period is from May to June.[1–4]

Main phytoconstituents	Ref.
amino acids (asparagine, glutamine, aspartic acid, glutamic acid, arginine, alanine, glycine, threonine, cysteine)	6–8
alkaloids	
flavonoids (kaempherol, hyperoside, hesperedide, diosgenin)	
terpenoids (cumene, ionone, spathulenol, caryophyllene, phytol)	
sterols (β-sitosterol)	
fatty acids (palmitic, linoleic, oleic, palmitoleic, stearic, α-linolenic, myristic acid)	
phenolic acids (p-coumaric acid, gallic acid, ferulic, vanillic acid)	
sulphur compounds (glutamyl peptides, sulfoxides, thiosulphinates, (poly)sulfides, alliin, methiin, isoalliin, propiin	
nucleoside	
carbohydrates	
aldehydes&ketones	

hydrocarbons
miscellaneous (carotenoids)

Medicinal Use

The vegetable product *Allii ursini bulbus et folium* is used for its hypocholesterolemic, uricosuric,[1,6] antiscorbutic,[4] detoxifying, diuretic, antiseptic, bactericidal, antitoxic, stomachic, carminative, febrifuge, vermifuge, expectorant, hypotensive, hemostatic, digestive disorder-alleviating, and tonic properties.[7,9,10]

Biological Activity

■ antioxidant, antimicrobial, antimycotic, cytotoxic, antiproliferative, cardioprotective[7]

It is used in digestive conditions (helminthiasis), metabolic disorders (gout), cardiovascular issues (hypertension, atherosclerosis), and renal problems (cystitis, nephritis).[6]

Other Uses

■ culinary (pesto, soups, gnocchi, risotto, ravioli, spice for cheese, salads)

■ dietary supplement[7]

■ household disinfectant

■ insecticide[10]

Allium victorialis L. (Alpine leek)

Scientific name: *Allium victorialis* L.
Common name: Victory onion, Alpine leek, Alpine broad-leaf allium
Romanian name: Ceapă de munte, Ai de munte
Taxonomic classification
Kingdom: *Plantae*
Subkingdom: *Cormobionta*
Phylum: *Spermatophyta*
Subphylum: *Magnoliophytina (Angiospermae)*
Class: *Liliatae (Liliopsida, Monocotyledonatae)*
Subclass: *Liliidae*
Order: *Liliales*
Family: *Alliaceae*
Genus: *Allium*
Species: *victorialis* L. (Figure 1.2)[2-3]

SPREAD

This species is sporadically distributed from the spruce forest zone to the subalpine zone, through meadows, grassy areas, rocky terrain, or peat bogs.[2-4]

DESCRIPTION

It is a perennial herbaceous plant, with a height of 30–60 cm, and it has a cylindrical or cylinder-conical bulb, usually solitary, occasionally 2–3, positioned at the extremity of an obliquely ascending rhizome, with a length of 3–10 cm. The bulb is composed of intersecting reticulate fibers and has brown or brownish-gray tunics. The stem is vertical, angular at the superior part, and cylindrical at the base. There are 2–4 leaves, shortly petiolate, with sheaths bearing linear, violet striations, encircling 1/3–3/4 of the stem. The inferior leaves are sessile, while the superior ones have short petioles, are flat, smooth, elongated-lanceolate, with obtuse tips and attenuated bases into the petioles. The flowers are greenish-white or white, fragrant, with a perigone initially campanulate, later expanded, and tepals shorter than the stamens, measuring 4–6 mm in length. The spathe is membranous, whitish, equal to or shorter than the inflorescence. The capsule is globular-trilobed, with broadly obcordate valves, and contains spherical seeds. The flowering period is from July to August.[2-4]

Figure 1.2 *Allium victorialis* L. (Alpine leek).

Source: 11.

Main phytoconstituents   Ref.
amino acids (alanine)   [12–15]
flavonoids (quercetin, astragalin, icariside, siringin, kaempherol, roseoside, rutin,
    allivictoside A, allivictoside B, allivictoside C, allivictoside D, allivictoside E, allivictoside
    F, allivictoside G, allivictoside H)
terpenoids (β-amyrin)
sterols (β-sitosterol)
alkaloids
phenolic acids (ferulic acid, p-coumaric acid)
saponins
sulfur compounds (methyl allyl disulfide, diallyl disulfide, S-methylthiosulfinate, methyl
    allyl trisulfide, 1,4-dithiane)
aldehydes & ketones
hydrocarbons
miscellaneous (alliumonoate)

Medicinal Use

■ skin problems (wounds, ulcers, eczema), gynecological disorders (fibroids, cysts, endometriosis,
    infertility, dysmenorrhea, menorrhagia), diuretic, antianemic and antidiabetic, antihemorrhagic,
    antispasmodic, regulation of thyroid, and reproductive hormones[12–15]

Biological Activity

■ antibacterial, antioxidant, anticancer, hepatoprotective, neuroprotective, antifungal, antiviral,
    gastroprotective, anti-inflammatory, wound healing, antiproliferative, enzyme inhibitory
    activity, gynecological disorders[12–15]

Other Uses

- ornamental
- culinary use (salads, pesto)[12–15]

REFERENCES

1. Bejenaru C., Mogoșanu G.D., Bejenaru L.E., Popescu H. 2020. *Botanică farmaceutică. Cormobionta*, ediția a III-a. Ed. Sitech, Craiova.

2. Ciocârlan V. 2009. *Flora ilustrată a României. Pteridophyta et Spermatophyta*. Ediția a 3-a revizuită și adăugită. Ed. Ceres, București.

3. Sârbu I., Ștefan N., Oprea A. 2013. *Plante vasculare din România. Determinator ilustrat de teren*. Ed. Victor B Victor, București.

4. Săvulescu T. (ed). 1966. *Flora RPR*, vol. XI. Ed. Academiei RPR, București.

5. https://dryades.units.it/floritaly/index.php?procedure=taxon_page&tipo=all&id=6901

6. Bejenaru L.E., Mogoșanu G.D., Bejenaru C., Popescu H. 2015. *Farmacognozie-Fitoterapie*. Vol. I, ediția a III-a. Ed. Sitech, Craiova.

7. Sobolewska D., Podolak I., Makowska-Wąs J. 2015. *Allium ursinum*: botanical, phytochemical and pharmacological overview. *Phytochemistry Reviews* 14(1):81–97.

8. Stupar A., Šaric L., Vidovic S., Bajic A., Kolarov V., Šaric B. 2022. Antibacterial potential of *Allium ursinum* extract prepared by the green extraction method. *Microorganisms* 10: 1358

9. www.remediilenaturii.ro/remediu_fitoterapie.php?n=Leurd%C4%83&ID=233

10. https://pfaf.org/user/plant.aspx?latinname=Allium+ursinum

11. https://dryades.units.it/floritaly/index.php?procedure=taxon_page&tipo=all&id=6892

12. Lee K.T., Choi J.H., Kim D.H., Son K.H., Kim W.B., Kwon S.H., Park H.J. 2001. Constitutents and the antitumor principle of *Allium victorialis* var. *platyphyllum*. *Archives of Pharmacal Research* 24(1):44–50.

13. Khan S., Kazmi M.H., Inamullah F., Itrat F. 2019. A review on phytochemical constituents and pharmacological properties of *Allium victorialis*. *International Journal of Scientific&Engineering Research* 10(6):278–288.

14. Kyung-Tae L., Jaesoon C., Dae-Hyun K., Kun-Ho S., Won-Bae K., Sang-Hyuk K., Hee-Juhn P. 2001. Constituents and the antitumor principle of *Allium victorialis var. platyphyllum*. *Archives of Pharmacal Research* 24:44–50.

15. Young S.K., Jung D.H., Lee I.S., Choi S.J., Yu S.Y., Ku S.K., Kim M.H., Kim J.S. 2013. Effects of *Allium victorialis* leaf extracts and its single compounds on aldose reductase, advanced products and TGF-β1 expression in mesangial cells. *BMC Complementary and Alternate Medicine* 13: 251.

2 *Amaryllidaceae* Juss. Family

Cornelia Bejenaru[1], Antonia Radu[1], George Dan Mogoşanu[2],
Ludovic Everard Bejenaru[2], Andrei Biţă[2], and Adina-Elena Segneanu[3]*
[1] Department of Pharmaceutical Botany, Faculty of Pharmacy, University of Medicine and Pharmacy of Craiova, 2 Petru Rareş Street, Craiova, Dolj County, Romania
[2] Department of Pharmacognosy & Phytotherapy, Faculty of Pharmacy, University of Medicine and Pharmacy of Craiova, 2 Petru Rareş Street, Craiova, Dolj County, Romania
[3] Institute for Advanced Environmental Research–West University of Timisoara (ICAM–WUT), 4 Oituz Street, Timisoara, Romania

BOTANICAL DESCRIPTION

Perennial plants with bulbs, rhizomes, bulbotubers, and thickened roots that grow in warm and temperate regions. The stem is scapiform, and the basal leaves are linear or elliptical. The flowers are of type 3, hermaphroditic, and actinomorphic or zygomorphic. At the base of the flowers, there are hypsophylls. The flowers have 6 petaloid tepals in 2 verticils, either free or united. In some species, a coronule (paracorolla) is present between the inner whorl of tepals and the androecium, which can be free at the base or fused into a tube. The androecium consists of 6 stamens (3+3), and the gynoecium is tricarpellary and trilocular, with the ovary located inferiorly, exhibiting central, rarely parietal, placentation, and anatropous ovules arranged in two rows in each locule. The fruit is a loculicidal capsule or berry. The seeds are rounded or flat, with or without appendages.[1-4]

Many species from this family are appreciated for their ornamental qualities, which is why they are cultivated. In Romania, there are 5 genera with a total of 17 species, with only 6 of them being spontaneous.[2]

The following medicinal plants from the spontaneous flora of Romania are presented, belonging to this family, on which chemical or biological screening were carried out.

Galanthus nivalis L. (Common snowdrop)

Scientific name: *Galanthus nivalis* L.
Common name: Common snowdrop
Romanian name: Ghiocei
Taxonomic classification
Kingdom: *Plantae*
Subkingdom: *Cormobionta*
Phylum: *Spermatophyta*
Subphylum: *Magnoliophytina (Angiospermae)*
Class: *Liliatae (Liliopsida, Monocotyledonatae)*
Subclass: *Liliidae*
Order: *Liliales*
Family: *Amaryllidaceae*
Genus: *Galanthus*
Species: *nivalis* L. (Figure 2.1)[1-3]

SPREAD

It grows spontaneously in forests, shrublands, and meadows across a range of habitats from the lowlands to the mountains.[1-4]

DESCRIPTION

It is a small, perennial herbaceous species (10–30 cm in height) with tunicate bulbs, which are ovoid or globose, covered in brown tunics, from which two linear basal leaves are formed, that are flat with a slightly pronounced keel, and a scapiform, compressed stem, bearing a single actinomorphic flower with two protective bracts (spathe): one long spathe with two green keels and white, membranous edges, and a short spathe covered by the long one. The floral envelope is a white petaloid perigone, dialytpetalous, with the internal tepals smaller, each with a green spot under

DOI: 10.1201/9781003270515-3

Figure 2.1 *Galanthus nivalis* L. (Common snowdrop).

Source: 5.

the emarginate tip. The stamens have short filaments and anthers with an acuminate, membranous appendage at the tip. The capsule is ovoid, with 1–5 seeds in each locule. The perfoliation is flat. It blooms in February–March. The genus name derives from the Greek words "gala" (milk) and "anthos" (flower).[1-4]

Main phytoconstituents   Ref.

alkaloids (galanthamine, lycoramine, nivalidine, galanthine, incartine, galanthusine,
   hippeastrine, graciline, crinine, haemanthamine, tazettine, narciclasine)   [6-8]

flavonoids (quercetin, isorhamnetin, isoquercitrin, quercetin, kaempherol, hyperoside, rutin)

terpenoids

sterols

fatty acids

phenolic acids (chlorogenic acid, vanillic acid, caffeic acid, gallic acid, syringic acid,
   p-coumaric acid, ferulic acid, anisic acid, rosmarinic acid, salicylic acid, cinnamic acid)

tannins

Medicinal Use

■ pain, migraine, headache, Alzheimer's disease, hypertension, heart diseases, neurological
   disorders, dermatological disorders (skin pigmentation on face and hands, ephelides),
   panaritium [9-10]

Biological Activity

The plant is a source for obtaining galantamine, which is used in the treatment of myasthenia gravis, myopathies, and polyneuropathies.[1]

■ antioxidant, antibacterial, antifungal, antiparasitic (malaria), antiviral, analgesic, antitumor,
   cytotoxic, anti-inflammatory, antiaging[9-10]

Other uses

- ornamental
- insecticidal
- dye (yellow) for natural fibers
- cosmetic (antiage products)
- melliferous[9–10]

Leucojum aestivum L. (Summer snowflake)

Scientific name: *Leucojum aestivum* L.
Common name: Summer snowflake, Loddon lily
Romanian name: Ghiocei de baltă
Taxonomic classification
Kingdom: *Plantae*
Subkingdom: *Cormobionta*
Phylum: *Spermatophyta*
Subphylum: *Magnoliophytina (Angiospermae)*
Class: *Liliatae (Liliopsida, Monocotyledonatae)*
Subclass: *Liliidae*
Order: *Liliales*
Family: *Amaryllidaceae*
Genus: *Leucojum*
Species: *aestivum* L. (Figure 2.2)[2–3]

Figure 2.2 *Leucojum aestivum* L. (Summer snowflake).
Source: 11.

SPREAD

The species is sporadically distributed from the sylvosteppe zone to oak forests, occurring in groves, reed beds, swamps, meadows, and wet forests along the banks of water bodies.[2-4]

DESCRIPTION

It is a perennial herbaceous plant, reaching a height of 30–60 cm, with an ovoid bulb covered by membranous, whitish or grayish tunics. The 5–7(9) leaves are broadly linear, shiny, concave canaliculate, often exceeding the compressed, scapiform stem. The spathe is lanceolate and membranous, with 2 green keels. The flowers, measuring 14–17 cm in length, are grouped 2–8 in an umbelliform inflorescence. The flowers are campanulate and white, with broadly lanceolate tepals that have a yellow-greenish spot at the tip. The seeds are black, without a strophiole. The flowering period is from May to June.[2-3]

Main phytoconstituents	Ref.
alkaloids (galanthamine, lycoramine, nivalidine, galanthine, incartine, galanthusine, hippeastrine, graciline, crinine, haemanthamine, tazettine, narciclasine)	[12]
flavonoids (naringenin, quercetin, isorhamnetin, isoquercitrin, quercetin, kaempherol, hyperoside, rutin)	
terpenoids	
sterols	
fatty acids	
phenolic acids (rosmarinic acid, p-coumaric acid, syringic acid, gallic acid, gentisic acid)	

Medicinal Use

- poliomyelitis, dystrophy, myasthenia, myopathy, and paralysis in newborns, neuritis, neurodermatitis[13]

Biological Activity

- antibacterial, antiretroviral, cytostatic, anticancer[14]

Other Uses

- ornamental

- cosmetic (antiaging products)[12-13]

Sternbergia colchiciflora Waldst. et Kit. (Autumn crocus)

Scientific name: *Sternbergia colchiciflora* Waldst. et Kit.
Common name: Autumn crocus, Slender sternbergia
Romanian name: Ghiocel de toamnă[15]
Taxonomic classification
Kingdom: *Plantae*
Subkingdom: *Cormobionta*
Phylum: *Spermatophyta*
Subphylum: *Magnoliophytina (Angiospermae)*
Class: *Liliatae (Liliopsida, Monocotyledonatae)*
Subclass: *Liliidae*
Order: *Liliales*
Family: *Amaryllidaceae*
Genus: *Sternbergia*
Species: *colchiciflora* Waldst. et Kit. (Figure 2.3)[2-3]

SPREAD

The species is rarely found from the lowland area up to the sessile oak forest zone, in dry, rocky meadows, forest clearings, and rocky slopes. It can be found in the counties of Dolj, Mehedinți, Caraș-Severin, Giurgiu, Tulcea, Buzău, and Prahova.[2-4]

Figure 2.3 *Sternbergia colchiciflora* Waldst. et Kit. (Autumn crocus).

Source: 16.

DESCRIPTION

It is a perennial herbaceous plant that blooms in the autumn, while its leaves and fruit develop in the spring. It has an ovoid bulb covered in brown tunics, and typically, it has 4–6 leaves surrounded at the base by an aphyllous sheath, smooth or ciliately scabrous on the edges, which is usually erect. The stem is a scape, ranging in height from 1–6 cm, compressed and short during the flowering phase, and accrescent in the fruiting phase, producing a globose fruit at ground level. The seeds are brown with pronounced appendages. The flowering period is from September to October, and fruiting occurs from February to April.[2–4]

Main phytoconstituents   Ref.
amino acids   [17–20]
alkaloids (galanthamine, lycoramine, nivalidine, galanthine, incartine, galanthusine,
    hippeastrine, graciline, crinine, haemanthamine, tazettine, narciclasine)
phenolic acids
flavonoids
terpenoids
sterols
fatty acids
miscellaneous

Medicinal Activity

■ antidiabetic, anthelmintic[19–20]

Biological Activity

■ antioxidant, anti-inflammatory, antidiabetic, antimicrobial, antiviral, antitumor[19–21]

Other Uses

■ ornamental[19–20]

REFERENCES

1. Bejenaru C., Mogoșanu G.D., Bejenaru L.E., Popescu H. 2020. *Botanică farmaceutică. Cormobionta*, ediția a III-a. Ed. Sitech, Craiova.

2. Ciocârlan V. 2009. *Flora ilustrată a României. Pteridophyta et Spermatophyta.* Ediția a 3-a revizuită și adăugită. Ed. Ceres, București.

3. Sârbu I., Ștefan N., Oprea A. 2013. *Plante vasculare din România. Determinator ilustrat de teren.* Ed. Victor B Victor, București.

4. Săvulescu T. (ed). 1966. *Flora RPR*, vol. XI. Ed. Academiei RPR, București.

5. https://dryades.units.it/valerio/index.php?procedure=taxon_page&id=6989&num=1764

6. Benedec D., Oniga I., Hanganu D. et al. 2018. Sources for developing new medicinal products: biochemical investigations on alcoholic extracts obtained from aerial parts of some Romanian *Amaryllidaceae* species. *BMC Complementary and Alternate Medicine* 18:226.

7. Berkov S., Codina C., Bastida J. 2012. The genus *Galanthus:* A source of bioactive compounds. Phytochemicals. In: Rao A.V. (ed). Phytochemicals – A Global Perspective of Their Role in Nutrition and Health. InTech Europe, Rjeka, Croatia, pp. 235–254.

8. Bulduk I., Karafakıoğlu Y. 2019. Evaluation of galantamine, phenolics, flavonoids and antioxidant content of *Galanthus* species in Turkey. *International Journal of Biochemistry Research & Review* 25(1):1–12.

9. www.remediilenaturii.ro/remediu_fitoterapie.php?n=Ghiocel&ID=225

10. https://powo.science.kew.org/taxon/urn:lsid:ipni.org:names:64496-11

11. https://dryades.units.it/lamone/index.php?procedure=taxon_page&id=6986&num=1763

12. Al-Faris H.D.H., Bulduk I., Kahraman A. 2019. Biochemical and micro-morphoanatomical investigations on *Leucojum aestivum L. Notulae Botanicae Horti Agrobotanici Cluj-Napoca* 47(4):1382–1393.

13. Demir A. 2014. Medical resource value appraisal for *Leucojum aestivum* in Turkey. *American Journal of Alzheimer's Disease & Other Dementias* 29(5):448–451.

14. Georgiev V., Ivanov I., Berkov S., Ilieva M., Georgiev M., Gocheva T., Pavlov A. 2012. Galanthamine production by *Leucojum aestivum L.* shoot culture in a modified bubble column bioreactor with internal sections. *Engineering in Life Sciences* 12(5):534–543.

15. www.salvaeco.org/plante/page/sternbergia_colchiciflora.php

16. https://link.springer.com/article/10.1007/s42977-020-00018-4

17. Can Ağca A., Yazgan Ekici A.N., Yılmaz Sarıaltın S., Çoban T., Saltan Işcan G., Sever Yılmaz B. 2021. Antioxidant, anti-inflammatory and antidiabetic activity of two *Sternbergia taxons* from Turkey. *South African Journal of Botany* 136:105–109.

18. Acikara Ö.B., Yilmaz B.S., Yazgan D., Işcan G.S. 2019. Quantification of galantamine in *Sternbergia species* by high performance liquid chromatography. *Turkish Journal Of Pharmaceutical Sciences* 16(1):32–36.

19. Unver N., Irem Kaya G., Tansel Ozturk H. 2005. Antimicrobial activity of *Sternbergia sicula* and *Sternbergia lutea. Fitoterapia* 76(2):226–229.

20. Iscan G.S., Sarialtin S.Y., Yazgan A., Yilmaz B.S., Agca A.C., Coban T. 2017. The antioxidant, anti-inflammatory and antidiabetic activities of *Sternbergia lutea ssp. lutea* and *Sternbergia lutea ssp. sicula. Toxicology Letters* 280:S89.

21. Georgiev V., Ivanov I., Pavlov A. 2020. Recent progress in Amaryllidaceae biotechnology. *Molecules* 25(20):4670.

3 *Apiaceae* (Lindl.) (*Umbelliferae* Juss.) Family

Cornelia Bejenaru[1], Antonia Radu[1], George Dan Mogoşanu[2],
Ludovic Everard Bejenaru[2], Andrei Biţă[2], and Adina-Elena Segneanu[3*]

[1] Department of Pharmaceutical Botany, Faculty of Pharmacy, University of Medicine and Pharmacy of Craiova, 2 Petru Rareş Street, Craiova, Dolj County, Romania
[2] Department of Pharmacognosy & Phytotherapy, Faculty of Pharmacy, University of Medicine and Pharmacy of Craiova, 2 Petru Rareş Street, Craiova, Dolj County, Romania
[3] Institute for Advanced Environmental Research–West University of Timisoara (ICAM–WUT), 4 Oituz Street, Timisoara, Romania

BOTANICAL DESCRIPTION

Herbaceous annual, biennial, or perennial plants, less commonly woody, with taproots or rhizomes. The stem is finely striped to grooved, fistulous. The leaves are without stipules, arranged alternately, with pinnately divided (rarely palmate or round) lamina and well-developed, channelled or ventricose-inflated sheaths. The flowers are grouped in compound umbel inflorescences; rarely, inflorescences are simple umbels or capitula. The bracts at the base of the umbel are called involucres, and those at the base of the umbellets form the involucel. The flowers are small, hermaphroditic, actinomorphic or slightly zygomorphic, pentamerous in type. The floral envelope consists of 5 small, dentiform sepals, and 5 free petals inserted at the base of the stylopodium, entire, emarginate, or with a sharp tip. The androecium consists of 5 equal stamens, alternating with the petals, and the gynoecium is bicarpellate syncarpous, with an inferior bilocular ovary, continued with 2 styles at the base of which there is a nectariferous disk (stylopodium). The fruit is a dicaryopsis composed of 2 mericarps, which are attached to a central axis (carpophore), often bifid. Each mericarp (semi-fruit) has a flat internal (ventral) face and a convex and longitudinally ribbed external (dorsal) face. The external face of the mericarp has 5 longitudinal ribs (3 dorsal and 2 lateral), alternating with 5 grooves called vittae. Secondary ribs may appear longitudinally at the level of the vittae. The pericarp contains conducting fascicles (along the ribs) and secretory canals (in the vittae). The seed has an endosperm rich in oil and aleurone grains.[1–4]

A highly diverse family, comprising 150 genera that encompass around 3,000 species, with Romania hosting 57 genera and 131 species.[1,5–6]

The most representative medicinal plants from the traditional Romanian medicine were selected in order to describe in detail.

Sanicula europaea L. (Wood sanicle)

Scientific name: *Sanicula europaea* L.
Common name: Wood sanicle
Romanian name: Sânişoară
Taxonomic classification
Kingdom: *Plantae*
Subkingdom: *Cormobionta*
Phylum: *Spermatophyta*
Subphylum: *Magnoliophytina (Angiospermae)*
Class: *Magnoliatae (Magnoliopsida, Dicotyledonatae)*
Subclass: *Rosidae*
Order: *Apiales (Umbellales)*
Family: *Apiaceae (Umbelliferae)*
Subfamily: *Saniculoideae*
Genus: *Sanicula*
Species: *europaea* L. (Figure 3.1)[1–3]

SPREAD

Often found in shady forests and thickets, on moist and humus-rich soils, ranging from the lowland to the mountainous regions.[1,4]

Figure 3.1 *Sanicula europaea* L. (Wood sanicle).

Source: 7.

DESCRIPTION

The genus name derives from the Latin word "sanare," which means "to heal," and the Romans used it for wound healing.[4]

It is a perennial herbaceous plant with a short, vertical rhizome covered with petiole remnants and a stem that is either leafless or has a few reduced leaves, glabrous, and reaches a height of 20–50 cm. The leaves are palmately 3–7 lobed, with basal leaves being long-petiolate and glossy, while stem leaves are short-petiolate or sessile. The flowers are sessile and form simple capituliform umbels grouped in cymes with small involucral bracts. The corolla has obcordate petals that can be white or reddish. The dicaryopsis is covered with soft, uncinate prickles. The flowering period spans from May to June.[1,2,4]

Main phytoconstituents	Ref
terpenoids (selinene, caryophyllene, α-pinene, nonane, limonene, longifolene, cymene, β-bisabolene, cadinene, saniculoside N, saniculoside R1)	8–9
flavonoids	
saponin (saniculasaponins, sandrosaponin)	
sterols	
fatty acids (palmitic acid)	
phenolic acids (rosmarinic acid, caffeic acid)	
hydrocarbons	
aldehydes & ketones	

Medicinal Use

The aerial part of the plant (*Saniculae herba*) is used as an astringent, cicatrizant, and anti-inflammatory.[1]

Biological Activity

- antioxidant, antiexsudative, antimycotic, and antiviral (e.g., anti-HIV) activities[9–11]

Other Uses

Leaves and tender shoots can be prepared by cooking, but they should not be consumed in excessive amounts due to their saponin content. This plant is considered a last resort for food during times of famine.[12]

Eryngium planum L. (Sea holly)

Scientific name: *Eryngium planum* L.
Common name: Sea holly
Romanian name: Scai vânăt
Taxonomic classification
Kingdom: *Plantae*
Subkingdom: *Cormobionta*
Phylum: *Spermatophyta*
Subphylum: *Magnoliophytina (Angiospermae)*
Class: *Magnoliatae (Magnoliopsida, Dicotyledonatae)*
Subclass: *Rosidae*
Order: *Apiales (Umbellales)*
Family: *Apiaceae (Umbelliferae)*
Subfamily: *Saniculoideae*
Genus: *Eryngium*
Species: *planum* L. (Figure 3.2)[1–3]

SPREAD

Frequently found from the lowlands up to the beech forest zone, in meadows, forest edges, riverbanks, sandy areas, pastures, clearings, and ruderal sites.[1–4]

Figure 3.2 *Eryngium planum* L. (Sea holly).
Source: 13.

DESCRIPTION

A perennial herbaceous species, equipped with a fusiform rhizome covered by the petioles of leaves from the previous years, and a stem reaching 30–60 cm in height, branching into 3–5 bluish branches in the superior part. Basal leaves are long-petiolate, entire, ovate, with a cordate base and serrated setiform margins. Inferior stem leaves are short-petiolate, while superior stem leaves are sessile, amplexicaul, 3–5 palmately divided, with dentate lobes. The capitulum is bluish, ovoidal or subglobose, surrounded by lanceolate involucral bracts, narrow and rigid, exceeding the inflorescence. It blooms in the months of July and August.[1-4]

Main phytoconstituents   Ref

terpenoids (germacrene-D, α-muurolene, curcumene, eryngiumgenins, erynginol A,
   steganogenin)   [14-15]

flavonoids (rutin, kaempferol, quercetin, daidzein, kaempferol, isorhamnetin, genistein,
   luteolin, apigenin)

saponin (saniculasaponin II and III)

sterols (cholesterol, brassicasterol, campesterol, stigmasterol

fatty acids

phenolic acids (rosmarinic acid, caffeic acid, chlorogenic acid, p-coumaric acid, ferulic acid,
   sinapic acid)

hydrocarbons

aldehydes & ketones

Medicinal Use

The aerial part harvested during flowering (*Eryngii plani herba*) has antitussive action and can be used to combat convulsive coughing.[1]

Biological Activity

■   antioxidant, anti-inflammatory, amebicidal, antimicrobial, antidiabetic, antitumor acivities[14,16-19]

Other Uses

This species can be used to attract wildlife.[20]

Carum carvi L. (Caraway)

Scientific name: *Carum carvi* L.
Common name: Caraway
Romanian name: Chimen
Taxonomic classification
Kingdom: *Plantae*
Subkingdom: *Cormobionta*
Phylum: *Spermatophyta*
Subphylum: *Magnoliophytina (Angiospermae)*
Class: *Magnoliatae (Magnoliopsida, Dicotyledonatae)*
Subclass: *Rosidae*
Order: *Apiales (Umbellales)*
Family: *Apiaceae (Umbelliferae)*
Subfamily: *Apioideae*
Genus: *Carum*
Species: *carvi* L. (Figure 3.3)[1-3]

SPREAD

Commonly found from the oak zone to the beech zone, in pastures, nutrient-rich areas, meadows, forest edges, and ruderal sites.[1-4] Cultivated as a spice, aromatic, or medicinal plant.[1]

DESCRIPTION

A biennial or perennial herbaceous plant with a thick taproot, sometimes branched, and a grooved, fistulous stem. The superior part of the stem is branched, sparsely foliated, and it can reach a height

Figure 3.3 *Carum carvi* L. (Caraway).

Source: 21.

of 30–50 (even 100) cm. The leaves are 2–3 times pinnatisect, with lanceolate or linear laciniae. Basal leaves are long-petiolate, while stem leaves are sessile. The flowers, which can be white or pink, are grouped in compound umbels and lack both involucres and involucels. The mericarps are ovate, with narrowed tips, slightly arcuate, and have 5 prominent main ribs. Flowering occurs in May to June.[1–4]

Main phytoconstituents	**Ref**
terpenoids (carvone, dihydrocarvone, α-pinene,limonene, γ-terpinene, linalool, *p*-cymene, linalool)	22–25

coumarins (herniarin, scopoletol)
alkaloids
flavonoids (quercetin, kaempferol)
glycosides
sterols (*β*-sitosterol)
fatty acids (linoleic acid, oleic acid, palmitic acid myristic acid, lauric acid, stearic acid, linolenic acid)
phenolic acids
carbohydrates (mannans)
hydrocarbons
lignins
aldehydes & ketones

Medicinal Use

Carvi fructus has the following actions and uses: gastrointestinal spasmolytic and carminative, especially useful for infants and young children; cholagogue in the treatment of hepatic insufficiency; expectorant in bronchitis; galactagogue.[4]

Biological Activity

Research indicates that this plant possesses various beneficial properties, including appetite suppression, alleviating oral mucositis, managing irritable bowel syndrome, addressing hypothyroidism, reducing blood pressure, acting as a vasodilator and cardiac modulator, combating tuberculosis, exhibiting anticancer potential, mitigating colitis, managing cardiovascular disease, addressing obesity, showcasing antimicrobial properties, controlling hyperlipidemia, regulating blood sugar levels, antiulcerogenic, antiproliferative and countering oxidative stress. This plant can be effectively utilized in the treatment and prevention of numerous health conditions, with minimal to no side effects.[23-26]

Toxicity

At high doses, *Carvi aetheroleum* is neurotoxic and abortifacient.[22]

Other Uses

Caraway seeds, known for their pungent taste, are employed in various pharmaceutical applications. They serve as a flavoring agent in medicinal confectionery, helping mask the taste of certain formulations. Additionally, *Carvi aetheroleum*, rich in aromatic compounds, finds application in the flavoring of pharmaceutical products like lozenges and syrups, enhancing their palatability, being also used in the perfume industry.[27]

Angelica archangelica L. (Garden angelica)

Scientific name: *Angelica archangelica* L.
Common name: Garden angelica
Romanian name: Angelică
Taxonomic classification
Kingdom: *Plantae*
Subkingdom: *Cormobionta*
Phylum: *Spermatophyta*
Subphylum: *Magnoliophytina (Angiospermae)*
Class: *Magnoliatae (Magnoliopsida, Dicotyledonatae)*
Subclass: *Rosidae*
Order: *Apiales (Umbellales)*
Family: *Apiaceae (Umbelliferae)*
Subfamily: *Apioideae*
Genus: *Angelica*
Species: *archangelica* L. (Figure 3.4)[1-3]

SPREAD

Sporadic in the mountainous region, along streams, in peat bogs, on the banks of water bodies, and at forest edges.[1-4]

DESCRIPTION

A robust biennial or perennial herbaceous species, reaching heights of up to 2 meters, with a thick napiform rhizome that becomes multicapitate at maturity and contains a white to yellowish sap. The stem is finely striped, erect, cylindrical, branched, and glabrous, with a very thick base and large glabrous leaves. The leaves are 2–3 times pinnatisect, with large ovate ultimate segments, featuring a narrow cartilaginous margin and an uneven doubly serrated edge. The leaf sheath is swollen and well-developed. The flowers are whitish-green, arranged in large compound umbels, lacking involucres and involucels. The mericarp has broad lateral ribs. It blooms in July to August.[1,4]

Main phytoconstituents   Ref

coumarins (angelicin, archangelicin, biakangelicin, decursinol angelate, osthenol, imperatorin, umbelliferone, isoimperatorin, decursin, umbelliprenin, lomatin, marmesin, bergapten, xanthotoxin, oxypeucedanin, osthole, pimpinellin)   [22,29-38]
flavonoids (rutin, myricetin, hesperidin, vanillin, oleuropein, resveratrol, apigenin, naringenin)

Figure 3.4 *Angelica archangelica* L. (Garden angelica).

Source: 28.

terpenoids (ligustilide, bisabolol, cadinol, phellandrene, elemene, pinene, limonene, sabinene, cymene, terpinene, myrcene, caryophyllene, eudesmol, thujene, germacrene, camphene, nothoapiole, cyclosativene, cubenene, α-copaene, δ-cadinene, β-barbatene, α-humulene, α-murolene, β-bisabolene)
sterols (xanthogonine, xanthogalol, xanthalin, ostruthol, isooxyposedan, β-sitosterol)
fatty acids (linoleic acid, oleic acid)
phenolic acids (ferulic acid, benzoic acid, ellagic acid, gallic acid, pyrogallol, chlorogenic acid, caffeic acid, syringic acid)
saponins
furanocoumarin glycosides (apterin, marmesinin)
tannins
hydrocarbons
aldehydes & ketones
others (butylidene phthalide, β-eudesmol, spathulenol)

Medicinal Use

The vegetable product *Angelicae radix* is used for neuroses, gastric ulcers, exhaustion, bronchial asthma, influenza, anorexia, nausea, indigestion, aerophagia, and abdominal bloating. It is also included in the composition of some bitter-type beverages.[1,22]

Biological Activity

These vegetable products have various pharmacological actions, including digestive, bitter tonic, stomachic, sedative, carminative, spasmolytic, anticonvulsant, muscle relaxant, antibacterial, diaphoretic,[1,22] antioxidant, antimicrobial, anti-inflammatory, antifungal, antiviral, antitumor, insecticidal, radioprotective, hepatoprotective, gastroprotective, and uterotonic activities;[36] the angelic acid in the volatile oil also exhibits nervous sedative and spasmolytic effects, similar to 2-methylbutenol found in hop cones, while archangelicin acts as a coronary vasodilator.[1,22]

Other Uses

- culinary

- gin preparation

- preservation of jams, fruits

- flavoring various alcoholic beverages

- perfumery[35,39]

Athamanta turbith (L.) Brot. (Candy carrot)

Scientific name: *Athamanta turbith* (L.) Brot.
Common name: Candy carrot
Romanian name: Brie
Taxonomic classification
Kingdom: *Plantae*
Subkingdom: *Cormobionta*
Phylum: *Spermatophyta*
Subphylum: *Magnoliophytina (Angiospermae)*
Class: *Magnoliatae (Magnoliopsida, Dicotyledonatae)*
Subclass: *Rosidae*
Order: *Apiales (Umbellales)*
Family: *Apiaceae (Umbelliferae)*
Subfamily: *Apioideae*
Genus: *Athamanta*
Species: *turbith* (L.) Brot. (Figure 3.5)[2–3]
Subspecies: – *hungarica* (Borbás) Tutin (*A. hungarica* Borbás)[2–3]

Figure 3.5 *Athamanta turbith* (L.) Brot. (Candy carrot).
Source: 40.

SPREAD

A rare species found in the oak to fir forest zone, on sunny limestone rocks. It is an endemic species of Romania, found in the Meridional Carpathian Mountains.[2–4]

DESCRIPTION

A perennial herbaceous plant with a cylindrical, thick, brown rhizome covered by the sheaths of the leaves from the previous year. The stem is initially ascending, then erect, branching from the base, cylindrical, unevenly finely striped, finely and densely grayish pubescent, with numerous leaves, and reaches a height of 10–50 cm. The leaves are triangular, 3–4 (5) times pinnately divided, with distanced, filiform or narrowly linear folioles, featuring entire margins, terminated by a mucron, and are either glabrous or sparsely short-pubescent. Basal leaves are long-petiolate and vaginate, stem leaves are sessile, with membranous sheaths, and those in the superior part are small and slightly divided. The flowers are grouped in compound umbels, without involucres, and with an involucel consisting of short, lanceolate, broadly membranous, pubescent, and ciliate on the margins folioles. The elongated ellipsoidal fruits are brown, with blunt, slightly pronounced ribs, covered in small, dense setiform hairs. The flowering period is from May to July.[2–4]

Main phytoconstituents	Ref
flavonoids (kaempferol, quercetin, avicularin, luteolin)	41–44

terpenoids (oleanolic acid, ursolic acid, hederagenin, p-cymene, thymol, camphene, carvacrol, pinene, myrcene, phellandrene, terpinene, p-cymene, limonene, terpinene, humulene, murolene, terpinolene, elemene, caryophyllene, bicyclogermacrene, amorphene, myristicin, germacrene B, spathulenol)

sterols

organic acid

fatty acids (myristic, pentadecanoic, palmitic, palmitoleic, stearic, petroselinic, oleic, linoleic, a-linolenic, arachidic, behenic acid)

phenolic acids (chlorogenic acid)

saponide

Carbohydrates

others (dodecanal, hexanal, apiole)

Medicinal Use

stomach tonic, inflammation of the upper respiratory tract (gingivitis, throat), hemostatic[36–39]

Biological Activity

antibacterial, astringent, tonic digestive, antidiarrheic, anti-inflammatory[36–39]

Other Uses

■ tea substitute.[36–39]

Heracleum palmatum Baumg. (Cow parsnip)

Scientific name: *Heracleum palmatum* Baumg. (*H. sphondylium* L. subsp. *transsilvanicum* (Schur) Brummit)
Common name: Cow parsnip, Hogweed
Romanian name: Talpa ursului
Taxonomic classification
Kingdom: *Plantae*
Subkingdom: *Cormobionta*
Phylum: *Spermatophyta*
Subphylum: *Magnoliophytina (Angiospermae)*
Class: *Magnoliatae (Magnoliopsida, Dicotyledonatae)*
Subclass: *Rosidae*
Order: *Apiales (Umbellales)*
Family: *Apiaceae (Umbelliferae)*
Subfamily: *Apioideae*

Figure 3.6 *Heracleum palmatum* Baumg. (Cow parsnip).

Source: 45.

Genus: *Heracleum*
Species: *palmatum* Baumg. (Figure 3.6)[2-3]

SPREAD

This species is an endemic plant found in the Carpathian Mountains, sporadically distributed from the beech forest zone up to the subalpine zone, growing in rocky valleys, coniferous forests, near streams, and forest clearings.[2-4]

DESCRIPTION

A robust, perennial herbaceous plant with a blackish-brown rhizome and an erect, strongly ridged, fistulous stem, branching, sparsely short-hairy, and reaching a height of 100–250 cm. Basal leaves are large, measuring 50 cm in length and 30 cm in width, deeply cordate, entire or palmately lobed to palmately partite, with 7–9 lobulate lobes. Stem leaves are sessile, round, with a prominently swollen sheath, and lanate-hairy on the external surface. The flowers have whitish to yellowish petals, grouped in compound umbels without involucres and with an involucel consisting of numerous folioles. The fruits are long-pedicellate, flattened dorsiventrally, and yellow or brown. It blooms from June to July.[2-4]

Main phytoconstituents	Ref
flavonoids (rhamnetin, quercetin, astragalin, apigenin, rutin, isoquercetin, hyperoside, isorhamnetin, pelargonidin, catechin, cyanidin, sexangularetin, scutellarin, baicalin)	[46-47]
terpenoids (squalene, geranylacetone, amyrin, urkish acid, oleanolic acid)	
organic acid (malonic acid, azelaic acid, succinic acid, benzoic acid, oxalic acid. Citric acid, pimelic acid)	

fatty acids (myristic acid, tricosanoic acid, behenic acid, palmitic acid, lignoceric acid palmitoleic acid, lauric acid, margaric acid)

phenolic acids (gallic acid; chlorogenic acid, caffeic acid, ferulic acid, vanillic acid, syringic acid, gentisic acid, coumaric acid)

sterols (β-sitosterol)

carbohydrates: (glucose, sorbitol, fructose)

hydrocarbons (heptacosane, nonacosane)

aldehydes & ketones (hexenal, nonanal, dodecanal, dodecan-2-one)

other (parasorbic acid, eriobofuran, noreriobofuran)

Medicinal Use

disorders of the respiratory tract (fever, infections, colds, flu), rheumatism, gout, cancer, diabetes, diuretic, cholagogue, hemostatic[46–47]

Biological Activity

antioxidant, anti-inflammatory, antibacterial, antitumoral, analgesic, antiviral, hepatoprotective cardioprotective, antiparasitic, antirheumatic, vasorelaxant, antiatherogenic[46–47]

Other Uses

- culinary (jam, jellies, conserves, marmalades, vinegar, wines, spirits, confectionery, ketchup, pies, soups, urkish delight)

- coffee substitute

- woodcraft (baskets, cartwheels, kitchen utensils, crates, furniture, hoops for barrels)

- cosmetics (face-mask)

- black dye

- decorative [46–47]

Heracleum sphondylium L. (Common hogweed)

Scientific name: *Heracleum sphondylium* L.
Common name: Common hogweed, Hogweed
Romanian name: Brânca ursului
Taxonomic classification
Kingdom: *Plantae*
Subkingdom: *Cormobionta*
Phylum: *Spermatophyta*
Subphylum: *Magnoliophytina (Angiospermae)*
Class: *Magnoliatae (Magnoliopsida, Dicotyledonatae)*
Subclass: *Rosidae*
Order: *Apiales (Umbellales)*
Family: *Apiaceae (Umbelliferae)*
Subfamily: *Apioideae*
Genus: *Heracleum sphondylium* L.
Species: *sphondylium* L.[2–3]
Subspecies: – *elegans* (Crantz) Schűbl. Et G. Martens (subsp. *montanum* (Schleich. Ex Gaudin) Briq.) has the stem leaves provided with 3 segments and a pair of lateral segments, that, along with the terminal segment, are three-divided. The flowers have white petals, with the external ones being radiate. It is rarely found from the beech forest zone up to the subalpine zone;[2,3]

- *sibiricum* (L.) Simonk. (*H. flavescens* Willd.) is found from the oak forests zone up to the fir forest zone and has flowers with 2 mm long petals, slightly emarginate, non-radiate, greenish-yellow in color;[2,3]

- *sphondylium* is widespread from the oak forest zone up to the fir forest zone and has white, occasionally pink, unequal petals, with the external ones deeply emarginate, ranging from 3–10 mm in length, and they are radiate (Figure 3.7).[2,3]

Figure 3.7 *Heracleum sphondylium* L. (Common hogweed).

Source: 48.

SPREAD

A frequently encountered species, from the oak forest zone up to the subalpine zone, growing in shrublands, meadows, pastures, grassy slopes, forest clearings, and forest edges.[2-4]

DESCRIPTION

A biennial or perennial species with a ringed, fusiform, branched, thick, yellowish rhizome and an erect, fistulous stem, highly ridged, covered with patent, setiform hairs, and reaching a height of 50–150 cm. The leaves are variable pinnate, but may occasionally be lobed or divided. The vagina of stem leaves is slightly swollen and rough-hairy or glabrous. It lacks an involucre, and the involucel has numerous lanceolate to linear-subulate folioles. It blooms from June to September.[2-4]

Main phytoconstituents	Ref
terpenoid (β-amyrin, phytol, linalool, erpineol, nerolidol, β-ionone, lupeone, betulinic acid, taraxerol, germanicol, oleanolic acid, ursolic acid)	49–50

flavonoids (quercetin, apigenin, isorhamnetin, hyperoside, avicularin, vitexin, vicenin, sparin, luteolin, kaempferol, astragalin)

organic acid (ascorbic acid, succinic acid, fumaric acid)

fatty acids (linoleic acid, palmitic acid, arachidic acid, behenic acid, tricosanoic acid, lignoceric acid)

phenolic acids (ferulic acid, cinnamic acid, caffeic acid, coumaric acid, gentisic acid, vanilic acid, syringic acid)

sterols (beta-sitosterol, stigmasterol)

saponins

Medicinal Use

insomnia, toothache, stomach pains, menstrual cramps, premenstrual syndrome, hyperactivity, neuralgia, rheumatism, cough, asthma, malaria [49–50]

Biological Activity

anti-inflammatory, antioxidant, antimicrobial, antibacterial, antitumoral, antifungal, analgesic, neuroprotective [49–50]

Other Uses

■ ornamental. [49–50]

REFERENCES

1. Bejenaru C., Mogoşanu G.D., Bejenaru L. E., Popescu H. 2020. *Botanică farmaceutică. Cormobionta*, ediția a III-a. Ed. Sitech, Craiova.

2. Ciocârlan V. *Flora ilustrată a României. Pteridophyta et Spermatophyta.* Ediția a 3-a revizuită și adăugită. 2009. Ed. Ceres, București.

3. Sârbu I., Ştefan N., Oprea A. 2013. *Plante vasculare din România, Determinator ilustrat de teren.* Ed. Victor B Victor, București.

4. Săvulescu T. (ed). 1958. *Flora RPR*, vol. VI. Ed. Academiei RPR, București.

5. Oroian S., Samarghitan M. 2018. Flora from Fagaras area (Mures County) as potential sourse of medicinal plants. *Acta Biologica Marasiensis* 1(1):60–70.

6. Linnell J.D.C., Kaltenborn B., Bredin Y., Gjershaug J.O. 2016. *Biodiversity assessment of the Fagaras Mountains.* Romania – NINA Report 1236. 86 pp., Trondheim, Norwegian Institute for Nature Research, ISBN: 978-82-426-2876-3.

7. https://dryades.units.it/Roma/index.php?procedure=taxon_page&id=3411&num=4970

8. Pavlović M., Kovačević N., Tzakou O., Couladis M. 2006. Essential oil composition of *Sanicula europaea* L. *Flavour and Fragrance Journal* 21(4): 687–689.

9. Talag A.H.M. 2016. Phytochemical investigation and biological activities of *Sanicula europaea* and *Teucrium davaeanum.* Isolation and identification of some constituents of *Sanicula europaea* and *Teucrium davaeanum* and evaluation of the antioxidant activity of ethanolic extracts of both plants and cytotoxic activity of some isolated compounds *PhD Thesis*, Faculty of Life Sciences, University of Bradford, England.

10. Karagöz A., Arda N., Gören N., Nagata K., Kuru A. 1999. Antiviral activity of *Sanicula europaea* L. extracts on multiplication of human parainfluenza virus type 2. *Phytotherapy Research* 13: 436–438.

11. Hiller K. 2005. *Sanicula europaea* L. Sanikel. *Zeitschrift für Phytotherapie* 26(5). doi: 10.1055/s-2005-922111.

12. https://pfaf.org/user/Plant.aspx?LatinName=Sanicula+europaea

13. https://dryades.units.it/floritaly/index.php?procedure=taxon_page&tipo=all&id=11714

14. Paun G., Neagu E, Moroeanu V., Albu C., Savin S., Radu G.L. 2019. Chemical and bioactivity evaluation of *Eryngium planum* and *Cnicus benedictus* polyphenolic-rich extracts. *BioMed Research International* vol. 2019, Article ID 3692605, 10 pages.

15. Erdem S.A., Nabavi S.F., Orhan I.E., Daglia M., Izadi M., Nabavi S.M. 2015. Blessings in disguise: A review of phytochemical composition and antimicrobial activity of plants belonging to the genus *Eryngium. DARU Journal of Pharmaceutical Sciences* 23:53.

16. Derda M. et al. 2013. The evaluation of the amebicidal activity of *Eryngium planum* extracts. *Acta Poloniae Pharmaceutica* 70(6):1027–1034.

17. Grytsyk A., Gnatoyko K. 2022. Research of acute toxicity and anti-inflammatory activity of extracts of *Eryngium planum*. *SSP Modern Pharmacy and Medicine* 2(1):1–12.

18. Arykbayeva A.B., Ustenova G.O., Sharipov K.O., Beissebayeva U.T., Kaukhova I.E., Myrzabayeva A., Gemejiyeva N.G. 2023. Determination of chemical composition and antimicrobial activity of the CO_2 extract *of Eryngium planum L. International Journal of Biomaterials* vol. 2023, Article ID 4702607, 11 pages.

19. Kikowska M., Piotrowska-Kempisty H., Kucińska M. et al. 2023. Saponin fractions from *Eryngium planum* L. induce apoptosis in ovarian SKOV-3 cancer cells. *Plants* 12(13):2485.

20. https://pfaf.org/user/Plant.aspx?LatinName=Eryngium+planum

21. https://dryades.units.it/dolomitifriulane/index.php?procedure=taxon_page&id=3594&num=2512

22. Mogoşanu G.D., Bejenaru L.E., Bejenaru C., Popescu H. 2015. *Farmacognozie-Fitoterapie*, Vol. II, ediția a III-a. Ed. Sitech, Craiova.

23. Egyptian Herbal Monograph. 2022. Carum carvi, 236–240, EDA, Egypt.

24. Bhushan B. 2021. Phytochemistry and remedial properties of *Carum carvi* Seeds. *World Journal of Pharmaceutical and Life Science* 7(1):87–92.

25. Joshi R.K., Soulimani R. 2020. Ethno-medicinal and phytochemical potential of *Carum carvi* Linn. and *Cuminum cyminum*: A review. *International Journal of Pharmacognosy and Life Science* 1(1):33–37.

26. Sepide M., Kiani S. 2016. Pharmacological activities of *Carum carvi L. Pharma Letter* 8(6):135–138.

27. https://pfaf.org/user/Plant.aspx?LatinName=Carum+carvi

28. https://dryades.units.it/floritaly/index.php?procedure=taxon_page&tipo=all&id=3609

29. Bone K., Mills S. 2013. *Principles of herbal treatment, in principles and practice of phytotherapy.* (Second Edition), Ed. Churchill Livingstone: 83–88, ISBN 9780443069925.

30. Aćimović M.G., Pavlović S.Đ., Varga A.O., Filipović V.M., Cvetković M.T., Stanković J.M., Cabarkapa I.S. 2017. Chemical composition and antibacterial activity of *Angelica archangelica* root essential oil. *Natural Product Communications* 12(2):205–206.

31. Bhat Z.A, Kumar D, Shah M.Y. 2011. *Angelica archangelica* Linn. is an angel on earth for the treatment of diseases. *International Journal of Nutrition, Pharmacology, Neurological Diseases* 1:36–50.

32. Kumar D., Bhat Z.A., Kumar V., Chashoo I.A., Khan N.A., Shah M.Y. 2011. Pharmacognostical and phytochemical evaluation of *Angelica archangelica* Linn. *International Journal of Drug Development and Research* 3(3):173–188.

33. Nivinskienė O., Butkienė R., Mockutė D. 2005. The chemical composition of the essential oil of *Angelica archangelica* L. roots growing wild in Lithuania. *Journal of Essential Oil Research* 17(4):373–377.

34. Romm A., Ganora L., Hoffmann D., Yarnell E., Abascal K., Coven M. 2010. Fundamental principles of herbal medicine. In: Romm A., Hardy M.L., Mills S. (eds). Botanical medicine for women's health. Churchill Livingstone, London, UK, pp. 24–74, ISBN 9780443072772.

35. Sowndhararajan K., Deepa P., Kim M., Park S.J., Kim S. 2017. A Review of the composition of the essential oils and biological activities of *Angelica species*. *Scientia Pharmaceutica* 85(3):33.

36. Kaur A., Bhatti R. 2021. Understanding the phytochemistry and molecular insights to the pharmacology of *Angelica archangelica* L. (garden angelica) and its bioactive components. *Phytotherapy Research* 35(11):5961–5979.

37. Alkan Türkuçar S., Aktaş Karaçelik A., Karaköse M. 2021. Phenolic compounds, essential oil composition, and antioxidant activity of *Angelica purpurascens* (Avé-Lall.) Gill. *Turkish Journal of Chemistry* 45(3):956–966.

38. Aćimović M.G., Pavlović, S.Đ., Varga A.O., Filipović V.M., Cvetković M.T., Stanković J.M., Čabarkapa I.S. 2017. Chemical composition and antibacterial activity of *Angelica archangelica* root essential oil. *Natural Product Communications* 12(2):205–206.

39. http://prepar.ro/condimente/angelica.html

40. https://dryades.units.it/duino_en/index.php?procedure=taxon_page&id=3521&num=2503

41. Tomic A., Petrovic S., Drobac M., Božić D. 2008. Composition and antimicrobial activity of the fruit essential oils of two *Athamanta turbith subspecies*. *Chemistry of Natural Compounds* 44(6):789–791.

42. Tomić A., Petrović S., Pavlović M., Trajkovski B., Milenković M., Vučićević D., Niketić M. 2009. Antimicrobial and antioxidant properties of methanol extracts of two *Athamanta turbith* subspecies. *Pharmaceutical Biology* 47(4):314–319.

43. Tomic, A., Petrovic, S., Pucarevic, M. et al. 2006. Fatty acid composition of two *Athamanta turbith subspecies*. *Chemistry of Natural Compounds* 42:391–393.

44. Tomić, A., Petrović, S., Pavlović, M., Tzakou, O., Couladis, M., Milenković, M., Lakušić, B. 2009. Composition and antimicrobial activity of the rhizome essential oils of two *Athamanta turbith* subspecies. *Journal of Essential Oil Research* 21(3):276–279.

45. https://spontana.org/species.php?id=1201

46. Hosseinzadeh Z., Ramazani A., Razzaghi-Asl N. 2019. Plants of the genus *Heracleum* as a source of coumarin and furanocoumarin. *Journal of Chemical Reviews* 1(2):78–98.

47. Gîrd C.E., Costea T., Duţu L.E., Popescu M.L., Nencu I. 2017. Phytochemical and phytobiological research regarding roots of *Heracleum sphondylium* L. and *H. palmatum* Baumg. *Trends in Toxicology and Related Sciences* 1(1):1–15.

48. https://dryades.units.it/floritaly/index.php?procedure=taxon_page&tipo=all&id=3642

49. Benedec D., Hanganu D., Filip L., Oniga I., Tiperciuc B., Olah N.K., Gheldiu A.M., Raita O., Vlase L. 2017. Chemical, antioxidant and antibacterial studies of Romanian *Heracleum sphondylium*. *Farmacia* 65:252–256.

50. Sheppard A.W. 1991. *Heracleum sphondylium* L. *Journal of Ecology*, 79(1):235–258.

4 *Asteraceae* Martinov (*Compositae* Adans.) Family

Cornelia Bejenaru[1], Antonia Radu[1], George Dan Mogoşanu[2],
Ludovic Everard Bejenaru[2], Andrei Biţă[2], and Adina-Elena Segneanu[3]

[1] Department of Pharmaceutical Botany, Faculty of Pharmacy, University of Medicine and Pharmacy of Craiova, 2 Petru Rareş Street, Craiova, Dolj County, Romania
[2] Department of Pharmacognosy & Phytotherapy, Faculty of Pharmacy, University of Medicine and Pharmacy of Craiova, 2 Petru Rareş Street, Craiova, Dolj County, Romania
[3] Institute for Advanced Environmental Research–West University of Timisoara (ICAM–WUT), 4 Oituz Street, Timisoara, Romania

BOTANICAL DESCRIPTION

Plants that are generally herbaceous, annual, biennial, or perennial, rarely subshrubs, exceptionally trees (some species in tropical areas). They have a well-developed radicular system, which can include tuberized roots or adventitious buds. Underground stems may be present, and the aboveground stem can be simple, scapiform, or branched. The leaves are unstipulated, simple (entire or pinnately divided) or pinnately compound. Leaf arrangement is alternate or opposite, often in a basal rosette. The flowers are grouped in a specific racemose inflorescence (calathidium or anthodium). The receptacle of the inflorescence can be flat, convex, elongated, conical, globular, compact, or hollow, and it can be hairy or glabrous, depending on the species. Its size also varies. At the base of the inflorescence, there are involucral bracts that vary in size, shape, or consistency. The flowers are pentamerous, hermaphroditic or unisexual, actinomorphic or zygomorphic. Inflorescences typically have both types of flowers, both hermaphroditic and unisexual. The floral envelope consists of a reduced calyx (often formed by the persistent pappus on the fruit) and a gamopetalous corolla. The corolla can either be a tube terminated with 5 (rarely 8) teeth, characteristic of actinomorphic tubular flowers, or it can consist of a tube that continues with a long, linear extension on one side (ligule), terminated with 3–5 teeth, characteristic of zygomorphic ligulate flowers. The androecium consists of 5 stamens attached to the corolla tube by filaments, with long, introrse anthers, either fused or adherent, forming a tube through the middle of which the style passes (sinantherous androecium). The gynoecium consists of 2 united carpels, with a uniovulate ovary (anatropous ovule) positioned inferiorly, continuing into a long, cylindrical style with 2 stigmas at the tip, equipped with collecting hairs. At the base of the style, there are nectarines that secrete abundant nectar rich in sugar. The fruit is an achene (cypsela), which can have a persistent pappus, sometimes with setae or a coronule. Each fruit contains a single seed without endosperm. The number of species in this family is approximately 20,000, grouped into around 1,000 genera, making it the most well-represented family in *Dicotyledonatae*. The distribution of plants in this family covers a very large geographic and climatic range, from tropical to arctic regions, with the majority of species growing spontaneously in steppe areas. In Romania, there are 320 species belonging to 95 genera, ranging from the lowland to the mountainous regions. Due to the nectar produced by these plants, pollination is primarily entomophilous. Fruits are dispersed away from their place of origin either by wind or animals. In the case of anemochory, the presence of the pappus aids in dissemination, while in zoochory, the fruits often have retrorse spines. Among the many species within this family, only a relatively small number have practical significance, from an ornamental, industrial, culinary, or medicinal point of view.[1-4]

The species of this family are subdivided in two subfamilies: *Asteroideae* (*Tubuliflorae*) and *Cichorioideae* (*Liguliflorae*).[1-3]

Plants with therapeutic properties demonstrated by recent research on metabolites profile and biological activities were selected.

Eupatorium cannabinum L. (Hemp-agrimony)

Scientific name: *Eupatorium cannabinum* L.
Common name: Hemp-agrimony, Holy rope
Romanian name: Cânepa codrului
Taxonomic classification
Kingdom: *Plantae*
Subkingdom: *Cormobionta*

DOI: 10.1201/9781003270515-5

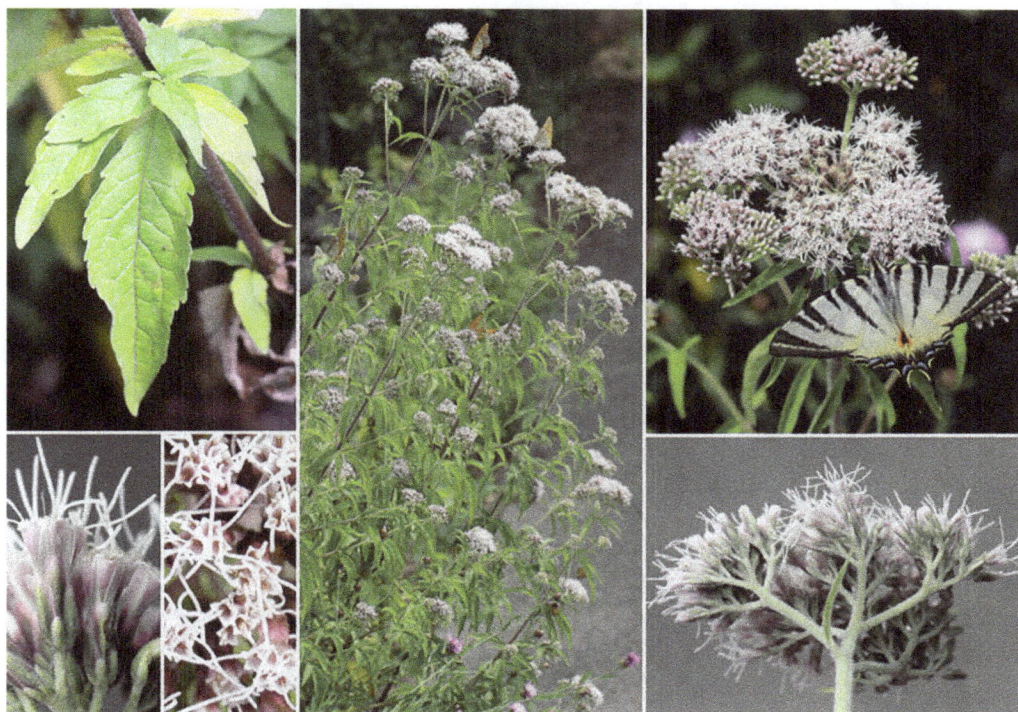

Figure 4.1 *Eupatorium cannabinum* L. (Hemp-agrimony).

Source: 5.

Phylum: *Spermatophyta*
Subphylum: *Magnoliophytina (Angiospermae)*
Class: *Magnoliatae (Magnoliopsida, Dicotyledonatae)*
Subclass: *Asteridae*
Order: *Asterales (Compositales)*
Family: *Asteraceae (Compositae)*
Subfamily: *Asteroideae (Tubuliflorae)*
Genus: *Eupatorium*
Species: *cannabinum* L. (Figure 4.1)[1–3]

SPREAD

The species is common and widespread, often found in wet areas, along waterbanks, in thickets, clearings within forests, and ruderal places, ranging from the lowlands to the mountainous regions.[1–4]

DESCRIPTION

Perennial herbaceous plant, with a cylindrical, tuberous root and a tall stem reaching 50–150 cm in height. The stem is cylindrical, shortly hairy, and can be simple or branched. The leaves are petiolate, palmately compound, with 3–5 lanceolate folioles that have serrated margins. The arrangement of leaves on the stem is opposite. The flowers are reddish, hermaphroditic, tubular, 5-toothed, and form small anthodia (with 4–6 flowers) with involucral folioles that are either glabrous or finely pubescent, unequal, and arranged in imbricate series of 2–3. The anthodia are grouped in dense, compound corymb-type inflorescences. The fruit is a penta-edged achene, truncate at the top, and sparsely glandular-hairy. It blooms from July to September.[1–4]

Main phytoconstituents Ref.
6–9
pyrrolizidine alkaloids (supinine, echinatine, lycopsamine)
flavonoids (eupatorine, quercetin, rutin, eupatilin, exadeceno, astragalin, kaempferol)
phenolic acids (caffeic acid, chlorogenic acid)
fatty acids (palmitic acid)
terpenoids (germacrene D, α-terpinene, phellandrene, thymol, cymene, elemol, β-bisabolene, spathulenol, humulene, farnesene, phytol, α-zingiberene, elemol, linalool, carvacrol, 2-carene, limonene, cadinol)
hydrocarbons (hexadecane, exadecenoi, octadecane)
benzofurans (euparone, euparin)
carbohydrates
others (indole, stevioside)

Medicinal Use

- choleretic, laxative, depurative, diuretic, expectorant, febrifuge, emetic, would healing, hypocholesterolemic, tonic, anthelmintic[6–9]

The rhizomes with roots (*Eupatorii cannabini rhizoma cum radicibus*) and the leaves (*Eupatorii cannabini folium*) are used in the treatment of hepato-biliary diseases, acute gastritis, and urinary bladder inflammation.[1]

Biological Activity

antioxidant, antimicrobial, antitumor, anti-inflammatory, hepatoprotective, immuno-modulatory[6–9]

Other Uses

- insect repellent

- ornamental value

- cosmetic use (skin conditioner)

- dyes (blue and black)[6–9]

Achillea millefolium L. (Common yarrow)

Scientific name: *Achillea millefolium* L.
Common name: Common yarrow
Romanian name: Coada şoricelului
Taxonomic classification
Kingdom: *Plantae*
Subkingdom: *Cormobionta*
Phylum: *Spermatophyta*
Subphylum: *Magnoliophytina (Angiospermae)*
Class: *Magnoliatae (Magnoliopsida, Dicotyledonatae)*
Subclass: *Asteridae*
Order: *Asterales (Compositales)*
Family: *Asteraceae (Compositae)*
Subfamily: *Asteroideae (Tubuliflorae)*
Genus: *Achillea*
Species: *millefolium* L.[1–3]
Subspecies: – *millefolium* has sparsely pubescent leaves, hypsofilous involucral bracts with a brown stripe, narrow on the margins and white ligules. Commonly found from oak forests, to the sessile oak forest level, towards the subalpine level, in shrublands and meadows.[2,3]

- *sudetica* (Opiz) Weiss (*A. sudetica* Opiz) has pubescent leaves, hypsofilous bracts with a broad brown-black stripe, and reddish ligules. It is sporadically present from the spruce forest level to the subalpine level on rocky soils and meadows (Figure 4.2).[2,3]

Figure 4.2 *Achillea millefolium* L. (Common yarrow).

Source: 10.

SPREAD

It spontaneously grows in forest clearings, glades, shaded and relatively moist areas, pastures, meadows, and grassy places, ranging from the hilly region to the subalpine zone.[1-4]

Description

A perennial herbaceous plant with a creeping, oblique, or horizontal rhizome and stolons. The stem is glabrous or moderately hairy, erect or ascending, striated, simple or branching towards the top, reaching heights of 10–60 cm. The leaves are alternate, lanceolate, 2(3)–pinnatisect, green, uniformly hairy, or subglabrous, with an untoothed rachis (1 mm wide). The calathides are small, elongated ovoid, forming corymbose inflorescences (corymb with calathides). The ligulate flowers are white or reddish, with obovate ligules, numbering 5–6. The phyllaries are squamiform, membranous, imbricate, and have a distinct brown margin. The achenes are long and lack a pappus. The flowering period is from June to August.[1-4]

Main phytoconstituents The volatile oil contains 6–18% azulene (chamazulene, S-guaiazulene), 10% 1,8-cineole (eucalyptol), borneol, and camphor. According to F.R.X., the medicinal product must contain at least 0.2% blue-colored volatile oil.[11]   Ref.

alkaloids (betonicine, stachydrine, trigonelline)   [12-15]
coumarins
amino acids
flavonoids (apigenin, luteolin, quercitin, kaempferol, isorhamnetin)
sesquiterpene lactones (achillin, achillicin, austricin, balchanolide, millefin, millefolide, terpinene)
triterpenoids (beta-amyrin, thujone, taraxasterol, methanol, pinene, limonene, sabinene, camphene, myrcene, borneol, terpineol, carveol, germacrene D, cadinene)
phenolic acids (caffeic acid, salicylic acid, chlorogenic acids)
fatty acids (linoleic acid, palmitic acid, oleic acid, myristic acid)

sterols (stigmasterol, cholesterol, campesterol, tocopherols)
carboxylic acids (oxalic, quinic, malic, citric, shikimic acid)
carbohydrates (fructose, glucose, sucrose, trehalose, raffinose)
tannins
organic acids (formic acid, acetic acid, hexadecenoic acid)
other: hydrocarbons, phenols, alcohols, esters, aldehydes, ketones

Medicinal use

Common yarrow flowers are harvested to obtain the vegetable product known as *Millefolii flos*, and through hydrodistillation, blue-colored volatile oil is obtained (*Millefolii aetheroleum*). *Millefolii flos* has stomachic, hemostatic, and anti-inflammatory (due to azulenes) actions, along with bitter tonic, choleretic, expectorant, carminative, and antispasmodic properties.[1,11]

The vegetal product is recommended for wound healing properties, in diabetes, gastrointestinal disorders (stomach pain, bloating, irritable bowel), hepato-biliary diseases, gynecological disorders (amenorrhea), mild infections, and for depurative effect.[12–15]

Biological Activity

The vegetable product *Millefolii flos* has bitter tonic, choleretic, expectorant, carminative, antispasmodic, aromatic,[1,11] antioxidant, antitumoral, antimicrobial, astringent, anti-inflammatory, sedative, antifungal, estrogenic, diuretic, anisecretory (digestive), immunomodulatory, and anti-inflammatory properties.[12–14]

Millefolii aetheroleum is administered for conditions such as loss of appetite, dyspepsia, gastralgia, epistaxis, pulmonary and hemorrhoidal bleeding, metrorrhagia (as a hemostatic agent), as well as in cases of cholecystitis, dry cough, inflammatory states, and nocturnal enuresis (urinary incontinence). Additionally, the volatile oil is used as an antifungal agent in certain toothpaste formulations.[1,11]

Other Uses

- cosmetic use (moisturizing creams, shampoo)

- culinary: leaves are used in salads, soups, omelets and other egg dishes, seasoning dishes, infusion[16]

Artemisia absinthium L. (Absinthe)

Scientific name: *Artemisia absinthium* L.
Common name: Absinthe, Wormwood
Romanian name: Pelin
Taxonomic classification
Kingdom: *Plantae*
Subkingdom: *Cormobionta*
Phylum: *Spermatophyta*
Subphylum: *Magnoliophytina (Angiospermae)*
Class: *Magnoliatae (Magnoliopsida, Dicotyledonatae)*
Subclass: *Asteridae*
Order: *Asterales (Compositales)*
Family: *Asteraceae (Compositae)*
Subfamily: *Asteroideae (Tubuliflorae)*
Genus: *Artemisia*
Species: *absinthium* L. (Figure 4.3)[1–3]

SPREAD

Species of Eurasian origin, with a specific scent, commonly found in ruderal areas, shrublands, and along roadsides, ranging from lowlands to the beech forest zone.[2,3]

DESCRIPTION

Perennial herbaceous species, with a woody, strongly branched root, and an erect, cylindrical, woody stem. The stem is finely striped, gray pubescent with long, erect branches, reaching a height of

Figure 4.3 *Artemisia absinthium* L. (Absinthe).

Source: 17.

60–120 cm. Its leaves are uniformly white-pubescent on both sides, with a longer than wide lamina, bipinnatisect (in the inferior part of the stem) or bipinnatipartite (in the superior part of the stem), with segments measuring 2–4 mm in width. The numerous anthodia are globose, nodding, short-pedicellate, with a diameter of approximately 3 mm, and their involucral bracts are broadly ovate-subrotund, biseriate, with the internal ones membranous. The flowers in the internal part are hermaphroditic, while those in the external part are female and have a reduced corolla. Flowering occurs from July to September.[1–4]

Main phytoconstituents

Absinthii aetheroleum, which has a green color, is mainly composed of 35–45% thujone (yellow), chamazulene (blue), and thujol (9%). The officinal vegetable product according to F.R. X should contain at least 0.3% volatile oil.[11]   Ref

terpenoids (camphene, myrcene, thujone, artemisia ketone, cymene, thujene, cadinene, nerol, limonene, terpinene, borneol, camphor)   18–23

sterols (ergosterol, α- tocopherol, β sitosterol, stigmasterol)

fatty acids (palmitic acid, stearic acid, palmitoleic acid, oleic acid, docosadienoic acid, linoleic acid, eicosadienoic acid)

flavonoids (artemitin, santin, casticin, quercetin, apigenin, kaempferol, naringin, rutin, catechin, myricetin, morin, naringenin)

sesquiterpene lactones (absinthin, arabsin, artemetin, matricin, absilactone, eudesmanolide, anabsinthin, artemolin artabin, ketopelenolide)

phenolic acids (chlorogenic acid, caffeic acid, ferulic acid, vanillic acid, syringic acid, sinapic acid, p-coumaric acid, salicylic acid)

coumarins (coumarin, herniarin)

organic acids (succinic acid, cinnamic acid, malic acid)

azulene (chamazulene, azulene)

bitter principles (artamaridinin, artamarin, artamarinin, quebrachitol)

other (trimethoxybezoic acid, fraganol, benzeneacetaldehyde, benzene)

Medicinal Use

asthma, gynecological conditions, liver disease, anthelmintic, intestinal cramps, kidney stones, headaches, insomnia, digestive tonic[18–24]

Biological Activity

anti-inflammatory, antioxidant, antimicrobial, antiseptic, neuroprotective, carminative, antifungal, antimicrobial, analgesic, anthelmintic, antimalarial, hepatoprotective, chloretic, antidepressant, anti-inflammatory, narcotic, immunomodulant[18–24]

Absinthe is a bitter and aromatic tonic, recommended in cases of loss of appetite and gastroenteritis due to its strong choleretic action, which is similar to that of artichoke. Its polymethoxylated flavonoids (such as artemisinin) have spasmolytic properties.[11]

Toxicity

Short courses of preparations containing wormwood are recommended due to its emmenagogue and neurotoxic effects, primarily attributed to thujone and absinthin. In high doses, these compounds can lead to central nervous system disturbances, motor and sensory impairments, nerve degeneration, and potentially fatal effects.

Intoxication with the French drink known as absinthe ("absinthism") presents symptoms similar to alcoholism, including epileptiform seizures that can lead to exitus. Currently, absinthe is no longer prepared with wormwood and has been banned in France under the Law Against Absinthe (1915) due to causing a widespread national intoxication. At the European level, regulations have been imposed on the production and distribution of beverages flavored with volatile oils rich in ketones (such as wormwood, hyssop, anise, and caraway).[11]

Other Uses

- cosmetic use: shampoos, face serums, masks, essences, tonics, moisturizing creams[23]

- preparation of alcoholic beverages: vermouth, wine, liqueur[23]

- "bouquet-garni" for: flavoring pork, pheasant, turkey dishes, vegetable dishes, soups[23]

Artemisia annua L. (Sweet sagewort)

Scientific name: *Artemisia annua* L.
Common name: Sweet sagewort
Romanian name: Năfurică, Măturicea raiului
Taxonomic classification
Kingdom: *Plantae*
Subkingdom: *Cormobionta*
Phylum: *Spermatophyta*
Subphylum: *Magnoliophytina (Angiospermae)*
Class: *Magnoliatae (Magnoliopsida, Dicotyledonatae)*
Subclass: *Asteridae*
Order: *Asterales (Compositales)*
Family: *Asteraceae (Compositae)*
Subfamily: *Asteroideae (Tubuliflorae)*
Genus: *Artemisia*
Species: *annua* L. (Figure 4.4)[1–3]

SPREAD

Species of Eurasian origin, sporadic in ruderal areas, flood-prone areas, and gravel bars, ranging from the lowlands to the beech forest zone.[1–4]

DESCRIPTION

An annual herbaceous species, growing to a height of 50–150 cm, with an erect, branched, and glabrous stem, and alternately arranged leaves. The leaf blade is 2–3 times pinnatisect, with the ultimate segments being linear-lanceolate. The inflorescence is a panicle with nutant, globose, small, pedicellate anthodia. It blooms from July to September.[1–4]

Figure 4.4 *Artemisia annua* L. (Sweet sagewort).

Source: 25.

Main phytoconstituents Ref

flavonoids (artemisinin, quercetin, luteolin, rutin, apigenin, rhamnetin, chrysosplenetin, acacetin, casticin, taxifolin, chrysosplenol D, quercetagetin) [26]

amino acids (glycine, valine, leucine, threonine, serine, arginine, phenylalanine, histidine, aspartic acid)

terpenoids (artemisine, artemisia ketone, artemisitene, spathulenol, artesimic acid, camphene, beta-amyrin, cadinene, myrcene, linalool, terpinene, camphor, nerol, sabinol, eugenol, borneol, geraniol, germacrene D, selinene, taraxasterone)

coumarins (scopoletin, coumarin, exadecen, tomentin A, scoparone)

phenolic acids (caffeic acid, coumaric acid, cinnamic acid, salicylic acid)

fatty acids (oleanic acid, myristic acid, palmitic acid)

sterols (beta-sitosterol, stigmasterol, daucosterol)

Medicinal Use

antimalarial, antihemorrhagic, digestive, stomachic, antipyretic, tonic of the nervous system, hypoglycemic[26]

Biological Activity

antioxidant, antidiabetic, antibacterial, antiviral, antitumor, immunomodulatory, antiparasitic, antibacterial, antifungal[26]

Other Uses

- herbicidal
- cosmetics and perfumes
- ornamental

- culinary
- nutraceuticals
- wine[26]

Matricaria recutita L. (*M. chamomilla* L. p.p.; *Chamomilla recutita* (L.) Rauschert) (Chamomile)

Scientific name: *Matricaria recutita* L.
Common name: Chamomile
Romanian name: Mușețel
Taxonomic classification
Kingdom: *Plantae*
Subkingdom: *Cormobionta*
Phylum: *Spermatophyta*
Subphylum: *Magnoliophytina (Angiospermae)*
Class: *Magnoliatae (Magnoliopsida, Dicotyledonatae)*
Subclass: *Asteridae*
Order: *Asterales (Compositales)*
Family: *Asteraceae (Compositae)*
Subfamily: *Asteroideae (Tubuliflorae)*
Genus: *Matricaria*
Species: *recutita* L. (Figure 4.5)[1–3]

SPREAD
Eurasian species, commonly found in meadows, ruderal or marshy areas, vineyards, and acacia plantations, ranging from lowlands to mountain regions.[1–4]

Figure 4.5 *Matricaria recutita* L. (*M. chamomilla* L. p.p.; *Chamomilla recutita* (L.) Rauschert) (Chamomile).

Source: 27.

DESCRIPTION

It is an annual herbaceous plant, measuring 10–50 cm in height, with a specific, pleasant, and intense odor. The root is fusiform, strongly branched, and the stem is erect or ascending, glabrous, striated, and tufted-branched. The leaves are 2–3 times pinnatisect, with linear, green, and glabrous laciniae. The anthodia have a diameter of 1.5–2 cm, and the receptacle is conical, hollow inside, without persistent bracts. The central flowers are yellow, numerous, small, tubular, and 5-dentate, while the marginal ones are ligulate and white, numbering 12. The achenes are long, cylindrical, slightly curved, with a weakly developed coronule at the tip.[1,4]

Main phytoconstituents	Ref.

flavonoids (apigenin, luteolin, quercetin, rutin, naringenin, patuletin, hispidulin) [28–31]
coumarins (scopoletin, herniarin, umbelliferone)
phenolic acids (caffeic acid, chlorogenic acid)
terpenoids (matricin, farsenene, sabinene, chamazulene, bisabolol, limonene, spathulenol, germacrene D, linalool, nerol, geraniol, β-elemene, β-caryophyllene, humulene, nerol, camphor, geraniol, phytol, artemisia ketone)
fatty acids (tocopherol)
sterols (campesterol, stigmasterol, β-sitosterol, cholesterol)
Hydrocarbons

Medicinal Use

Flowers are harvested from chamomile, yielding the medicinal product known as *Chamomillae flos*. From the flowers, through steam distillation, a blue-colored (due to azulenes) volatile oil with a specific scent is obtained. *Chamomillae flos* has antihistaminic, epithelizing, carminative, antiseptic, emollient, demulcent, stomachic, sedative, mildly sudorific, antipruritic, and antispasmodic properties.[1,11,28–31]

The volatile oil is a strong anti-inflammatory (antiflogistic) and antiulcer agent by activating the biosynthesis of protective prostaglandins in the gastric mucosa. Infusion of *Chamomillae flos* is not recommended for evening consumption because it has an excitatory effect on the central nervous system and may cause insomnia.[1]

Biological Activity

■ antioxidant, antibacterial, anti-inflammatory, antipruritic, antiseptic, antispasmodic, immunomodulator, antihyperglycemic[28–31]

Other Uses

■ beverages (liqueur)

■ cosmetic products (soap, shampoo, moisturizer creams, lotions, toothpaste),

■ flavoring for sanitation products (paper handkerchiefs, wipes)

■ yellow dyes[28–31]

Arnica montana L. (Wolf's bane)

Scientific name: *Arnica montana* L.
Common name: Wolf's bane, Mountain tobacco, Mountain arnica
Romanian name: Arnică, Podbal de munte
Taxonomic classification
Kingdom: *Plantae*
Subkingdom: *Cormobionta*
Phylum: *Spermatophyta*
Subphylum: *Magnoliophytina (Angiospermae)*
Class: *Magnoliatae (Magnoliopsida, Dicotyledonatae)*
Subclass: *Asteridae*
Order: *Asterales (Compositales)*
Family: *Asteraceae (Compositae)*

Figure 4.6 *Arnica montana* L. (Wolf's bane).

Source: 32.

Subfamily: *Asteroideae* (*Tubuliflorae*)
Genus: *Arnica*
Species: *montana* L. (Figure 4.6)[1-3]

SPREAD

The species is sporadically distributed from the beech zone to the spruce zone, through meadows, pastures, shrubs, and grasslands.[2-4] It is widespread throughout the Carpathian range.[33]

DESCRIPTION

Perennial herbaceous plant, with a thick, cylindrical, horizontal rhizome with many fibrous roots. The stem reaches a height of 20–50 cm, is cylindrical, erect, solitary, simple, or with 1–2 branches. The basal rosette leaves are elliptical and rough; the stem leaves are smaller and arranged in 2–3 pairs in an opposite fashion. The anthodia, with a diameter of 4–8 cm, are terminal, with yellow ligulate flowers arranged marginally and tubular flowers in the center, which are yellow-orange.[1] The fruits are long achenes, hairy, with an 8 mm pappus. It blooms from June to August.[1-4,34]

Main phytoconstituents   Ref.

alkaloids (tussilagine, pyrrolidineacetic acid)   35–37
terpenoid (helenalin, farnesene, germacrene, carriophyllene, chamissonoid, phellandrene, limonene, champhene, carvacrol, cymene, sabinene, myrcene, cumene)
flavonoids (quercetin, patuletin, kampferol, luteolin, apigenin, hispidulin, eupafolin)
carotenoids
coumarins (umbelliferone, scopoletin)
fatty acids
phenolic acids (cynarin, chlorogenic acid, caffeic acid)
polysaccharides (inulin)

hydrocarbons
aldehydes and ketones
organic acids (angelic acid, formic acid, fumaric acid, valerianic acid, tanic acid, succinic acid)
sterols
tannins
lignans

Medicinal Use

From the medicinal product *Arnicae flos* (arnica flowers), a tincture is prepared, which has anti-inflammatory, antiechimotic, cicatrizing, and fracture healing stimulation actions. It should not be applied to open wounds due to its irritating effect.[1] Wolf's bane is used for various purposes, including the treatment of traumatic injuries, pain, sores, bruises, bronchitis, uterine bleeding, indigestion, skin inflammations, fever, colds, rheumatism, cardiovascular diseases, hair stimulation, arthralgia, arthritis, arteriosclerosis, anemia, car or sea sickness.[35–37]

Biological Activity

It is used in the therapy of phlebitis, hematomas, thrombosis, sprains (joint twisting), injuries, insect stings.[11] Wolf's bane is also reported to have various other properties, including analgesic, antioxidant, cytotoxic, wound healing, antibacterial, antitumor, anti-inflammatory, antifungal, immunomodulatory, antibruising, sedative, hypopigmentation, antiplatelet, and uterotonic effects.[35–36]

Other Uses

■ cosmetic (shampoos, hair tonics, antidandruff products, bath products, perfumes, skin-conditioners, cremes, ointments)

■ beverages

■ candy

■ baked products

■ gelatins

■ puddings[35–37]

Arctium lappa L. (Bardane)

Scientific name: *Arctium lappa* L.
Common name: Bardane, Beggar's buttons, Burdock, Great burdock, Snake's rhubarb, Thorny burr
Romanian name: Brustur
Taxonomic classification
Kingdom: *Plantae*
Subkingdom: *Cormobionta*
Phylum: *Spermatophyta*
Subphylum: *Magnoliophytina (Angiospermae)*
Class: *Magnoliatae (Magnoliopsida, Dicotyledonatae)*
Subclass: *Asteridae*
Order: *Asterales (Compositales)*
Family: *Asteraceae (Compositae)*
Subfamily: *Asteroideae (Tubuliflorae)*
Genus: *Arctium*
Species: *lappa* L. (Figure 4.7)[1–3]

SPREAD

Eurasian species, common in ruderal areas, along waterways, floodplains, from the lowlands to the mountains.[1–4]

Figure 4.7 *Arctium lappa* L. (Bardane).

Source: 38.

DESCRIPTION

The plant is a biennial with a fleshy taproot and leafy stem, reaching a height of 1–1.5 m. The stem is erect, sulcate, and extensively branched. The basal leaves, which appear in the first year of vegetation, are large, long-petiolated, with an ovate-cordate blade, acute or obtuse apex, and distanced dentate margins. They are glabrous on the superior side, dark green, and densely hairy on the dorsal side. The anthodia are long-pedunculate, with a diameter of 2–3 cm, arranged in a wide corymb, with purple flowers, elongated campanulate, measuring 2–2.5 cm in length. The involucre is glabrous or subglabrous (the same length as the flowers) and finely pointed and uncinate. The plants bloom in the months of July and August.[1-4]

Main phytoconstituents   Ref.
flavonoids (quercetin, luteolin, rutin, rhamnetin, astragalin)   [39-42]
terpenoids (amyrin, ursollic acid, arctiopicrin, taraxasterol, taraxasterol, lupeol, phytol, geraniol, linalool, tymol, carrophillene, cadinene, elemene, squalene, myrcene, guajen)
phenolic acids (chlorogenic acid, caffeic acid, coumaric acid, salicylic acid)
fatty acids (docosanoic acid, hexadecenoic acid, pentadecanic acid, linoleic acid, stearic acid)
organic acids (benzoic acid, succinic acid)
sterols (beta-sitosterol, stigmasterol, campesterol, daucosterol)
lignans (arctigenin, arctiin, lappaol, artignan, neoarctin)
carbohydrates (fructose, inulin, rhamnose, glucose, galactose, xylan, arabinan, sorbitol, mannitol)
aldehydes (hexanal, pentanal, octanal, decanal)

Medicinal Use

The therapeutic use of this species includes both the root (*Bardanae radix*) and the leaves (*Bardanae folium*). *Bardanae radix* has antibacterial, antifungal, choleretic, bitter tonic, depurative (laxative, diuretic, diaphoretic), and anticancer properties.[1,11] It is also used for wound healing (dermatitis, eczema, wounds, ulcers), combating fatigue, treating rheumatism, managing diabetes, promoting diuresis, detoxification, and managing hypertension.[39–42] The leaves (*Bardanae folium*) have emollient, antipruritic, and trophic effects. They are used topically in the treatment of certain dermatological conditions and insect stings.[1,11]

Biological Activity

It is used in hepatobiliary diseases (cholecystitis, biliary lithiasis), dermatological conditions (acne, furunculosis, seborrhea, alopecia, varicose ulcers), oral conditions (stomatitis, oral aphthae), urinary conditions (cystitis)[1,8], and has antioxidant, antimicrobial, antipyretic, diuretic, diaphoretic, hypoglycemic, anti-inflammatory, hepatoprotective, gastroprotective, antimutagenic, and antitumor properties.[39–42]

Other Uses

- food, nutraceutical

- cosmetic uses (shampoo, moisturizer creams, antiacne creams and lotions, etc.)[39–42]

- the root is traditionally consumed in Japan under the name *gobo*[1,11]

Tanacetum corymbosum (L.) Sch.-Bip. (*Chrysanthemum corymbosum* L.) (Corymbose tansy)

Scientific name: *Tanacetum corymbosum* (L.) Sch.-Bip.
Common name: Corymbose tansy
Romanian name: Năpraznic
Taxonomic classification
Kingdom: *Plantae*
Subkingdom: *Cormobionta*
Phylum: *Spermatophyta*
Subphylum: *Magnoliophytina (Angiospermae)*
Class: *Magnoliatae (Magnoliopsida, Dicotyledonatae)*
Subclass: *Asteridae*
Order: *Asterales (Compositales)*
Family: *Asteraceae (Compositae)*
Subfamily: *Asteroideae (Tubuliflorae)*
Genus: *Tanacetum*
Species: *corymbosum* (L.) Sch.-Bip.[2–3]
Subspecies: – *corymbosum* has hypsophilous involucral bracts with narrow, light brown margins and ligules of 10–15 mm. It is distributed from the lowlands to the beech zone (Figure 4.8).[2,3]

- – *subcorymbosum* (Schur) Pawl. (subsp. *clusii* (Fisch. Ex Rchb.) Heywood) has hypsophilous involucral bracts with broad, blackish-brown margins and ligules of 15–20 mm. It is present in the spruce zone up to the subalpine zone.[2,3]

SPREAD

The species is distributed from the steppe zone up to the spruce zone, through shrublands, glades, forest edges, forest clearings, ruderal areas, and grassy rocks.[2–4]

DESCRIPTION

Perennial herbaceous plant, with a knotty, oblique or horizontal rhizome, and an erect stem, finely edged, hairy, reaching a height of 30–80 cm. The leaves are glabrous on the superior surface and glabrescent on the inferior surface, simple or bipinnatisect, with 7–12 pairs of lanceolate laciniae. The basal and inferior leaves are petiolate, while the superior ones are sessile. The anthodia, numbering 3–10, are grouped in a loose corymb. The central tubular flowers are yellow, and the ligulate ones are white, measuring 10–20 mm in length and arranged on the edge of the anthodium. The fruits are achenes with 5 edges and have a membranous coronule at the top. This plant blooms from June to August.[2–4]

Figure 4.8 *Tanacetum corymbosum* (L.) Sch.-Bip. (*Chrysanthemum corymbosum* L.) (Corymbose tansy).
Source: 43.

Main phytoconstituents   Ref.
terpenoids (cadinene, tujene, pinocasrvone, phellandrene, cymene, sabinene)   44–48
phenolic acids (chlorogenic acid)
flavonoids (casticin, exadeceno, hispidulin, luteolin, apigenin)
sterols (sigmasterol, campesterol, ergosterol, β-sitosterol)

Medicinal Use

digestive disorders (gastritis), anthelmintic[44]

Biological Activity

antimicrobial, antiviral, tonic, exadecen, carminative, antispasmodic, anticoagulant, anti-inflammatory, antifibrinolytic, cytotoxic activities[44,49]

Other uses

■ cosmetics

■ insecticides

■ balsams, dyes

■ food preservatives

■ flavouring agents[44]

Carduus nutans L. (Nodding thistle)

Scientific name: *Carduus nutans* L.
Common name: Nodding thistle

Figure 4.9 *Carduus nutans* L. (Nodding thistle).

Source: 50.

Romanian name: Ciulin
Taxonomic classification
Kingdom: *Plantae*
Subkingdom: *Cormobionta*
Phylum: *Spermatophyta*
Subphylum: *Magnoliophytina (Angiospermae)*
Class: *Magnoliatae (Magnoliopsida, Dicotyledonatae)*
Subclass: *Asteridae*
Order: *Asterales (Compositales)*
Family: *Asteraceae (Compositae)*
Subfamily: *Asteroideae (Tubuliflorae)*
Genus: *Carduus*
Species: *nutans* L.[3]
Subspecies: – *nutans* (Figure 4.9)

SPREAD
Species found from the plains to the beech forest level, in rocky pastures, uncultivated areas, plantations, shrubs, road edges, and ruderal locations.[3-4]

DESCRIPTION
Biennial herbaceous plant with a taproot and an erect stem, solitary or branched, foliate, spiny-winged, reaching a height of 30–100 cm. Leaves in the basal rosette are large, oblanceolate, glabrous or only sparsely hairy on the inferior surface, pinnatipartite, with ovate, three-lobed segments, each lobe ending in yellow spines and small spines between them. The basal rosette disappears after flowering. Stem leaves are decurrent on the stem and form lobed, spiny wings. Branches bear a single, long-peduncled anthodium, densely tomentose, with a diameter of 3.5–6 cm. Involucral

61

folioles are 2–5 mm, gradually narrowing towards the apex and ending in a spine. The flowers are red with a long corolla, deeply lobed, and exserted from the involucre. Achenes are long, about 4–5 mm, finely transversely rugose. Blooms from June to August.[3-4]

Main phytoconstituents   Ref.
[51-54]

amino acids (arginine, threonine, tyrosine, proline, tryptophan, aspartic acid, isoleucine, phenyl alanine)

fatty acids (oleic acid, stearic acid, linoleic acid, palmitic acid, arachidic acid, dodecanoic acid)

terpenoids (farnesene, geraniol, caryophyllene)

hydrocarbons (hexadecane, dodecane, tricosane, pentacosane, heptacosane)

phenols (eugenol, 4-vinyl guaiacol)

aldehydes and ketones (nonanal, tetradecanal, p-methoxyacetophenone)

flavonoids (apigenin, isorhamnetin. Luteolin, kempferol, rutin, tilianin)

phenolic acids (chlorogenic acid, ferulic acid)

Medicinal Use

febrifuge, prevention of atherosclerosis, stimulate liver function, detoxifying, hypertension, myalgia, antihemorroidal, cardiotonic, diuretic[51-54]

Biological Activity

hepatoprotective, immunostimulant, antioxidant, neurodegeneration protective, cardiovascular protective, antitumoral[51-56]

Other Uses

- raw material for paper preparation
- culinary (roasted) or chow[57]

Cirsium arvense (L.) Scop. (Canada thistle)

Scientific name: *Cirsium arvense* (L.) Scop.
Common name: Canada thistle
Romanian name: Pălămidă
Taxonomic classification
Kingdom: *Plantae*
Subkingdom: *Cormobionta*
Phylum: *Spermatophyta*
Subphylum: *Magnoliophytina (Angiospermae)*
Class: *Magnoliatae (Magnoliopsida, Dicotyledonatae)*
Subclass: *Asteridae*
Order: *Asterales (Compositales)*
Family: *Asteraceae (Compositae)*
Subfamily: *Asteroideae (Tubuliflorae)*
Genus: *Cirsium*
Species: *arvense* (L.) Scop. (Figure 4.10)[2-3]

SPREAD

Common species, frequently distributed from the lowlands to the spruce forest level, in fields, forest clearings, riverside areas, ruderal sites, shrubs, and meadows.[2-4]

Description

Perennial herbaceous plant, with stolons and long, horizontal rhizomes deep in the soil. The stem is cylindrical, erect, costate, glabrous, with long, erect branches and a height of 50–150 cm. Basal leaves disappear before flowering, stem leaves are lanceolate, inferior ones attenuated in the short petiole, and middle and superior ones sessile, entire, lobed or pinnatifid. At the tips of the lobes, there are more vigorous spines, and between them, the leaves are weakly spiny-setiform. They are glabrous on the superior surface, sometimes sparsely hairy, and on the inferior surface, they are

Figure 4.10 *Cirsium arvense* L. Scop. (Canada thistle).

Source: 58.

densely tomentose or almost white. The inflorescence is a corymb with pedunculate calathides, ovoid-globose, 8–10 mm in diameter, and involucral folioles are ovate-lanceolate, imbricate, terminated with a very short, patent spinelet. The flowers are reddish-violet, longer than the involucre. Achenes have a dirty white pappus. The flowering period is from June to August.[2–4]

Main phytoconstituents Ref
[59–62]
fatty acids (tocopherol, octadecanoic acid, exadecenoic acid)
flavonoids (acacetin, morin, cirsimarin, luteolin, kaempferol, rutin)
phenolic acids (chlorogenic acid, caffeic acid, vanillic acid, syringic acid)
coumarins (scopoletin)
sterols (sitosterol, β amyrin)
terpenoids (linalool, camphor, lupeol, citronellol, phytol, bisabolol, cadinene, selinene)
Tannins

Medicinal Use

peptic ulcer, leukemia, diuretic, astringent, hepatoprotective, calcium deficiencies/avitaminosis[59–63]

Biological Activity

antifungal, antioxidant, antimicrobial, antitumoral, antidiabetic, anthelmintic[63]

Other Uses

- salad

- puree

- coffee substitute[63]

Cichorium intybus L. (Common chicory)

Scientific name: *Cichorium intybus* L.
Common name: Common chicory
Romanian name: Cicoare
Taxonomic classification
Kingdom: *Plantae*
Subkingdom: *Cormobionta*
Phylum: *Spermatophyta*
Subphylum: *Magnoliophytina (Angiospermae)*
Class: *Magnoliatae (Magnoliopsida, Dicotyledonatae)*
Subclass: *Asteridae*
Order: *Asterales (Compositales)*
Family: *Asteraceae (Compositae)*
Subfamily: *Cichorioideae (Liguliflorae)*
Genus: *Cichorium*
Species: *intybus* L.[1-3]
Subspecies: – *intybus* – Wild chicory (Figure 4.11)[2]

  – *sativum* (DC.) Janch. – Cultivated chicory[2]

SPREAD

Frequently found from lowland to mountainous areas, in meadows, ruderal and cultivated areas, riverbanks, meadows, and pastures.[1-3,65]

Figure 4.11 *Cichorium intybus* L. (Common chicory).
Source: 64.

Description

Perennial herbaceous plant with a vertical, thick root and an erect stem that can reach up to 1 meter in height. The stem is green, branched, sulcate, and may be either glabrous or roughly hairy. The basal leaves, arranged in a rosette, are oblanceolate, runcinate, and attenuated into petioles, while the stem leaves are lanceolate, sessile, and cordate at the base, with rough hairs or becoming glabrous. The anthodia are terminal and axillary, solitary or grouped in pairs or threes, with hermaphroditic, ligulate flowers that are blue in color. The achenes have a very short pappus along their superior edge, resembling small teeth. Flowering occurs from July to September.[1-3,65]

Main phytoconstituents   Ref.

fatty acids (eicosanoic acid, tetradecanoic acid, octadecanoic acid, heptadecanoic acid, octadecadienoic acid)   [66-68]

sesquiterpene lactones (germacranolide, guajanolide, lactucin, lactucopicrin)

terpenoids (β–amyrin, caryophyllene, lupeol, thymol, terpineol, taraxasterol, linalool, camphor)

coumarins (scopoletin, umbelliferone)

sterols (campesterol, β-sitosterol, stigmasterol)

phenolic acids (chlorogenic acid, caffeic acid, malic acid, cichoric acids)

organic acids (succinic acid, quinic acid, shikimic acid, oxalic acid, caftaric acid)

anthocyanins (cyanidin, malvidin, pelargonidin, delphinidin)

flavonoids (quercetin, ladanetin, spicoside, apigenin, kaempferol)

carbohydrates (inulin, fructose, glucose)

hydrocarbons (octacosane, hexadecane, eicosane)

amino acids (arginine, histidine, leucine, lysine, cysteine, phenylalanine, threonine, valine, methionine)   [69]

Medical Uses

diuretic, laxative, diabetes, febrifuge, hepatoprotective, choleretic, slow digestion, bloating, depurative, spasmolytic, antiarthritis, antiseptic, arteriosclerosis, cholesterol, jaundice, kidney disorders[66,70-74]

Biological Activity

The entire plant (*Cichorii radix et herba*) has properties that include being a bitter tonic, depurative, stomachic, and hypoglycemic.[1] Additionally, it exhibits antimicrobial, analgesic, anti-inflammatory, photoprotective, hepatoprotective, antidiabetic, hypolipidemic, antioxidant, antifungal, and bone-strengthening effects.[40-73]

Other Uses

- salads

- coffee substitutes

- honey species[66,70-74]

Hieracium pilosella L. (Mouse-ear hawkweed)

Scientific name: *Hieracium pilosella* L. (*Pilosella officinarum* Vaill.)
Common name: Mouse-ear hawkweed
Romanian name: Vulturică, Culcușul vacii[65]
Taxonomic classification
Kingdom: *Plantae*
Subkingdom: *Cormobionta*
Phylum: *Spermatophyta*
Subphylum: *Magnoliophytina (Angiospermae)*
Class: *Magnoliatae (Magnoliopsida, Dicotyledonatae)*
Subclass: *Asteridae*
Order: *Asterales (Compositales)*
Family: *Asteraceae (Compositae)*
Subfamily: *Cichorioideae (Liguliflorae)*
Genus: *Hieracium*
Species: *pilosella* L. (Figure 4.12)[2-3]

Figure 4.12 *Hieracium pilosella* L. (Mouse-ear hawkweed).

Source: 75.

SPREAD

Common species throughout the country, from oak forests to the spruce zone, in sunny meadows, fallow lands, and forest clearings.[2,3,65]

DESCRIPTION

Perennial herbaceous species, with a thick, horizontal rhizome and numerous long, slender above-ground stolons, simple or branched, densely covered with long hairs. The leaves are small and spaced apart, and the scape can be solitary or rarely 2–5, each bearing an anthodium with or without dense, stellate hairs or sometimes with a glandular, gray tomentose appearance. The basal rosette leaves are coriaceous, lanceolate, acute, entire, rarely distant and finely denticulate, green on the superior surface, densely long and setiform hairy, while the inferior surface is gray or white-pressed tomentose. The anthodium is globular, with hypsofilous involucral bracts that may have variable hairiness and a stellate-hairy or glabrous margin. The flowers are yellow, and the ligules are red-striped on the external surface. The flowering period is from May to August.[2,3,65]

Main phytoconstituents

flavonoids (apigenin, luteolin, kaempferol, isoetin)
coumarins (umbelliferone)
phenolic acids (caffeic acid, chlorogenic acid)
fatty acids (palmitic acid)
terpenoids (beta-amyrin, taraxerol, lupeol)
sterols (campesterol, stigmasterol, β-sitosterol, cholesterol, fecosterol)

Ref.
[76–77]

Medicinal Use

diuretic, depurative, astringent, hypertension, wound healing, renal inflammations (cystitis, nephritis, urolithiasis), rheumatism, expectorant, respiratory diseases (asthma, bronchitis)[76–77]

Biological Activity

antioxidant, antitumor, anti-inflammatory, antiseptic, hepatoprotective, immunomodulatory[76-77]

REFERENCES

1. Bejenaru C., Mogoşanu G.D., Bejenaru L.E., Popescu H. 2020. *Botanică farmaceutică. Cormobionta*, ediția a III-a. Ed. Sitech, Craiova.

2. Ciocârlan V. *Flora ilustrată a României. Pteridophyta et Spermatophyta.* Ediția a 3-a revizuită și adăugită. 2009. Ed. Ceres, București.

3. Sârbu I., Ştefan N., Oprea A. 2013. *Plante vasculare din România, Determinator ilustrat de teren.* Ed. Victor B Victor, București.

4. Săvulescu T. (ed). 1964. *Flora RPR*, vol. IX. Ed. Academiei RPR, București.

5. https://dryades.units.it/lamone/index.php?procedure=taxon_page&id=5394&num=1208

6. Judzentiene A., Garjonyte R., Budiene J. 2016. Variability, toxicity, and antioxidant activity of *Eupatorium cannabinum* (hemp agrimony) essential oils. *Pharmaceutical Biology* 54(6):945–953.

7. Grigore A., Neagu G., Dobre N., Albulescu A. 2018. Evaluation of antiproliferative and protective effects of *Eupatorium cannabinum* L. extracts. *Turkish Journal of Biology* 42:341–351.

8. Ionita L., Grigore A., Pirvu L., Draghici E., Bubueanu1 C., Ionita C., Panteli M., Dobre N. 2013. Pharmacological activity of an *Eupatorium cannabinum* L. extract. *Romanian Biotechnological Letters* 18(6):8779–8786.

9. Tiţă M.G., Lupuleasa D., Mogoşanu G.D. 2012. Histo-anatomical researches on the vegetative organs of *Eupatorium cannabinum* L. species. *Farmacia* 60(5):621–633.

10. https://dryades.units.it/floritaly/index.php?procedure=taxon_page&tipo=all&id=5644

11. Mogoşanu G.D., Bejenaru L.E., Bejenaru C., Popescu H. 2015. *Farmacognozie-Fitoterapie*, Vol. II, ediția a III-a. Ed. Sitech, Craiova.

12. Astafyeva O., Sukhenko l., Kurashov E., Krylova J., Egorov M., Bataeva Y., Baimukhambetova A. 2018. Essential oil-bearing plants, chemical composition and antibacterial properties of *Achillea micrantha. Indian Journal of Pharmaceutical Sciences* 80(3):434–441.

13. Dias M.I., Barros L., Dueñas M., Pereira E., Carvalho A.M., Alves R.C., Oliveira M.B., Santos-Buelga C, Ferreira I.C. 2013. Chemical composition of wild and commercial *Achillea millefolium L.* and bioactivity of the methanolic extract, infusion and decoction. *Food Chemistry* 141(4):4152–4160.

14. Falconieri D., Piras A., Porcedda S., Marongiu B., Gonçalves M.J., Cabral C., Cavaleiro C., Salgueiro L. 2011. Chemical composition and biological activity of the volatile extracts of *Achillea millefolium. Natural Product Communications* 6(10):1527–1530.

15. Chandler R.F., Hooper S.N., Harvey M.J. 1982. Ethnobotany and phytochemistry of yarrow, *Achillea millefolium, Compositae. Economic Botany Springer on behalf of New York Botanical Garden Press* 36(2):203–223.

16. Simon, J.E., Chadwick A.F., Craker L.E. 1984. *Herbs: an Indexed Bibliography*. 1971–1980. The Scientific Literature on Selected Herbs, and Aromatic and Medicinal Plants of the Temperate Zone. Archon Books, 770 pp., Hamden, CT.

17. https://dryades.units.it/euganei/index.php?procedure=taxon_page&id=5706&num=1274

18. Bailen M., Julio L.F., Diaz C.E., Sanz J., Martínez-Díaz R.A., Cabrera R., Burillo J., Gonzalez-Coloma A. 2013. Chemical composition and biological effects of essential oils from *Artemisia absinthium L.* cultivated under different environmental conditions. *Industrial Crops and Products* 49:102–107.

19. Bhat R.R., et al. 2019. Chemical composition and biological uses of *Artemisia absinthium* (Wormwood). In: Ozturk M., Hakeem K. (eds) *Plant and Human Health*, vol. 3. Springer, Cham, Switzerland, pp. 37–64.

20. El-Saber Batiha G., Olatunde A., El-Mleeh Amany, Hetta H.F., Al-Rejaie S., Alghamdi S., Zahoor M., Beshbishy A.M., Murata T., Zaragoza-Bastida A., Rivero-Perez N. 2020. Bioactive compounds, pharmacological actions and pharmacokinetics of Wormwood (*Artemisia absinthium*). *Antibiotics (Basel)* 9(6):353.

21. Msaada K., Salem N., Bachrouch O., Bousselmi S., Tammar S., Alfaify A., Al Sane K., Ben Ammar W., Azeiz S., Brahim A.H., Hammami M., Selmi S., Limam F., Marzouk B. 2015. Chemical composition and antioxidant and antimicrobial activities of wormwood (*Artemisia absinthium* L.) essential oils and phenolics. *Journal of Chemistry*, vol. 2015, Article ID 804658, 12 pp.

22. Al-Snafi A.E. Pharmacological and Nutritional Importance of *Artemisia*. In: *Medicinal and aromatic plants of the world*. Encyclopedia of Life Support Systems (EOLSS), United Nations Educational, Scientific and Cultural Organization (UNESCO), Paris, France, pp. 123–186.

23. Szopa A., Pajor J., Klin P., Rzepiela A., Elansary H.O., Al-Mana F.A., Mattar M.A., Ekiert H. 2020. *Artemisia absinthium* L. – Importance in the history of medicine, the latest advances in phytochemistry and therapeutical, cosmetological and culinary uses. *Plants* 9(9):1063.

24. Hashimi A., Siraj, M., Ahmed Y., Siddiqui A., Jahangir U., Scholar M. 2019. One for all – *Artemisia absinthium* (Afsanteen) "A Potent unani drug." *Tang (Humanitas Medicine)* 9(4):e5.

25. https://dryades.units.it/euganei/index.php?procedure=taxon_page&id=5724&num=1270

26. Segneanu A.E., Marin C.N., Ghirlea I.O.-F., Feier C.V.I., Muntean C., Grozescu, I. 2021. *Artemisia annua* growing wild in Romania – A metabolite profile approach to target a drug delivery system based on magnetite nanoparticles. *Plants* 10:2245.

27. https://dryades.units.it/euganei/index.php?procedure=taxon_page&id=5666&num=1255

28. Singh O, Khanam Z, Misra N, Srivastava MK. 2011. Chamomile (*Matricaria chamomilla* L.): An overview. *Pharmacognosy Review* 5(9):82–95.

29. Gupta V., Mittal P., Bansal P., Khokra S.L, Kaushik D. 2010. Pharmacological potential of *Matricaria recutita*-A review. *International Journal of Pharmaceutical Sciences and Drug Research* 2(1):12–16.

30. Singh O., Khanam Z., Misra N., Srivastava M.K. 2011. *Chamomile* (*Matricaria chamomilla* L.): An overview, *Pharmacognosy Reviews* 5(9):82–95.

31. Stanojevic L.P., Marjanovic-Balaban Z.R., Kalaba V.D., Stanojevic J.S., Cvetkovic D.J. 2016. Chemical composition, antioxidant and antimicrobial activity of chamomile flowers essential oil (*Matricaria chamomilla* L.). *TEOP* 19(8): 2017–2028.

32. https://dryades.units.it/dolomitifriulane/index.php?procedure=taxon_page&id=5745&num=3816

33. Maftei D.E. 2019. An overview on the importance, spread, sustainable cultivation and preservation of the mountain *Arnica*. *Studii şi Cercetări-Universitatea "Vasile Alecsandri" din Bacău* 28(1): 119–121.

34. Waizel-Bucay J., de Lourdes C-J.M. 2014. *Arnica montana L.,* relevant European medicinal plant -Research Note. *Revista Mexicana de Ciencias Forestales* 5(25):99–109.

35. Kriplani P., Guarve K., Baghael U.S. 2017. *Arnica montana* L. – a plant of healing: review. *Journal of Pharmacy and Pharmacology* 69: 925–945.

36. Sugier D, Sugier P, Jakubowicz-Gil J, Winiarczyk K, Kowalski R. 2019. Essential oil from *Arnica montana* L. Achenes: chemical characteristics and anticancer activity. *Molecules* 24(22):4158.

37. Kriplani, P., Guarve, K. and Baghael, U.S. 2017. *Arnica montana* L. – a plant of healing: review. *Journal of Pharmacy and Pharmacology* 69:925–945.

38. https://dryades.units.it/Roma/index.php?procedure=taxon_page&id=5837&num=1284

39. Wang D., Bădărau A.S., Swamy M.K., Shaw S., Maggi F., da Silva L.E., López V., Yeung A.W.K., Mocan A., Atanasov A.G. 2019. *Arctium* species secondary metabolites chemodiversity and bioactivities. *Frontiers in Plant Science* 10:834.

40. Al-Snafi A.E. 2014. The pharmacological importance and chemical constituents of *Arctium lappa*. A review. *International Journal for Pharmaceutical Research Scholars* 3(1):663–670.

41. Azizov U.M., Khadzhieva U.A., Rakhimov D.A. et al. 2012. Chemical composition of dry extract of *Arctium lappa* roots. *Chemistry of Natural Compounds* 47:1038–1039.

42. Gao Q., Yang M., Zuo Z. 2018. Overview of the anti-inflammatory effects, pharmaco-kinetic properties and clinical efficacies of arctigenin and arctin from *Arctium lappa* L. *Acta Pharmacologica Sinica* 39(5):787–801.

43. https://dryades.units.it/rosandra_en/index.php?procedure=taxon_page&id=5694&num=1266

44. Ivănescu B., Pop C.E., Vlase L., Corciovă A., Gherghel D., Vochita G., Tuchiluş C., Mardari C., Teodor C.M. 2021. Cytotoxic effect of chloroform extracts from *Tanacetum vulgare, T. macrophyllum* and *T. corymbosum* on Hela, A375 and V79 cell lines. *Farmacia* 69(1):12–20.

45. Sowa P., Marcincáková D., Miłek M., Sidor E., Legáth J., Dzugan M. 2020. Analysis of cytotoxicity of selected *Asteraceae* plant extracts in real time, their antioxidant properties and polyphenolic profile. *Molecules* 25:5517.

46. Vilhelmova N., Simeonova L., Nikolova N., Pavlova E., Gospodinova Z., Antov G., Galabov A., Nikolova I. 2020. Antiviral, cytotoxic and antioxidant effects of *Tanacetum vulgare* L. crude extract *in vitro*. *Folia Medica* 62(1):172–179.

47. Thomas O.O. 1989. Phytochemistry of the leaf and flower oils of *Tanacetum corymbosum*. *Fitoterapia* 60(3):225–228.

48. Kumar V., Tyagi D. 2013. Chemical composition and biological activities of essential oils of genus *Tanacetum* – a review. *Journal of Pharmacognosy and Phytochemistry* 2(3):159–163.

49. Zinicovscaia I., Ciocarlan A., Lupascu L. et al. 2019. Chemical analysis of *Tanacetum corymbosum* (L.) Sch. *Bip. using neutron activation analysis. Journal of Radioanalytical and Nuclear Chemistry* 321:349–354.

50. https://dryades.units.it/Roma/index.php?procedure=taxon_page&id=5845&num=7114

51. Isik G., Yücel E. 2017. Fatty acid composition of *Carduus nutans* seeds. *International Journal of Agriculture and Environmental Research* 3(1):19–25.

52. Formisano C., Rigano D., Senatore F., De Feo V., Bruno M., Rosselli S. 2007. Composition and allelopathic effect of essential oils of two thistles: *Cirsium creticum* (Lam.) D.'Urv. ssp. *triumfetti* (Lacaita) Werner and *Carduus nutans* L. *Journal of Plant Interactions* 2(2):115–120.

53. Bain J.F., Desrochers A.M. 1988. Flavonoids of *Carduus nutans* and *C. acanthoides*. *Biochemical Systematics and Ecology* 16(3):265–268.

54. Maciel L.S., Marengo A., Rubiolo P., Leito I., Herodes K. 2021. Derivatization-targeted analysis of amino compounds in plant extracts in neutral loss acquisition mode by liquid chromatography-tandem mass spectrometry. *Journal of Chromatography A*, 1656:462555.

55. Zhelev I., Dimitrova-Dyulgerova I., Belkinova D., Mladenov R. 2013. Content of phenolic compounds in the genus *Carduus* L. from Bulgaria. *Ecologia Balkanica* 5(2):13–21.

56. Zheleva-Dimitrova D., Zhelev I., Dimitrova-Dyulgerova I. 2011. Antioxidant activity of some *Carduus* species growing in Bulgaria. *Free Radicals and Antioxidants* 1(4).

57. www.invasiveplantatlas.org/subject.html?sub=3011

58. https://dryades.units.it/gallignano/index.php?procedure=taxon_page&id=5918&num=26962

59. Amiri N., Yadegari M., Hamedi B. 2018. Essential oil composition of *Cirsium arvense* L. produced in different climate and soil properties. *Records of Natural Products* 12:251–262.

60. Dehjurian A., Lari J., Motavalizadehkakhky A. 2017. Anti-bacterial activity of extract and the chemical composition of essential oils in *Cirsium arvense* from Iran. *Journal of Essential Oil Bearing Plants* 20(4):1162–1166.

61. Khan Z., Ali F., Khan S., Ali I. 2011. Phytochemical study on the constituents from *Cirsium arvense*. *Mediterranean Journal of Chemistry* 2: 64–69.

62. Demirtas I., Tüfekçi A., Aksit H., Gül F. 2018. Determination of phenolic profile of *Cirsium arvense* (L.) Scop. subsp. *vestitum* (Wimmer et Grab.) Petrak plant. *Research Journal of Pharmaceutical, Biological and Chemical Sciences* 1:33–36.

63. Khan Haq U.Z., Khan S., Chen Y., Wan P. 2013. *In vitro* antimicrobial activity of the chemical constituents of *Cirsium arvense* (L). Scop. *Journal of Medicinal Plant Research* 7:1894–1898.

64. https://dryades.units.it/floritaly/index.php?procedure=taxon_page&tipo=all&id=6152

65. Săvulescu T. (ed). 1965. *Flora RPR*, vol. X. Ed. Academiei RPR, București.

66. Janda K, Gutowska I, Geszke-Moritz M, Jakubczyk K. 2021. The Common Cichory (*Cichorium intybus* L.) as a source of extracts with health-promoting properties-A review. *Molecules* 26(6):1814.

67. Street R.A., Sidana J., Prinsloo G. 2013. *Cichorium intybus*: Traditional uses, phytochemistry, pharmacology, and toxicology. *Evidence-Based Complementary and Alternative Medicine*, vol. 2013, Article ID 579319, 13 pages.

68. Nwafor I.C., Shale K., Achilonu M.C. 2017. Chemical composition and nutritive benefits of chicory (*Cichorium intybus*) as an ideal complementary and/or alternative livestock feed supplement. *Scientific World Journal* 7343928.

69. Ying G. W., Gui L. J. 2012. Chicory seeds: a potential source of nutrition for food and feed. *Journal of Animal & Plant Sciences* 13(2):1736–1746.

70. Chinyelu N.I., Shale K., Chilaka A.M. 2017. Chemical composition and nutritive benefits of chicory (*Cichorium intybus*) as an ideal complementary and/or alternative livestock feed supplement. *Scientific World Journal*, ID 7343928, 11 pp.

71. Haji Akber A., Xue-lei X., Dan T. 2020. Chemical constituents and their pharmacological activities of plants from *Cichorium* genus. *Chinese Herbal Medicines* 12(3):224–236.

72. Saeed M., Abd El-Hack M., Alagawany M., Dhama K. 2017. Chicory (*Cichorium intybus*) herb: chemical composition, pharmacology, nutritional and healthical applications. *International Journal of Pharmacology* 13(4):351–360.

73. Street R.A., Sidana J., Prinsloo G. 2013. *Cichorium intybus*: Traditional uses, phytochemistry, pharmacology, and toxicology. *Evidence-Based Complementary and Alternative Medicine*, Article ID 579319, 13 pages.

74. Akbar S. 2020. *Cichorium intybus* L. (*Asteraceae/Compositae*). In: *Handbook of 200 Medicinal Plants*. Springer, Cham, Switzerland, pp. 609–621.

75. https://dryades.units.it/floritaly/index.php?procedure=taxon_page&tipo=all&id=6391

76. Gawrońska-Grzywacz M., Krzaczek T. 2007. Identification and determination of triterpenoids in *Hieracium pilosella* L. *Journal of Separation Science* 30: 746–750.

77. Gawrońska-Grzywacz M., Krzaczek T., Los R., Malm A., Cyranka M., Rzeski W. 2011. Biological activity of new flavonoid from *Hieracium pilosella* L. *Central European Journal of Biology* 6:397–404.

5 *Boraginaceae* Adans. Family

Cornelia Bejenaru[1], Antonia Radu[1], George Dan Mogoşanu[2],
Ludovic Everard Bejenaru[2], Andrei Biţă[2], and Adina-Elena Segneanu[3]

[1] Department of Pharmaceutical Botany, Faculty of Pharmacy, University of Medicine and Pharmacy of Craiova, 2 Petru Rareş Street, Craiova, Dolj County, Romania
[2] Department of Pharmacognosy & Phytotherapy, Faculty of Pharmacy, University of Medicine and Pharmacy of Craiova, 2 Petru Rareş Street, Craiova, Dolj County, Romania
[3] Institute for Advanced Environmental Research–West University of Timisoara (ICAM–WUT), 4 Oituz Street, Timisoara, Romania

BOTANICAL DESCRIPTION

This family groups together herbaceous plants, rarely subshrubs, with harsh tector hairs and entire, non-stipulate leaves arranged alternately. The flowers adhere to a pattern of 5, are hermaphroditic, actinomorphic, and rarely zygomorphic, grouped in scorpioid cymes. The calyx is gamosepalous, persistent, with 5 teeth or divisions, and the corolla is gamopetalous and can be tubular, infundibuliform, or hypocrateriform (protrusions called fornices are found on the corolla, which close the tube of the corolla). The fornices can be scaly, papillose or hairy, and vaulted. The androecium consists of stamens attached to the tube of the corolla, and the gynoecium has a bicarpellate and tetralocular ovary and a gynobasic style (more rarely starting from the tip of the ovary). The fruit is formed from 2 or 4 achenes or nutlets (mericarpic fruit), more rarely being a drupe.[1–4]

The family encompasses 90 genera, with approximately 1600 species distributed globally, except for the Antarctic region.[4]

The *Boraginaceae* family is divided into 2 subfamilies: the *Heliotropioideae* subfamily, which includes 2 genera found in the flora of Romania: *Argusia* Boehm., *Heliotropium* L., and the *Boraginoideae* subfamily with 18 genera in the flora of our country: *Cerinthe* L., *Lithospermum* L., *Onosma* L., *Alkanna* Tausch, *Echium* L., *Myosotis* L., *Pulmonaria* L., *Nonea* Medik., *Symphytum* L., *Anchusa* L., *Lycopsis* L., *Rochelia* Rchb., *Eritrichium* Schrad. ex Gaudin, *Lappula* Moench, *Asperugo* L., *Cynoglossum* L., *Rindera* Pall., *Omphalodes* Mill.[2–3]

Symphytum officinale L. (Comfrey)

Scientific name: *Symphytum officinale* L.
Common name: Comfrey
Romanian name: Tătăneasă
Taxonomic classification
Kingdom: *Plantae*
Subkingdom: *Cormobionta*
Phylum: *Spermatophyta*
Subphylum: *Magnoliophytina (Angiospermae)*
Class: *Magnoliatae (Magnoliopsida, Dicotyledonatae)*
Subclass: *Asteridae*
Order: *Boraginales*
Family: *Boraginaceae*
Subfamily: *Boraginoideae*
Genus: *Symphytum*
Species: *officinale* L.[1–3]
Subspecies: – *officinale* (*S. officinale* var. *purpureum* Pers.) features violet-purple flowers that are 1–2 cm in length (Figure 5.1);[2–3]

- *bohemicum* (F.W. Schmidt) Celak. has smaller, yellowish flowers;[2–3]

- *uliginosum* (A. Kern.) Nyman (*S. tanaicense* Steven) is a perennial plant, with a height of 10–120 cm, has a flowering period from May to July. It is sporadically found from the plains area to the sessile oak forest zone in moist farmlands and swampy meadows.[3]

DOI: 10.1201/9781003270515-6

Figure 5.1 *Symphytum officinale* L. (Comfrey).

Source: 5.

SPREAD

The plant is commonly found in wet meadows, backwaters, river meadows, damp hayfields, moist farmlands, and ditches from the sylvosteppe area up to the beech tree zone.[1-4]

DESCRIPTION

Perennial plant, tall (50–120 cm), densely setose-hairy, with a branched, short, and thick rhizome. The aerial stem is erect, branched, roughly hairy, with setiform hairs that are slightly curved upwards and have a length of 1–2 mm, edged at the base and winged-ridged at the top. The basal leaves are petiolate, ovate, large, while the stem leaves are broadly decurrent, dark green in color and roughly hairy on the superior surface, and pale with a white midrib, roughly hairy and with reticulate and prominent secondary veins. The pedicellate flowers, which are red-violet in color, form terminal cincinni. The flowers are actinomorphic, gamopetalous, with a cylindrical-campanulate corolla featuring triangular fornices, with an exerted style and an elongated connective above the anther. The fruits are shiny, smooth, with a well-developed strophiole, surrounded by a finely ribbed and long-dentate attachment ring. The comfrey blooms from May to August.[1-2,4]

Main Phytoconstituents

Symphyti radix contains: 0.5–0.8% allantoin; 4–6.5% catechol-type tannin; 1–3% asparagine; approx. 0.5% carotenoids; 0.2–0.5% pyrrolizidine alkaloids, primarily lycopsamine, intermedine, symphytine, symviridine, echimidine, and corresponding N-oxides; triterpenoid saponosides, mono- and bidesmosides with aglycones hederagenin and oleanolic acid; polyuronides (ozuronic mucilages); homogeneous polyholosides (starch, inulin); depsides (lithospermic acid, rosmarinic acid); glycoproteins; polyphenolcarboxylic acids; silicic acid.[6]

Main phytoconstituents   Ref.

alkaloids (intermedine, lasiocarpine, heliosupine, lycopsamine, echiumine, myoscorpine,
    consolidin, consolicin, symviridine, symphytine, intermedine *N*-oxide, lycopsamine
    N-oxide, 7-acetyl intermedine, 7-acetyl lycopsamine, 7- acetyl intermedine *N*-oxide,
    uplandicine *N*-oxide, asperumine, *N*-oxide, myoscorpine *N*-oxide, echiumine *N*-oxide,
    symphytine *N*-oxide, symviridine N-oxide, heliosupine N-oxide, asperumine *N*-oxide)   7–9

terpenoids (symphytoxide A, isobauerenol)

flavonoids (quercetin, kaempferol,

saponin (symphytoxide A, cauloside D, leontoside A, leontoside B, leontoside D)

sterols (β-sitosterol, stigmasterol)

fatty acids (Γ-linolenic acid)

phenolic acids (p-hydroxybenzoic acid, hydrocaffeic acid, chlorogenic acid, rosmarinic acid)

carbohydrates (glucuronic acid, mannose, rhamnose, xylose, fructan, glucose, fructose)

tannins (pyrocatechol)

miscellaneous (allantoin, choline, delphinidin, cyanidin, malvidin)

Medicinal Use

The vegetable product *Symphyti radix* syn. *Consolidae radix* (comfrey root) exhibits demulcent,
cicatrizing, and regenerative actions on damaged tissues (due to the presence of allantoin,
polyholosides), antipruritic effects (mucilages), anti-inflammatory properties (polyphenolcarboxylic
acids, glycoproteins), and stimulates the formation of callus on fractured bones (allantoin). The
herbal product is contraindicated during pregnancy and lactation.[1]

Biological Activity

The vegetable product *Symphyti radix* is used in locomotor disorders (bruises, dislocations,
arthritis, fractures), especially in periosteal and joint inflammations; in vascular conditions
(thrombophlebitis, hematomas); in furuncles, due to its cicatrizing action.

In Romanian ethnopharmacology, comfrey root is used for gingivitis, periodontitis, laryngitis,
gastritis, gastrointestinal ulcers, diarrhea, pleurisy, and rheumatism.[6]

Other Uses

■ culinary; tea; biomass; compost; gum[10]

Symphytum cordatum Waldst. et Kit. (Black burdock)

Scientific name: *Symphytum cordatum* Waldst. et Kit.
Common name: Black burdock
Romanian name: Brustur negru
Taxonomic classification
Kingdom: *Plantae*
Subkingdom: *Cormobionta*
Phylum: *Spermatophyta*
Subphylum: *Magnoliophytina (Angiospermae)*
Class: *Magnoliatae (Magnoliopsida, Dicotyledonatae)*
Subclass: *Asteridae*
Order: *Boraginales*
Family: *Boraginaceae*
Subfamily: *Boraginoideae*
Genus: *Symphytum*
Species: *cordatum* Waldst. et Kit. (Figure 5.2)[2–3]

SPREAD

The species is commonly found from the beech zone to the spruce zone in shady forests and
shrubs.[2–4]

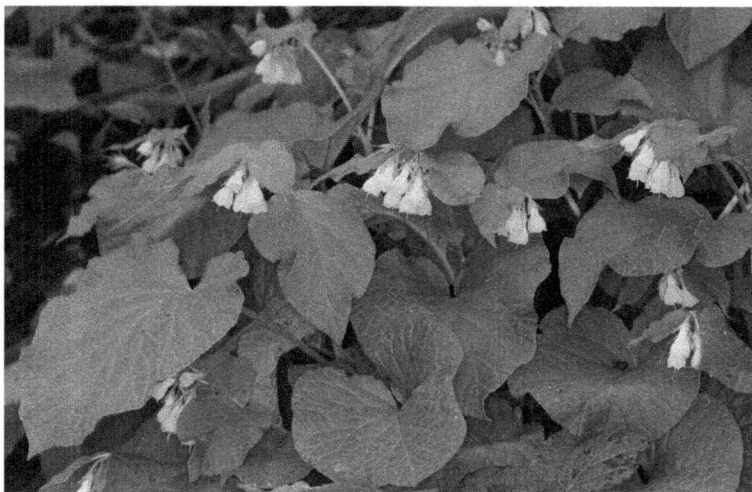

Figure 5.2 *Symphytum cordatum* Waldst. et Kit. (Black burdock).

Source: 11.

DESCRIPTION

Perennial herbaceous plant, with a cylindrical, thick, fleshy, horizontal, branched rhizome that generates 1–2 basal leaves in the first year of vegetation, and in the following year one or more stems. The flowering stem is 20–60 cm tall, simple, unbranched, glabrescent, with sparse, setiform, and reflexed hairs. The basal leaves are long petiolate, up to 25 cm in length and more than 10 cm in width, deeply cordate, shortly acuminate. The stem leaves (2–3) are broadly ovate, cordate or truncated at the base, shortly petiolate or sessile. The bracteal leaves are ovate-lanceolate. The flowers have shortly pubescent pedicels, but also with longer dispersed hairs. The flowers are grouped in pauciflorous scorpioid cymes and have a whitish-yellow corolla, with triangular-lanceolate fornices and twice as long as the calyx. The fruits are elongated cylindrical, roughly reticulated, and with a dentate ring around the strophiole. The flowering period is from May to June.[2–4]

Main phytoconstituents	Ref.
amino acids	7,12–13

coumarins

alkaloids (echimidine, symviridine, symphytine, intermedine, lasiocarpine, lycopsamine)

flavonoids (quercetin, astragalin, hyperoside, myricetin, vicenin, acacetin, hesperetin, nicotiflorin, apigenin, kaempherol, luteolin)

terpenoids

sterols

fatty acids (palmitic acid, stearic acid, oleic acid, linoleic acid, linolenic acid, linolenic acid, stearidonic acid)

phenolic acids (danshensu, chlorogenic acid, ferulic acid, rosmarinic acid, p-coumaric acid, yunnaneic acid E, shmobashiric acid C, monardic acid A, lithospermic acid, salvianolic acid H, pulmitric acid A, pulmitric acid B, cichonic acid)

lignnans (globoidnans A and B, pulmonarioides A and B)

tannins

saponins

anthocyans (delphinidin, petunidin)

carbohydrates

miscellaneous (allantoin)

Medicinal Use

- wound healing, astringent, antiexudative, hematomas and thrombophlebitis, tissue regeneration, arthritis, gout, osteoarticular pain[7,12–13]

Biological Activity

- antioxidant, antibacterial, analgesic, antifungal, anti-inflammatory, antimicrobial, hepatoprotective [7,12–13]

Other Uses

- melliferous
- ornamental [7,12–13]

Cynoglossum officinale L. (Houndstongue)

Scientific name: *Cynoglossum officinale* L.
Common name: Houndstongue
Romanian name: Limba câinelui
Taxonomic classification
Kingdom: *Plantae*
Subkingdom: *Cormobionta*
Phylum: *Spermatophyta*
Subphylum: *Magnoliophytina (Angiospermae)*
Class: *Magnoliatae (Magnoliopsida, Dicotyledonatae)*
Subclass: *Asteridae*
Order: *Boraginales*
Family: *Boraginaceae*
Subfamily: *Boraginoideae*
Genus: *Cynoglossum*
Species: *officinale* L. (Figure 5.3)[1–3]

SPREAD

The plant is common throughout the country, found along forest edges, roadside verges, forest clearings, acacia plantations, riverbanks, shrublands, seedbeds, meadows, and degraded pastures.[1–4]

DESCRIPTION

The genus name is derived from the Greek words *kyon* – dog, and *glossa* – tongue, as the plant's leaves are elongated in the shape of a tongue.[4]

It is a biennial plant, with a greyish appearance and a mouse-like smell. It has a taproot, brown in color, and fleshy. The stem is erect, with numerous leaves; the basal part retains the remnants of the leaves from the previous year and features long, patent, soft hairs, reaching a height of 30–80 cm. The basal leaves are long-petiolate, ranging from elongated elliptic to broadly elliptic shapes, with an acute apex, silky tomentose on both sides with appressed hairs. The stem leaves are lanceolate, semi-amplexicaul, sessile, featuring soft, short, and appressed hairs on both surfaces, more prominently hairy on the inferior side. The pedicellate flowers, numerous and lacking bracts, are grouped in panicles with multifloral cincinni, with one or two arranged terminally, the others axillary. The corolla is broadly infundibuliform, weakly reticulate-veined, measuring 5–9 mm in length, initially violet in color, then turning to a dark red-purple hue. The fornices in the throat of the corolla are reddish, dilated and papillose at the tip, with the stamens inserted in the superior part of the corolla tube. The fruits are flat or slightly concave achenes on the external surface, covered with glochidiate hairs. The flowering period is from May to July.[1–4,15]

Figure 5.3 *Cynoglossum officinale* L. (Houndstongue).

Source: 14.

Main phytoconstituents	Ref.
	16–20

amino acids

coumarins

alkaloids (echinatine, heliosupine, heliotrine, rinderine, cyanustine, cynaustraline)

flavonoids (quercetin, astragalin, hyperoside, myricetin, vicenin, acacetin, hesperetin, nicotiflorin, apigenin, kaempherol, luteolin)

terpenoids

sterols

fatty acids (palmitic acid, stearic acid, oleic acid, linoleic acid, linolenic acid, linolenic acid, and stearidonic acid)

phenolic acids (danshensu, chlorogenic acid, ferulic acid, rosmarinic acid, p-coumaric acid, yunnaneic acid E, shmobashiric acid C, monardic acid A, lithospermic acid, salvianolic acid H, cichonic acid)

tannins

anthocyans (delphinidin, petunidin,

miscellaneous (allantoin)

Medicinal Use

■ diuretic, respiratory disorder (bronchitis, pulmonary catarrh) dysentery, intestinal pain, hemorrhages, hemorrhoids, thrombophlebitis, skin diseases (eczema, wounds, skin ulcers), infections (abscesses), emollient, dandruff[17–21]

Biological Activity

■ antioxidant, antidiabetic, antihyperlipidaemic, antifertility, antitumor, antiseptic, analgesic, anti-inflammatory, hepatoprotective, antihemorrhagic[17–21]

Other Uses

- culinary (leaves)
- food dye (flowers)
- dye[17]

Pulmonaria officinalis L. (Lungwort)

Scientific name: *Pulmonaria officinalis* L.
Common name: Lungwort, Mary's tears
Romanian name: Mierea ursului, Cuscrişor
Taxonomic classification
Kingdom: *Plantae*
Subkingdom: *Cormobionta*
Phylum: *Spermatophyta*
Subphylum: *Magnoliophytina (Angiospermae)*
Class: *Magnoliatae (Magnoliopsida, Dicotyledonatae)*
Subclass: *Asteridae*
Order: *Boraginales*
Family: *Boraginaceae*
Subfamily: *Boraginoideae*
Genus: *Pulmonaria*
Species: *officinalis* L. (Figure 5.4)[1–3]

SPREAD

This species is commonly found in the forests from the oak zone up to the beech zone, in meadows, woods, and forest edges, as well as in shrubs and glades.[1–4]

Figure 5.4 *Pulmonaria officinalis* L. (Lungwort).

Source: 22.

DESCRIPTION

Perennial plant, densely roughly hairy, with a thick rhizome of 3–6 mm, 15 cm long, branched, creeping. The stem is 15–30 cm tall and has 4–5 leaves. The basal leaves, which persist through the winter, are arranged in a rosette, are long-petiolate, erect, elliptical or cordate, suddenly narrowed at the base, with whitish or light green spots, and densely roughly hairy. The inferior stem leaves are spatulate, narrowing into a winged petiole, while the middle ones are elongated oval to elliptical, abruptly narrowing towards the base. The actinomorphic flowers are red initially, later becoming blue-violet, and are grouped in scorpioid cymes. After flowering, the calyx is 1.5–2.5 times longer than it is wide, expanding into a funnel shape. At the throat of the corolla, there are five tufts of long hairs. The stamens and style are enclosed within the corolla. The fruit is a smooth tetra-achene. The flowering period is from March to May.[1–4,23]

Main phytoconstituents   Ref.
amino acids   [24–25]
alkaloids (lycopsamine)
flavonoids (quercetin, astragalin, hyperoside, myricetin, vicenin, acacetin, hesperetin, nicotiflorin, apigenin, kaempherol, luteolin)
terpenoids
sterols
fatty acids (palmitic acid, stearic acid, oleic acid, linoleic acid, linolenic acid, linolenic acid, stearidonic acid)
phenolic acids (danshensu, chlorogenic acid, ferulic acid, rosmarinic acid, p-coumaric acid, yunnaneic acid E, shmobashiric acid C, monardic acid A, lithospermic acid, salvianolic acid H, pulmitric acis A and B, cichonic acid)
lignnans (globoidnans A and B, pulmonariosides A and B)
tannins
anthocyans (delphinidin, petunidin)

Medicinal Use

The leaves (*Pulmonariae folium*) have properties such as diaphoretic, expectorant, vitaminizing,[1] promoting wound healing, and exhibiting skin whitening, diuretic and hemostatic activities.[24]

Biological Activity

- antioxidant, antiseptic, anti-inflammatory, antianemic, anticoagulant, antineurodegenerative, anticonvulsant, immunomodulant[26]

Other Uses

- culinary (salads)

- melliferous[27]

Anchusa officinalis L. (Bugloss)

Scientific name: *Anchusa officinalis* L.
Common name: Common bugloss, Alkanet
Romanian name: Miruță,[1–3] Limba boului[4]
Taxonomic classification
Kingdom: *Plantae*
Subkingdom: *Cormobionta*
Phylum: *Spermatophyta*
Subphylum: *Magnoliophytina (Angiospermae)*
Class: *Magnoliatae (Magnoliopsida, Dicotyledonatae)*
Subclass: *Asteridae*
Order: *Boraginales*
Family: *Boraginaceae*
Subfamily: *Boraginoideae*
Genus: *Anchusa*

Figure 5.5 *Anchusa officinalis* L. (Bugloss).

Source: 28.

Species: *officinalis* L.[1-3]
Subspecies: – *officinalis* features lanceolate bracts, the fruiting calyx with laciniae longer than the tube, and the stigma exerting from the calyx.[2-3] The fruits are dispersed-verrucose, with spaced tubercles (Figure 5.5);[3]

- – *procera* (Besser) Ciocârlan[2-3] (*A. procera* Besser)[3] has wide triangular bracts, the fruiting calyx with laciniae equal to or shorter than the tube, and the stigma not exerting from the calyx.[2-3] The fruits are densely verrucose, with tangent, close tubercles, and initially scorpoid inflorescence.[3]

SPREAD

Common throughout the country, from the plains to the mountainous areas, along the edges of cultivated fields, alongside railway tracks, near roads, on sandy areas, and in meadows.[1-4]

DESCRIPTION

A perennial species, occasionally biennial. The plant grows robustly (can reach a height of 70 cm), with a rough hairy appearance, due to the presence of setiform hairs that cover it and point in all directions. The leaves are wide, elongated-lanceolate; the inferior ones narrow into the petiole, while the middle and superior ones are rounded at the base, sessile, and covered with hairs. The flowers are small, violet in color, rarely white or pink, grouped in cincinni. The calyx has laciniae that are fused only at the base, and the corolla has laciniae that are shorter than the tube, which features fornices. The stamens (attached to the corolla tube) and the style are enclosed within the corolla tube. The fruit is a reticulated and rugose achene. It blooms from May to July.[1-4]

Main phytoconstituents	Ref.
	29–32

amino acids (glutamine, aspartic acid, glutamic acid, threonic acid)
coumarins (isofraxidin)
alkaloids (lycopsamine)
flavonoids (quercetin, naringenin, rutin, apigenin, kaempherol, luteolin, syringin, clinopodic acid A, gingerol)
terpenoids
sterols
fatty acids (myristic acid, palmitic acid, linoleic acid, stearic acid, arachidic acid, behenic acid, lignoceric acid
phenolic acids (sinaptic acid, chlorogenic acid, ferulic acid, rosmarinic acid, p-coumaric acid, salvianolic acid A, salvianolic acid E)
saponins
tannins (gallocatechin, epigallocatechin, catechin, epicatechin)
miscellaneous (malic acid, allantoin, embellin)

Medicinal Use

■ expectorant, emollient, bruises, rheumatic pain, diuretic, diarrhea, cough, skin inflammation, insect bites, nervousness, depression, antiscorbutic, diaphoretic, antigout, cicatrizing, expectorant, digestive, sedative[33–34]

Biological Activity

■ antibiotic, anticancer, anti-inflammatory[29–32]

Other Uses

■ melliferous

■ ornamental

■ culinary (young shoots, flowers and leaves taste similar to spinach)

■ musk fragrance (dried leaves)[35]

Echium vulgare L. (Common viper's bugloss)

Scientific name: *Echium vulgare* L.
Common name: Common viper's bugloss
Romanian name: Limba şarpelui
Taxonomic classification
Kingdom: *Plantae*
Subkingdom: *Cormobionta*
Phylum: *Spermatophyta*
Subphylum: *Magnoliophytina (Angiospermae)*
Class: *Magnoliatae (Magnoliopsida, Dicotyledonatae)*
Subclass: *Asteridae*
Order: *Boraginales*
Family: *Boraginaceae*
Subfamily: *Boraginoideae*
Genus: *Echium*
Species: *vulgare* L. (Figure 5.6)[1–3]

SPREAD

The plant is commonly found in ruderal places, along roadsides, wastelands, railway embankments, meadows, shrublands, and thickets from the plains up to the beech forest zone.[1–4]

Figure 5.6 *Echium vulgare* L. (Common viper's bugloss).

Source: 36.

DESCRIPTION

A biennial plant, with rough hairs and a gray appearance. The black taproot is thick, measuring 6–15 mm. The stem is cylindrical, erect, covered with short, soft hairs, and among these, there are also rigid, long, setiform, and tuberculate hairs, with a red or purple base. The leaves are linear-lanceolate, acute, with the inferior ones arranged in a basal rosette and attenuated into the petiole, while the stem leaves are sessile, with a single vein, and alternately arranged. The flowers are zygomorphic, grouped in simple, unbranched, erect cincinni, which become arched when mature. The corolla has a wide tube, slightly curved and bilabiate, pubescent, with the tube shorter than the calyx, blue in color, with exserted stamens, with long, reddish filaments, curved upwards, and at the base fused with the corolla tube and having blue anthers, elliptical, dorsifixed. The style is long, thin, pubescent, longer than the stamens, and ends with a bilobed stigma. The fruit is a tetra-achene. The blooming period is from June to August.[1-4]

Main phytoconstituents   Ref.

amino acids (lysine, tyrosine, glutamic acid, aspartic acids, phenylalanine, glycine)   [37-38]

alkaloids (echimidine, acetylechimidine, uplandicine, 9-*O*-angelylretronecine, echiuplatine, leptanthine, echimiplatine, echivulgarine, vulgarine, 7-*O*-acetylvulgarine)

flavonoids (quercetin, naringenin, rutin, apigenin, kaempherol, luteolin, myricetin)

terpenoids

sterols (β-sitosterol, stigmasterol)

fatty acids (palmitic acid, oleic acid, linoleic acid, nervonic acid)

phenolic acids (caffeic acid, gallic acid, chlorogenic acid, ferulic acid, rosmarinic acid, vanillic acid, salicylic acid, p-coumaric acid, catechol, catechin)

saponins

tannins

naphthoquinones (acetylshikonin, deoxyshikonin, isovalerylshikonin, shikonin, 3,3-dimethylacrylshikonin)
anthocyanins (delphinidin, cyanidin)

Medicinal Use

■ antiseptic, antispasmodic, calming, diaphoretic, emollient, wound healing, antidiarrheal, depurative, diuretic[39–40]

Biological Activity

■ anti-inflammatory, antioxidant, antibacterial, antiviral, anxiolytic, antimicrobial, antiage, antitumoral, anti-hyperlipidemia[37,41]

Other Uses

■ ornamental

■ melliferous[37,40–41]

Echium italicum L. (Italian viper's bugloss)

Scientific name: *Echium italicum* L.
Common name: Italian viper's bugloss
Romanian name: Coada vacii
Taxonomic classification
Kingdom: *Plantae*
Subkingdom: *Cormobionta*
Phylum: *Spermatophyta*
Subphylum: *Magnoliophytina (Angiospermae)*
Class: *Magnoliatae (Magnoliopsida, Dicotyledonatae)*
Subclass: *Asteridae*
Order: *Boraginales*
Family: *Boraginaceae*
Subfamily: *Boraginoideae*
Genus: *Echium*
Species: *italicum* L. (Figure 5.7)[2–3]

SPREAD
Sporadically found from the plains area up to the hilly region in ruderal places, meadows, stabilized dunes, sandy areas, and slopes.[2–4]

DESCRIPTION
A biennial herbaceous plant, with a greyish appearance, robust, very hispid hairy. The taproot is 10–12 mm thick and dark red in color. The stem is erect, rigid, slightly angular, sometimes branching from the inferior part, covered with long, spiky, rigid, patent setaceous hairs, white or yellow in color, arranged on small tubercles, but also with short and soft hairs. The basal leaves are lanceolate, the inferior stem leaves are linear-lanceolate, attenuated into the petiole, while the middle and superior ones are decreasing in size. They are densely setiform hairy on the superior surface, and hairy only on the median vein on the inferior surface. The flowers with short pedicels are grouped in scorpioid cymes, arched at maturity, with the inferior ones having 2–3 branches. The flowers have an infundibuliform corolla, setiform hairy, yellowish-white in color, rarely flesh-colored, with unequal lobes and a very narrow tube. The stamens have glabrous filaments, 3 of them are longer than the corolla, and the anthers are pale pink. The gynoecium has a short, pubescent style, and the fruits are yellowish, slightly shiny. It blooms from June to August.[2–4]

Main phytoconstituents	Ref.
amino acids	37,43–44
coumarins	

Figure 5.7 *Echium italicum* L. (Italian viper's bugloss).

Source: 42.

alkaloids
flavonoids (quercetin, naringenin, rutin, apigenin, kaempherol, luteolin)
terpenoids
sterols (stigmasterol, brassicasterol)
fatty acids (palmitic acid, oleic acid, linoleic acid, stearic acid, oleic acid, stearidonic acid)
phenolic acids (caffeic acid, syringic acid, synapsis acid, chlorogenic acid, ferulic acid,
  rosmarinic acid, vanillic acid, p-coumaric acid)
naphthoquinones (isobutyrylshikonin, propionylshikonin, angelylshikonin, shikonin,
  3,3-dimethylacrylshikonin)
saponins
tannins
anthocyanins (delphinidin, cyanidin)

Medicinal Use

- anxiolytic, sedative, insecticidal, wound healing, depurative, diaphoretic, diuretic[37,43–44]

The ethnomedicinal history of the *Echium* species can be traced back to 300 B.C. in the
Mediterranean area. Reports indicate various species have been used as folk medicine in the region,
utilised predominately for their sedative, anti-inflammatory, antioxidant, and anxiolytic properties,
treating ailments including fissures of the hands, general abrasions, and even snakebites.[37]

Biological Activity

- antioxidant, antimicrobial, analgesic, anti-inflammatory[37,43–44]

Other Uses

- culinary (young shoots)[45]

REFERENCES

1. Bejenaru C., Mogoșanu G.D., Bejenaru L.E., Popescu H. 2020. *Botanică farmaceutică. Cormobionta*, ediția a III-a. Ed. Sitech, Craiova.

2. Ciocârlan V. 2009. Flora ilustrată a României. Pteridophyta et Spermatophyta. ediția a 3-a revizuită și adăugită. Ed. Ceres, București.

3. Sârbu I., Ștefan N., Oprea A. 2013. *Plante vasculare din România, Determinator ilustrat de teren.* Ed. Victor B Victor, București.

4. Săvulescu T. (ed). 1960. *Flora RPR*, vol. VII. Ed. Academiei RPR, București.

5. https://dryades.units.it/FVG/index.php?procedure=taxon_page&id=4318&num=1048

6. Mogoșanu G.D., Bejenaru L.E., Bejenaru C., Popescu H. 2015. *Farmacognozie-Fitoterapie*, Vol. II, ediția a III-a. Ed. Sitech, Craiova.

7. Salehi B., Sharopov F., Tumer T. B., Ozleyen A., Rodríguez-Pérez C. et al. 2019. *Symphytum* species: A comprehensive review on chemical composition, *food applications and phytopharmacology. Molecules* 24(12):2272.

8. EMA/HMPC/572844/2009 Committee on Herbal Medicinal Products (HMPC). 2011. Assessment report on *Symphytum officinale* L., radix.

9. Aftab K., Shaheen F., Mohammad F.V., Noorwala M., Ahmad V.U. 1996. Phyto-pharmacology of saponins from *Symphytum officinale* L. In: Waller G.R., Yamasaki K. (eds) *Saponins Used in Traditional and Modern Medicine. Advances in Experimental Medicine and Biology*, vol 404. Springer, Boston, MA, pp. 429–442.

10. https://pfaf.org/user/Plant.aspx?LatinName=Symphytum+officinale

11. https://spontana.org/species.php?id=19

12. Dresler S., Szymczak G., Wójcik M. 2017. Comparison of some secondary metabolite content in the seventeen species of the *Boraginaceae* family. *Pharmaceutical Biology*, 55(1):691–695.

13. Bujor O. 2003. *The guide of medicinal and aromatic plants from A to Z.* Ed. Fiat Lux, Bucuresti, ISBN 973-9250-68-8.

14. https://dryades.units.it/lamone/index.php?procedure=taxon_page&id=4376&num=3589

15. Zouhar K. 2002. *Cynoglossum officinale.* Fire Effects Information System [online], U.S. Department of Agriculture, Forest Service, Rocky Mountain Research Station, Fire Sciences Laboratory.

16. www.iucngisd.org/gisd/species.php?sc=1193

17. Joshi K. 2016. *Cynoglossum* L.: A review on phytochemistry and chemotherapeutic potential. *Journal of Pharmacognosy and Phytochemistry* 5(4):32–39.

18. Kopp T., Abdel-Tawab M., Mizaikof B. 2020. Extracting and analyzing pyrrolizidine alkaloids in medicinal plants: A review. *Toxins* 12:320.

19. Maslennikov P.V., Chupakhina G.N., Skrypnik L.N. 2014. The content of phenolic compounds in medicinal plants of a botanical garden (Kaliningrad Oblast). *Biology Bulletin* 41(2):133–138.

20. Pfister J.A., Molyneux R.J., Baker D.C. 1992. Pyrrolizidine alkaloid content of houndstongue (*Cynoglossum officinale* L.). *Journal of Range Management* 45:254–256.

21. www.sfatulmedicului.ro/plante-medicinale/limba-cainelui-cynoglossum-officinalis_14553

22. https://dryades.units.it/euganei/index.php?procedure=taxon_page&id=4306&num=1046

23. Meeus S., Brys R., Honnay O., Jacquemyn H. 2013. Biological flora of the British Isles*: Pulmonaria officinalis. Journal of Ecology* 101(5):1353–1368.

24. Chauhan S., Jaiswal V., Cho Y.I., Lee H.J. 2022. Biological activities and phytochemicals of lungworts (genus *Pulmonaria*) focusing on *Pulmonaria officinalis. Applied Sciences* 12:667.

25. Haberer W., Witte L., Hartmann T., Dobler S. 2002. Pyrrolizidine Alkaloids in *Pulmonaria obscura. Planta Medica* 68(5):480–482.

26. https://csgohap.ru/en/medunica-lekarstvennaya-pulmonaria-officinalis-medunica-lecheb nye-svoistva/

27. https://plants.ces.ncsu.edu/plants/pulmonaria-officinalis/

28. https://dryades.units.it/duino/index.php?procedure=taxon_page&id=4331&num=6616

29. Tsiokanos E., Cartabia A., Tsafantakis N., Lalaymia I., Termentzi A., Miguel M., Declerck S., Fokialakis N. 2022. The metabolic profile of *Anchusa officinalis L.* differs according to its associated arbuscular mycorrhizal fungi. *Metabolites* 12:57.

30. Boskovic I., Đukić D.A., Maskovic P., Mandić L., Perovic S. 2018. Phytochemical composition and antimicrobial, antioxidant and cytotoxic activities of *Anchusa officinalis* L. *extracts. Biologia* 73:1035–1041.

31. Svirska S., Grytsyk A. 2018. Investigation of tannins in *Anchusa officinalis* L. *The Pharma Innovation Journal* 7(4): 758–761.

32. Svirska S.P. 2021. The study of fatty acids in *Anchusa officinalis* L. herb of the Ukrainian flora. *News of Pharmacy* 2(102):3–7.

33. http://mediplantepirus.med.uoi.gr/pharmacology_en/plant_details.php?id=66

34. https://agroromania.manager.ro/articole/plante-medicinale/limba-boului-anchusa-offi cinalis-planta-melifera-si-medicinala-invaziva-in-culturile-de-lucerna-435.html

35. www.worldfloraonline.org/taxon/wfo-0000533707

36. https://dryades.units.it/euganei/index.php?procedure=taxon_page&id=4294&num=1043

37. Jin J., Boersch M., Nagarajan A., Davey A.K., Zunk M. 2020. Antioxidant properties and reported ethnomedicinal use of the genus *Echium* (*Boraginaceae*). *Antioxidants (Basel)* 9(8):722.

38. Kapusterynska A.R., Hamada V.R., Krvavych A.S., Konechna R.T., Kurka M.S., Novikov V.P. 2020. Investigation of the extract's composition of viper's bugloss (*Echium vulgare*). *Ukrainica Bioorganica Acta* 15(1):42–46.

39. www.naturalherbs.ro/ro/project/iarba-sarpelui-echium-vulgare/

40. www.revistagalenus.ro/fitofarmacie/echium-vulgare-l-iarba-sarpelui/

41. Wang W., Jin J., Xu H., Shi Y., Boersch M., Yin Y. 2022. Comparative analysis of the main medicinal substances and applications of *Echium vulgare* L. and *Echium plantagineum* L.: A review. *Journal of Ethnopharmacology* 285:114894.

42. https://dryades.units.it/floritaly/index.php?procedure=taxon_page&tipo=all&id=4289

43. Bošković I.D., Đukić D. A., Mašković P.Z., Mandić G. 2017. Phytochemical composition and biological activity of *Echium italicum* L. plant extracts. *Bulgarian Chemical Communications* 49(4):836–845.

44. Al-Snafi A.E. 2017. Pharmacological and Therapeutic Importance of *Echium italicum*-A Review. *Indo American Journal of Pharmaceutical Sciences* 4(02):394–398.

45. https://upwikiro.top/wiki/Echium

6 *Caprifoliaceae* Adans. Family

Cornelia Bejenaru[1], Antonia Radu[1], George Dan Mogoşanu[2],
Ludovic Everard Bejenaru[2], Andrei Biţa[2], and Adina-Elena Segneanu[3]
[1] Department of Pharmaceutical Botany, Faculty of Pharmacy, University of Medicine
and Pharmacy of Craiova, 2 Petru Rareş Street, Craiova, Dolj County, Romania
[2] Department of Pharmacognosy & Phytotherapy, Faculty of Pharmacy, University of
Medicine and Pharmacy of Craiova, 2 Petru Rareş Street, Craiova, Dolj County, Romania
[3] Institute for Advanced Environmental Research–West University of
Timisoara (ICAM–WUT), 4 Oituz Street, Timisoara, Romania

BOTANICAL DESCRIPTION

In this family, woody plants are predominant, which grow in shrub form (although there are rarely perennial herbaceous plants and trees included). The leaves lack stipules, are simple (entire or incised) or compound, and are arranged opposite. The flowers are hermaphroditic, pentamerous, and can be actinomorphic or zygomorphic. The calyx is gamosepalous, and the corolla is gamopetalous. The corolla can be rotate, infundibuliform, campanulate, or bilabiate. There are 5 stamens inserted at the base of the corolla tube, with ovoid, bilocular anthers. The gynoecium consists of 2–5 carpels, with 2–5 locules, each containing 1 or numerous ovules. The ovary is inferior. The fruit can be a berry, drupe, achene, or capsule. The seeds contain horny or fleshy albumen.[1–4]

The family comprises 12 genera found throughout the world. In Romania, there are 3 genera in the spontaneous flora: *Sambucus* L., *Viburnum* L., and *Lonicera* L.[1–3]

Sambucus L.

In this genus, there are woody or herbaceous plants with imparipinnate compound leaves and hermaphroditic, actinomorphic flowers with a rotate corolla, grouped in cymose inflorescences. The fruit is a berry-like drupe with 1–3 seeds.[1,4]

Sambucus nigra L. (Black elder)

Scientific name: *Sambucus nigra* L.
Common name: Elder, Elderberry, Black elder, European elder, European elderberry
Romanian name: Soc negru
Taxonomic classification
Kingdom: *Plantae*
Subkingdom: *Cormobionta*
Phylum: *Spermatophyta*
Subphylum: *Magnoliophytina (Angiospermae)*
Class: *Magnoliatae (Magnoliopsida, Dicotyledonatae)*
Subclass: *Asteridae*
Order: *Dipsacales*
Family: *Caprifoliaceae*
Genus: *Sambucus*
Species: *nigra* L.[1–3]
Variations: – *nigra* has folioles with entire margins (Figure 6.1);[2]

   laciniata L. has pinnately lobed to pinnately dissected folioles. It is found in Banat, Muntenia, Transylvania, and Dobrogea.[2]

SPREAD

A common species found from the steppe zone to the beech forest zone, through forests, shrubs or thickets, clearings, forest edges, and meadows.[1–4]

DESCRIPTION

Shrub, 2–7 m in height, with soft white pith and fissured rhytidome. The stems are brown and have numerous prominent lenticels, while the shoots are vigorous and erect. The leaves

DOI: 10.1201/9781003270515-7

Figure 6.1 *Sambucus nigra* L. (Black elder).

Source: 5.

are imparipinnate compound with 5–7 short-petiolate, elliptic or ovate-elliptic, acute folioles, irregularly sharply serrate along the margins, and sparse hairs on the veins on the inferior surface. The flowers are small, pleasantly and strongly scented, grouped in corymbiform or umbelliform cymose inflorescences, flat or terminal, with a diameter of 12–20 cm. The flowers have a white to yellowish-white corolla with 5 rounded lobes and yellow anthers. The fruit is a black berry, globular, glossy, with red juice, and the fruiting peduncles are purple to violet at maturity. It blooms from May to July.[1-4]

Main Phytoconstituents

Sambuci flos contains: cyanogenic heterosides, with the main representative being (S)-sambunigroside (sambunigrin or benzaldehyde-cyanohydrin-O-glucoside); flavonoids (rutin); saponins; catechic tannin; mucilage; volatile oil; polyphenolcarboxylic acids.
   The nauseating smell of fresh elderflowers is due to aliphatic amines.[6]

Main phytoconstituents   Ref.
amino acids (glutamic acid, alanine, leucine, aspartic acid, tyrosine)   [6-9]
terpenoids (linalool, nerol, geranial, β-amyrin, camphor, pinene, α-farnesene, oleanoic acid,
   citronellol, myrcenol, ursolic acid, δ-cadinene, β-bergamotene, citral, fenchone, germacrene
   D, α-humulene, eugenol, limonene, β-elemene, phytoene, α-terpineol, β-caryophyllene,
   α-phellandrene, safranal, eucalyptol, β-bourbonene, myrtenol, α-cubebene, β-cyclocitral,
   verbenone, β-ocimene, β-damascenone, β-ionone, aromadendrene, calamenene)
coumarins (esculetin)
iridoids (sambunigrin, prunasin)
flavonoids (quercetin, kaempferol, naringenin, myricetin, astragalin, isorhamnetin, morin, rutin)
tannins (catechin, epicatechin, ellagic acid)
sterols (β-sitosterol, campesterol, stigmasterol)

organic acids (citric and malic acid, tartaric acid, shikimic acid, fumaric acid, quinic acid, acetic acid, valeric acid, benzoic acid)

alcohols (3-hexen-1-ol, 1-hexanol, 2-phenyl-ethanol)

carbohydrates (glucose, sucrose, fructose, arabinose, rhamnose, xylose, mannose, galactose, glucuronic acid, galacturonic acid)

fatty acids (palmitic acid, myristic acid, stearic acid, oleic acid, capric acid, lauric acid, lignoceric acid, arachidic acid, cerotic acid)

phenolic acids (chlorogenic acid, gallic acid, protocatechuic acid, gentisic acid, p-hydroxybenzoic acid, vanillic acid, caffeic acid, ferulic acid, sinapic acid, p-coumaric acid, salicylic acid, neochlorogenic acid)

anthocyanins (cyanidin, chrysanthemin, peonidin, callistephin, pelargonidin)

aldehydes & ketones (hexanal, benzaldehyde, heptanal, pentanal, 3-hydroxy-2-butanone, phenylacetaldehyde, nonanal)

saponins

amines (ethylamine, *n*-propylamine, *i*-propylamine, *i*-butylamine)

Medicinal Uses

Elder flowers are harvested when 75% of the inflorescences are open. They are dried to create the herbal product known as *Sambuci flos*. This product has the following actions: diuretic and sudorific, antitussive, expectorant, emollient, immunostimulant, mild laxative, and antirheumatic.[1]

Biological Activity

It is used in the treatment of influenza conditions, respiratory viral infections, and other conditions accompanied by feverish symptoms.[6] Due to its mild laxative and diuretic actions, the product *Sambuci flos* is used in weight-loss regimens.

Externally, elderflowers can be administered in the form of hot baths or poultices for conditions like furuncles, abscesses, burns, blisters, and rheumatism.[6]

Other Uses

- culinary

- insect repellent

- cosmetic

- dye

- fungicide

- microscope slides[10]

Sambucus racemosa L. (American red elder)

Scientific name: *Sambucus racemosa* L.
Common name: American red elder
Romanian name: Soc roşu
Taxonomic classification
Kingdom: *Plantae*
Subkingdom: *Cormobionta*
Phylum: *Spermatophyta*
Subphylum: *Magnoliophytina (Angiospermae)*
Class: *Magnoliatae (Magnoliopsida, Dicotyledonatae)*
Subclass: *Asteridae*
Order: *Dipsacales*
Family: *Caprifoliaceae*
Genus: *Sambucus*
Species: *racemosa* L. (Figure 6.2)[1–3]

Figure 6.2 *Sambucus racemosa* L. (American red elder).

Source: 11.

SPREAD

The species is frequently distributed from the beech zone to the fir zone through clearings and forest edges, as well as shrubs.[1-4]

DESCRIPTION

Shrub, 1–4 m tall, with reddish-brown pith and light brown stems. The leaves are imparipinnate compound with 1–3 pairs of elliptic to ovate-lanceolate folioles, with an elongated, sharp apex, acute serrate margins, glabrous or dispersely hairy on the inferior side, along the veins. The flowers appear shortly before or simultaneously with the leaves, have a 4 mm wide greenish-yellow corolla, with deciduous petals, and yellow anthers. They are grouped in an ovoid to globose inflorescence of panicle type. The fruit is a red berry. Blooms in April-May.[1-4]

Main phytoconstituents　　　　　　　　　　　　　　　　　　　　　Ref.
amino acids　　　　　　　　　　　　　　　　　　　　　　　　　　　12–14
terpenoids
coumarins (esculetin)
iridoids
flavonoids (quercetin, vitexin, hyperoside, isorhamnetin, rutin)
tannins (epicatechin, catechin)
sterols (β-sitosterol, campesterol, stigmasterol)
organic acids (citric and malic acid, tartaric acid, shikimic acid, fumaric acid, quinic acid, acetic acid, valeric acid, benzoic acid)
glycosides
alcohols
carbohydrates
fatty acids

phenolic acids (gallic acid, caffeic acid, caftaric acid, p-coumaric acid)
anthocyanins (cyanidin)
aldehydes & ketones

Medicinal Uses

It has been employed in traditional medicine for addressing a range of conditions, including bone fractures, rheumatism, diabetes, respiratory and pulmonary issues, skin disorders, inflammatory conditions, diarrhea, and more.[14]

Biological Activity

antiviral, antioxidant, antimicrobial, antidiabetic, anti-inflammatory, antidepressant, analgesic, antigiardial, immunomodulatory, scolicidal, antiulcerogenic, antiradical, bone-protective, antiglycemic, antiosteoporotic, hypolipidemic, antiglycation, and wound-healing properties[14–19]

Other Uses

■ ornamental by its red fruits[4]

Sambucus ebulus L. (Dwarf elder)

Scientific name: *Sambucus ebulus* L.
Common name: Dwarf elder, Dwarf elderberry
Romanian name: Boz
Taxonomic classification
Kingdom: *Plantae*
Subkingdom: *Cormobionta*
Phylum: *Spermatophyta*
Subphylum: *Magnoliophytina (Angiospermae)*
Class: *Magnoliatae (Magnoliopsida, Dicotyledonatae)*
Subclass: *Asteridae*
Order: *Dipsacales*
Family: *Caprifoliaceae*
Genus: *Sambucus*
Species: *ebulus* L. (Figure 6.3)[1–3]

SPREAD

The species is common from the lowlands to the mountainous areas, through ruderalized places, fields, uncultivated areas, near fences, and pastures.[1–4]

DESCRIPTION

Perennial species, 0.5–2 m in height, with a thick, creeping rhizome and an herbaceous, straight, glabrous or dispersely hairy, sulcate, unbranched stem. Leaves are imparipinnate compound, with 7–11 ovate-lanceolate folioles, acute apex, sharply serrate margin, glabrous or sparsely hairy on the inferior surface, and prominent, ovate, green stipules with serrate margins. Actinomorphic flowers, grouped in multiflower corymb-like cymose inflorescences, have the corolla with a white interior and a reddish exterior, and stamens with red-purple anthers, later turning black. The fruits are black berries. The entire plant has an unpleasant odor. It blooms from June to August.[1–4]

Main phytoconstituents	Ref.
amino acids (cysteine, serine, alanine, proline, threonine, methionine, isoleucine, aspartic acid, valine, arginine, phenylalanine, tyrosine, lysine, histidine, tryptophan, glycine)	20–24
terpenoids (β-bisabolene, germacrene D, geraniol,α-cubebene, amyrin, eugenol, citronellal)	
iridoids (ebulin, ebulitin, sambunigrin, prunasin)	
flavonoids (quercetin, kaemferol, rutin)	
tannins (epicatechin, catechin)	
sterols (β-sitosterol)	
organic acids (isovaleric acid, valeric acid, quinic acid, benzoic acid)	
glycosides	

Figure 6.3 *Sambucus ebulus* L. (Dwarf elder).
Source: 14.

alcohols
carbohydrates (galactose, glucose, sucrose,
fatty acids (palmitic acid, linoleic acid, stearic acid, myristic acid)
phenolic acids (chlorogenic acid, acid, caffeic acid, acetic acid, p-coumaric acid, ferrulic acid)
anthocyanins (cyanidin)
aldehydes & ketones
stillbens (piceid, resveratrol)

Medicinal Uses

The underground part of the plant (*Sambuci ebuli radix*) has antimicrobial, diuretic, sudorific, and purgative actions.[1,4]

Biological Activity

■ hydropsy, rheumatism[4]

Other Uses

■ culinary

■ tea

■ dye

■ ink

■ repellent[25]

Viburnum opulus L. (European cranberrybush)

Scientific name: *Viburnum opulus* L.
Common name: European cranberrybush, Guelder rose, Cramp bark
Romanian name: Călin
Taxonomic classification
Kingdom: *Plantae*
Subkingdom: *Cormobionta*
Phylum: *Spermatophyta*
Subphylum: *Magnoliophytina (Angiospermae)*
Class: *Magnoliatae (Magnoliopsida, Dicotyledonatae)*
Subclass: *Asteridae*
Order: *Dipsacales*
Family: *Caprifoliaceae*
Genus: *Viburnum*
Species: *opulus* L. (Figure 6.4)[1-3]

SPREAD
Frequently occurring shrub from the lowlands to the beech forest level, in thickets, floodplains, riversides, forest edges, and clearings.[1-4]

DESCRIPTION
Shrub of 1.5–4 m in height, with smooth, glabrous, grayish branches. On the long branches and shoots, the leaves are broadly ovate and trilobed, while on the short branches, they are pentalobed, with acuminate, deeply dentate lobes, rounded or truncate at the base, smooth and green on the

Figure 6.4 *Viburnum opulus* L. (European cranberrybush).
Source: 26.

superior surface, glabrous, and light green, pubescent along the veins on the inferior surface. The petioles are long, channelled, with filiform stipules and 2–4 large, prominent, disciform glands in the vecinity of the leaf lamina. The flowers, white or white-reddish, form dense umbelliform cymose inflorescences, with numerous flowers, the marginal ones being sterile, radiate, and larger. The fruit is a red drupe. Flowering occurs in May–June.[1-4]

Main phytoconstituents   Ref.
amino acid (leucine, glutamic acid, aspartic acid, arginine acid)   [27–30]
iridoids (secologanate, arbutin, viburtioside, opulus I-IV)
terpenoids (squalene, phytol, carvone, linalool, ursolic acid, caryophyllene, germacrene D)
flavonoids (quercetin, rutin, isorhamnetin, luteolin, hyperoside)
coumarins (scopoletin)
tannins (catechin, epicatechin, ellagic acid)
sterols (campesterol, stigmasterol, β-sitosterol)
fatty acids (oleic acid, linoleic acid, palmitic acid, capric acid)
phenolic acids (gallic acid, vanillic acid, chlorogenic acid, p-coumaric acid, neochlorogenic acid, homogentisic acid)
organic acids (malic acid, citric acid, quinic acid, succinic acid, ascorbic acid, oxalic acid, tartaric acid, fumaric acid, ferulic acid, syringic acid, 3-methyl-butanoic acid)
anthocyans (cyanidin)
saponins
carotenoids
aldehydes&ketones (2-octanone,
alcohols&phenols (hexanol, pentanol, butanol, phenol, cresol, syringol)
hydrocarbons (pentadecane, heptane)
miscellaneous (gambiriin, cinchonain I-IV)

Medicinal Uses

For centuries, the bark, leaves, and fruits of this plant have been employed for traditional medicinal purposes. One of the most notable medicinal applications is the impact of the plant's bark on the uterus, a practice observed in both Europe and America. Moreover, particular attention is given to the fruit due to its noteworthy antiurolithiatic properties.[31]

Biological Activity

■ antimicrobial, antioxidant, antinociceptive, anti-inflammatory, antiobesity, antiurolithiatic, antidiabetic, osteogenic, cardioprotective, and cytoprotective properties[31–36]

Other Uses

■ culinary

■ dye

■ ink

■ hedge[37]

Lonicera xylosteum L. (European fly honeysuckle)

Scientific name: *Lonicera xylosteum* L.
Common name: European fly honeysuckle
Romanian name: Caprifoi
Taxonomic classification
Kingdom: *Plantae*
Subkingdom: *Cormobionta*
Phylum: *Spermatophyta*
Subphylum: *Magnoliophytina (Angiospermae)*
Class: *Magnoliatae (Magnoliopsida, Dicotyledonatae)*
Subclass: *Asteridae*

Figure 6.5 *Lonicera xylosteum* L. (European fly honeysuckle).

Source: 38.

Order: *Dipsacales*
Family: *Caprifoliaceae*
Genus: *Lonicera*
Species: *xylosteum* L. (Figure 6.5)[1-3]

SPREAD

A species commonly found in mountainous and submontane forests, as well as in forests, forest edges, clearings, shrublands, and on limestone rocks.[1-4]

DESCRIPTION

Erect shrub, pubescent, 1–2 m in height, with fistulous, short pubescent, grayish branches. Leaves are short-petiolate, ovate-elliptic, 3–7 cm long, with a rounded or abruptly narrowed base, acute apex, entire margin, and densely appressed hairy on both surfaces. Flowers are zygomorphic, with a bilabiate, yellowish-white, pubescent at the exterior corolla, grouped in pairs on a common peduncle, 15–20 mm long, axillary. The peduncle of the inflorescence is of the same length as the flowers. The calyx is deciduous. The fruit is a red or yellow, glossy berry. Flowering occurs in May–June.[1-4]

Main phytoconstituents	Ref.
amino acids	[39-41]
coumarins	
alkaloids (xylostosidine)	
glycosides	
iridoids	
flavonoids	
terpenoids (ursolic acid, pomolic acid)	

sterols
fatty acids
phenolic acids (chlorogenic acid, caffeic acid, ferulic acid)
lignnans
tannins
saponins
organic acids

Medicinal Uses

expectorant, laxative, respiratory diseases, febrifuge, sudorific, diaphoretic, astringent, diuretic, would healing[39–41]

Biological Activity

antioxidant, anti-inflammatory, antitumoral, antispasmodic[39–41]

Other Uses

■ culinary (salads, jellies, teas)

■ ornamental[39–41]

REFERENCES

1. Bejenaru C., Mogoșanu G.D., Bejenaru L.E., Popescu H. 2020. *Botanică farmaceutică. Cormobionta*, ediția a III-a. Ed. Sitech, Craiova.

2. Ciocârlan V. 2009. *Flora ilustrată a României. Pteridophyta et Spermatophyta*. Ediția a 3-a revizuită și adăugită. Ed. Ceres, București.

3. Sârbu I., Ștefan N., Oprea A. 2013. *Plante vasculare din România, Determinator ilustrat de teren. Ed.* Victor B Victor, București.

4. Săvulescu T. (ed). 1961. *Flora RPR*, vol. VIII. Ed. Academiei RPR, București.

5. https://dryades.units.it/ampezzosauris/index.php?procedure=taxon_page&id= 5142&num=3422

6. Bejenaru L.E., Mogoșanu G.D., Bejenaru C., Popescu H. 2015. *Farmacognozie-Fitoterapie*. Vol. I, ediția a III-a. Ed. Sitech, Craiova.

7. Ferreira, S.S., Silva, A.M., Nunes, F.M. 2020. *Sambucus nigra* L. fruits and flowers: chemical composition and related bioactivities. *Food Reviews International* 38(6):1237–1265.

8. Szablewski T., Świerk D., Krejpcio Z., Suchowilska E., Tomczyk Ł. 2021. *Sambucus nigra* extracts–natural antioxidants and antimicrobial compounds. *Molecules* 26(10):2910.

9. Pascariu O. 2022. Bioactive compounds from elderberry: extraction, health benefits, and food applications. *Processes* 10(11):2288.

10. https://pfaf.org/user/plant.aspx?LatinName=Sambucus+nigra

11. https://dryades.units.it/dolomitifriulane/index.php?procedure=taxon_page&id= 5143&num=3924

12. Miljković V. 2019. *Sambucus nigra* and *Sambucus racemosa* fruit: a schematic review on chemical characterization. *Chemia Naissensis – Review* 2(2):1–25.

13. Miljkovic V., Đorđević B., Arsic B., Markovic, Nikolic M., Todorovic L., et.al. 2020. *Sambucus racemosa* L. fruit extracts obtained with conventional and deep eutectic solvents. *Facta Universitatis – Series Physics Chemistry and Technology* 18:89–97.

14. Waswa E. N., Li J., Mkala E. M., Wanga V. O., Mutinda E. S., Nanjala C. et.al. 2022. Ethnobotany, phytochemistry, pharmacology, and toxicology of the genus *Sambucus* L. (*Viburnaceae*). *Journal of Ethnopharmacology* 292:115102.

15. Mlinaric A., Kreft S., Umek A., Strukelj B. 2000. Screening of selected plant extracts for *in vitro* inhibitory activity on HIV-1 reverse transcriptase (HIV-1 RT). *Die Pharmazie* 55(1):75–77.

16. Masuda T., Inouchi T., Fujimoto A., Shingai Y., Inai M., Nakamura M., Imai S. 2012. Radical scavenging activity of spring mountain herbs in the Shikoku mountain area and identification of antiradical constituents by simple HPLC detection and LC-MS methods. *Bioscience, Biotechnology, and Biochemistry* 76(4):705–711.

17. Miljković V.M., Đorđević B.S., Arsić B., Marković M.S., Nikolić L.B., Todorović Z.B., Nikolić G.S. 2020. *Sambucus racemosa* L. fruit extracts obtained with conventional and deep eutectic solvents. *Facta Universitatis–Series Physics, Chemistry and Technology* 18(2):89–97.

18. Đelić G., Mašković P., Simić Z., Pavlović M., Timotijević S. 2020. Antioxidative activity and adoption of metals in species *Sambucus racemosa* L. *Savetovanje o Biotehnologiji* 25(1):35–42.

19. https://pfaf.org/User/Plant.aspx?LatinName=Sambucus+ebulus

20. Kaya Y., Haji E.K., Arvas Y.E., Aksoy H.M. 2019. *Sambucus ebulus* L.: Past, present and future. *Proceedings of the 2nd International Conference on Biosciences and Medical Engineering (ICBME2019): Towards innovative research and cross-disciplinary collaborations, AIP Conf. Proc.* 2155:020030.

21. Chirigiu L., Chirigiu R.G., Tircomnicu V. et al. 2011. GC-MS analysis of chemical composition of *Sambucus ebulus* leaves. *Chemistry of Natural Compounds* 47:126–127.

22. Shokrzadeh M., Saeedi Saravi S.S. 2010. The chemistry, pharmacology and clinical properties of *Sambucus ebulus*: A review. *Journal of Medicinal Plants Research* 4(2):095–103.

23. Kiselova-Kaneva Y., Galunska B., Nikolova M., Dincheva I., Badjakov I. 2022. High resolution LC-MS/MS characterization of polyphenolic composition and evaluation of antioxidant activity of *Sambucus ebulus* fruit tea traditionally used in Bulgaria as a functional food. *Food Chemistry* 367:130759.

24. Tasinov O., Dincheva I., Badjakov I., Kiselova-Kaneva Y., Galunska B., Nogueiras R., Ivanova D. 2021. Phytochemical composition, anti-inflammatory and ER stress-reducing potential of *Sambucus ebulus* L. fruit extract. *Plants* 10(11):2446.

25. https://pfaf.org/User/Plant.aspx?LatinName=Sambucus+ebulus

26. https://dryades.units.it/valerio/index.php?procedure=taxon_page&id=5144&num=3425

27. Kajszczak D., Zakłos-Szyda M., Podsędek A. 2020. *Viburnum opulus* L. – A review of phytochemistry and biological effects. *Nutrients* 12(11):3398.

28. Bock K., Jensen S.R., Nielsen B.J., Norn V. 1978. Iridoid allosides from *Viburnum opulus*. *Phytochemistry* 17(4):753–757.

29. Dienaitė L, Pukalskienė M, Pereira CV, Matias AA, Venskutonis PR. 2020. Valorization of European cranberry bush (*Viburnum opulus* L.) berry pomace extracts isolated with pressurized ethanol and water by assessing their phytochemical composition, antioxidant, and antiproliferative activities. *Foods* 9(10):1413.

30. Liu D., Bai M., Zhao P., Huang X. 2022. Research review on the chemical constituents and pharmacological effects of *Viburnum opulus*[J]. *Asian Journal of Traditional Medicines* 17(2):65–71.

31. Şeker K.G., İlgün S. 2022. *Viburnum opulus* L. In: *Novel drug targets with traditional herbal medicines: scientific and clinical evidence* (pp. 569–583). Springer International Publishing, Cham.

32. Altun M.L., Saltan Çitoğlu G., Sever Yılmaz B., Özbek H. 2009. Antinociceptive and anti-inflammatory activities of *Viburnum opulus*. *Pharmaceutical Biology* 47(7):653–658.

33. Cesonienė L., Daubaras R., Kraujalytė V., Venskutonis P.R., Šarkinas A. 2014. Antimicrobial activity of *Viburnum opulus* fruit juices and extracts. *Journal für Verbraucherschutz und Lebensmittelsicherheit* 9:129–132.

34. Barak T. H., Celep E., İnan Y., Yesilada E. 2019. Influence of in vitro human digestion on the bioavailability of phenolic content and antioxidant activity of *Viburnum opulus* L. (European cranberry) fruit extracts. *Industrial Crops and Products* 131:62–69.

35. Podsędek A., Zakłos-Szyda M., Polka D., Sosnowska D. 2020. Effects of *Viburnum opulus* fruit extracts on adipogenesis of 3T3-L1 cells and lipase activity. *Journal of Functional Foods* 73:104111.

36. İlhan M., Ergene B., Süntar I., Özbilgin S., Saltan Çitoğlu G., Demirel M. A., Küpeli Akkol E. 2014. Preclinical evaluation of antiurolithiatic activity of *Viburnum opulus* L. on sodium oxalate-induced urolithiasis rat model. *Evidence-Based Complementary and Alternative Medicine* 2014:578103.

37. https://pfaf.org/user/Plant.aspx?LatinName=Viburnum+opulus

38. https://dryades.units.it/ampezzosauris/index.php?procedure=taxon_page&id=5152&num=4412

39. Padmanabhan P., Correa-Betanzo J., Paliyath G. 2016. Berries and related fruits. In: Caballero M., Finglas P.M., Toldrá F. (eds). *Encyclopedia of Food and Health*. Academic Press–Elsevier, London, UK, pp. 364–371.

40. Cordell G.A. 1999. The monoterpene alkaloids. In: Cordell G.A. (ed). *The Alkaloids: Chemistry and Biology*, vol. 53. Academic Press–Elsevier, London, UK, pp. 261–376.

41. Popović Z., Smiljanić M., Kostić M., Nikić P., Janković S. 2014. Wild flora and its usage in traditional phytotherapy (Deliblato Sands, Serbia, South East Europe). *Indian Journal of Traditional Knowledge* 13(1):9–35.

7 *Caryophyllaceae* Juss. Family

Cornelia Bejenaru[1], Antonia Radu[1], George Dan Mogoşanu[2],
Ludovic Everard Bejenaru[2], Andrei Biţă[2], and Adina-Elena Segneanu[3]
[1] Department of Pharmaceutical Botany, Faculty of Pharmacy, University of Medicine and Pharmacy of Craiova, 2 Petru Rareş Street, Craiova, Dolj County, Romania
[2] Department of Pharmacognosy & Phytotherapy, Faculty of Pharmacy, University of Medicine and Pharmacy of Craiova, 2 Petru Rareş Street, Craiova, Dolj County, Romania
[3] Institute for Advanced Environmental Research–West University of Timisoara (ICAM–WUT), 4 Oituz Street, Timisoara, Romania

BOTANICAL DESCRIPTION

The family comprises herbaceous plants, which can be annual, biennial, or perennial, with opposite, simple, entire leaves, usually without stipules. The flowers are arranged in dichasial cymose inflorescences and are actinomorphic, bisexual (rarely unisexual), with a double floral envelope, usually 4–5-merous, gamosepalous, and dialypetalous (sometimes the corolla is completely reduced). The androecium has twice the number of stamens compared to the floral envelope elements, while the gynoecium is superior, unilocular, uni- or pluriovulate, septate, and arranged on an extension of the axis called the gynophore, which becomes a carpophore at maturity. The fruit is a dehiscent capsule, opening through valves or teeth, rarely being a berry or achene, with campylotropous ovules. The seeds are round, pear-shaped, or reniform, with a visible funiculus, attached to a central placenta. The embryo is bent, rarely straight, with narrow cotyledons.[1–3]

The family is subdivided into two subfamilies: Subfam. *Caryophylloideae*, which includes 21 genera in the spontaneous flora of Romania, and Subfam. *Paronychioideae*, which comprises 5 genera spread throughout the spontaneous flora of our country.[2,4]

The genus *Herniaria* is represented in the spontaneous flora of Romania by the following species: *Herniaria glabra* L., *Herniaria hirsuta* L., and *Herniaria incana* Lam.[2,4]

Herniaria glabra L. (Smooth rupturewort)

Scientific name: *Herniaria glabra* L.
Common name: Smooth rupturewort
Romanian name: Feciorică
Taxonomic classification
Kingdom: *Plantae*
Subkingdom: *Cormobionta*
Phylum: *Spermatophyta*
Subphylum: *Magnoliophytina (Angiospermae)*
Class: *Magnoliatae (Magnoliopsida, Dicotyledonatae)*
Subclass: *Caryophyllidae*
Order: *Caryophyllales*
Family: *Caryophyllaceae*
Subfam. *Paronychioideae*
Genus: *Herniaria*
Species: *glabra* L. (Figure 7.1)[1,2,4]

SPREAD

The species is sporadic from the oak forest zone up to the beech zone, occurring in dry and sandy places, meadows, landslides, and stony areas.[2,3,4]

DESCRIPTION

It is an annual, biennial, or perennial herbaceous plant, creeping, small, with a height ranging from 5 to 15 cm. It is glabrous and branched, forming dense circular tufts. The leaves are ovate or elliptical with laciniate margins and stipules, small and arranged in an opposite pattern. The small greenish flowers are clustered in glomerule cymes, located axillary to the leaves. The fruits are

DOI: 10.1201/9781003270515-8

Figure 7.1 *Herniaria glabra* L. (Smooth rupturewort).

Source: 5.

ovate, rough, and monospermous, longer than the calyx. The shiny seeds are lentiform, smooth, and black in color. The plant blooms from June to July.[1,3,4]

Main phytoconstituents   Ref.
amino acids (alanine, glutamic acid, phenylalanine, histidine, leucine, asparagine, isoleucine, glycine, threonine)   6–7
coumarins (herniarin umbelliferon)
flavonoids (quercetin, isorhamnetin, luteolin, rutin)
terpenoids (medicagenic acid, glabrosides B and C)
sterols
fatty acids
phenolic acids (gallic acid, caffeic acid, chlorogenic acid)
tannins

Medicinal Use

The vegetable product *Herniariae herba* has antiseptic, anti-inflammatory, antilithiasic, and uricosuric properties.[1]

Biological Activity

Studies concluded that long-term oral consumption of saponins extracted from *Herniaria glabra* can lead to a reduction in arterial blood pressure (hypertension) and influence the renal function (the transport of salt and water in renal tubules).[8]

Moreover, *Herniaria glabra* exhibits antimicrobial and bactericidal activity in urinary tract infections treatment (stronger than *Vaccinium vitis-idaea*). Additionally, it has profound litholytic effect on gallstones, being also efficient in urolithiasis. Furthermore, it demonstrated antiscalant properties and neuroprotective effects. Applied topically, it possesses skin antiaging properties and antipigmentation activity caused by the direct inhibition of elastase, collagenase, and tyrosinase enzymes.[9]

101

Other Uses

■ ornamental (gardens)[9]

In the spontaneous flora of Romania, the genus *Gypsophila* includes the following species: *Gypsophyla paniculata* L., *Gypsophyla muralis* L., *Gypsophyla petraea* (Baumg.) Rchb., *Gypsophyla pallasii* Ikonn. (*G. glomerata* auct. non Pall.), *Gypsophyla perfoliata* L. (*G. trichotoma* Wender.), *Gypsophyla collina* Steven ex Ser. (*G. fastigiata* L. var. *leioclados* Borbás).[2,4]

Gypsophila paniculata L. (Baby's breath)

Scientific name: *Gypsophila paniculata* L.
Common name: Baby's breath
Romanian name: Ipcărige
Taxonomic classification
Kingdom: *Plantae*
Subkingdom: *Cormobionta*
Phylum: *Spermatophyta*
Subphylum: *Magnoliophytina (Angiospermae)*
Class: *Magnoliatae (Magnoliopsida, Dicotyledonatae)*
Order: *Caryophyllales*
Family: *Caryophyllaceae*
Subfam. *Caryophylloideae*
Genus: *Gypsophila*
Species: *paniculata* L. (Figure 7.2)[1,2,4]

SPREAD

The species is sporadic and can be found in sandy and rocky places, rocky coasts, along railways, in the plains and hilly areas.[1–4]

DESCRIPTION

The perennial herbaceous plant has a strongly branched stem from the base, mostly glabrous (finely pubescent only in the inferior part), and reaches a height of 50–90 cm. The lanceolate leaves are acute, uninervous, and narrow towards the apex. The flowers have pedicels that are 2–3 times

Figure 7.2 *Gypsophila paniculata* L. (Baby's breath).

Source: 10.

the length of the calyx and are without involucres, forming a lax, multifloral cymose inflorescence. The floral envelope consists of a gamosepalous calyx with membranous commissural nerves and a white or pale reddish corolla, about 3–4 mm long (twice the length of the calyx). There are 10 stamens and a superior ovary, unilocular, and multi-ovulate. The fruit is a capsule, longer than the calyx. The seeds are flattened and verrucose. The plant blooms from June to September.[1–4]

Main phytoconstituents   Ref.
amino acids   [11–15]
flavonoids
terpenoids (germacrene-D, cymene, ocimene, gypsogenin, limonene, terpinene, myristicin
sterols
fatty acids (palmitic acid)
phenolic acids (p-coumaric acid, syringic acid, dihydroferulic acid)
saponins
tannins
phytoecdysteroids

Medicinal Use

After harvesting and drying, the roots and rhizomes form the medicinal product known as *Saponariae albae radix*.[1] This vegetable product has expectorant, diuretic, and depurative properties.[11]

Toxicity

Due to the saponins, sometimes there are toxic phenomena that are difficult to tolerate (hemolysis), both through an irritating action on mucous membranes, especially when taken orally, and through external contact with the powdered vegetable material (causing sneezing and irritation).[11]

Biological Activity

- Antioxidant, antimicrobial, and antidiabetic activity.[12,16–18]
- It was revealed that saponins derived from *Gypsophila paniculata* substantially enhance the immunospecific cytotoxic efficacy of immunotoxins against human lymphoma.[19]

Other Uses

- cleansing: The roots are employed for cleansing both hair and clothing.

- ornamental: Grown for decorative purposes in floral compositions.

- Saponin supplier: It serves as a saponin source with diverse commercial uses, including the production of photographic film and hemolytic laboratory reagents. Its detergent attribute renders it valuable for crafting soap and shampoo.[20]

The genus *Saponaria* is represented in the flora of Romania by 4 species: *Saponaria officinalis* L., *Saponaria pumilio* (L.) Fenzl. ex A. Braun, *Saponaria bellidifolia* Sm., and *Saponaria glutinosa* M. Bieb. [2,4]

Saponaria officinalis L. (Common soapwort)

Scientific name: *Saponaria officinalis* L.
Common name: Common soapwort, Bouncing-bet, Crow soap, Soapweed
Romanian name: Odogaci, Săpunariţă
Taxonomic classification
Kingdom: *Plantae*
Subkingdom: *Cormobionta*
Phylum: *Spermatophyta*
Subphylum: *Magnoliophytina (Angiospermae)*
Class: *Magnoliatae (Magnoliopsida, Dicotyledonatae)*
Order: *Caryophyllales*
Family: *Caryophyllaceae*
Subfam. *Caryophylloideae*
Genus: *Saponaria*
Species: *officinalis* L. (Figure 7.3)[1,2,4]

Figure 7.3 *Saponaria officinalis* L. (Common soapwort).

Source: 21.

SPREAD

The plant grows on the banks of water bodies, in meadows, along fences, roads, and cultivated areas, from the steppe zone up to the beech zone.[2,3]

DESCRIPTION

The perennial plant has a cylindrical, creeping rhizome, strongly branched, with both fertile and sterile shoots. The aerial stem is erect, reaching a height of 30–70 cm, and slightly branched in the superior part. The leaves are elliptical, tapering, trinerved, glabrous, and have a rough margin. The fragrant flowers are grouped in multifloral paniculate dichasial inflorescences. The floral envelope consists of a cylindrical calyx, about 2 cm long, and a corolla with white or pink petals, approximately 1 cm long, with an emarginate apex. The fruits are capsules that open through 4 or 5 unequal teeth. The seeds are numerous, reniform, slightly flattened, and black. The plant blooms from June to September.[1,3]

Main phytoconstituents	Ref.
amino acids	22–23
alkaloids	
flavonoids	
Saponins	

terpenoids (patchouli alcohol, hederagenin, phytol, gypsogenic acid, quillaic acid, hydroxyhederagenin, gypsogenin, vaccaroside D, dianchinenoside B, saponarioside)
sterols
fatty acids
phenolic acids
glycosides
tannins
Aldehydes & ketones

Medicinal Use

The vegetable product *Saponariae rubrae radix* contains triterpenic saponins with expectorant, diuretic, and depurative effects.[9]

Biological Activity

– Antioxidant, antimicrobial, anticancer, antifungal, and insecticidal properties.[22,24–27]
– Saporin, extracted from *Saponaria officinalis*, is a protein toxin utilized as a modeling tool for neurodegenerative diseases, studying the roles of specific brain cell subpopulations, and therapeutic applications. Conjugates based on saporin have been employed to investigate cellular mechanisms associated with sleep, general anesthesia, epilepsy, pain, as well as the progression of Parkinson's and Alzheimer's diseases.[28,29]

Other Uses

■ Extracts from *Saponariae radix* are used in the textile industry and for foam fire extinguishers.[3]

The genus *Lychnis* is represented in the spontaneous flora of our country by the following species: *Lychnis flos-cuculi* L., *Lychnis coronaria* (L.) Desr., *Lychnis viscaria* L. (*V. vulgaris* Bernh.).[2,4]

Lychnis flos-cuculi L. (Ragged-Robin)

Scientific name: *Lychnis flos-cuculi* L.
Common name: Ragged-Robin
Romanian name: Floarea cucului
Taxonomic classification
Kingdom: *Plantae*
Subkingdom: *Cormobionta*
Phylum: *Spermatophyta*
Subphylum: *Magnoliophytina (Angiospermae)*
Class: *Magnoliatae (Magnoliopsida, Dicotyledonatae)*
Order: *Caryophyllales*
Subfam. *Caryophylloideae*
Family: *Caryophyllaceae*
Genus: *Lychnis*
Species: *flos-cuculi* L. (Figure 7.4)[1,2,4]

SPREAD

Frequently found in damp meadows and marshes, in moist soils, at the edges of ponds, from lowlands to mountains.[1–4]

DESCRIPTION

It is a perennial herbaceous plant, reaching a height of 30–60 cm, with both flowering and sterile shoots. The floral axes are erect, angular, sparsely short-hairy, and sticky under the nodes, sometimes reddish in color. The elongated-spatulate leaves are acute (inferior ones) or linear-lanceolate (superior ones), arranged oppositely. The flowers, grouped in inflorescences (dichasia), have a tubular-campanulate calyx, glabrous, green or pink with 10 wide, dark-colored ribs. The corolla is pink, composed of deeply 4-cleft petals with linear laciniae, palmately divergent, either simple or involute. The coronula is deeply bidentate or dentate on the external face. The fruit is a

Figure 7.4 *Lychnis flos-cuculi* L. (Ragged-Robin).

Source: 30.

capsule that opens through 5 short, acute teeth, with numerous brown, verrucose-aculeate seeds. Flowering occurs from May to August.[1-4]

Main phytoconstituents   Ref.
amino acids   [31]
Flavonoids (apigenin, luteolin
terpenoids (oleanoic acid, gypsogenin, quillaic acid, gypsogenic acid, viticosterone E)
sterols
fatty acids
phenolic acids (ferulic acid,
phytoecdysteroids
tannins
Aldehydes & ketones

Medicinal Use

In traditional medicine, Ragged-Robin has been referenced as a treatment for migraines and abdominal discomfort, and as having attributes related to wound healing.[31-32]

Biological Activity

Lychnis flos-cuculi extracts exhibit antioxidant, bactericidal, fungistatic, antimitotic, hemorheological, neuroprotective, and anti-inflammatory properties, along with blood-thinning characteristics. Moreover, there are documented observations suggesting that this species can enhance the excitability of uterine smooth muscle.[31-32]

Other Uses

■ The root contains saponins, which serve as a soap alternative suitable for laundering clothes, hair, and more. Extraction involves boiling the roots in water. Care should be taken not to overboil the roots, as this could lead to the breakdown of the saponins.[33]

REFERENCES

1. Bejenaru C., Mogoșanu G.D., Bejenaru L.E., Popescu H. 2020. *Botanică farmaceutică. Cormobionta*, ediția a III-a. Ed. Sitech, Craiova.

2. Ciocârlan V. 2009. *Flora ilustrată a României. Pteridophyta et Spermatophyta*. Ediția a 3-a revizuită și adăugită. Ed. Ceres, București.

3. Săvulescu T. (ed). 1953. *Flora RPR*, vol. II. Ed. Academiei RPR, București.

4. Sârbu I., Ștefan N., Oprea A. 2013. *Plante vasculare din România, Determinator ilustrat de teren.* Ed. Victor B Victor, București.

5. https://dryades.units.it/rosandra_en/index.php?procedure=taxon_page&id=759&num=5252

6. Ammor K., Bousta D., Jennan S., Bennani B., Chaqroune A., Mahjoubi F. 2018. Phytochemical screening, polyphenols content, antioxidant power, and antibacterial activity of *Herniaria hirsuta* from Morocco. *Scientific World Journal* 2018:7470384.

7. Al-Snafi A.E., 2018. Pharmacological importance of *Herniaria glabra* and *Herniaria hirsuta* – A review. *Indo American Journal of Pharmaceutical Sciences* 5(04):2167–2175.

8. Rhiouani H., Settaf A., Lyoussi B., et al. 1999. Effects of saponins from *Herniaria glabra* on blood pressure and renal function in spontaneously hypertensive rats. *Therapie* 54(6):735–739.

9. Kozachok S., Pecio Ł., Orhan I.E., Deniz F.S.S., Marchyshyn S., Oleszek W. 2020. Reinvestigation of *Herniaria glabra* L. saponins and their biological activity. *Phytochemistry* 169:112162.

10. https://bcinvasives.ca/invasives/babys-breath/

11. Mogoșanu G.D., Bejenaru L.E., Bejenaru C., Popescu H. 2015. *Farmacognozie-Fitoterapie*. Vol. II, ediția a III-a. Ed. Sitech, Craiova.

12. Jin C., Sun D. 2021. Gas chromatography-mass spectrometry analysis of natural products in *Gypsophila paniculate*. *HortScience* 56(10):1–4.

13. Chandra S., Rawat D.S. 2015. Medicinal plants of the family *Caryophyllaceae*: A review of ethno-medicinal uses and pharmacological properties. *Integrative Medicine Research* 4(3):123–131.

14. Shafaghat A., Shafaghatlonbar M. 2011. Antimicrobial activity and chemical constituents of the essential oils from flower, leaf and stem of *Gypsophila bicolor* from Iran. *Natural Product Communications* 6(2):275–276.

15. Chou S.C., Everngam M.C., Beck J.J. 2008. Allelochemical phenolic acids from *Gypsophila paniculata*. *Journal of Undergraduate Chemistry Research* 7(1):1.

16. Yao S., Ma L., Luo J.G., Wang J., Kong L.Y. 2010. Triterpenoid saponins from the roots of *Gypsophila paniculata*. *Chinese Journal of Natural Medicines* 8:28–33.

17. Korchowiec B., Korchowiec J., Kwiecińska K., Gevrenova R., Bouguet-Bonnet S., Deng C., Henry M., Rogalska E. 2022. The molecular bases of the interaction between a saponin from the roots of *Gypsophila paniculata* L. and model lipid membranes. *International Journal of Molecular Sciences* 23:3397.

18. Kołodziej B., Sęczyk Ł., Sugier D., Kędzia B., Chernetskyy M., Gevrenova R., Henry M. 2019. Determination of the yield, saponin content and profile, antimicrobial and antioxidant activities of three *Gypsophila* species. *Industrial Crops and Products* 138:111422.

19. Flavell D., Holmes S., Gibbs E., Fuchs H., Bachran C., Bachran D., Melzig M., Flavell S. 2010. Saponins from *Gypsophila paniculata* L. significantly potentiate the immunospecific cytotoxic activity of anti-CD19 and CD38 saporin-based immunotoxins for a human lymphoma cell line. *Cancer Research* 70(8 Suppl):5624.

20. www.stuartxchange.org/BabysBreath

21. www.misin.msu.edu/facts/detail/?project=misin&id=56&cname=Bouncingbet

22. Chandra S., Rawat D.S., Bhatt A. 2021. Phytochemistry and pharmacological activities of *Saponaria officinalis* L.: A review . *Notulae Scientia Biologicae* 13(1):10809.

23. Petrović G.M., Ilić M.D., Stankov-Jovanović V.P., Stojanović G.S., Jovanović S.C. 2017. Phytochemical analysis of *Saponaria officinalis* L. shoots and flowers essential oils. *Natural Product Research* 32(3):331–334.

24. Sengul M., Ercisli S., Yildiz H., Gungor N., Kavaz A., Cetin B. 2011. Antioxidant, antimicrobial activity and total phenolic content within the aerial parts of *Artemisia absinthum*, *Artemisia santonicum* and *Saponaria officinalis*. *Iranian Journal of Pharmaceutical Research* 10(1):49–56.

25. Lu Y., Van D., Deibert L., Bishop G., Balsevich J. 2015. Antiproliferative quillaic acid and gypsogenin saponins from *Saponaria officinalis* L. roots. *Phytochemistry* 113:108–120.

26. Tănase M.A.; Marinescu M.; Oancea P.; Răducan A., Mihaescu C.I., Alexandrescu E., Nistor C.L., Jinga L.I., Dițu L.M., Petcu C., et al. 2021. Antibacterial and photocatalytic properties of ZnO nanoparticles obtained from chemical *versus Saponaria officinalis* extract-mediated synthesis. *Molecules* 26:2072.

27. Sadowska B., Budzyńska A., Więckowska-Szakiel M., Paszkiewicz M., Stochmal A., Moniuszko-Szajwaj B., Kowalczyk M., Różalska B. 2014. New pharmacological properties of *Medicago sativa* and *Saponaria officinalis* saponin-rich fractions addressed to *Candida albicans*. *Journal of Medical Microbiology* 63(Pt 8):1076–1086.

28. Bolshakov A.P., Stepanichev M.Y., Dobryakova Y.V., Spivak Y.S., Markevich V.A. 2020. Saporin from *Saponaria officinalis* as a tool for experimental research, modeling, and therapy in neuroscience. *Toxins* 12:546.

29. Turmagambetova A.S., Alexyuk P.G., Bogoyavlenskiy A.P. et al. 2017. Adjuvant activity of saponins from Kazakhstani plants on the immune responses to subunit influenza vaccine. *Archives of Virology* 162:3817–3826.

30. https://dryades.units.it/ampezzosauris/index.php?procedure=taxon_page&id=795&num=781

31. Maliński M.P., Kikowska M.A., Soluch A., Kowalczyk M., Stochmal A., Thiem B. 2021. Phytochemical screening, phenolic compounds and antioxidant activity of biomass from *Lychnis flos-cuculi* L. *in vitro* cultures and intact plants. *Plants* 10(2):Article No. 206.

32. Maliński M.P., Tomczykowa M., Tomczyk M., Thiem B. 2014. Ragged Robin (*Lychnis flos-cuculi*) – a plant with potential medicinal value. *Revista Brasileira de Farmacognosia* 24(6):722–730.

33. https://pfaf.org/user/plant.aspx?LatinName=Lychnis+flos-cuculi

8 *Cistaceae* Adans. Family

Cornelia Bejenaru[1], Antonia Radu[1], George Dan Mogoşanu[2],
Ludovic Everard Bejenaru[2], Andrei Biţă[2], and Adina-Elena Segneanu[3]
[1] Department of Pharmaceutical Botany, Faculty of Pharmacy, University of Medicine and Pharmacy of Craiova, 2 Petru Rareş Street, Craiova, Dolj County, Romania
[2] Department of Pharmacognosy & Phytotherapy, Faculty of Pharmacy, University of Medicine and Pharmacy of Craiova, 2 Petru Rareş Street, Craiova, Dolj County, Romania
[3] Institute for Advanced Environmental Research–West University of Timisoara (ICAM–WUT), 4 Oituz Street, Timisoara, Romania

BOTANICAL DESCRIPTION

The family comprises approximately 160 species, predominantly spread across the Mediterranean region. In the Romanian flora, two genera are found[1]: *Helianthemum* and *Fumana*. The genus *Helianthemum* includes six species: *Helianthemum salicifolium* (L.) Mill., *Helianthemum nummularium* (L.) Mill., *Helianthemum canum* (L.) Baumg., *Helianthemum alpestre* (Jacq.) DC., and *Helianthemum rupifragum* A. Kern., while the genus *Fumana* is represented by a single species, *Fumana procumbens* (Dunal) Gren. et Godr.[2-3]

Shrubs, dwarf subshrubs, or annual herbaceous plants with erect or procumbent stems, simple leaves, entire, sometimes with revolute margins, often with stipules, arranged opposite or alternate. The flowers are grouped into simple terminal racemes, panicles, or simple cymes, sometimes solitary, actinomorphic, hermaphrodite, with a 3–5-merous floral envelope, with a double perianth, dialisepalous and persistent calyx, and a dialypetalous corolla in white, yellow, or red color. The androecium consists of numerous stamens with free filaments, and the gynoecium is tricarpellary, with a superior ovary, unilocular, with numerous ovules. The fruit is a loculicidal capsule, many-seeded, woody or fleshy. The seeds are subrounded or polyhedral with floury endosperm, and a twisted or curved embryo.[1-4]

Helianthemum nummularium (L.) Mill. (Common rock-rose)

Scientific name: ***Helianthemum nummularium*** (L.) Mill.
Common name: Common rock-rose
Romanian name: Iarba moroiului[5]
Taxonomic classification
Kingdom: *Plantae*
Subkingdom: *Cormobionta*
Phylum: *Spermatophyta*
Subphylum: *Magnoliophytina (Angiospermae)*
Class: *Magnoliatae (Magnoliopsida, Dicotyledonatae)*
Subclass: *Dilleniidae*
Order: *Violales*
Family: *Cistaceae*
Genus: *Helianthemum*
Species: *nummularium* (L.) Mill. (Figure 8.1)[1-3]

SPREAD

The plants are common in arid, sunny places, grassy rocky areas, or sandy locations, sparse bushes, from the plain region up to the alpine level.[1-4]

DESCRIPTION

Dwarf subshrub (5–50 cm), strongly branched, with ascending or prostrate stems, with lignified lower part, brown or gray-brown, divided into transversal cicatricial segments of the leaf insertions, and the upper part greenish-gray, tomentose. Stipulate leaves, with ovate-lanceolate, entire blades, white or gray-tomentose on the lower surface, with prominent midrib, flat or revolute margins, arranged opposite. The flowers, numbering from 2 to 15, are grouped in a scorpioid cyme, with densely tomentose floral pedicels as long as or longer than the calyx, reflexed at fruiting. Sepals are externally finely tomentose, while the petals are yellow,

DOI: 10.1201/9781003270515-9

Figure 8.1 *Helianthemum nummularium* (L.) Mill.

Source: 6.

approximately 1 cm long, with a darker spot, citron-yellow, or orange at the base. The androecium consists of numerous stamens, and the gynoecium has an oval-spherical ovary covered with dense clusters of hairs, a style twice as long as the ovary with a thickened tip, and the stigma longer than the stamens. The fruit is a short-hairy, round-oval, three-cornered capsule, of a dark-yellow color. The seeds are polyhedral-round or oval, with a warty seed coat, brown in color. Flowering occurs from May to July.[1–4]

The species *Helianthemum nummularium* (L.) Mill. presents three subspecies:

Helianthemum nummularium (L.) Mill. subsp. *nummularium* (incl. subsp. *tomentosum* (Scop.) Schinz et Thell.) with gray or white-tomentose leaves, especially on the lower surface. It is widespread in oak forests up to the fir forest zone.[2,3]

Helianthemum nummularium (L.) Mill. subsp. *obscurum* (Celak.) Holub (*H. hirsutum* (Thuill.) Mérat; *H. nummularium* subsp. *ovatum* (Viv.) Schinz et Thell.) has hairy sepals between the veins. It is found from the oak forest zone up to the fir forest zone.[2]

Helianthemum nummularium (L.) Mill. subsp. *grandiflorum* (Scop.) Schinz et Thell. has glabrous sepals between the veins. It is distributed from the beech forest zone up to the subalpine zone.[2,3]

Main Phytoconstituents

The main phytoconstituents found in *Helianthemum nummularium* can vary depending on the part of the plant analyzed and its specific location.

Main phytoconstituents	Ref.
flavonoids (quercetin, astragalin, kaempherol, rutin, tiloside)	7–10
terpenoids (cineole, nerol, carvone, camphor, spathulenol, piperitone, geraniol, β-damascenone, tymol, α-terpineol, safranal, β-ionone, linalool, phytol)	
sterols (sitosterol, daucosterol)	
carbohydrates (galactose, arabinose)	
fatty acids (linoleic acid)	
phenolic acids (chlorogenic acid, gallic acid, caffeic acid)	
tannins (catechin, epicatechin)	
organic acids (p-hydroxybenzoic acid)	
glycosides (plantainoside)	
esters	
hydrocarbons (tetradecane, pentadecane, heneicosane)	
Aldehydes & ketones (nonanal, decanal, 2-undecanone)	

Medicinal Use

In some traditional medicine systems, certain parts of the *Helianthemum nummularium* plant, including leaves, stems, and flowers, have been used for various medicinal purposes. The species was historically used in herbal remedies for a range of ailments, including mild respiratory conditions, inflammations, and digestive issues. Moreover, the plant was traditionally used for its astringent and antimicrobial properties, which have been beneficial in treating wounds and skin conditions.[11]

Biological Activity

Some of the reported biological activities of *Helianthemum nummularium* include:

■ antioxidant activity: The species contains various phenolic compounds, including flavonoids and tannins, which exhibit antioxidant properties. These compounds can scavenge free radicals and protect cells from oxidative stress.[9] Additionally, they act as protective agents against UV radiation.[12]

■ antimicrobial potential: *Helianthemum nummularium* extracts have demonstrated antimicrobial activity against certain bacteria and fungi.[13] This suggests a potential role in inhibiting the growth of pathogenic microorganisms, such as *Staphylococcus aureus*.[13]

Other Uses

Research findings suggest that *Helianthemum nummularium* is present in a herbal medicine known as "Bach flower essences," which is described as a remedy used in times of stress and fear to provide assistance and relief.[14]

Fumana procumbens (Dunal) Gren. et Godr

Scientific name: *Fumana procumbens* (Dunal) Gren. et Godr
Common name: Sprawling needle sunrose
Romanian name: Iarba osului[5]
Taxonomic classification
Kingdom: *Plantae*
Subkingdom: *Cormobionta*
Phylum: *Spermatophyta*
Subphylum: *Magnoliophytina (Angiospermae)*
Class: *Magnoliatae (Magnoliopsida, Dicotyledonatae)*
Subclass: *Dilleniidae*
Order: *Violales*
Family: *Cistaceae*
Genus: *Fumana*
Species: *procumbens* (Dunal) Gren. et Godr (Figure 8.2)[2-3]

SPREAD

Distributed from the steppe zone up to the beech forest zone, through sandy or rocky substrate meadows, on sandy and calcareous hills and slopes, arid, open, and stabilized dunes.[2-4]

DESCRIPTION

Semi-shrub with a height of 6–15 (20) cm, having a bush-like appearance, strongly branched, with procumbent or ascending, thin, short-hairy branches, lignified at the base. Leaves are arranged alternately, sessile, without stipules, linear, acicular, very narrow, with three rigid edges, pointed tip, ciliate or rough margins, and rolled back. The flowers are solitary, rarely grouped in 3–4, with floral pedicels of the same length as the calyx, arched-reflexed, non-glandular. The calyx consists of two cycles of sepals, the outer ones are green, linear-lanceolate, pointed, with ciliate margins, while the ones from the inner cycle are broad-ovate, pointed, glabrous, or slightly pubescent, reddish, with 4–5 prominent nerves, lighter in color. The corolla is composed of yellow petals with a basal spot. The external stamens are sterile. The gynoecium is tricarpellary. The fruits are spherical, glossy capsules of dark yellow color, provided with 3 glabrous edges or a few scattered hairs towards the top. The seeds are large, oval, black, or chestnut, with three shiny, smooth edges. Flowering occurs from May to July.[1-4]

Figure 8.2 *Fumana procumbens* (Dunal) Gren. Et Godr.

Source: 15.

The species *Fumana procumbens* (Dunal) Gren. et Godr has two subspecies:
Fumana procumbens (Dunal) Gren. et Godr subsp. *procumbens* has linear sepals in the outer cycle, about half the length of those in the inner cycle. The fruit is a capsule with erect-patent valves. It is sporadically distributed on rocky slopes and rocky areas in the Banat, Oltenia, Transylvania, and Muntenia regions.[2-3]
Fumana procumbens (Dunal) Gren. et Godr subsp. *sabulosa* Ciocârlan has narrow-elliptic sepals in the outer cycle, about 1/3 the length of those in the inner cycle. The fruit valves are horizontal when open. It is an endemic species in Romania. It is rarely found on sandy soils in the Danube Delta.[2,3]

Main phytoconstituents Ref.
flavonoids (quercetin, myricitrin, quercitrin, genkwanol A, stelleranol, dihydrodaphnodorin
   A-B, daphnodorin, naringenin, rhododendron, iriflophenone) [16-17]
terpenoids (myrcene, limonene, p-cymene, terpinolene, isomenthone, selinene, α-terpineol,
   geraniol, β-farnesene, viridiflorene. carvacrol, caryophyllene, β-selinene ϒ-muurolene linalool)
sterols
fatty acids
phenolic acids (gallic acid, p-hydroxybenzoic acid)
tannins (epicatechin, gallocatechin, epigallocatechin)
organic acids
glycosides
esters (3-hexenyl benzoate)
hydrocarbons (hexadecane, pentadecane, heptadecane)
aldehydes&ketones (nonanaI, decanal, 2-undecanone)
miscellaneous

Medicinal Use

Some of the reported traditional medicinal uses of *Fumana procumbens* include antimicrobial, antiulcerogenic, antidiarrheal, antirheumatic, and vasodilator activities.[18]

Biological Activity

cytotoxic and antioxidant activity[19]

Other Uses

- ecological studies about climate and environmental changes[19]

REFERENCES

1. Bejenaru C., Mogoșanu G.D., Bejenaru L.E., Popescu H. 2020. *Botanică farmaceutică. Cormobionta*. Ediția a III-a. Ed. Sitech, Craiova.

2. Ciocârlan V. 2009. *Flora ilustrată a României. Pteridophyta et Spermatophyta*. Ediția a 3-a revizuită și adăugită. Ed. Ceres, București.

3. Sârbu I., Ștefan N., Oprea A. 2013. *Plante vasculare din România, Determinator ilustrat de teren*. Ed. Victor B Victor, București.

4. Săvulescu T. (ed). 1955. *Flora RPR*, vol. III. Ed. Academiei RPR, București.

5. Drăgulescu C. 2018. *Dicționar de Fitonime Românești*. Ediția a 5-a completată, Editura Universității "Lucian Blaga" Sibiu.

6. https://plantcaretoday.com/helianthemum-nummularium.html

7. Agostini M., Hininger-Favier I., Marcourt L., Boucherle B., Gao B., Hybertson B.M., Bose S.K., et.al. 2020. Phytochemical and biological investigation of *Helianthemum nummularium*, a high-altitude growing alpine plant overrepresented in ungulates diets. *Planta Medica* 86(16):1185–1190.

8. Javidnia K., Nasiri A., Miri R., Jamalian A. 2007. Composition of the essential oil of *Helianthemum kahiricum* Del. from Iran. *Journal of Essential Oil Research* 19(1):52–53.

9. Pirvu L., Nicu A. 2017. Polyphenols content, antioxidant and antimicrobial activity of ethanol extracts from the aerial part of rock rose (*Helianthemum nummularium*) species. *Journal of Agricultural Science and Technology A* 7:61–67.

10. Benabdelaziz I., Haba H., Lavaud C., Harakat D., Benkhaled M. 2015. Lignans and other constituents from *Helianthemum sessiliflorum*. *Records of Natural Products* 9:342–348.

11. Airy G., Garnatje T., Ibáñez N., López-Pujol J., Nualart N., Vallès J. 2017. Medicinal plant uses and names from the herbarium of *Francesc Bolòs (1773–1844)*. *Journal of Ethnopharmacology* 204:142–168.

12. Corradi E., Abbet C., Gafner F., Hamburger M., Potterat O. 2013. Screening of alpine plant extracts as protective agents against UV-induced skin damage. *Planta Medica* 79:PC3.

13. Schnurr E., Barros I., Chaves H., Jesus J., et.al. 2018. Corrosion resistance and anti-biofilm effect of rock rose remedy: A potential preventive measure in implant therapy. *EC Microbiology* 14:148–159.

14. Halberstein R.A., Sirkin A., Ojeda-Vaz M.M. 2010. When less is better. A comparison of Bach® Flower remedies and homeopathy. *Annals of Epidemiology* 20(4):298–307.

15. www.pavelkaalpines.cz/Photos/Turkey2010/fumanaprocumbensgoremeturkey.html

16. Gürbüz P., Doğan Ş.D. 2017. Biflavonoids from *Fumana procumbens* (Dunal) Gren. & Godr. *Biochemical Systematics and Ecology* 74:57–59.

17. Laraoui H., Haba H., Long C., Benkhaled M. 2019. A new flavanone sulfonate and other phenolic compounds from *Fumana montana*. *Biochemical Systematics and Ecology* 86:103927.

18. Emerce E., Gurbuz P., Dogan S., Kadioglu E., Süntar I. 2019. Cytotoxic activity-guided isolation studies on *Fumana procumbens* (Dunal) Gren. & Godr. *Records of Natural Products* 13:189–198.

19. Dahlgren J.P., Bengtsson K., Ehrlén J. 2016. The demography of climate-driven and density-regulated population dynamics in a perennial plant. *Ecology* 97(4):899–907.

9 *Corylaceae* Mirb. Family

Cornelia Bejenaru[1], Antonia Radu[1], George Dan Mogoşanu[2],
Ludovic Everard Bejenaru[2], Andrei Biţă[2], and Adina-Elena Segneanu[3]
[1] Department of Pharmaceutical Botany, Faculty of Pharmacy, University of Medicine and Pharmacy of Craiova, 2 Petru Rareş Street, Craiova, Dolj County, Romania
[2] Department of Pharmacognosy & Phytotherapy, Faculty of Pharmacy, University of Medicine and Pharmacy of Craiova, 2 Petru Rareş Street, Craiova, Dolj County, Romania
[3] Institute for Advanced Environmental Research–West University of Timisoara (ICAM–WUT), 4 Oituz Street, Timisoara, Romania

BOTANICAL DESCRIPTION

The *Corylaceae* family is represented in the Romanian flora by two genera with the following species: *Corylus avellana* L., *Corylus colurna* L., *Carpinus betulus* L., and *Carpinus orientalis* Mill.[1–3]

This family includes woody monoecious plants, with simple, pinnately-veined leaves, either singly or doubly serrated, deciduous, petiolate, arranged alternately. The leaves have free, deciduous stipules. The male flowers are naked and grouped in simple, pendulous catkins. The female flowers have a reduced perianth, a bicarpellary and bilocular ovary, and are grouped in compound catkins. The fruit is an achene accompanied by an involucre composed of the bracts and bracteoles at the base of the flowers. Seeds are without endosperm and cotyledons rich in oil.[1–4]

Scientific name: *Corylus avellana* L. (Hazel)
Common name: Common hazel
Romanian name: Alun
Taxonomic classification
Kingdom: *Plantae*
Subkingdom: *Cormobionta*
Phylum: *Spermatophyta*
Subphylum: *Magnoliophytina (Angiospermae)*
Class: *Magnoliatae (Magnoliopsida, Dicotyledonatae)*
Subclass: *Hamamelidae*
Order: *Fagales*
Family: *Corylaceae*
Genus: *Corylus* L.
Species: *avellana* L. (Figure 9.1)[1–3]

DESCRIPTION OF THE GENUS

Corylus L. (Hazel)

The plants are woody with ovoid buds, often weakly compressed, covered by imbricated obovate scales. The rhytidome is thin and remains smooth for a longer time, while the annual stems are flexible. The leaves are ovate, petiolate, doubly serrated, with pinnate venation, arranged alternately, and provided with deciduous stipules. The plant blooms before leafing out. Male inflorescences are cylindrical and pendulous, while female inflorescences are globular, resembling buds. The fruits, called hazelnuts, are grouped in clusters of 1–4 and are accompanied by a long, laciniate cup. The seeds are exalbuminous and have fleshy cotyledons, rich in oil.[1,4]

SPREAD

This species is commonly distributed throughout deciduous forests from the plain to the middle montane zone, reaching its maximum vegetation at the oak forest level, between 700–800 (1000) m above sea level.[1,4]

DESCRIPTION

Shrub with a maximum height of 6 m, having annual stems of yellowish-gray matte color covered with reddish glandular trichomes and elongated, whitish lenticels. The leaves are obovate, hairy, petiolate, with a sharply pointed tip, irregularly doubly serrated margins, and arranged alternately.

DOI: 10.1201/9781003270515-10

Figure 9.1 *Corylus avellana* L. (Common hazel).

Source: 5.

The stipules are hairy and deciduous. Male flowers are grouped in catkins, 2–4 per short stems. Female flowers are grouped in 2–4 within an involucre divided into ovate, irregularly laciniate segments, with a length equal to that of the fruit.[1,4]

Main Phytoconstituents

Corylus avellana contains various phytoconstituents that contribute to its medicinal properties. The main phytochemicals found in this species include:

1. Tannins, a class of polyphenolic compounds, are present in significant amounts in the leaves of *Corylus avellana*. These compounds are known for their astringent properties and are thought to contribute to the plant's hemostatic actions. Tannins have also been reported to exhibit antioxidant and anti-inflammatory effects, further supporting the plant's potential medicinal uses.[6]

2. Flavonoids: *Corylus avellana* is recognized as a valuable source of flavonoids such as quercetin, kaempferol, and myricetin. Flavonoids are known for their diverse bioactivities, including antioxidant, anti-inflammatory, and vasoprotective properties, which may be relevant to the plant's medicinal use as an anti-inflammatory and venotonic agent.[7]

3. Phenolic acids: The plant also contains various phenolic acids, including ellagic acid and caffeic acid. These phenolic compounds exhibit antioxidant properties, further contributing to *Corylus avellana*'s potential health benefits.[8]

4. Fatty acids and oils: The seeds of *Corylus avellana*, commonly known as hazelnuts, are a rich source of fatty acids, particularly oleic acid. These monounsaturated fats, along with other constituents present in the oils, may have implications for cardiovascular health.[9]

5. Proteins and other constituents: The plant also contains proteins, minerals, vitamins (especially vitamin E), and other bioactive compounds that may contribute to its overall medicinal properties.[10–11]

These phytoconstituents work synergistically to provide *Corylus avellana* with its potential therapeutic applications, including anti-inflammatory, hemostatic, and venotonic properties.[12]

Main phytoconstituents   Ref.

amino acids (alanine, serine, arginine, lysine, histidine, threonine, tyrosine, phenylalanine, cystine, valine, methionine, isoleucine, aspartic acid, glutamic acid, glycine, proline)   [12–16]

flavonoids (quercetin, rutin, apigenin, hyperoside, kaempferol, luteolin, chrysine, naringenin, myricetin)

diarylheptanoids (hirsutenone, giffonin J, giffonin V, giffonin B-D, giffonin F, alnusone, giffonin H-I, giffonin L-P, giffonin T, giffonin U, carpinontriol B, oregonin)

terpenoids (pinene, linalool, limonene, β-caryophyllene, camphene, δ-3-carene, α-thujene, γ-terpinene, sabinene, α-terpinolene, β-phellandrene, p-cymene, α-damascone, β-damascenone)

sterols (β-sitosterol, cholesterol, ergosterol, stigmasterol)

fatty acids (oleic acid, lauric acid, stearic acid, myristic acid, palmitic acid, arahidic acid, behemic acid, lignoceric acid)

phenolic acids (ferulic acid, chlorogenic acid, ferrulic acid, vanillic acid p-hydroxybenzoic acid, salicylic acid, protocatechuic acid, shikimic acid, gallic acid, isovanillic acid)

carbohydrates (sucrose, stachyose, raffinose, glucose fructose, myo-inositol)

taxanes (paclitaxel, baccatin III, paclitaxel C, cephalomannine, 7-epipaclitaxel, taxinine M)

lignans (balanophonin, ceplignan, ficusal, ent-cedrusin)

tannins (catechin, epicatechin, gallocatechin, ellagic acid, epigallocatechin)

alcohols (dihydroconiferyl alcohol)

aldehydes & ketones (4-hydroxybenzaldehyde, 5-methyl-(E)-2-hepten-4-one)

miscellaneous (restraverol, veratroylglycol, phloretin)

Medicinal Use

Corylus avellana has several medicinal uses attributed to its therapeutic properties. The medicinal products obtained from this species are: Hazel leaves (*Coryli folium*), Hazel bark (*Coryli cortex*) and Hazelnuts (*Coryli nux*).

Traditionally, herbal medicine employs *Coryli folium* for its anti-inflammatory, hemostatic, and venotonic actions. These medicinal properties make it valuable in the treatment of various conditions. The hemostatic effects, induced by presence of tannins in *Corylus avellana*, are beneficial in controlling bleeding, which can be useful in wound management and stopping minor bleeding. In addition, *Corylus avellana* is known for its venotonic activity, which may support the improvement of venous circulation and the reduction of venous insufficiency symptoms. Flavonoids, particularly rutin, present in *Coryli folium*, are considered responsible for its venotonic effect. The anti-inflammatory properties are beneficial in reducing inflammation and managing conditions related to inflammation, such as arthritis.[17] Other medicinal purposes of *Corylus avellana* include:

- Antioxidant properties: *Corylus avellana* extracts contain various phytochemicals, such as flavonoids and phenolic compounds, which possess antioxidant activity. These antioxidants can help neutralize free radicals and reduce oxidative stress in the body, potentially contributing to overall health and well-being.[18]

- Cardiovascular health: hazelnuts are a good source of heart-healthy monounsaturated fats and other beneficial nutrients like vitamin E and magnesium. Consuming hazelnuts as part of a balanced diet may contribute to improved cardiovascular health by reducing cholesterol levels and supporting blood vessel function.[19]

- Skin health: Hazelnut oil, extracted from the nuts, is commonly used in skincare products due to its moisturizing and nourishing properties. It may help soothe and protect the skin, making it a popular ingredient in cosmetics and natural skincare remedies.[20]

- Neuroprotective effects: Some research suggests that certain compounds found in *Corylus avellana* may have neuroprotective properties, potentially benefiting brain health and cognitive function.[21]

Biological Activity

Corylus avellana exhibits various biological activities that have been studied for their potential therapeutic applications. Scientific research has shown that it possess anti-inflammatory, antioxidant, and antimicrobial activities, which are attributed to the presence of phytochemicals, such as flavonoids and phenolic compounds.[22]

The antioxidant activity is particularly relevant due to the presence of flavonoids like quercetin and kaempferol. Antioxidants play a crucial role in neutralizing free radicals and protecting cells from oxidative stress, potentially benefiting overall health.[23]

Additionally, the antimicrobial properties of *Corylus avellana* are noteworthy. Extracts from various parts of the plant have demonstrated inhibitory effects against certain pathogenic microorganisms, which may have implications for their use in traditional medicine.[24]

Furthermore, *Corylus avellana* extracts have been investigated for their potential as antidiabetic agents. Studies have shown that these extracts can inhibit carbohydrate-hydrolyzing enzymes, contributing to the regulation of blood glucose levels.[15]

Moreover, the plant has been evaluated for its antihyperlipidemic effects. Research suggests that *Corylus avellana* extracts can help reduce lipid levels, providing potential benefits in managing hyperlipidemia or obesity.[25]

In a 2023 study, researchers demonstrated the hepatoprotective effects of *Corylus avellana* gemmotherapy bud extract in a liver fibrosis model using diabetic mice.[26]

Other Uses

- wood is suitable for turning, while the stems are used for weaving[4]

- culinary use[4]

Corylus colurna L. (Turkish hazel)

Scientific name: *Corylus colurna* L.
Common name: Turkish Hazel
Romanian name: Alun turcesc
Taxonomic classification
Kingdom: *Plantae*
Subkingdom: *Cormobionta*
Phylum: *Spermatophyta*
Subphylum: *Magnoliophytina (Angiospermae)*
Class: *Magnoliatae (Magnoliopsida, Dicotyledonatae)*
Subclass: *Hamamelidae*
Order: *Fagales*
Family: *Corylaceae*
Genus: *Corylus* L.
Species: *colurna* L. (Figure 9.2)[1–3]

SPREAD

The plant is sporadically distributed in the Banat region and thrives on rocky, calcareous soils.[1,4]

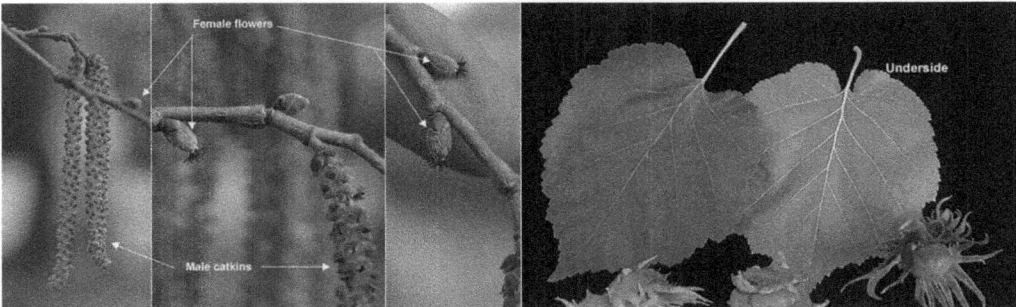

Figure 9.2 *Corylus colurna* L. (Turkish hazel).

Source: 27.

119

DESCRIPTION

A tree reaching 20 meters in height, with initially smooth bark, later covered by thick, grayish-yellow rhytidome that exfoliates into irregular, small plates. The leaves are orbicular, ovate-cordate, acuminate, doubly serrated, and have short petioles. The involucre, with long and rigid lobes, is much longer than the fruit.[1,4]

Main phytoconstituents   Ref.

amino acids (alanine, serine, arginine, lysine, histidine, threonine, tyrosine, phenylalanine, cystine, valine, methionine, isoleucine, aspartic acid, glutamic acid, glycine, proline)   28–29

coumarins (coumarin)

flavonoids (quercetin, apigenin, kaempferol, luteolin, myricetin)

diarylheptanoids (hirsutenone, oregonin)

terpenoids

sterols (β-sitosterol, cholesterol, ergosterol, stigmasterol, corylusterol)

fatty acids (oleic acid, lauric acid, stearic acid, myristic acid, palmitic acid, arahidic acid, behemic acid, lignoceric acid)

phenolic acids (caffeic acid, p-coumaric acid, gallic acid)

carbohydrates

taxanes

lignans

tannins (catechin, epicatechin, ellagic acid)

alcohols

aldehydes & ketones

miscellaneous

Medicinal Use

Corylus colurna extracts have been used in traditional medicine for a long time for the treatment of various conditions, such as phlebitis, varicose veins, haemorrhoidal symptoms, and eczema.[30]

Biological Activity

Studies have indicated that extracts from the bark of *Corylus colurna* possess antioxidant and anti-inflammatory activities. These properties are attributed to the presence of bioactive compounds such as flavonoids and phenolic acids.[29]

Furthermore, *Corylus colurna* has shown potential as an antimicrobial agent. Extracts derived from various parts of the plant have displayed inhibitory effects against specific pathogenic microorganisms, suggesting potential applications in traditional medicine.[31]

Additionally, the plant has been investigated for its potential hepatoprotective effects. Research suggests that *Corylus colurna* extracts can help protect the liver against oxidative damage, providing potential benefits in maintaining liver health.[32]

Other Uses

- wood is suitable for turning and carving[4]

- culinary use (large, tasty fruits)[4]

Carpinus betulus L. (European hornbeam)

Scientific name: *Carpinus betulus* L.
Common name: European hornbeam
Romanian name: Carpen
Taxonomic classification
Kingdom: *Plantae*
Subkingdom: *Cormobionta*
Phylum: *Spermatophyta*
Subphylum: *Magnoliophytina (Angiospermae)*
Class: *Magnoliatae (Magnoliopsida, Dicotyledonatae)*
Subclass: *Hamamelidae*
Order: *Fagales*

Family: *Corylaceae*
Genus: *Carpinus* L.
Species: *betulus* L. (Figure 9.3)[1–3]

DESCRIPTION OF THE GENUS *CARPINUS* L.

Plants that bear both male and female flowers, grouped in catkins. The buds are fusiform, with numerous scales. The leaves are simple, alternate, ovate, acute, and doubly serrated. The fruiting involucre is flat and often lobed. Flowering and leafing occur simultaneously.[1,4]

SPREAD

Widely distributed in forests from the plain to the lower montane zone.[4]

DESCRIPTION

A tree reaching 25 meters in height, with a pyramidal crown, smooth and thin rhytidome, whitish-gray in color. The leaves are obovate-oblong, longer than 4 cm, doubly serrated, and acute or acuminate. The fruiting involucre is three-lobed.[1,4]

Main Phytoconstituents

Carpinus betulus contains various phytoconstituents that contribute to its medicinal properties. Scientific studies have reported the presence of tannins in *Carpinus betulus*, which contribute to the plant's astringent properties and have antioxidant effects. It is also a rich source of flavonoids, such as quercetin and kaempferol, that are known for their diverse bioactivities, including antioxidant and anti-inflammatory properties. Moreover, this plant also contains phenolic acids, including caffeic acid and chlorogenic acid, with antioxidant effects. *Carpinus betulus* is reported to contain various terpenoids, such as β-sitosterol and α-amyrin. Terpenoids are a class of secondary metabolites that have been studied for their potential pharmacological activities.[34]

Figure 9.3 *Carpinus betulus* L. (European hornbeam).

Source: 33.

121

Main phytoconstituents   Ref.
amino acids   34–36
coumarins
flavonoids (quercetin, apigenin, kaempferol, luteolin, isorhamnetin, myricetin, rutin. afzelin)
diarylheptanoids (carpinontriol A-B, giffonin U, giffonin X, asuarinondiol)
terpenoids (betulinic acid, betulonic acid,)
sterols
fatty acids (linoleic acid)
phenolic acids (gallic acid, gentisic acid, vanillic acid, protocatechuic acid, p-coumaric acid, o-coumaric acid)
carbohydrates
taxanes
lignans (aviculin)
tannins (catechin, epicatechin, epigallocatechin, gallocatechin, ellagic acid)
alcohols
aldehydes & ketones (p-hydroxybenzaldehyde, syringaldehyde)
miscellaneous (eucaglobulin)

Medicinal Use

Carpinus betulus has been traditionally used in herbal medicine for various medicinal purposes, such as astringent and antiseptic,[37] therefore scientific research has explored its potential therapeutic applications.

Biological Activity

Studies have reported that *Carpinus betulus* extracts exhibit antioxidant and antimicrobial effects due to the presence of phenolic compounds. Certain investigations revealed the antimicrobial and antioxidant activity of gemmotherapic remedies from *Carpinus betulus* employed in the treatment of respiratory diseases.[38]

Phytochemical analysis of *Carpinus betulus* leaf extract resulted in the bioguided isolation of pheophorbide a, identified as the compound responsible for the anticancer properties found in the young leaves of *Carpinus betulus*.[35]

Other Uses

- Ornamental, in the form of hedges[4]
- forestry species[4]

Carpinus orientalis Mill. (Oriental hornbeam)

Scientific name: *Carpinus orientalis* Mill.
Common name: Oriental hornbeam
Romanian name: Cărpiniță
Taxonomic classification
Kingdom: *Plantae*
Subkingdom: *Cormobionta*
Phylum: *Spermatophyta*
Subphylum: *Magnoliophytina (Angiospermae)*
Class: *Magnoliatae (Magnoliopsida, Dicotyledonatae)*
Subclass: *Hamamelidae*
Order: *Fagales*
Family: *Corylaceae*
Genus: *Carpinus*
Species: *orientalis* Mill. (Figure 9.4)[1–3]

SPREAD

Sub-Mediterranean species, sporadic on sunny coasts from plains and hills.[1,4]

Figure 9.4 *Carpinus orientalis* Mill. (Oriental hornbeam).

Source: 39.

DESCRIPTION

A shrub reaching 7 meters in height, with smooth, grayish rhytidome. The leaves are smaller than 4 cm, ovate or ovate-oblong, with an acute apex and doubly serrated margin. The fruiting involucre is unlobed.[1,4]

Main phytoconstituents	Ref.
amino acids	40–41

amino acids
coumarins
flavonoids (quercetin, apigenin, kaempferol, luteolin, isorhamnetin, myricetin, rutin. afzelin)
diarylheptanoids
terpenoids (betulinic acid, betulonic acid)
sterols
fatty acids (palmitic, stearic, oleic, linoleic, gadoleic acid, myristic acid, pentadecanoic acid, margaric acid, margoleic acid)
phenolic acids (gallic acid, gentisic acid, vanillic acid, protocatechuic acid, p-coumaric acid, o-coumaric acid)
carbohydrates
taxanes
lignans (aviculin)
tannins (catechin, epicatechin, epigallocatechin, gallocatechin, ellagic acid)
alcohols
aldehydes & ketones
miscellaneous

Medicinal Use

■ eczema, psoriasis, and vitiligo[42]

Biological Activity

■ antimicrobial, immunomodulant[42]

Other Uses

■ forestry tree used for coastal stabilization[1]
■ etnoveterinary use (wolf bite)[43]

REFERENCES

1. Bejenaru C., Mogoșanu G.D., Bejenaru L.E., Popescu H. 2020. *Botanică farmaceutică. Cormobionta*. Ediția a III-a. Ed. Sitech, Craiova.

2. Ciocârlan V. 2009. *Flora ilustrată a României. Pteridophyta et Spermatophyta*. Ediția a 3-a revizuită și adăugită. Ed. Ceres, București.

3. Sârbu I., Ștefan N., Oprea A. 2013. *Plante vasculare din România, Determinator ilustrat de teren*. Ed. Victor B Victor, București.

4. Săvulescu T. (ed). 1952. *Flora RPR*, vol. I. Ed. Academiei RPR, București.

5. www.tehnicaagricola.md/index.php?pag=news&id=581&rid=701&l=ro

6. Amarowicz R., Dykes G.A., Pegg R.B. 2008. Antibacterial activity of tannin constituents from *Phaseolus vulgaris, Fagoypyrum esculentum, Corylus avellana* and *Juglans nigra. Fitoterapia* 79(3):217–219.

7. Ivanović S., Avramović N., Dojčinović B., Trifunović S., Novaković M., Tešević V., Mandić B. 2020. Chemical composition, total phenols and flavonoids contents and antioxidant activity as nutritive potential of roasted hazelnut skins (*Corylus avellana* L.). *Foods* 9:430.

8. Amaral J.S., Ferreres F., Andrade P.B., Valentão P., Pinheiro C., Santos A., Seabra R. 2005. Phenolic profile of hazelnut (*Corylus avellana* L.) leaves cultivars grown in Portugal. *Natural Product Research* 19(2):157–163.

9. Karaosmanoğlu H., Üstün N. 2019. Variations in fatty acid composition and oxidative stability of hazelnut (*Corylus avellana* L.) varieties stored by traditional method. *Grasas y Aceites* 70:288.

10. Ceylan F.D., Yılmaz H., Adrar N., Günal Köroğlu D., Gultekin Subasi B., Capanoglu E. 2022. Interactions between hazelnut (*Corylus avellana* L.) protein and phenolics and *in vitro* gastrointestinal digestibility. *Separations* 9:406.

11. Poșta D.S., Radulov I., Cocan I., Berbecea A.A., Alexa E., et al. 2022. Hazelnuts (*Corylus avellana* L.) from spontaneous flora of the west part of Romania: A source of nutrients for locals. *Agronomy* 12:214.

12. Bottone A., Cerulli A., D'Urso G., Masullo M., Montoro P., Napolitano A., Piacente S. 2019. Plant specialized metabolites in hazelnut (*Corylus avellana*) kernel and byproducts: an update on chemistry, biological activity, and analytical aspects. *Planta Medica* 85(11–12):840–855.

13. Oliveira I., Sousa A., Morais J.S., Ferreira I.C.F.R., Bento A., Estevinho L., Pereira J.A. 2008. Chemical composition, and antioxidant and antimicrobial activities of three hazelnut (*Corylus avellana* L.) cultivars. *Food and Chemical Toxicology* 46(5):1801–1807.

14. Shataer, D., Abdulla, R., Ma, Q.L. et al. 2020. Chemical composition of extract of *Corylus avellana* shells. *Chemistry of Natural Compounds* 56:338–340.

15. Masullo M., Lauro G., Cerulli A., Bifulco G., Piacente S. 2022. *Corylus avellana*: a source of diarylheptanoids with α-glucosidase inhibitory activity evaluated by in vitro and in silico studies. *Frontiers in Plant Science* 13:805660.

16. Squara S., Stilo F., Cialiè Rosso M., Liberto E., Spigolon N., et.al. 2022. *Corylus avellana* L. aroma blueprint: potent odorants signatures in the volatilome of high quality hazelnuts. *Frontiers in Plant Science* 13:840028.

17. Guiné R.P.F., Reis Correia P.M. 2020. Hazelnut: a valuable resource. *International Journal of Food Engineering* 6(2):67–72.

18. Alasalvar C., Karamac M., Amarowicz R., Shahidi F. 2006. Antioxidant and antiradical activities in extracts of hazelnut kernel (*Corylus avellana* L.) and hazelnut green leafy cover. *Journal of Agricultural and Food Chemistry* 54:4826e4832.

19. Tey S.L., Brown R.C., Chisholm A.W., Delahunty C.M., Gray A.R., Williams S.M. 2011. Effects of different forms of hazelnuts on blood lipids and α-tocopherol concentrations in mildly hypercholesterolemic individuals. *European Journal of Clinical Nutrition* 65(1):117–124.

20. Michalak M., Kiełtyka-Dadasiewicz A. 2019. Nut oils and their dietetic and cosmetic significance: a review. *Journal of Oleo Science* 68(2):111–120.

21. Bahaeddin Z., Yans A., Khodagholi F., Hajimehdipoor H., Sahranavard S. 2017. Hazelnut and neuroprotection: improved memory and hindered anxiety in response to intra-hippocampal A-beta injection. *Nutritional Neuroscience* 20(6):317–326.

22. Shataer D., Li J., Duan X.M., Liu L., Xin X.L., Aisa H.A. 2021. Chemical composition of the hazelnut kernel (*Corylus avellana* L.) and its anti-inflammatory, antimicrobial, and antioxidant activities. *Journal of Agricultural and Food Chemistry* 69(14):4111–4119.

23. Oliveira I., Sousa A., Morais J.S., Ferreira I.C., Bento A., Estevinho L., Pereira J.A., 2008. Chemical composition, and antioxidant and antimicrobial activities of three hazelnut (*Corylus avellana* L.) cultivars. *Food and Chemical Toxicology* 46(5):1801–1807.

24. Cappelli G., Giovannini D., Basso A.L., et al. 2018. A *Corylus avellana* L. extract enhances human macrophage bactericidal response against *Staphylococcus aureus* by increasing the expression of anti-inflammatory and iron metabolism genes. *Journal of Functional Foods* 45:499–511.

25. Mollica A., Gokhan Z., Azzurra S., Ferrante C., et al. 2018. Nutraceutical potential of *Corylus avellana* daily supplements for obesity and related dysmetabolism. *Journal of Functional Foods* 47:562–574.

26. Balta C., Herman H., Ciceu A., Mladin B., Rosu M., Sasu A., et al. 2023. Phytochemical profiling and anti-fibrotic activities of the gemmotherapy bud extract of *Corylus avellana* in a model of liver fibrosis on diabetic mice. *Biomedicines* 11:1771.

27. https://landscapeplants.oregonstate.edu/plants/corylus-colurna

28. Sharma M. 2010. Phytochemical investigation of fruits of *Corylus colurna* Linn. *Journal of Phytology* 2:89–100.

29. Riethmüller E., Tóth G., Alberti A., Sonati M., Kéry A. 2014. Antioxidant activity and phenolic composition of *Corylus colurna*. *Natural Product Communications* 9(5):679–682.

30. Riethmüller E., Könczöl A., Szakál D., Végh K., Balogh G.T., Kéry Á., 2016. HPLC-DPPH screening method for evaluation of antioxidant compounds in *Corylus* species. *Natural Product Communications* 11(5):641–644.

31. Ozgur C., Sahin M.D., Avaz S. 2013. Antibacterial activity of *Corylus colurna* L. (*Betulaceae*) and *Prunus divaricata* Ledep. subsp. divaricata (*Rosaceae*) from Usak. *Turkey. Bulgarian Journal of Agricultural Science* 19:1204–1207.

32. Riethmüller E., Tóth G., Alberti A., Végh K., Béni S., Balogh G.T., Kéry A. 2015. Occurence of diarylheptanoids in *Corylus* species native to Hungary. *Acta Pharmaceutica Hungarica* 85(1):29–38.

33. https://dryades.units.it/Roma/index.php?procedure=taxon_page&id=253&num=2160

34. Hofmann T., Nebehaj E., Albert L. 2016. Antioxidant properties and detailed polyphenol profiling of European hornbeam (*Carpinus betulus* L.) leaves by multiple antioxidant capacity assays and high-performance liquid chromatography/multistage electrospray mass spectrometry. *Industrial Crops and Products* 87:340–349.

35. Cieckiewicz E., Angenot L., Gras T., Kiss R., Frédérich M. 2012. Potential anticancer activity of young *Carpinus betulus* leaves. *Phytomedicine* 19(3–4):278–283.

36. Felegyi-Tótha C.A., Garádia Z., Darcsia A., et.al. 2022. Isolation and quantification of diarylheptanoids from European hornbeam (*Carpinus betulus* L.) and HPLC-ESI-MS/MS characterization of its antioxidative phenolics. *Journal of Pharmaceutical and Biomedical Analysis* 210:114554.

37. https://suntsanatos.ro/carpen-32.html

38. Orodan M., Vodnar D.C., Toiu A.M., et al. 2016. Phytochemical analysis, antimicrobial and antioxidant effect of some gemmotherapic remedies used in respiratory diseases. *Farmacia* 64(2):224–230.

39. https://dryades.units.it/rosandra_en/index.php?procedure=taxon_page&id=254&num=2159

40. Kokten K., Ozel H.B. 2020. Fatty acid composition of *Carpinus orientalis* collected from different locations in Turkey. *Chemistry of Natural Compounds* 56:899–901.

41. Jeon J.I., Chang C.S., Chen Z.D., Park T.Y. 2007. Systematic aspects of foliar flavonoids in subsect. *Carpinus* (*Carpinus, Betulaceae*). *Biochemical Systematics and Ecology*, 35(9):606–613.

42. Erarslan Z.B., Genc G.E., Kultur S. 2020. Medicinal plants traditionally used to treat skin diseases in Turkey – eczema, psoriasis, vitiligo. *Journal of Faculty of Pharmacy of Ankara University* 44(1):137–166.

43. Senkardes I., Dogan A., Emre G. 2022. An ethnobotanical study of medicinal plants in Taşköprü (Kastamonu–Turkey). *Frontiers in Pharmacology* 13:984065.

10 *Cupressaceae* Gray Family

Cornelia Bejenaru¹, Antonia Radu¹, George Dan Mogoşanu²,
Ludovic Everard Bejenaru², Andrei Biţă², and Adina-Elena Segneanu³
¹ Department of Pharmaceutical Botany, Faculty of Pharmacy, University of Medicine
and Pharmacy of Craiova, 2 Petru Rareş Street, Craiova, Dolj County, Romania
² Department of Pharmacognosy & Phytotherapy, Faculty of Pharmacy, University of
Medicine and Pharmacy of Craiova, 2 Petru Rareş Street, Craiova, Dolj County, Romania
³ Institute for Advanced Environmental Research–West University of
Timisoara (ICAM–WUT), 4 Oituz Street, Timisoara, Romania

BOTANICAL DESCRIPTION

Coniferous trees or shrubs, with scale-like leaves that are persistent and arranged opposite or acicular, subulate, arranged in whorls. The flowers are unisexual, small, terminal or axillary, borne on short shoots, either on the same plant (monoecious) or on separate plants (dioecious – in *Juniperus*). Male flowers are arranged terminally or axillary, and their structure includes stamens with short filaments and shield-shaped anthers horizontally positioned, each bearing 3–5 pollen sacs on the lower margin. Female flowers form small, rounded cones with a few carpels completely fused with bracts. Each carpel contains 2–20 atropous ovules (erect and adnate only at their carpel base). Both stamens and carpels are arranged opposite, except in *Juniperus*, where they are arranged in whorls. After flower development, a strobile is formed surrounded by fleshy or woody scaly bracts, either free or fused, valvate, or imbricate.[1-4]

The *Cupressaceae* family is represented in the Romanian flora by the genus *Juniperus*.

The genus *Juniperus* is represented in the spontaneous flora of our country by the following species: *Juniperus communis* L. (Common juniper), *Juniperus sibirica* Lodd. in Burgsd. (*J. communis* subsp. *alpina* (Suter) Celak; *J. nana* Willd. nom. illeg.) (Siberian juniper), and *Juniperus sabina* L. (Savin juniper). [1-3]

Juniperus L. (Common Juniper)

This genus includes dioecious trees or shrubs with leaves arranged opposite or in whorls of 3. The male cones are yellow and consist of numerous stamens arranged in opposite or whorled fashion, positioned in groups of 3 stamens. The female cones are spherical, axillary, borne on short peduncles covered with scales or terminal at the tips of the branches, and composed of 2 to numerous scales. Usually, they are formed from 3 carpels (each with one ovule) which, after fertilization, become fleshy and fuse together, adhering to each other and forming a false fruit (false berries called juniper "berries"), containing 1 to 12 seeds.[1-4]

Juniperus communis L. (Common juniper)

Scientific name: *Juniperus communis* L.
Common name: Common juniper
Romanian name: Ienupăr
Taxonomic classification
Kingdom: *Plantae*
Subkingdom: *Cormobionta*
Phylum: *Spermatophyta*
Subphylum: *Pinophytina (Gymnospermae)*
Class: *Pinatae (Pinopsida, Coniferopsida)*
Order: *Pinales*
Family: *Cupressaceae*
Genus: *Juniperus*
Species: *communis* L. (Figure 10.1)[1-3]

SPREAD

The shrub is widespread in the hilly and submontane regions, occurring either isolated, in clusters, or as thickets across meadows, edges or sparse forest areas, clearings, pastures, and even in swamps.[1-4]

DOI: 10.1201/9781003270515-11

Figure 10.1 *Juniperus communis* L. (Common juniper).

Source: 5.

DESCRIPTION

This shrub has an upright growth habit, reaching a height of 3 meters. Its stems are ascending, and the young, three-edged shoots are brown in color. The leaves, arranged in whorls of three, are acicular, subulate-linear, rigid, straight and sharply pointed, measuring up to 1.5 cm in length. They gradually narrow and sharpen at the tip, and on the upper surface, they have a slight channel and a broad white stripe, while on the lower surface, they are obtusely keeled and spread out widely. The whorls of leaves are spaced 3–10 mm apart. The female cones are short-stalked, round, initially green, then turning blackish-blue and frosted, with a diameter of under 1 cm, resembling berries. The seeds, usually three but sometimes only one or two, mature in the second or even third year.[1-4]

This species presents two varieties:

Juniperus communis L. var. *communis* is an erect shrub with leaves measuring 10–15 (20) mm in length and spaced 5–10 mm apart between whorls. Its distribution area is in the hilly and montane regions.[1-3]

Juniperus communis L. var. *intermedia* (Schur) Sanio is a shrub reaching up to 1 meter in height, with acicular leaves measuring 7–10 mm in length. These leaves are straight or slightly curved towards the stem, narrow-lanceolate, and flat on the upper surface. The distance between the whorls of leaves is 3–6 mm. This variety is widespread in the montane and subalpine regions, forming extensive thickets or clusters.[1-3]

Main Phytoconstituents

Juniperi fructus contains approximately 0.5–2% volatile oil; 5–30% carbohydrates; 5% tannin; biflavonoids (cupressuflavone, amentoflavone, bilobetin); anthocyanins; proanthocyanidols; a bitter principle (juniperine); diterpenoids (communic acid); organic acids; pectin; and resins. For the officinal product, the Romanian Pharmacopoeia X specifies a minimum content of 0.8% volatile oil.

The volatile oil comprises approximately 3% α-terpineol, with a violet-like odor; terpinolene; α- and β-pinene, in significant proportion.[6]

Main phytoconstituents	Ref.
amino acids	7–8

flavonoids (quercetin, quercitrin, apigenin, rutin, luteolin, scutellarein, nepetin, amentoflavone, bilobetin, cupressuflavone, hinokiflavone, biflavones, isocryptomerin amentoflavone, sciadopitysin)

coumarins (umbelliferone)

terpenoids (communic acid, *β*-pinene, isocupressic acid, cedrene, *α*-pinene, sabinene, myrcene, camphene, limonene, cadinene, juniper, junicedral, linalool)

sterols (β-sitosterol, campesterol)

fatty acids (palmitic acid, oleic acid, linoleic acid, juniperonic acid, stearic acid, gondoic acid)

phenolic acids

tannins

organic acids (malic acid, formic acid, acetic acid, ascorbic acid)

carbohydrates

resins

hydrocarbons

aldehydes&ketones

miscellaneous

Medicinal Use

The mature pseudoberries (cones) form the herbal product *Juniperi fructus,* used as a saluretic diuretic and carminative (F.R. X). Both the pseudoberries and the volatile oil obtained by hydrodistillation from them (*Juniperi aetheroleum*) have antiseptic, stomachic, and expectorant properties.[5] Under the name of "Juniper berry," *Juniperi fructus* is used as a flavoring agent in gin-type beverages.[1]

Biological Activity

antioxidant, hepatoprotective, anti-inflammatory, analgesic, antibacterial, antifungal, antimalarial, neuroprotective, anticataleptic, antihypercholesterolemic activity.[6–8]

Toxicity; Contraindications

The volatile oil is irritating, causing redness of the skin (rubefacient), and can even act as a blistering agent (causing irritation of the nephrons, with hematuria and albuminuria). For this reason, patients susceptible to kidney inflammatory conditions are advised not to use juniper oil.[6]

Juniperus sibirica Lodd. in Burgsd. (*J. communis* subsp. *alpina* (Suter) Celak; *J. nana* Willd. nom. illeg.)

Scientific name: *Juniperus sibirica* Lodd. in Burgsd. (*J. communis* subsp. *alpina* (Suter) Celak; *J. nana* Willd. nom. illeg.)
Common name: Siberian juniper
Romanian name: Ienupăr pitic
Taxonomic classification
Kingdom: *Plantae*
Subkingdom: *Cormobionta*
Phylum: *Spermatophyta*
Subphylum: *Pinophytina* (*Gymnospermae*)
Class: *Pinatae* (*Pinopsida, Coniferopsida*)
Order: *Pinales*
Family: *Cupressaceae*
Genus: *Juniperus*
Species: *sibirica* Lodd. in Burgsd. (Figure 10.2)[1–3]

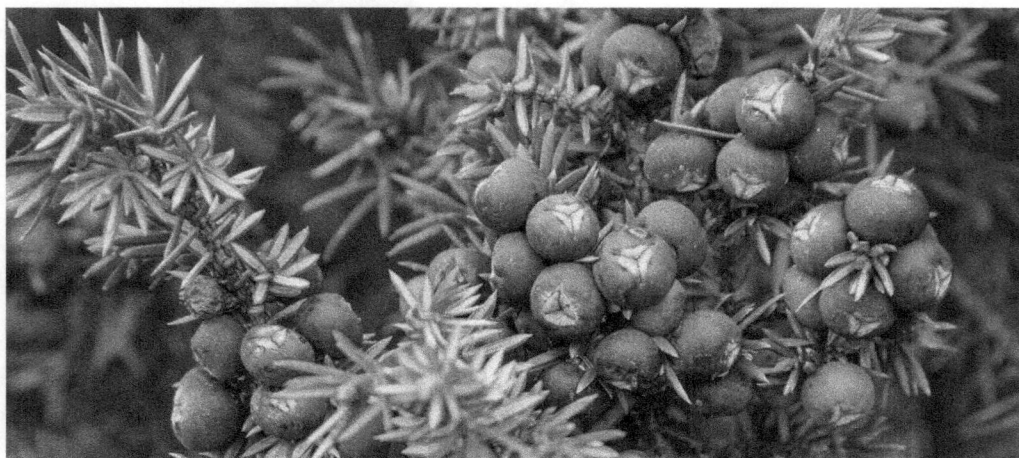

Figure 10.2 *Juniperus sibirica* Lodd. in Burgsd. (*J. communis* subsp. *alpina* (Suter) Celak; *J. nana* Willd. nom. illeg.).

Source: 8.

SPREAD

Frequently found in the alpine zone (reaching up to 2000 m altitude), where it forms shrubs and can become invasive in meadows. In the subalpine zone of the Carpathian Mountains and the Apuseni Mountains, it forms dense thickets in mountain clearings and pastures, especially on south-facing slopes. It can be found in the Ciucului, Gutin, Țibleș, Gurghiului, Harghita, Domogled, and Poiana Ruscă Mountains. In the form of peats, it is found in the Sebeș and Gilău Mountains.[1–4]

DESCRIPTION

A small shrub (0.5 m) with trailing branches. It has short and thick three-edged shoots, irregularly curved. The acicular leaves are 4–8 mm long, linear-lanceolate, abruptly tapering into a sharp tip, and bent towards the branches, but non-prickly. The leaves are distinctly concave on the upper surface and rounded on the lower surface, dark green, shiny, arranged in groups of three in close whorls (1–3 mm between whorls). The pseudoberries are spherical or ovoid, with a diameter of 7–10 mm, and have the same maturation period as juniper. [1,2,4]

Main phytoconstituents   Ref.
amino acids   9,10–11
flavonoids (apigenin, luteolin, rutin, catechin, cosmosiin, catechin)
coumarins
terpenoids (cimene, pinene, caryophyllene, cubebene, germacrene D, linalool, cadinene,
     cedrol, myrcene, sabinene, and limonene)
sterols (β-sitosterol)
fatty acids (palmitic acid, oleic acid, stearic, linoleic acid, juniperonic acid)
phenolic acids
tannins
organic acids
carbohydrates
resins
hydrocarbons
aldehydes&ketones
miscellaneous

Medicinal Use

The pseudoberries have similar uses to those of the species *J. communis*.
  Traditionally, different parts of the plant have been used for various therapeutic purposes:

- antiseptic and disinfectant: The essential oil obtained from *Juniperus sibirica* has strong antiseptic and disinfectant properties. It has been used to cleanse wounds and treat skin infections.

- diuretic and detoxifying: *Juniperus sibirica* is known for its diuretic properties, which means it can increase urine production and promote the elimination of toxins from the body.

- digestive aid: The plant has been used as a digestive aid to improve digestion and alleviate digestive issues like bloating, gas, and indigestion.

- respiratory health: In traditional medicine, *Juniperus sibirica* has been used to treat respiratory conditions like bronchitis, coughs, and congestion.

- rheumatism and arthritis: The essential oil of *Juniperus sibirica* is believed to have anti-inflammatory properties and has been used in folk medicine to alleviate symptoms of rheumatism and arthritis.[10,12,13]

Biological Activity

- antioxidant, antibacterial and anti-inflammatory, analgesic agent, diuretic, anthelmintic, hypoglycaemic, hypotensive, and abortifacient[10,14]

Juniperus sabina L. (Savin juniper)

Scientific name: *Juniperus sabina* L.
Common name: Savin juniper
Romanian name: Cetină de negi
Taxonomic classification
Kingdom: *Plantae*
Subkingdom: *Cormobionta*
Phylum: *Spermatophyta*
Subphylum: *Pinophytina* (*Gymnospermae*)
Class: *Pinatae* (*Pinopsida*, *Coniferopsida*)
Order: *Pinales*
Family: *Cupressaceae*
Genus: *Juniperus*
Species: *sabina* L. (Figure 10.3)[1–3]

SPREAD

A sporadic species found in the subalpine and lower alpine regions in several places in Banat and Transylvania. It is localized on limestone rocks and scree slopes, forming shrubby growths. [1,4]

DESCRIPTION

A 3-meter tall shrub with prostrate, highly branched stems and slender, edged young shoots. The leaves are obtuse, scaly, small (1 mm), ovate-rhombic, arranged oppositely in four longitudinal rows, closely pressed against the stem. On the lower surface, they are convex and each contains a secretory gland. When crushed, the leaves emit a characteristic, unpleasant odor. The plant can also have acicular leaves, arranged in whorls of 3. The flowers can be unisexual monoecious or dioecious. The female flowers, located terminally, are initially erect and then become pendulous. The spherical or globular-ovate pseudoberries, with a diameter of 5–7 mm, consist of 4 scales. They are pedunculate, nodding, blackish-blue, frosted, and contain 1–3 seeds. The seeds mature in the autumn of the first year of vegetation or in the following spring. They are toxic due to the content of sabinol.[1–4]

Figure 10.3 *Juniperus sabina* L. (Savin juniper).

Source: 15.

Main phytoconstituents	Ref.
	9,16–18

amino acids
flavonoids (apigenin, luteolin, rutin, catechin, cosmosiin, catechin, taxifolin, naringenin)
coumarins
terpenoids (pinene, cimene, linalool, caryophyllene, cubebene, cadinene, germacrene D, cedrol, totarol, ferruginol, myrcene, sabinene, and limonene)
sterols (β-sitosterol)
fatty acids (juniperonic acid, palmitic acid, oleic acid, linoleic acid, stearic acid)
phenolic acids (gallic acid, p-coumaric acid)
tannins
organic acids
carbohydrates
resins

Medicinal Use

The entire plant is extremely toxic, and ingestion can lead to death.[1] The leaves of *Juniperus sabina* were used in traditional medicine for the treatment of rheumatism and arthritic pain.[14,19]

Biological Activity

■ antibacterial effect[14]

■ weak antimicrobial activity on *Bacillus subtilis*, *Candida albicans*, *Escherichia coli*, and *Staphylococcus aureus*[18]

■ antioxidant, analgesic, antidiabetic, antiviral, antibacterial, hypotensive, anti-inflammatory, antitumoral abortifacient, antinociceptive effect[17,19–20]

Other Uses

■ decorative shrub suitable for rocky areas[4]

■ insecticidal effect[21]

REFERENCES

1. Bejenaru C., Mogoșanu G.D., Bejenaru L.E., Popescu H. 2020. *Botanică farmaceutică. Cormobionta*, ediția a III-a. Ed. Sitech, Craiova.

2. Ciocârlan V. 2009. *Flora ilustrată a României. Pteridophyta et Spermatophyta.* Ediția a 3-a revizuită și adăugită. Ed. Ceres, București.

3. Sârbu I., Ștefan N., Oprea A. 2013. *Plante vasculare din România, Determinator ilustrat de teren.* Ed. Victor B Victor, București.

4. Săvulescu T. (ed). 1952. *Flora RPR*, vol. I. Ed. Academiei RPR, București.

5. www.monaconatureencyclopedia.com/juniperus-communis-/?lang=en

6. Mogoșanu G.D., Bejenaru L.E., Bejenaru C., Popescu H. 2015. *Farmacognozie-Fitoterapie*, Vol. II, ediția a III-a. Ed. Sitech, Craiova.

7. Bais S., Gill N.S., Rana N., Shandil S. 2014. A phytopharmacological review on a medicinal plant: *Juniperus communis*. *International Scholarly Research Notices* 2014, Article ID 634723, 6 pages.

8. https://naturasiberica.ee/en/ingredient/juniper/ns-juniperus-sibirica-featured-image-2/

9. Güvenç A., Küçükboyacı N., Gören A.C. 2012. Fatty acid composition of *Juniperus* species (*Juniperus* section) native to Turkey. *Natural Product Communications* 7(7):919–922.

10. Lesjak M.M., Beara I.N., Orčić D.Z., Anačkov G.T., Balog K.J., Francišković M.M., Mimica-Dukić N.M. 2011. *Juniperus sibirica* Burgsdorf. as a novel source of antioxidant and anti-inflammatory agents. *Food Chemistry* 124(3):850–856.

11. Olech M., Nowak R., Ivanova D., Tashev A., Boyadzhieva S., Kalotova G., Angelov G., Gawlik-Dziki U. 2020. LC-ESI-MS/MS-MRM profiling of polyphenols and antioxidant activity evaluation of junipers of different origin. *Applied Sciences-Basel* 10(24):8921.

12. https://naturasiberica.ee/en/ingredient/juniper/ns-juniperus-sibirica-featured-image-2/

13. Bojor O. 2003. *Ghidul Plantelor Medicinale și Aromatice de la A la Z.* Ed. Fiat Lux, București.

14. Nikolić B., Vasilijević, B., Mitić-Ćulafić D, Marija L., Branka, V.G., Neda M. Knežević-Vukčević J. 2016. Screening of the antibacterial effect of *Juniperus sibirica* and *Juniperus sabina* essential oils in a microtitre platebased MIC assay. *Botanica Serbica* 40(1):43–48.

15. www.infoflora.ch/en/flora/juniperus-sabina.html

16. Sahin Yaglioglu A., Eser F. 2017. Screening of some *Juniperus* extracts for the phenolic compounds and their antiproliferative activities. *South African Journal of Botany* 113:29–33.

17. Kavaz D., Faraj R.E., 2023. Investigation of composition, antioxidant, antimicrobial and cytotoxic characteristics from *Juniperus sabina* and *Ferula communis* extracts. *Scientific Reports* 13(1):1–13.

18. Asili J., Emami S.A., Rahimizadeh M., Fazly-Bazzaz B.S., Hassanzadeh M.K. 2010. Chemical and antimicrobial studies of *Juniperus sabina* L. and *Juniperus foetidissima* Willd. essential oils. *Journal of Essential Oil Bearing Plants* 13(1):25–36.

19. Zhao J., Maitituersun A., Li C., Li Q., Xu F., Liu T. 2018. Evaluation on analgesic and anti-inflammatory activities of total flavonoids from *Juniperus sabina*. *Evidence-Based Complementary and Alternative Medicine* 2018:7965306.

20. Orhan N., Deliorman Orhan D., Gökbulut A., Aslan M., Ergun F. 2017. Comparative analysis of chemical profile, antioxidant, *in-vitro* and *in-vivo* antidiabetic activities of *Juniperus foetidissima* Willd. and *Juniperus sabina* L. *Iran Journal of Pharmacy Research* 16(Suppl):64–74.

21. Gao R., Gao C., Tian X., Yu X., Di X., Xiao H., Zhang X. 2004. Insecticidal activity of deoxypodophyllotoxin, isolated from *Juniperus sabina* L., and related lignans against larvae of *Pieris rapae* L. *Pest Management Science* 60:1131–1136.

11 *Dipsacaceae* Juss. Family

Cornelia Bejenaru[1], Antonia Radu[1], George Dan Mogoşanu[2],
Ludovic Everard Bejenaru[2], Andrei Biţă[2], and Adina-Elena Segneanu[3]

[1] Department of Pharmaceutical Botany, Faculty of Pharmacy, University of Medicine and Pharmacy of Craiova, 2 Petru Rareş Street, Craiova, Dolj County, Romania
[2] Department of Pharmacognosy & Phytotherapy, Faculty of Pharmacy, University of Medicine and Pharmacy of Craiova, 2 Petru Rareş Street, Craiova, Dolj County, Romania
[3] Institute for Advanced Environmental Research–West University of Timisoara (ICAM–WUT), 4 Oituz Street, Timisoara, Romania

BOTANICAL DESCRIPTION

The *Dipsacaceae* family encompasses 11 genera, with 160 species distributed across North Africa, Asia, and Europe. In the spontaneous flora of Romania, 6 genera are found: *Dipsacus, Knautia, Scabiosa, Cephalaria, Succisa,* and *Succisella*.[1,2]

It includes herbaceous plants with opposite leaves, without stipules. The flowers are hermaphroditic, zygomorphic, and sessile, arranged on a common receptacle in a capitulum-type inflorescence surrounded by involucral bracts at the base. The flowers may or may not have bracts resembling scales, and the common receptacle can be squamous or not. The flowers have an external calyx adherent to the fruit, sulcate, ribbed, or angular, and an internal gamosepalous calyx, persistent, cyathiform (saucer-like), entire, lobed, or aristate, sometimes composed of a tuft of hairs. The corolla is gamopetalous, bilaterally symmetrical, tubular, sub-bilabiate, with 4–5 lobes. The androecium consists of 4 stamens with bilocular, introrse anthers, inserted on the corolla tube alternately with its lobes. The gynoecium has an inferior ovary, unilocular, uniovulate, with a filiform style and an entire or bilobed stigma. The fruit is an achene accompanied by a persistent calyx. The seeds have fleshy endosperm.[1,3]

The genus *Dipsacus* is represented in the spontaneous flora of our country by the following species: *Dipsacus pilosus* L. (*Virga pilosa* (L.) Hill, *Cephalaria pilosa* (L.) Gren.), *Dipsacus strigosus* Willd. (*Virga strigosa* (Roem. et Schult.) Holub), *Dipsacus hmelinii* M. Bieb., *Dipsacus fullonum* L. (*D. sylvestris* Huds.), and *Dipsacus laciniatus* L.[1,2]

Dipsacus fullonum L. (Fuller's teasel)

Scientific name: *Dipsacus fullonum* L. (*D. sylvestris* Huds.)
Common name: Fuller's teasel, Wild teasel
Romanian name: Varga ciobanului
Taxonomic classification
Kingdom: *Plantae*
Subkingdom: *Cormobionta*
Phylum: *Spermatophyta*
Subphylum: *Magnoliophytina (Angiospermae)*
Class: *Magnoliatae (Magnoliopsida, Dicotyledonatae)*
Subclass: *Asteridae*
Order: *Dipsacales*
Family: *Dipsacaceae*
Genus: *Dipsacus*
Species: *fullonum* L. (Figure 11.1)[1-2]

SPREAD

Frequent throughout the country, from the sylvosteppe zone up to the beech forest level, in ruderal places, meadows, riverbanks, and shrublands.[2-3]

Description

A biennial plant with an erect, vigorous stem, prominently ridged, spiny, and branching towards the superior part, with unicapitate branches, reaching a height of 50–150 cm. The leaves are sessile, crenate-serrate at the base, with glabrous or sparsely aculeate margins. The flowers are grouped in ovoid-cylindrical capitula, with linear-subulate involucral bracts that are arched, ascending, and

Figure 11.1 *Dipsacus fullonum* L. (Fuller's teasel).

Source: 4.

longer than the inflorescence. The receptacle squamae are elongated-obovate, flexible, aristate, and straight, longer than the flowers. The external calyx is sessile, tetragonal, ending in a membranous limb, while the internal calyx is gamosepalous, with a concave, tetragonal, ciliate limb. The corolla is tetralobed, lilac in color, occasionally white, with 4 stamens inserted on the tube. The gynoecium has a bilobed stigma. The fruit retains the persistent calyx. The flowering period is from July to August.[1-3]

Main phytoconstituents Ref.
amino acids [5-6]
iridoids (oganic acid, loganin, sweroside, cantleyoside, and sylvestroside III)
flavonoids (hyperoside, kaempferol, quercetin, orientin, Saponaretin, apigenin, luteolin)
terpenoids (cymene, trujone, terpinen-4-ol, phytol, α-santonin, squalene)
sterols (campesterol, stigmasterol)
fatty acids (α-tocopherol)
phenolic acids (chlorogenic acid, neochlorogenic acid, caffeic acid,)
saponins

Medicinal Uses

Lyme disease, osteoporosis, diuretic, jaundice, would healing, diaphoretic, fibromyalgia, bone fracture, stomachic, antipyretic, laxative, skin disorders (acne, dermatitis, eczema, wounds, freckles, warts, fistula), cancer, Alzheimer's disease[5,7-10]

Biological Activity

antimicrobial, antioxidant, antiacetylcholinesterase, antitumoral[5]

Other Uses

- dye (blue as indigo substitute)
- ornamental
- winter food resource for birds
- culinary (smoothie, cooked or raw, tea, vinegar, tinctures)
- skin cosmetics[11-13]
- this plant was used for smoothing or pressing cloth fibers[5]

Dipsacus laciniatus L. (Cutleaf teasel)

Scientific name: *Dipsacus laciniatus* L.
Common name: Cutleaf teasel
Romanian name: Scaete
Taxonomic classification
Kingdom: *Plantae*
Subkingdom: *Cormobionta*
Phylum: *Spermatophyta*
Subphylum: *Magnoliophytina (Angiospermae)*
Class: *Magnoliatae (Magnoliopsida, Dicotyledonatae)*
Subclass: *Asteridae*
Order: *Dipsacales*
Family: *Dipsacaceae*
Genus: *Dipsacus*
Species: *laciniatus* L. (Figure 11.2)[1-2]

Figure 11.2 *Dipsacus laciniatus* L. (Cutleaf teasel).
Source: 14.

SPREAD

It is frequently found from the oak forest zone to the beech forest level, in shrublands, along the edges of forests and water bodies, in clearings, and in ruderal areas.[1-3]

Description

A biennial plant with an erect, cylindrical, ridged, sulcate, glabrous, spiny stem, reaching a height of 50–120 (150) cm. The basal leaves, arranged in a rosette, are lobed-crenate, lyrate, and densely setaceously ciliate along the margins, while the inferior surface is setaceously hairy, and the veins are aculeate. The stem leaves are pinnatifid or pinnatisect, connate, forming a reservoir for collecting water from precipitation. The flowers are grouped in ovoid-cylindrical capitula, with involucral bracts at most the length of the inflorescence, lanceolate or lanceolate-subulate, rigid, and pointing upwards. The corolla is pale violet or white. The flowering period is from June to August.[1-3]

Main phytoconstituents	Ref
	15-17

amino acids
iridoids (sylvestroside I-IV)
flavonoids (hyperoside, vanillin, verbacoside, taxifolin, hesperidin, apigenin, luteolin, quercetin)
terpenoids (ursolic acid, oleanoic acid, phytol, linalool)
sterols
fatty acids
phenolic acids (chlorogenic acid, caffeic acid, ferrulic acid, vanillic acid, gallic acid, p-coumaric, sinaptic acid)
saponins (asperosaponin V)
tanins (catechin, epicatechin)
carbohydrates (glucose, mannose, galactose, arabinose, rhamnose)

Medicinal use

Dipsacus laciniatus has been employed as medicinal remedy to address various conditions such as Alzheimer's disease, Lyme disease, fibromyalgia. Traditionally, the aerial parts were utilized to prepare tea for treating colds, while decoctions made from its roots and aerial parts were employed to address diabetes and cancer.[18-19]

Biological Activity

Research on *Dipsacus laciniatus* has revealed numerous biological activities, including antioxidant properties, antimicrobial activity, cytoprotective abilities, inhibition of HIV-1 reverse transcriptase, anticholinesterase, cytotoxic, and antinociceptive effects. It has also been documented to contain neuroprotective compounds that could be used in the treatment or prevention of neurodegenerative diseases. These neuroprotective agents may have a suppressing effect on neuroinflammation, leading to a reduction in neuronal damage.[18-19]

Other Uses

Dipsacus laciniatus is cultivated for its charming purple blooms, its ability to attract butterflies, and its role as a natural predator of agricultural pests.[18]

Scabiosa columbaria L. (Small scabious)

Scientific name: *Scabiosa columbaria* L.
Common name: Small scabious, Dwarf pincushion flower
Romanian name: Văduviță[20]
Taxonomic classification
Kingdom: *Plantae*
Subkingdom: *Cormobionta*
Phylum: *Spermatophyta*
Subphylum: *Magnoliophytina (Angiospermae)*
Class: *Magnoliatae (Magnoliopsida, Dicotyledonatae)*
Subclass: *Asteridae*
Order: *Dipsacales*

Figure 11.3 *Scabiosa columbaria* L. (Small scabious).

Source: 21.

Family: *Dipsacaceae*
Genus: *Scabiosa columbaria* L.
Species: *columbaria* L.[1-2]
Subspecies: – *pseudobanatica* (Schur) Jáv. et Csapody-Ung.; the involucral bracts are longer than or equal to the flower, and they are provided with long hairs;[1]

- *banatica* (Waldst. et Kit.) Diklić (*S. banatica* Waldst. et Kit.); the involucral bracts are shorter than the flower, with short hairs, and the flowers are reddish. The middle stem leaves have a terminal segment of the same width as the lateral ones;[2]

- *columbaria*; the middle stem leaves have a terminal segment that is larger and broader than the lateral ones, and the flowers are blue-violet (Figure 11.3).[1-2]

SPREAD

The species is sporadically distributed from the oak forest level to the spruce forest level, on rocky soils, in shrublands, meadows, grassy rocky areas, forest edges, dry slopes, and rocky places.[1-3]

Scabiosa columbaria is native to Yugoslavia and Romania (The Danube Gorge, Maramureş Mts), its native range is South-Eastern Carpathians.[22]

DESCRIPTION

Perennial herbaceous plant, reaching a height of 30–80 cm. The stem is glabrous or only finely hairy in the inferior part. The leaves in the basal rosette and those in the inferior part of the stem are dentate, lyrate, with the terminal lobe ranging from rhomboidal to pinnately divided. The leaves on the sterile shoots are elongated spatulate, rounded or obtuse, entire, rarely lyrate-sect, and can be hairy or glabrous. The middle and superior leaves are glabrous, pinnatisect, with linear-lanceolate laciniae, and the terminal segment is broader or linear-lanceolate. The involucral bracts are narrow-lanceolate, and the calyx setae are 3–6 times longer than the corolla, with a brown-blackish color.

The setae of the internal calyx are 2–4 times longer than those of the external calyx. The flowering period is from June to August.[1-3]

Main phytoconstituents	Ref.
	[9-11]

amino acids
coumarins
alkaloids
iridoids (secologanin, loganin)
diarylheptanoids (curcumene)
flavonoids (cathechin,
terpenoids (terpineol, thymol, eugenol, β-damascenone, ionone, cedrenol, linalool, caryophyllene, verbenene, palustrol)
sterols (sitosterol)
fatty acids (stearic acid, palmitic acid)
Esters
organic acids
phenolic acids (gallic acid, chlorogenic acid, caffeic acid)
miscellaneous (damascenone)

Medicinal Uses

diuretic, laxative, wound healing, antispasmodic, astringent, scabies, eye infections, heartburn, respiratory problems, female infertility, venereal diseases[23-25]

Biological Activity

antioxidant, anti-inflammatory, antimicrobial, antibacterial, antifungal, antiprotozoal anticancer[23]

Other Uses

- ornamental

- mellifera[23-25]

Knautia arvensis (L.) J. M. Coult. (Field scabiosa)

Scientific name: *Knautia arvensis* (L.) J. M. Coult.
Common name: Field scabiosa
Romanian name: Mușcata dracului,[1-2] Mușcatu dracului[3]
Taxonomic classification
Kingdom: *Plantae*
Subkingdom: *Cormobionta*
Phylum: *Spermatophyta*
Subphylum: *Magnoliophytina (Angiospermae)*
Class: *Magnoliatae (Magnoliopsida, Dicotyledonatae)*
Subclass: *Asteridae*
Order: *Dipsacales*
Family: *Dipsacaceae*
Genus: *Knautia*
Species: *arvensis* (L.) J. M. Coult. [1-2]
Subspecies: – *rosea* (Baumg.) Soó (*K. dumetorum* Heuff.) has undivided leaves, and the capitula have a diameter of less than 2 cm. The marginal flowers are weakly radiate, and the corolla is pink-lilac;[1]

- *pannonica* (Heuff.) O. Schwarz (*K. arvensis* var. *budensis* (Simonk.) Szabó) are gray or white-tomentose plants with deeply divided leaves having narrow segments. The superior surface of the leaves is densely hairy, while the inferior surface is gray-woolly tomentose. The capitula have a diameter of 2 cm;[1-2]

- *arvensis* rough-hairy or glabrescent plants, with stems having reddish-purple spots in the inferior part. Leaves may be divided or undivided, with broad segments. The capitula have a diameter of 2–4 cm (Figure 11.4).[1-2]

Figure 11.4 *Knautia arvensis* (L.) J. M. Coult. (Field scabiosa).

Source: 26.

SPREAD

This species is commonly distributed from the sylvosteppe zone up to the spruce forest level, in shrublands, arable lands, ruderal areas, meadows, clearings, and dry places.[1-3]

Description

Perennial herbaceous plant with an elongated, branched rhizome, an erect cylindrical stem, either simple or branched, with retrorse, non-glandular hairs in the inferior part and glabrous superior stem, but beneath the inflorescence, it bears stiff, reflexed, and glandular hairs at times. The basal and inferior leaves exhibit various shapes, elongated-lanceolate with acute tips, and sparsely dentate margins, decurrent. The basal leaves are green during flowering. The petiole and the inferior surface are setaceously hairy, occasionally glabrous. The stem leaves, more abundant in the inferior half of the stem, are deeply divided, lyrate, with spaced acute laciniae, entire, and the terminal lacinia is larger. The flowers, situated on a setaceous-hairy receptacle, are grouped in capitula with a diameter of 3–4 cm, and they have elongated ovate involucral bracts, acute, with the internal ones being thinner and shorter than the flowers. The external calyx is tetralobed, smooth, not ribbed, with 4 or more teeth, while the internal calyx has 8–10 setaceous teeth. The fruit is densely hairy. It blooms from June to August.[1-3]

Main phytoconstituents   Ref.
amino acids   [27-29]
coumarins
alkaloids
iridoids
flavonoids (hyperoside, kaempferol, isovitexin, apigenin, luteolin)
terpenoids (terpinen-3-ol, phytol, β-Amyrin, linalool)
sterols (sitosterol)

fatty acids (myristic acid, palmitic acid, palmitoleic acid, margaric acid, linolenic acid, stearic acid, docosahexaenoic acid)

phenolic acids (chlorogenic acid, cryptochlorogenic acid, gallic acid, salicylic acid, syringic acid, caffeic acid, rosmarinic acid, ferulic acid, p-coumaric acid)

hydrocarbons

tannins

Medicinal Uses

cough, sore throat, wound healing diuretic, astringent, diuretic, respiratory disorders (pharyngitis, cough, tuberculosis), scabies, burns, bruises[30-32]

Biological Activity

antioxidant, antimicrobial, anti-inflammatory[30-32]

Other Uses

■ cosmetic products[29]

Cephalaria uralensis (Murray) Roem. et Schult.

Scientific name: *Cephalaria uralensis* (Murray) Roem. et Schult.
Common name: Uralian scalehead
Romanian name: Sipică,[2] Sipică mare[20]
Taxonomic classification
Kingdom: *Plantae*
Subkingdom: *Cormobionta*
Phylum: *Spermatophyta*
Subphylum: *Magnoliophytina (Angiospermae)*
Class: *Magnoliatae (Magnoliopsida, Dicotyledonatae)*
Subclass: *Asteridae*
Order: *Dipsacales*
Family: *Dipsacaceae*
Genus: *Cephalaria*
Species: *uralensis* (Murray) Roem. et Schult.[1-2]
Subspecies: – *uralensis* has inferior leaves that are pinnatifid, with entire, oblong, or lanceolate segments;[2]

> – *multifida* (Roman) Roman et Beldie is rare, considered an endemic species in Romania, and it is cited in the Mehedinți county between Drobeta-Turnu Severin and Gura Văii, Dudașul Schelei and Gura Văii, and Gura Văii and Schela Cladovei. It has inferior leaves that are pinnatifid, with linear or linear-lanceolate segments, and 5–6 lobes (Figure 11.5).[2]

SPREAD

The species is sporadically distributed from the sylvosteppe zone to the oak forest level, in meadows, shrublands, on sunny rocky slopes.[1-2,34]

Description

Perennial herbaceous plant with a long, thick, multicapitate rhizome and a tall stem measuring 30–60 (100) cm, sometimes simple or branched. The base of the stem is arched and covered with hirsute, reversed hairs, while the superior part is glabrous. The leaves are coriaceous, obovate, with basal leaves decurrent into the petioles and being pinnatifid with 2–4 pairs of generally entire segments. The stem leaves are sessile, decurrent, amplexicaul, with lanceolate, ciliate segments, and the superior leaves are broadly linear. The flowers are whitish-yellow or pale yellow, weakly radiate, and have acuminate internal squamae of the receptacle, while the external ones are ovate. The external calyx has 4 long and 4 short, corniculate teeth. The fruit is pubescent. The flowering period is from July to September.[1-3,35]

Figure 11.5 *Cephalaria uralensis* (Murray) Roem. et Schult.

Source: 33.

Main phytoconstituents	Ref
amino acids	36-37

iridoids (loganic acid, loganin)

flavonoids (hyperoside, kaempferol, quercetin, isovitexin, nicotiflorin, isoorientin, swertiajaponin, apigenin, guiaverin, diosmetin, luteolin, nepetin)

terpenoids

sterols

fatty acids

phenolic acids (chlorogenic acid, neochlorogenic acid, caffeic acid, ferulic acid, p-coumaric acid)

hydrocarbons

antocyans (cyanidin)

Medicinal Uses

seborrheic skin diseases, acne, antipyretic, tonic, anticonvulsant, haemolysis[36,38]

Biological Activity

antioxidant, anti-inflammatory, antimicrobial, cytotoxic[36]

Other Uses

■ ornamental[38]

REFERENCES

1. Ciocârlan V. 2009. *Flora ilustrată a României. Pteridophyta et Spermatophyta.* Ediția a 3-a revizuită și adăugită. Ed. Ceres, București.

2. Sârbu I., Ștefan N., Oprea A. 2013. *Plante vasculare din România, Determinator ilustrat de teren.* Ed. Victor B Victor, București.

3. Săvulescu T. (ed). 1961. *Flora RPR*, vol. VIII. Ed. Academiei RPR, București.

4. https://dryades.units.it/Roma/index.php?procedure=taxon_page&id=5213&num=1696

5. Oszmiański J., Wojdyło A., Juszczyk P., Nowicka P. 2020. Roots and leaf extracts of *Dipsacus fullonum* L. and their biological activities. *Plants (Basel)* 9(1):78.

6. Saar-Reismaa P., Bragina O., Kuhtinskaja M., Reile I., Laanet P.R., Kulp M., Vaher M. 2022. Extraction and fractionation of bioactives from *Dipsacus fullonum* L. leaves and evaluation of their anti-*Borrelia* activity. *Pharmaceuticals* 15:87.

7. http://mediplantepirus.med.uoi.gr/pharmacology_en/plant_details.php?id=261

8. www.naturalmedicinalherbs.net/herbs/d/dipsacus-fullonum=teasel.php

9. www.botanical.com/botanical/mgmh/t/teazle09.html

10. https://sfatnaturist.ro/varga-ciobanului-o-planta-medicinala-cu-puteri-uimitoare/

11. https://healthembassy.co.uk/en/herbs/13-fuller-s-teasel-herb-dipsacus-fullonum-l.html

12. https://temperate.theferns.info/plant/Dipsacus+fullonumm

13. www.healthbenefitstimes.com/teasel/

14. https://dryades.units.it/floritaly/index.php?procedure=taxon_page&tipo=all&id=5214

15. Benabderrahim M.A., Sarikurkcu C., Elfalleh W., Ozer M.S. 2019. *Datura innoxia* and *Dipsacus laciniatus*: Biological activity and phenolic composition. *Biocatalysis and Agricultural Biotechnology* 19:101163.

16. Skała, E., & Szopa, A. *Dipsacus* and *Scabiosa species* – The source of specialized metabolites with high biological relevance: A review. *Molecules* 28(9):3754.

17. Kocsis Á., Szabó, L.F., Podányi B. 1993. New *bis*-iridoids from *Dipsacus laciniatus. Journal of Natural Products* 56(9):1486–1499.

18. Benabderrahim M.A., Sarikurkcu C., Elfalleh W., Sabih Ozer M. 2019. *Datura innoxia* and *Dipsacus laciniatus*: Biological activity and phenolic composition. *Biocatalysis and Agricultural Biotechnology* 19:101163.

19. Taşkin T., Akkilic Y., Celik B.O., Senkardes I. ve Ozakpinar O.B. 2020. Investigation of the biological activities of different extracts from *Dipsacus laciniatus* aerial. *Natural Products Journal* 10(1):15–19.

20. Drăgulescu C. 2018. *Dicționar de Fitonime Românești*. Ediția a 5-a completată. Editura Universității "Lucian Blaga," Sibiu.

21. https://dryades.units.it/dolomitibellunesi/index.php?procedure=taxon_page&id= 5253&num=4530

22. Hurdu B.I., Pușcaș M., Turtureanu P.D., Niketić M., Vonica Ghizela, Coldea Gh. 2012. *A critical evaluation of the Carpathian endemic plant taxa list from the Romanian Carpathians.* Contribuții Botanice Grădina Botanică "Alexandru Borza" Cluj-Napoca LVII:39–47.

23. Maroyi A. 2019. *Scabiosa columbaria*: A review of its medicinal uses, phytochemistry, and biological activities. *Asian Journal of Pharmaceutical and Clinical Research* 12(8):10–14.

24. www.naturplant.ro/knautia-arvensis-muscata-dracului/

25. www.botanical.com/botanical/mgmh/s/scafie29.html

26. https://dryades.units.it/floritaly/index.php?procedure=taxon_page&tipo=all&id=5240

27. Kopytko Y.F., Dargaeva T.D., Rendyuk T.D. 2020. Composition of the field *Scabious* (*Knautia arvensis* L.). *Pharmaceutical Chemistry Journal* 54(7):725–733.

28. Moldoch J., Szajwaj B., Masullo M., Pecio L., Oleszek W., Piacente, S., Stochmal A. 2011. Phenolic constituents of *Knautia arvensis* aerial parts. *Natural Product Communications* 6(11):1627–1630.

29. Marijan M., Jablan J., Jakupovic L., Jug M., Marguí E., Dalipi R., Sangiorgi E., Zovko Koncic M. 2022. Plants from urban parks as valuable cosmetic ingredients: green extraction, chemical composition and activity. *Agronomy* 12(1):204.

30. http://mediplantepirus.med.uoi.gr/pharmacology_en/plant_details.php?id=283

31. https://pfaf.org/user/Plant.aspx?LatinName=Knautia+arvensis

32. www.naturplant.ro/knautia-arvensis-muscata-dracului/

33. www.semanticscholar.org/paper/Biotechnological-Potential-of-Cephalaria-uralensis Chrz%C4%85szcz-Szewczyk/ffb359d9381749bc143e8902cd4aa149c7446393

34. Ciocârlan V. 2011. The variability of *Cephalaria uralensis* (Murray) Roem. et Schult. *Journal of Plant Development* 18(1):105–108.

35. Bojnanský V., Fargašová A. 2007. *Atlas of seeds and fruits of Central and East-European flora: The Carpathian Mountains region.* Springer Science & Business Media, Dordrecht, The Netherlands.

36. Chrząszcz M., Miazga-Karska M., Klimek K., Granica S., Tchórzewska D., Ginalska G., Szewczyk K. 2020. Extracts from *Cephalaria uralensis* (Murray) Roem. & Schult. and *Cephalaria gigantea* (Ledeb.) Bobrov as potential agents for treatment of acne vulgaris: Chemical characterization and *in vitro* biological evaluation. *Antioxidants* 9(9):796.

37. Szewczyk K.D.S. 2022. *Cephalaria* (*Caprifoliaceae*) genus. *Encyclopedia*. Available online: https://encyclopedia.pub/entry/11594 (accessed on 11 October 2022).

38. www.herbal-organic.com/en/herb/120330

12 *Elaeagnaceae* Adans. Family

Cornelia Bejenaru[1], Antonia Radu[1], George Dan Mogoşanu[2],
Ludovic Everard Bejenaru[2], Andrei Biţă[2], and Adina-Elena Segneanu[3]

[1] Department of Pharmaceutical Botany, Faculty of Pharmacy, University of Medicine and Pharmacy of Craiova, 2 Petru Rareş Street, Craiova, Dolj County, Romania
[2] Department of Pharmacognosy & Phytotherapy, Faculty of Pharmacy, University of Medicine and Pharmacy of Craiova, 2 Petru Rareş Street, Craiova, Dolj County, Romania
[3] Institute for Advanced Environmental Research–West University of Timisoara (ICAM–WUT), 4 Oituz Street, Timisoara, Romania

BOTANICAL DESCRIPTION

Woody plants (trees or spiny shrubs), covered with scaly or whitish stellate hairs on the entire aerial part, especially on the leaves. The leaves are simple, entire, arranged alternately or oppositely. The flowers are hermaphroditic, polygamous, or unisexual-dioecious, actinomorphic, with a simple floral envelope, arranged axillary, solitary, or in pauciflorus racemes. The perigone may have 2 or 4 (sometimes 6) sepaloid tepals; the androecium consists of 4–8 stamens inserted on the internal walls of the receptacle. The receptacle of male flowers has a plate or cup-like appearance. The receptacle of female and hermaphrodite flowers is large, cylindrical, long, tightly enveloping the ovary only at the base or entirely and becomes fleshy at maturity, accreting with the fruit, which has a drupaceous appearance, with membranous pericarp, surrounded by the persistent, fleshy receptacle. The seed has a woody tegument and is exalbuminous.[1-4]

The family includes 2 genera: *Hippophaë* L. and *Elaeagnus* L., which are found in the spontaneous flora of Romania.[2-3]

Hippophaë rhamnoides L. (Sea buckthorn)

Scientific name: *Hippophaë rhamnoides* L.
Common name: Sea buckthorn
Romanian name: Cătină albă, Cătină de râu
Taxonomic classification
Kingdom: *Plantae*
Subkingdom: *Cormobionta*
Phylum: *Spermatophyta*
Subphylum: *Magnoliophytina (Angiospermae)*
Class: *Magnoliatae (Magnoliopsida, Dicotyledonatae)*
Subclass: *Rosidae*
Order: *Elaeagnales*
Family: *Elaeagnaceae*
Genus: *Hippophaë*
Species: *rhamnoides* L.[1-3]
Subspecies: – *fluviatilis* Van Soest is a less spiny plant with elongated branches, 3–6 cm wide leaves, and ovoid seeds;

– *carpatica* Rousi has 5–8 (10) mm wide leaves, spherical fruits, and elongated-ellipsoidal seeds;

– *rhamnoides* has cylindrical fruits and ellipsoidal seeds (Figure 12.1).[2-3]

SPREAD

Frequently distributed from the seaside region up to the mountainous zone, along the shores, sandy and gravelly coastal areas, on rocky coasts, cliffs, and saliferous geological formations.[1-4]

DESCRIPTION

Spiny shrub (rarely a tree), unisexual-dioecious, reaching a height of 1–4 (10) m, with divergent branches and trailing shoots, dark reddish-brown. Annual branches are covered with numerous silvery-gray scaly hairs, while the two-year-old branches have lateral shoots terminated by a thorn. Buds are ovoid-spherical, densely covered with scaly-stellate, copper-colored hairs. Leaves are lanceolate, up to 1 cm wide, entire, very short-petiolate, with a whitish inferior surface, due to the

DOI: 10.1201/9781003270515-13

Figure 12.1 *Hippophaë rhamnoides* L. (Sea buckthorn).

Source: 5.

presence of tector hairs. Flowers are small, yellow in color, appearing before the leaves. Fruits are drupaceous, yellow-orange in color, ovoid, and contain hard seeds. Blooms in early spring (April–May). Cultivated for soil stabilization, as well as for food and medicinal purposes due to its fruit properties.[1-4]

Main phytoconstituents Ref.

amino acids (proline, cysteine, asparagine, aspargic acid, threonine, serine, glutamine, glycine, phenylalanine, valine) 6-13

alkaloids (hippophamide, 4-[(Z)-p-coumaroylamino]butan-1-ol, 4-[(E)-p-coumaroyl-amino] butan-1-ol)

flavonoids (quercitin, rutin, myricetin, catechin, epicatechin, syringetin, kaempferol, quercetin, luteolin, isorhamnetin, naringenin, gallocatechin, epigallocatechin)

terpenoids (oleanoic acid, ursolic acid, pomolic acid, dulcioic acid, corosolic acid uvaol, erythrodiol, linaool, germacrene B, phytol, β-amyrin, lupeol, cycloartenol, camphor, safranal, β-cyclocitral, maslinic acid, arjunolic acid, betulinic acid, betulin, lupeol, friedelan-3-ol, squalene)

sterols (cholesterol, β-sitosterol, campesterol, stigmasterol, citrostadienol, obtusifoliol, campesterol, isofucosterol, gramisterol, sitostanol)

fatty acids (palmitic acid, linolenic acid, alpha-linolenic acid, oleic acid, stearic acid, eicosanoic acid, behenic acid)

carbohydrates (galactose, fructose, xylose)

phenolic acids (gallic acid, chlorogenic acid, caffeic acid, ferrulic acid, p-cumaric acid, salicylic)

carotenoids (lutein, zeaxanthin, β-cryptoxanthin, *cis*-β-carotene, β-carotene)

organic acids

tannins (hippophaenin A, hippophaenin B, hippophaenin C, elaeagnatin A, pedunculagin, casuarinin, casuarictin, strictinin, isostrictinin, tellimagrandin I, stachyurin, castalagin, vescalagin, pterocarinin A)

saponins (hippophoside A, hippophoside B, hippophoside C, hippophoside D, hippophoside E, hippophoside F)
aldehydes&ketones
miscellaneous (vitamin C, vitamin D, vitamin A, vitamin B_1, vitamin B_6, vitamin K_1, vitamin E)

Medicinal Uses

Hippophaë fructus exhibits antiscorbutic and antioxidant action (ascorbic acid, vitamin E, flavonoids), anti-inflammatory properties (pentacyclic triterpenoid acids), as well as cicatrizing and antiulcer effects (fatty oil, carotenoids).[6]

- would healing, skin regeneration, digestive disorders (gastric, duodenal ulcers), gynecological diseases (vagina, cervix inflammation), rheumatism, antihemorrhagic, febrile diseases, high cholesterol[7-13]

Biologic Activity

- antioxidant, anti-inflammatory, antimicrobial, antibacterial, antiviral, antihyperlipidemic, antitumoral, anticoagulant, antiplatelet, antidiabetic, hepatoprotective, neuroprotective.[7-13]

Sea buckthorn fruits are administered in cases of scurvy and other avitaminosis (as a polyvitamin supplement), gastritis, gastroduodenal ulcers, dermatoses, mycoses, and ocular conditions (corneo-conjunctival burns, corneal grafts, phlyctenular keratoconjunctivitis, plastic and reconstructive surgery of the eyelids and conjunctival sac).

The fatty oil from the pericarp is used for burns, frostbites, and gynecological conditions.[6]

Other Uses

- culinary (sauces, jams)
- cosmetic products (antiaging and antiwrinkle serums and creams, UV protection lotions and creams, peeling products, baths salts, masks, hair removal products, shampoos, hair conditioners)
- ornamental[4]
- fixing degraded terrains[4]

Elaeagnus angustifolia L. (Russian olive)

Scientific name: *Elaeagnus angustifolia* L.
Common name: Russian olive, Oleaster
Romanian name: Sălcioară, Salcie mirositoare
Taxonomic classification
Kingdom: *Plantae*
Subkingdom: *Cormobionta*
Phylum: *Spermatophyta*
Subphylum: *Magnoliophytina (Angiospermae)*
Class: *Magnoliatae (Magnoliopsida, Dicotyledonatae)*
Subclass: *Rosidae*
Order: *Elaeagnales*
Family: *Elaeagnaceae*
Genus: *Elaeagnus*
Species: *angustifolia* L. (Figure 12.2)[2-3]

SPREAD

Subspontaneous species, cultivated for ornamental purposes and as protective hedges. Found in the Mediterranean region, Central Asia, and Western Asia.[2-4]

DESCRIPTION

It is a spiny tree or shrub, reaching a height of 3–7 m, with young branches that are silvery and densely covered with scaly hairs. The leaves have petioles that are 5–8 mm long, and they are

Figure 12.2 *Elaeagnus angustifolia* L. (Russian olive).

Source: 14.

elongated-lanceolate to linear-lanceolate, with acute or obtuse tips, and a broadly cuneate base. They measure 4–8 cm in length, have a dark green, glabrous superior surface, and a silver inferior surface covered in dense, silvery scaly hairs. The flowers are shortly pedunculate, occurring in groups of 1–3 in the inferior part of the branches, measuring 1 cm in length. They are silvery on the external surface and yellow on the internal surface, and they can be hermaphroditic or polygamous, emitting a pleasant fragrance. The tube of the receptacle is elongated, about the same length as the tepals, that form a four-lobed perigone. The fruit is ellipsoidal, measuring 1 cm in length, light yellow, covered in silvery scales, with a sweet, mealy pericarp. It blooms in June.[2-4]

Main phytoconstituents	Ref.
amino acids (proline, cysteine, asparagine, aspargic acid, threonine, serine, glutamine, glycine, alanine, valine)	15-20

alkaloids (elaeagnin, harman, dihydroharman, harmin, harmol)

flavonoids (rutin, catechin, epicatechin, gallocatechin, epigallocatechin, kaempferol, quercetin, luteolin, isorhamnetin)

terpenoids (farnesyl, limonene, nerolidole, squalene, theaspirane A, theaspirane B β-damascenone, ionone, cadinene, nerolidole B, spathulenol, germacrene D, phytol, caryophyllene)

sterols (β-sitosterol)

fatty acids (lauric acid, tridecanoic acid, myristic acid, pentadecanoic acid, palmitic acid, palmitoleic acid, heptadecanoic acid, linolenic acid, oleic, stearic acid, eicosanoic, docosanoic)

carbohydrates (fructose, glucose)

phenolic acids (caffeic acid, vanilic acid, ferrulic acid)

organic acids

tannins

Aldehydes & ketones

Medicinal Uses

- wound healing, tonic, antipyretic, astringent, analgesic, digestive disorders[15-20]

Biologic Activity

- antioxidant, anti-inflammatory, analgesic, antimicrobial, antibacterial, antifungal, antinociceptive, antiarthritic, gastroprotective, cardioprotective, hypolipidemic, insecticidal, neuroprotective, antimutagenic, antitumoral[15-20]

Other Uses

- culinary
- ornamental
- timber, furniture
- polluted soil remediation[21]

REFERENCES

1. Bejenaru C., Mogoșanu G.D., Bejenaru L.E., Popescu H. 2020. *Botanică farmaceutică. Cormobionta*, ediția a III-a. Ed. Sitech, Craiova.

2. Ciocârlan V. 2009. *Flora ilustrată a României. Pteridophyta et Spermatophyta*. Ediția a 3-a revizuită și adăugită. Ed. Ceres, București.

3. Sârbu I., Ștefan N., Oprea A. 2013. *Plante vasculare din România. Determinator ilustrat de teren*. Ed. Victor B Victor, București.

4. Săvulescu T. (ed). 1956. *Flora RPR*, vol. IV. Ed. Academiei RPR, București.

5. https://dryades.units.it/dolomitibellunesi/index.php?procedure=taxon_page&id= 3157&num=2164

6. Mogoșanu G.D., Bejenaru L.E., Bejenaru C., Popescu H. 2015. *Farmacognozie-Fitoterapie*, Vol. II, ediția a III-a. Ed. Sitech, Craiova.

7. Criste A., Urcan A.C., Bunea A., Pripon Furtuna F.R., Olah N.K., Madden R.H., Corcionivoschi N. 2020. Phytochemical composition and biological activity of berries and leaves from four Romanian sea Buckthorn (*Hippophae rhamnoides* L.) varieties. *Molecules* 25(5):1170.

8. Zuchowski J. 2023. Phytochemistry and pharmacology of sea buckthorn (*Elaeagnus rhamnoides*; syn. *Hippophae rhamnoides*): progress from 2010 to 2021. *Phytochemistry Reviews* 22(1):3–33.

9. Zielińska, A., Nowak, I. 2017. Abundance of active ingredients in sea-buckthorn oil. *Lipids in Health and Disease* 16:95.

10. Kukin T.P., Shcherbakov D.N., Gensh K.V., Tulysheva E.A., Salnikova O.I., Grazhdannikov A.E., Kolosova E. A. 2017. Bioactive components of sea buckthorn *Hippophae rhamnoides* L. foliage. *Russian Journal of Bioorganic Chemistry* 43(7):747–751.

11. Slynko N., Kuibida L., Tatarova L., Galitsyn G., Goryachkovskaya T., Peltek S. 2019. Essential oils from different parts of the sea buckthorn *Hippophae rhamnoides* L. *Advances in Bioscience and Biotechnology* 10:233–243.

12. Jaroszewska A., Biel W. 2017. Chemical composition and antioxidant activity of leaves of mycorrhized sea-buckthorn (*Hippophae rhamnoides* L.). *Chilean Journal of Agricultural Research* 77(2):155–162.

13. Kuhkheil A., Naghdi Badi H., Mehrafarin A., Abdoss V. 2017. Chemical constituents of seabuckthorn (*Hippophae rhamnoides* L.) fruit in populations of central Alborz Mountains in Iran. *Journal of Pharmacognosy (RJP)* 4(3):1–12.

14. https://dryades.units.it/saline/index.php?procedure=taxon_page&id=3158&num=3834

15. Hamidpour R., Hamidpour S., Hamidpour M., Shahlari M., Sohraby M., Shahlari N., Hamidpour R. 2016. Russian olive (*Elaeagnus angustifolia* L.): from a variety of traditional medicinal applications to its novel roles as active antioxidant, anti-inflammatory, anti-mutagenic and analgesic agent. *Journal of Traditional and Complementary Medicine* 7(1):24–29.

16. Torbati M, Asnaashari S, Heshmati Afshar F. 2016. Essential pil from flowers and leaves of *Elaeagnus angustifolia* (*Elaeagnaceae*): Composition, radical scavenging and general toxicity activities. *Advanced Pharmaceutical Bulletin* 6(2):163–9.

17. Artikova G.N., Ishimov U.Zh., Sobirova F.A., Genzhemuratova G.P., Bektursynova A.P., Matchanov A.D. 2020. Comprehensive study of the chemical composition of the plant *Elaeagnus angustifolia* L. *European Journal of Molecular & Clinical Medicine* 7(1):3320–3331.

18. Tehranizadeh Z.A., Baratian A., Hosseinzadeh H. 2016. Russian olive (*Elaeagnus angustifolia*) as a herbal healer. *Bioimpacts* 6(3):155–167.

19. Dincă L., Timiș-Gânsac V. 2020. Oleaster (*Eleagnus angustifolia* L.) – A constituent element of the biodiversity from Dobrogea Plateau. *Journal of Horticulture, Forestry and Biotechnology* 24(1):17–22.

20. Nazir N., Zahoor M., Nisar M. 2020. A review on traditional uses and pharmacological importance of genus *Elaeagnus* species. *The Botanical Review* 86:247–280.

21. www.cabidigitallibrary.org/doi/10.1079/cabicompendium.20717

13 *Equisetaceae* Michx. ex DC. Family

Cornelia Bejenaru[1], Antonia Radu[1], George Dan Mogoşanu[2],
Ludovic Everard Bejenaru[2], Andrei Biţă[2], and Adina-Elena Segneanu[3]

[1] Department of Pharmaceutical Botany, Faculty of Pharmacy, University of Medicine and Pharmacy of Craiova, 2 Petru Rareş Street, Craiova, Dolj County, Romania
[2] Department of Pharmacognosy & Phytotherapy, Faculty of Pharmacy, University of Medicine and Pharmacy of Craiova, 2 Petru Rareş Street, Craiova, Dolj County, Romania
[3] Institute for Advanced Environmental Research–West University of Timisoara (ICAM–WUT), 4 Oituz Street, Timisoara, Romania

BOTANICAL DESCRIPTION

The plants of this family are perennial herbaceous, with underground rhizomes equipped with tubers for storing reserve substances. The stems are articulated, hollow, and rigid due to the impregnation of the external tissue membrane (epidermis) with SiO_2. The internodes are ribbed and longitudinally grooved. At the nodes, there are reduced leaves (scales), arranged in whorls and fused together, forming a toothed sheath, with the number of sheath teeth equal to the number of stem ribs. Some species have sterile stems (do not produce sporophylls) and fertile stems. The stems are branched, with branches arranged in whorls. The sporophylls (modified leaves bearing sporangia) have a shield-like peltate shape and form terminal spikes with 5–12 sporangia each. The spores, which are formed in sac-like sporangia, are morphologically identical, while the prothalli formed from them are different (male and female). The spores are surrounded by two long, spiraled, hygroscopic threads that aid in their dispersal. The species in this family are isosporous and heteroprothallian.[1-3]

The species of the only genus (*Equisetum*) of this family are commonly known as Horsetails.[1] The genus *Equisetum* is represented in the spontaneous flora of Romania by 9 species: *Equisetum arvense* L., *Equisetum telmateia* Ehrh. (*E. maximum* auct.), *Equisetum sylvaticum* L., *Equisetum pratense* Ehrh., *Equisetum fluviatile* L. em Ehrh. (*E. limosum* L.), *Equisetum palustre* L., *Equisetum ramosissimum* Desf., *Equisetum variegatum* Schleich. ex Weber et D. Mohr, and *Equisetum hyemale* L.[2,4]

Equisetum arvense L. (Common horsetail)

Scientific name: Equisetum arvense L.
Common name: Common horsetail or Field horsetail
Romanian name: Coada calului
Taxonomic classification
Kingdom: *Plantae*
Subkingdom: *Cormobionta*
Phylum: *Pteridophyta*
Class: *Equisetatae* (*Equisetopsida*)
Order: *Equisetales*
Family: *Equisetaceae*
Genus: *Equisetum*
Species: *arvense* L. (Figure 13.1)[1-2,4]

SPREAD

This species is common from plains to mountains, on sandy terrain, in meadows, riverbanks, or cultivated fields, as well as in ravines and mossy sinkholes.[1,3]

DESCRIPTION

Perennial herbaceous plant, with a dark brownish rhizome and sometimes elongated-ovate or round tubers. It has two types of stems (sterile and fertile) with a height of 15–50 cm. The fertile stem lacks chlorophyll (it is brown in color), smooth, simple, succulent, and appears early in spring before the sterile one. It has spore-bearing spikes at the tip, measuring 1–4 cm in length, elongated-ovate or cylindrical with a blunt tip. On the fertile stems, there are 4–6 spaced sheaths, each with 6–12 teeth, sometimes up to 16. After the spores are dispersed, the fertile stem dries up. The sterile stem, green in color, has whirls of articulated branches at the nodes, appearing later, and exhibits 6–19 prominent and rough ribs.[1,3]

DOI: 10.1201/9781003270515-14

Figure 13.1 *Equisetum arvense* L. (Common horsetail).

Source: 5.

Main Phytoconstituents

The vegetable product *Equiseti herba* contains: 4–5% saponins, predominantly ecvisetonin, which, upon hydrolysis, releases the aglycone (ecvisetogenol) and the carbohydrate portion (arabinose and fructose); flavonoids; volatile oil; resins; organic acids; alkaloids (nicotine) in trace amounts; and 10% mineral salts, mainly (6–8%) silica dioxide. According to Romanian Pharmacopeea (F.R. X), the medicinal product must contain a minimum of 20% soluble substances.[6]

The vegetable product *Equiseti herba* represents the sterile stems, harvested in the months of July–August. It contains flavonoids, mineral salts, volatile oil, and is used as a diuretic, anti-inflammatory, antiedema, and a stimulant for collagen synthesis.[1]

Main phytoconstituents
alkaloids (equisetin, nicotine, palustrine, palustrinine)
flavonoids (quercetin, apigenin, kaempherol, isoquercitrin, genkwanin, onitin, luteolin, isovanillin)
terpenoids (tymol, farnesyl, ursolic acid, oleanolic acid, betulinic acid, germanicol, taraxerol, isobauerenol, geraniol, phytol)
sterols (campesterol, sitosterol, cholesterol)
saponins (ecvisetonin)
phenolic acids (chlorogenic acid, caffeic acid)
glycosides (coniferin)
organic acids (tartaric acid, silicic acid)

Ref.
6-10

Biological Activity

The vegetable product exhibits diuretic action (Romanian Pharmacopeea (F.R. X)) due to its content of saponins, flavonoids, mineral salts, and volatile oil. Additionally, it demonstrates anti-inflammatory and antiedema effects, while also promoting the strengthening of connective tissue by stimulating collagen synthesis and providing silica dioxide. In phytotherapy, it is considered a source of assimilable silicon, similar to *Cynosbati fructus* (Jean Valnet). *Equiseti herba* is used in the form of infusions, compresses, and baths for genitourinary diseases, edema, polyarthritis, osteoarthritis, fractures, frostbite, and eczema. The juice obtained from the fresh plant has a hemostatic effect in cases of epistaxis and hemorrhoids.[6]

Other Uses

The fertile stems, which bear strobili, can be cooked and consumed similarly to asparagus, known as "tsukushi" in Japan. In the Pacific Northwest, Native Americans eat the young shoots of this plant raw, but caution must be exercised. *E. arvense* contains thiaminase, an enzyme that metabolizes the B vitamin, thiamine, which can potentially lead to thiamine deficiency and associated liver damage if consumed chronically.[11-14]

Equisetum telmateia Ehrh. (*E. maximum* auct.)

Scientific name: Equisetum telmateia Ehrh. (E. maximum auct.)
Common name: Giant horsetail or Great horsetail
Romanian name: coada-calului mare
Taxonomic classification
Kingdom: *Plantae*
Subkingdom: *Cormobionta*
Phylum: *Pteridophyta*
Class: *Equisetatae (Equisetopsida)*
Order: *Equisetales*
Family: *Equisetaceae*
Genus: *Equisetum*
Species: *telmateia* Ehrh. (Figure 13.2)[1-2,4]

SPREAD

A common plant found from lowlands to mountains, in wet forests, meadows, and along waterways, as well as on moist rocky slopes and ditch edges.[1,3-4]

Figure 13.2 *Equisetum telmateia* Ehrh. (*E. maximum* auct.).

Source: 15.

DESCRIPTION

Perennial herbaceous plant, with a strong rhizome of dark brown-black color, and two types of stems (sterile and fertile), ranging in height from 30 to 100 cm. The fertile stem, lacking chlorophyll, appears first. It is non-branching, with a diameter of 2 cm, hollow, and has a fusiform spike measuring 4–8 cm in length. It bears 1–2 sterile rings at the base and numerous sheaths closely arranged, with each sheath having 20–30, sometimes up to 35 teeth. After spore dispersal, the fertile stem dries up and typically survives for around 7–8 days. The robust sterile stem has internodes of 10 cm in length, is smooth with faint stripes, and has numerous branches arranged in whorls. [1,3]

Main phytoconstituents   Ref.
alkaloids   10, 16-17
flavonoids (kaempherol, Isovanillin)
terpenoids (β-caryophyllene, α-ionone, limonene, farnesol, 1,8 cineole, β-thujone, phenanthrene, crocusatin A, β-cyclohomogeraniol, α-damascone, jasmone)
tannins (catechin)
fatty acids (myristic acid, palmitic acid)
sterols (stigmasterol)
saponins
esthers (benzyl benzoate)
phenolic acids (protocatechuic acid, p-hydroxy-benzoic, caffeic acid)
organic acids (tetradecanoic acid)
hydrocarbons (pentadecane, heptadecane, octadecane, eicosane)
aldehydes&keyones (2-octenal, nonanal, hexadecanol, 6-methyl-3,5-heptadien-2-one, 4-methylacetophenone, tetradecanal)
misceleous (shikimic acid, 2,6-dimethylcyclohexanol, 3-butylpyridine)

Medicinal Use

In traditional medicine, the aerial parts of *Equisetum telmateia* have been employed as a source of biologically active compounds for the treatment of inflammation, diarrhea, stomach-aches, eczema, and mouth infections.[16]

- expectorant, rheumatism, gastrointestinal disorders, diuretic, hypertension, bone fractures, genitourinary diseases, prostatitis[16-18]

Biological Activity

- antioxidant and antimicrobial activity[18]

Other Uses

This particular species was highly favored by coastal native peoples as a significant springtime vegetable. They would pick and consume the young fertile and vegetative shoots after removing the sheaths. However, it is important to note that this genus can be poisonous to both livestock and humans if consumed in large quantities. Some native groups used to harvest the tops of these plants, boil them, and drink the resulting liquid as a remedy for urinary ailments. Moreover, *E. telmateia*, like many other *Equisetum* species, was utilized as a medicinal plant for treating burns. The stems would be burned, and the ashes applied to the wounds. Additionally, the silicon dioxide crystals in *Equisetum* species made them excellent scouring tools. Native peoples would extensively use them for smoothing and polishing wood and soapstone. Even in modern times, hunters and outdoors enthusiasts still use them as scouring utensils to clean pots and pans.[19-21]

Equisetum palustre L. (Marsh horsetail)

Scientific name: *Equisetum palustre* L.
Common name: Marsh horsetail
Romanian name: Barba ursului de bahne
Taxonomic classification
Kingdom: *Plantae*
Subkingdom: *Cormobionta*

Figure 13.3 *Equisetum palustre* L. (Marsh horsetail).

Source: 22.

Phylum: *Pteridophyta*
Class: *Equisetatae (Equisetopsida)*
Order: *Equisetales*
Family: *Equisetaceae*
Genus: *Equisetum*
Species: *palustre* L. (Figure 13.3)[1-2,4]

SPREAD

The plant commonly grows from plains to mountains, in wet areas such as marshy meadows and peat bogs, riverbanks, wet orchards, and damp meadows.[1,3]

DESCRIPTION

Perennial herbaceous plant with a single type of stems, having with 6-8 deep ridges, and reaching a height of 10–40 cm. The stems have numerous assimilatory branches, erect or arching-erect, simple, with 5 ribs, and at the top, there are spore-bearing spikes measuring 15–30 mm in length, ovoid-oblong and obtuse. The stems are deeply ridged, soft, and dry up in the autumn. The sheaths have a number of teeth equal to the number of ridges on the stem.[1,3]

Main phytoconstituents   Ref.
alkaloids (palustridiene, palustrine, nicotine)   10,17,23-24
flavonoids (luteolin, apigenin, genkwanin, kaempferol, isovanillin)
terpenoids (limonene, α-farnesene, phytol, axerophthene, α-ionone, farnesol, 1,8 cineole,
     β-thujone, dihydroactinidiolide, nerolidol, phenanthrene, crocusatin A, α-damascone)
fatty acids (myristic acid, palmitic acid)
sterols (stigmasterol)
esthers (benzyl salicylate)

phenols (phenol, 2,4,6-trimethylphenol)
phenolic acids (homovanillic acid, ferulic acid)
organic acids (benzoic acid, phenylacetic acid, 2-Hexanoic acid)
hydrocarbons (5-methyltetradecane, heneicosane, docosane)
aldehydes&keyones (undecanal, phenylethanal, 2,3-octanedione, 2,4-decadienal, 4-(2,4,6-trimethylphenyl)-2-butanone, 2-octenal, nonanal, hexadecanol, 6-methyl-3,5-heptadien-2-one, 4-methylacetophenone, tetradecanal)
miscellaneous (3,4-dimethyl-5-pentyl-2(5H)-furanone, decanol)

Medicinal Use

It is not used in therapy, avoiding contamination of the product *Equiseti herba* with vegetable material from this species.[1]

The aerial parts of *Equisetum palustre* L. are used to treat peptic ulcer disease in the traditional medicine of Turkey.[14]

Biological Activity

antiulcerogenic and gastroprotective activity[25]

Other Uses

■ The stems, which contain 10% silica, serve as effective scouring tools for metal and can be used as fine sandpaper. They are also utilized as a polish for brass, hardwood, and other materials. When infused, the stems act as an efficient fungicide against mildew, mint rust, and blackspot on roses. Additionally, they can be used as a beneficial liquid feed.[26]

REFERENCES

1. Bejenaru C., Mogoşanu G.D., Bejenaru L.E., Popescu H. 2020. *Botanică farmaceutică. Cormobionta*, ediţia a III-a. Ed. Sitech, Craiova.

2. Ciocârlan V. 2009. *Flora ilustrată a României. Pteridophyta et Spermatophyta*. Ediţia a 3-a revizuită şi adăugită. Ed. Ceres, Bucureşti.

3. Săvulescu T. (ed). 1952. *Flora RPR*, vol. I. Ed. Academiei RPR, Bucureşti.

4. Sârbu I., Ştefan N., Oprea A. 2013. *Plante vasculare din România, Determinator ilustrat de teren*. Ed. Victor B Victor, Bucureşti.

5. https://naturallycuriouswithmaryholland.wordpress.com/2011/05/15/field-horsetail-equisetum-arvense/

6. Mogoşanu G.D., Bejenaru L.E., Bejenaru C., Popescu H. 2015. *Farmacognozie-Fitoterapie*, Vol. II, ediţia a III-a. Ed. Sitech, Craiova.

7. Mimica-Dukic N., Simin N., Cvejic J., Jovin E., Orcic D., Bozin B. 2008. Phenolic compounds in field horsetail (*Equisetum arvense* L.) as natural antioxidants. *Molecules* 13(7):1455–1464.

8. Makia R., Al-Sammarrae K.W., Al-Halbosiy M.M.F., Al-Mashhadani M.H. 2022. Phytochemistry of the genus *Equisetum* (*Equisetum arvense*). *GSC Biological and Pharmaceutical Sciences* 18(02): 283–289.

9. European Medicines Agency. 2016. EMA/HMPC/278089/2015-Committee on Herbal Medicinal Products (HMPC)-Assessment report on *Equisetum arvense* L., herba.

10. Boeing T., Tafarelo Moreno K.G., Junior A.G. 2021. Phytochemistry and pharmacology of the genus *Equisetum* (*Equisetaceae*): A narrative review of the species with therapeutic potential for kidney diseases. *Evidence-Based Complementary and Alternative Medicine: ECAM*, 2021:6658434.

11. Ashkenazi M., Jacob J. 2003. *Food culture in Japan. Westport.* Greenwood Press, Westport, CT, USA. ISBN 0-313-32438-7.

12. Gunther E. 1973. *Ethnobotany of western Washington: the knowledge and use of indigenous plants by native Americans* (Revised ed.). University of Washington Press, Seattle, WA, USA. ISBN 9780295952581.

13. Israelsen C.E., McKendrick S.S., Bagley C.V. 2010. *Poisonous plants and equine* (revised ed.). Utah State University, Logan, UT, USA.

14. Fabre B., Geay B., Beaufils P. 1993. Thiaminase activity in *Equisetum arvense* and its extracts. *Plantes Médicinales et Phytothérapie* 26(3):190–197.

15. www.infoflora.ch/en/flora/equisetum-telmateia.html

16. Yeganegi M., Tabatabaei Y.F., Mortazavi S.A., Asili J., Alizadeh Behbahani B., Beigbabaei A. 2018. *Equisetum telmateia* extracts: Chemical compositions, antioxidant activity and antimicrobial effect on the growth of some pathogenic strain causing poisoning and infection. *Microbial Pathogenesis* 116:62–67.

17. Milovanović V., Radulović N., Mitić V., Palić R., Stojanović G. 2008. Chemical composition of the essential oils of *Equisetum palustre* L. and *Equisetum telmateia* Ehrh. *Journal of Essential Oil Research* 20(4):310–314.

18. Radojevic I.D., Stankovic M.S., Stefanovic O.D., Topuzovic M.D., Comic L.R., Ostojic A.M. 2012. Great horsetail (*Equisetum telmateia* Ehrh.): active substances content and biological effects. *Experimental and Clinical Sciences Journal* 11:59–67.

19. Cooke S.S., ed. 1997. *A field guide to the common wetland plants of Western Washington & Northwest Oregon.* Seattle Audubon Society and Washington Native Plant Society. Seattle Audubon Society, Seattle, WA, USA.

20. Turner N.J., Thompson L.C., Thompson M.T., York A.Z. 1990. *Thompson ethnobotany: knowledge and usage of plants by the Thompson Indians of British Columbia.* Royal British Columbia Museum, Victoria, BC, USA.

21. Walters D.R., Keil D.J. 1996. *Vascular plant taxonomy.* Fourth edition. Kendall/Hunt Publishing, Dubuque, IO, USA.

22. www.infoflora.ch/en/flora/equisetum-palustre.html

23. Cramer L., Ernst L., Lubienski M., Papke U., Schiebel H.M., Jerz G., Beuerle T. 2015. Structural and quantitative analysis of *Equisetum* alkaloids. *Phytochemistry* 116:269–282.

24. Müller J., Puttich P.M., Beuerle T. 2020. Variation of the main alkaloid content in *Equisetum palustre* L. in the light of its ontogeny. *Toxins* 12(11):710.

25. Ilhan G., Yesilada E., Ito S. 2009. An anti-ulcerogenic flavonol diglucoside from *Equisetum palustre* L. *Journal of Ethnopharmacology* 121(3):360–365.

26. www.naturalmedicinalherbs.net/herbs/e/equisetum-palustre=marsh-horsetail.php

14 *Ericaceae* Juss. Family

Cornelia Bejenaru[1], Antonia Radu[1], George Dan Mogoşanu[2],
Ludovic Everard Bejenaru[2], Andrei Biţă[2], and Adina-Elena Segneanu[3]

[1] Department of Pharmaceutical Botany, Faculty of Pharmacy, University of Medicine and Pharmacy of Craiova, 2 Petru Rareş Street, Craiova, Dolj County, Romania
[2] Department of Pharmacognosy & Phytotherapy, Faculty of Pharmacy, University of Medicine and Pharmacy of Craiova, 2 Petru Rareş Street, Craiova, Dolj County, Romania
[3] Institute for Advanced Environmental Research-West University of Timisoara (ICAM–WUT), 4 Oituz Street, Timisoara, Romania

BOTANICAL DESCRIPTION

Shrubs or subshrubs widely distributed, especially in the temperate and cold regions of the Northern Hemisphere. These plants are mainly found in mountainous areas and have simple leaves without stipules. The leaves are entire, often persistent, arranged alternately, opposite, or in verticils. The hermaphrodite flowers, actinomorphic, of 4–5 type, gamosepalous and gamopetalous, are solitary or grouped in racemose inflorescences. The calyx and corolla are persistent, with petals brightly colored (sometimes deciduous). The corolla is cylindrical, urceolate, or campanulate, rarely dialypetalous. The stamens are double in number compared to petals, arranged on the edge of the receptacle, and have anthers that open through pores located at the tip. The flowers have a well-developed interstaminal disk and a superior gynoecium (in *Vaccinium* and *Oxycoccus*, an inferior gynoecium), with a multi-locular and multi-ovulate ovary. Placentation is central. The fruit is a berry, drupe, or capsule. The family is divided into three subfamilies: *Ericoideae*, *Arbutoideae*, and *Rhododendroideae*.[1-4]

In the spontaneous flora of Romania, there are 11 species belonging to 7 genera: *Bruckenthalia* Rchb., *Calluna* Salisb., *Rhododendron* L., *Loiseleuria* Desv., *Arctostaphylos* Adans., *Andromeda* L., and *Vaccinium* L.[2-3]

Vaccinium L.

This genus includes mountainous species that grow as shrubs (or subshrubs) with petiolate, simple leaves arranged alternately. The leaves have an ovate or elliptical blade and can be deciduous or persistent. The flowers, typically of 4 or 5 type, are usually grouped in racemes (terminal or axillary), occasionally solitary. The corolla of the flowers is white or reddish, urceolate or campanulate, and deciduous. The fruit is a berry.[1]

Vaccinium myrtillus L. (Bilberry)

Scientific name: *Vaccinium myrtillus* L.
Common name: Bilberry, Blaeberry, Wimberry, Whortleberry, European blueberry
Romanian name: Afin, Afin negru
Taxonomic classification
Kingdom: *Plantae*
Subkingdom: *Cormobionta*
Phylum: *Spermatophyta*
Subphylum: *Magnoliophytina (Angiospermae)*
Class: *Magnoliatae (Magnoliopsida, Dicotyledonatae)*
Subclass: *Dilleniidae*
Order: *Ericales*
Family: *Ericaceae*
Subfamily: *Arbutoideae*
Genus: *Vaccinium*
Species: *myrtillus* L. (Figure 14.1)[1-3]

SPREAD

The bilberry is a spontaneous species, forming extensive shrubs in mountain meadows, shrublands, and forest clearings, from the beech zone up to the subalpine zone, throughout the entire country.[1-4]

DOI: 10.1201/9781003270515-15

Figure 14.1 *Vaccinium myrtillus* L. (Bilberry).

Source: 5.

DESCRIPTION

This subshrub (or shrub) reaches a height of 15–50 cm. The stem is four-edged, highly branched, with green branches and sharp edges. The leaves are green on both sides, ovate, serrated, deciduous with a rounded base, very short-petiolate, with finely serrated margins and very prominent veining on the inferior surface. The flowers have a persistent calyx fused with the ovary, and the corolla is campanulate or urceolate, ranging from greenish to pale pink. The androecium consists of 8–10 stamens with yellowish-brown anthers, provided with spurs. The fruits are spherical berries, bluish-black in color, rimy, and contain crescent-shaped brown seeds. The flowering period is from May to June.[1-4]

Main phytoconstituents   Ref.

flavonoids (gallocatechin, astragalin, catechin, arbutin, rutin, quercetin, kaempherol, apigenin, isorhamnetin, chrysoeriol, hyperoside, luteolin, laricitrin, myricetin, syringetin, avicularine)

6-15

terpenoids (ursolic acid, amyrin, oleanolic acid, lupeol, uvaol,)

alkaloids (myrtine)

iridoids (asperuloside, vaccinoside, scandoside, geniposide, monotropein)

sterols (campesterol, cholesterol, sitosterol, stigmasterol, sitostanol, citrostadienol)

fatty acids (palmitic acid, stearic acid, oleic acid, linoleic acid, elaidic acid, arachidic acid, eicosanoic acid)

phenolic acids (chlorogenic acid, caffeic acid, gallic acid, ellagic acid, p-coumaric, ferrulic acid, gentisic acid, syringic acids)

antocynins (cyanidin, delphinidin, peonidin, malvidin, petunidin)

tannins (gallotannin, tannic acid, corilagin)

hydrocarbons (octadecane, nonadecane, eicosane)

carotenoids: (lutein, zeaxanthin, and β-carotene)

miscellaneous

Medicinal Use

Myrtilli folium is hypoglycemic,[6-7] antiseptic, astringent, antianemic, digestive disorders (biliary disorders, hepatitis, diarrhoea, dysentery, enterocolitis, nausea, vomiting, food poisoning), scurvy, respiratory inflammation (fever, coughs, tuberculosis), urinary disorders (inflammation and urinary tract infections, hematuria, kidney stones, enuresis, prostatitis, reduced uric acid), eye inflammation, edema, diabetes, abscesses.[10,14]

Biological Activity

- anti-inflammatory, sedative, antioxidant, analgesic, antimicrobial, immunomodulatory, antidiabetic, antiproliferative, hypoglycemic, hepatoprotective, cardioprotective [6-9,11]

Other Uses

- cosmetic (hair products, moisturizers, hand care products, deodorants, etc.)

- food industry (extruded products, bakeries, confectionary, food colorant, jam, ice creams, sauces, yogurts, beverages, liqueur)

- food supplements: vitamins, minerals and other antioxidants[12]

Vaccinium vitis-idaea L. (Lingonberry)

Scientific name: *Vaccinium vitis-idaea* L.
Common name: Lingonberry, Partridgeberry, Mountain cranberry, Cowberry
Romanian name: Merişor,[1-3] Coacăză de munte, Merişor de munte[4]
Taxonomic classification
Kingdom: *Plantae*
Subkingdom: *Cormobionta*
Phylum: *Spermatophyta*
Subphylum: *Magnoliophytina (Angiospermae)*
Class: *Magnoliatae (Magnoliopsida, Dicotyledonatae)*
Subclass: *Dilleniidae*
Order: *Ericales*
Family: *Ericaceae*
Subfamily: *Arbutoideae*
Genus: *Vaccinium*
Species: *vitis-idaea* L. (Figure 14.2)[1-3]

SPREAD

This species is commonly distributed from the beech zone up to the subalpine zone, through forest clearings, peat bogs, shrublands, and meadows.[1-4]

DESCRIPTION

This subshrub (or shrub) has an unridged, cylindrical, erect stem and is shorter in height (5–30 cm) compared to the bilberry. The leaves are leathery, rigid, short-petiolate, and persistent, with rare brown dots, having a shiny superior surface and a pale green inferior surface. The blade is ovate or elliptical, with entire and revolute margins. The flowers are white, tinged with red, have a faint scent, and are grouped in dense, nutant racemes. The corolla is campanulate-urceolate, with 5 acuminate laciniae turned outward. The stamens are not exerted, with hairy filaments at the base, anthers without spurs, and the ovary is inferior. The fruit is a red berry with crescent-shaped, reddish-brown seeds. The flowering period is from May to July.[1-4]

Main phytoconstituents   Ref.
flavonoids (hyperoside, quercetin, arbutin, avicularine, nicotiflorin isorhamnetin,
    kaempherol, astragalin, syringetin)   6,10-11,17-23
terpenoids (ursolic acid, amyrin, betulin, lupeol, taraxasterol, uvaol, friedelin, erythrodiol,
    oleanolic acid, α-terpineol, farnesene, linalool, menthane, caryophyllane, eudesmane, geraniol)
carbohydrates (glucose, galactose, xylose, rhamnose, mannose, fucose, arabinose)
iridoids (asperuloside, vaccinoside, scandoside, geniposide, monotropein)

Figure 14.2 *Vaccinium vitis-idaea* L. (Lingonberry).

Source: 16.

sterols (campesterol, sitosterol, stigmasterol)
fatty acids (palmitic acid)
phenolic acids (chlorogenic acid, caffeic acid, cinnamic acid, p-coumaric, ferrulic acid, gentisic acid, syringic acids)
antocynins (cyanidin, peonidin, petunidin)
tannins (cinnamtannin, proanthocyanidin)
hydrocarbons (heptacosane, pentacosane, nonacosane)
carotenoids: (lutein, zeaxanthin, and β-carotene)
organic acids (tartaric acid, citric acid, shikimic acid, fumaric acid)
miscellaneous (stilbene, resveratrol)

Medicinal use

- anticonvulsant, wound healing, eye inflammation, digestive, astringent, detoxifying, anticonvulsant, rheumatic diseases, antipyretic, respiratory disorders, urinary diseases, diuretic[11, 18-23]

Vitis idaeae folium also exhibits diuretic action (due to flavonoids and pentacyclic triterpenoid acids), anti-inflammatory properties (ursolic acid inhibits the serum complement), hemostatic effects, and antidiarrheal properties (attributed to tannins). Additionally, it has been noted for its contraceptive properties.[6]

Biological Activity

- antioxidant, antibacterial, antiviral, immunostimulant, antifungal analgesic, anti-inflammatory, antiproliferative antipyretic, diuretic, antidiabetic, neuroprotective, vasoprotective activities[10-11,18-23]

Indications: vesical catarrh, bladder irritation, cystitis, pyelonephritis, urethritis, leukorrhea, diarrhea, and hemorrhages.[6]

Other Uses

- culinary (compote, smoothie, syrup, game, liver dishes, jam)

- cosmetic industry (hair products, emollient, skin conditioning, skin protection, moisturizers, hand care products, deodorants, etc.)

- food industry (extruded products, bakeries, sauces, beverages, ice creams, yogurts, jam, liqueur, confectionary, food colorant)

- food supplements, vitamins, minerals and other antioxidants[11,18-23]

Arctostaphylos uva-ursi (L.) Spreng. (Common bearberry)

Scientific name: *Arctostaphylos uva-ursi* (L.) Spreng.
Common name: Common bearberry, Kinnikinnick
Romanian name: Strugurii ursului
Taxonomic classification
Kingdom: *Plantae*
Subkingdom: *Cormobionta*
Phylum: *Spermatophyta*
Subphylum: *Magnoliophytina (Angiospermae)*
Class: *Magnoliatae (Magnoliopsida, Dicotyledonatae)*
Subclass: *Dilleniidae*
Order: *Ericales*
Family: *Ericaceae*
Subfamily: *Arbutoideae*
Genus: *Arctostaphylos*
Species: *uva-ursi* (L.) Spreng. (Figure 14.3)[1-3]

Figure 14.3 *Arctostaphylos uva-ursi* (L.) Spreng. (Common bearberry).
Source: 24.

SPREAD

A glacial relict species in the flora of Romania, it is protected as a natural monument, sporadically growing in the subalpine zone, on dolomitic limestone and serpentine rocky outcrops, through forest clearings, dry, and sandy places.[1-4]

DESCRIPTION

A creeping shrub, strongly branched, reaching 20–100 cm in height. Young stems are pubescent. The leaves are simple, coriaceous, evergreen, firm, glabrous, with obtuse or slightly emarginate tips, and entire, non-revolute margins, very finely pubescent, arranged alternately, with an obovate, pale green blade and reticulate veins on the inferior surface. The flowers are of the 5-type, actinomorphic, and grouped in clusters of 3–12 in nutant racemes. The corolla is gamopetalous, urceolate, hairy on the interior, white with red edges, and deciduous. The gynoecium is superior. The fruit is a red berry with reniform, flattened seeds. It blooms from May to June.[1-4]

Main phytoconstituents   Ref. [25-29]

flavonoids (arbutin, quercetin, myricetin, kaempherol, hyperoside, corilagin, picein)

terpenoids (ursolic acid, betulinic acid, α-terpineol, uvaol, lupeol, geranyl acetone, linalool, selinene, thujone, camphor, menthone, borneol, nerol, cymenol, geraniol, damascenone, cedrene, caryophyllene, humulene, muurolene, cadinol, acorenone, amyrin, taraxenol, arbutin)

glycoside (ericolin)

sterols (sitosterol)

fatty acids (hexadecanoic acid, linoleic acid)

phenolic acids (ellagic acid, caffeic acid, gallic acid, p-coumaric, syringic acid, salicylic acid, ferrulic acid)

tannins (gallotannin, tannic acid, corilagin)

hydrocarbons (octadecane, nonadecane, eicosane)

aldehydes & ketones (furfural, hexanal)

other (ocimenol, vitispirane, allantoin, cyanidin and delphinidin)

Medicinal Use

The vegetable product *Uvae-ursi folium* has a wide range of properties and potential uses, including being antiseptic, diuretic, antibacterial, antilitiasic, hemostatic, antitumoral, antidiarrheal, antidiabetic,[1,6] dyslipidemia management, antimutagenic, astringent, sedative, effective for various skin diseases like wounds, ulcers, and eczema, and beneficial for urogenital diseases, myopia, retinal abnormalities, dark adaptation, increased capillary resistance, slowing the progression of lens capacity, neuroprotective, antispasmodic, antihemorrhagic, cardioprotective, and tonic.[25-29]

Biological Activity

■ antimicrobial, antibiotic, antiproliferative, antioxidant, diuretic, antiseptic[25-29]

Bearberry leaves also exhibit anti-inflammatory, antidiarrheal, hemostatic, and antilitiasic actions. They are used in cases of vesical catarrh, cystitis, cystopyelitis, urethritis, pyelonephritis, leukorrhea, diarrhea, hemorrhages, biliary lithiasis, and renal lithiasis.[1,6]

Precautions. Contraindications

Due to its high concentration, gallic tannin in *Uvae-ursi folium* can be an active or aggressive gastric irritant, by its astringent effect, leading to low tolerance (nausea, vomiting) in patients with gastric conditions. This plant product is contraindicated during pregnancy because it can act as a uterotonic, therefore affecting uterine contractions.[6]

Other Uses

■ cosmetic (antiage, sunscreen products, skin depigmentation solutions, hair tonic, hair loss control)

■ textile black dye

■ food preservation

■ beverages

- cakes
- fish or game dishes [25-29]

Rhododendron myrtifolium Schott et Kotschy (*R. kotschyi* Simonk.) (Rhododendron)

Scientific name: *Rhododendron myrtifolium* Schott et Kotschy
Common name: Rhododendron
Romanian name: Smirdar
Taxonomic classification
Kingdom: *Plantae*
Subkingdom: *Cormobionta*
Phylum: *Spermatophyta*
Subphylum: *Magnoliophytina (Angiospermae)*
Class: *Magnoliatae (Magnoliopsida, Dicotyledonatae)*
Subclass: *Dilleniidae*
Order: *Ericales*
Family: *Ericaceae*
Subfamily: *Rhododendroideae*
Genus: *Rhododendron*
Species: *myrtifolium* Schott et Kotschy (Figure 14.4)[1-3]

SPREAD

The species grows spontaneously in the mountain meadows of the alpine-subalpine zone of the Southern Carpathians, in rocky and grassy areas, on skeletal soils, forming large shrubs. It is protected as a natural monument.[1-4]

DESCRIPTION

This shrub, reaching a height of up to 50 cm, has long, weakly branched branches, with young stems covered in rusty, scaly glands. The leaves, 1–2 cm in length and 5–11 mm in width, are arranged alternately, especially towards the tips of the branches. They are tough, leathery, persistent, glabrous, elliptic, ovate-elliptic, or elongated-elliptic, with a narrowly cuneate base,

Figure 14.4 *Rhododendron myrtifolium* Schott et Kotschy (*R. kotschyi* Simonk.) (Rhododendron).
Source: 30.

with entire, revolute margins, and provided with rusty, scaly glands on the inferior surface. The flowers are of the 5-type and are grouped in clusters of 6–10 in terminal umbelliform racemes (the inflorescences resemble small bouquets). The flowers are large (1.5 cm), with a disciform calyx, zygomorphic, infundibuliform corolla with 5 laciniae, reddish-purple in color, deciduous, and 10 stamens with anthers without appendages, opening via pores. The gynoecium is superior, and the style is shorter than the ovary. The fruit is a dehiscent capsule through 5 valves and contains numerous seeds. It blooms from June to July.[1-4]

Main phytoconstituents Ref.
amino acids [31-35]
flavonoids (quercetin, myricetin, kaempherol, hyperoside, farrerol, luteolin, rutin, cathechin)
alkaloid hirsutine
terpenoids (germacrene, curzerene, and germacrone, spathulenol, pinene, camphene, sabinene, myrcene, phellandrene, ocimene, terpinene, cymene, thujone, elemene, linalool, selinene, curcumene, humulene, borneol, elemol, betulinic acid, avicularin, viridiflorol)
sterols (sitosterol, sigmasterol)
fatty acids (palmitic acid, stearic acid, arachidic acid, behenic acid, eicosanoic acid, heptadecanoic acid)
phenolic acids (caffeic acid, quinic acid, gallic acid, p-coumaric, syringic acid, salicylic acid, gallic acid, vanillic acid)
tannins
hydrocarbons (tetracosane, pentacosane)
anthocyans (azaleatin, peonidin cyanidin)

Medicinal Use

■ respiratory diseases (expectorant, cough, lung), hematuria, rheumatic disorders [31-35]

The flowers are used as a diaphoretic and expectorant when prepared as an infusion. The leaves have narcotic and aphrodisiac properties.

This species is seldom used in therapy due to its high toxicity, which can lead to progressive paralysis, including respiratory failure.

Honey obtained from the nectar of rhododendron flowers is also toxic.[1]

Biological Activity

■ cytotoxic, anti-inflammatory, antibacterial, analgesic, antiviral, hepatoprotective effects[31-35]

Other Uses

■ culinary (syrups, jam)[4,36-38]

REFERENCES

1. Bejenaru C., Mogoșanu G.D., Bejenaru L.E., Popescu H. 2020. *Botanică farmaceutică. Cormobionta*, ediția a III-a. Ed. Sitech, Craiova.

2. Ciocârlan V. *Flora ilustrată a României. Pteridophyta et Spermatophyta*. Ediția a 3-a revizuită și adăugită. 2009. Ed. Ceres, București.

3. Sârbu I., Ștefan N., Oprea A. 2013. *Plante vasculare din România, Determinator ilustrat de teren*. Ed. Victor B Victor, București.

4. Săvulescu T. (ed). 1960. *Flora RPR*, vol. VII. Ed. Academiei RPR, București.

5. https://dryades.units.it/euganei/index.php?procedure=taxon_page&id=3740&num=3889

6. Bejenaru L.E., Mogoșanu G.D., Bejenaru C., Popescu H. 2015. *Farmacognozie-Fitoterapie*. Vol. I, ediția a III-a. Ed. Sitech, Craiova.

7. Bojor O. 2003. *Ghidul plantelor medicinale şi aromatice de la A la Z*. Fditura FIAT LUX, Bucureşti.

8. Chu W.K., Cheung S.C.M., Lau R.A.W., Benzie I.F.F. 2011. *Bilberry (Vaccinium myrtillus L.) in herbal medicine: biomolecular and clinical aspects*. 2nd edition, CRC Press/Taylor & Francis, Boca Raton, FL, USA.

9. Elkiran O., Avşar C. 2020. Chemical composition and biological activities of the essential oil from the leaves of *Vaccinium myrtillus* L. *Bangladesh Botanical Society* 49(1):91–96.

10. Ștefănescu B.E., Călinoiu L.F., Ranga F., Fetea F., Mocan A., Vodnar D.C., Crişan G. 2020. Chemical composition and biological activities of the Nord-West Romanian wild bilberry (*Vaccinium myrtillus* L.) and Lingonberry (*Vaccinium vitis-idaea* L.) leaves. *Antioxidants* 9(6):495.

11. Tundis R., Tenuta M.C., Loizzo M.R., Bonesi M., Finetti F., Trabalzini L., Deguin B. 2021. *Vaccinium* species (*Ericaceae*): from chemical composition to bio-functional activities. *Applied Sciences* 11:5655.

12. Pires T.C.S.P., Caleja C., Santos-Buelga C., Barros L., Ferreira I.C.F.R. 2020. *Vaccinium myrtillus* L. fruits as a novel source of phenolic compounds with health benefits and industrial applications – a review. *Current Pharmaceutical Design* 26(16):1917–1928.

13. Szakiel A, Pączkowski C, Huttunen S., 2012. Triterpenoid content of berries and leaves of bilberry *Vaccinium myrtillus* from Finland and Poland. *Journal of Agricultural and Food Chemistry* 60(48):11839–11849.

14. Bunea A., Rugină D., Pintea A., Andrei S., Bunea C., Pop, R., Bele C. 2012. Carotenoid and fatty acid profiles of bilberries and cultivated blueberries from Romania. *Chemical Papers* 66(10):935–939.

15. Güder A., Gür M., Engin M.S. 2015. Antidiabetic and antioxidant properties of bilberry (*Vaccinium myrtillus* Linn.) fruit and their chemical composition. *Journal of Agricultural Science and Technology* 17(2):387–400.

16. https://dryades.units.it/dolomitifriulane/index.php?procedure=taxon_page&id=3736&num=3890

17. Roman I., Puică C., Toma V.A., 2014. The effect of *Vaccinium vitis idaea* L. extract administration on some biochemical bloodparameters in alcohol intoxicated rats. *Studia Universitatis Vasile Goldis Arad, Seria Stiintele Vietii* 24(4):363–367.

18. Alexeevich Shamilo A., Nikolaevna Bubenchikova V., Valentinovich Chernikov M., et.al.,. 2020. *Vaccinium vitis-idaea* L.: Chemical contents, pharmacological activities. *Pharmaceutical Sciences* 26(4):344–362.

19. Bujor O.C., Ginies C., Popa V.I., Dufour C. 2018. Phenolic compounds and antioxidant activity of lingonberry (*Vaccinium vitis-idaea* L.) leaf, stem and fruit at different harvest periods. *Food Chemistry* 252:356–365.

20. Kryvtsova M.V., Trush K., Eftimova J., Koščová J., Spivak M.J. 2019. Antimicrobial, antioxidant and some biochemical properties of *Vaccinium vitis-idaea* L. *Mikrobiolohichnyi Zhurnal* 81(3):40–52.

21. Raudone L., Vilkickyte G., Pitkauskaite L., Raudonis R., Vainoriene R., Motiekaityte V. 2019. Antioxidant activities of *Vaccinium vitis-idaea* L. leaves within cultivars and their phenolic compounds. *Molecules* 24(5):844.

22. Bubenchikova V., Chernikov M., Pozdnyakov D., Shamilov A., Garsiya E. 2020. *Vaccinium vitis-idaea* L.: chemical contents, pharmacological Activities. *Pharmaceutical Sciences* 26(4):344–362.

23. Ho K.Y., Huang J.S., Tsai C.C., Lin T.C., Hsu Y.F., Lin C.C. 1999. Antioxidant activity of tannin components from *Vaccinium vitis-idaea* L. *Journal of Pharmacy and Pharmacology* 51(9):1075–1078.

24. https://dryades.units.it/ampezzosauris/index.php?procedure=taxon_page&id=3731&num=3886

25. Kurkin V.A., Ryazanova T.K., Daeva E.D. et al. 2018. Constituents of *Arctostaphylos uva-ursi* leaves. *Chemistry of Natural Compounds* 54(2):278–280.

26. Radulović N., Blagojević P., Palić R. 2010. Comparative study of the leaf volatiles of *Arctostaphylos uva-ursi* (L.) Spreng. and *Vaccinium vitis-idaea* L. (*Ericaceae*). *Molecules* 15(9):6168–6185.

27. Sugier P., Sęczyk Ł., Sugier D., Krawczyk R., Wójcik M., Czarnecka J., Okoń S., Plak A. 2021. Chemical characteristics and antioxidant activity of *Arctostaphylos uva-ursi* L. Spreng. at the southern border of the geographical range of the species in Europe. *Molecules* 26(24):7692.

28. Chaika N., Koshovyi O., Ain R., Kireyev I., Zupanets, A., Odyntsova V. 2020. Phytochemical profile and pharmacological activity of the dry extract from *Arctostaphylos uva-ursi* leaves modified with phenylalanine. *ScienceRise: Pharmaceutical Science* 6(28):74–84.

29. Aronson J.K. 2016. *Meyler's Side Effects of Drugs: The International Encyclopedia of Adverse Drug Reactions and Interactions*. Sixteenth Ed., Seven Volume, ISBN 978-0-444-53716.

30. https://ukrbin.com/show_image.php?imageid=47317&big=1

31. Schepetkin I.A., Özek G., Özek T., Kirpotina L.N., Khlebnikov A.I., Quinn M.T. 2021. Chemical composition and immunomodulatory activity of essential oils from *Rhododendron albiflorum*. *Molecules* 26(12):3652.

32. Popescu R., Kopp B. 2013. The genus *Rhododendron*: An ethnopharmacological and toxicological review. *Journal of Ethnopharmacology* 147(1):42–62.

33. Tămaş M., Balica G., Ştefănescu C. 2020. Research on some plant species containing essential oils performed at University of Medicine Pharmacy "Iuliu Haţieganu" Cluj-Napoca. *Contributii Botanice* 55:141–151.

34. Carballeira N.M., Cartagena M., Tasdemir D. 2008. Fatty acid composition of Turkish *Rhododendron* species. *Journal of the American Oil Chemists' Society* 85(7):605–611.

35. Prokopiv A., Tkachenko H., Honcharenko V., Kurhaluk N., Nachychko V., Sosnovsky Y., Osadowski Z. 2019. Effect of *Rhododendron myrtifolium* Schott & Kotschy leaf extracts on HCl-induced hemolysis in human erythrocytesm. *Agr.bio.div. Impr. Nut., Health Life Qual. Agrobiodiversity for Improving Nutrition, Health and Life Quality* 3:394–403.

36. www.montaniarzi.ro/bujorul-de-munte-minunea-care-infloreste-in-iunie/

37. http://temperate.theferns.info/plant/Rhododendron+myrtifolium

38. www.plantpedia.ro/bujor-de-munte/

15 *Euphorbiaceae* Juss. Family

Cornelia Bejenaru[1], Antonia Radu[1], George Dan Mogoşanu[2],
Ludovic Everard Bejenaru[2], Andrei Biţă[2], and Adina-Elena Segneanu[3]

[1] Department of Pharmaceutical Botany, Faculty of Pharmacy, University of Medicine and Pharmacy of Craiova, 2 Petru Rareş Street, Craiova, Dolj County, Romania
[2] Department of Pharmacognosy & Phytotherapy, Faculty of Pharmacy, University of Medicine and Pharmacy of Craiova, 2 Petru Rareş Street, Craiova, Dolj County, Romania
[3] Institute for Advanced Environmental Research–West University of Timisoara (ICAM–WUT), 4 Oituz Street, Timisoara, Romania

BOTANICAL DESCRIPTION

Family with over 10,000 species, classified into about 300 genera, of which in the Romanian flora, only 38 spontaneous species belong to three genera: *Mercurialis*, *Chamaesyce*, and *Euphorbia*. The species of this family are distributed across all continents, with the majority being present in regions with a warm climate. These plants are herbaceous, annual, biennial, or perennial; in other regions, they can be woody, and at times, cactiform. The stems are thick, round, or angled, containing milky latex that is irritant, blistering, and toxic. The leaves are simple or compound, with entire or dentate margins, arranged alternately or opposite, reduced to scales or transformed into spines in plants with cladodes or phylloclades. The flowers are unisexual, mostly monoecious, rarely dioecious, actinomorphic, grouped in various cymose inflorescences, sometimes specific (cyathium). The floral envelope may be absent or simple, rarely forming a perianth. Male flowers have from 1 to numerous free stamens (rarely filaments are fused or branched), while female flowers have a superior gynoecium composed of 3 carpels with a trilocular ovary. The fruit is a capsule that opens into 3 valves, separating from the central axis into 2–6 monospermous nuts (more rarely a berry or drupe). The seeds, smooth or reticulate, often with a caruncle, are rich in fatty oil. The embryo is straight and has broad, flat cotyledons.[1-3]

The genus *Mercurialis* is represented in the spontaneous flora of our country by the following species: *Mercurialis annua* L., *Mercurialis perennis* L., *Mercurialis ovata* Sternb. et Hoppe, *Mercurialis paxii* (Graebn.) Rauschert.[2,4]

From the *Chamaesyce* genus, the following species are found in the spontaneous flora of Romania: *Chamaesyce nutans* (Lag.) Small (*Euphorbia nutans* Lag.), *Chamaesyce peplis* (L.) Prokh. (*Euphorbia peplus* L.), *Chamaesyce humifusa* (Willd. ex Schlecht.) Prokh. (*Euphorbia humifusa* Willd. ex Schlecht.), *Chamaesyce canescens* (L.) Prokh. (*C. vulgaris* Prokh. *Euphorbia chamaesyce* L.), *Chamaesyce maculata* (L.) Small (*Euphorbia maculata* L.; *E. supina* Rafin.).[2,4]

From the *Euphorbia* genus, the following species are present in the spontaneous flora of Romania: *Euphorbia dentata* Michx., *Euphorbia helioscopia* L., *Euphorbia stricta* L. (*E. serrulata* Thuill.), *Euphorbia platyphyllos* L., *Euphorbia villosa* Waldst. et Kit. ex Willd., *Euphorbia palustris* L., *Euphorbia carniolica* Jacq., *Euphorbia lingulata* Heuffel, *Euphorbia jacquinii* Fenzl. ex Boiss., *Euphorbia epithymoides* L. (*E. polychroma* A. Kern.), *Euphorbia carpatica* Woloszczak, *Euphorbia angulata* Jacq., *Euphorbia dulcis* L., *Euphorbia myrsinites* L., *Euphorbia exigua* L., *Euphorbia peplus* L., *Euphorbia falcata* L., *Euphorbia segetalis* L., *Euphorbia taurinensis* All. (*E. graeca* Boiss. et Sprun.), *Euphorbia amygdaloides* L., *Euphorbia paralias* L., *Euphorbia salicifolia* Host, *Euphorbia agraria* M. Bieb., *Euphorbia seguieriana* Necker (*E. gerardiana* Jacq.), *Euphorbia glareosa* Pall. ex M. Bieb., *Euphorbia lucida* Waldst. et Kit., *Euphorbia cyparissias* L., *Euphorbia esula* L., and *Euphorbia virgata* Waldst. et Kit.[2,4]

TOXICITY

Species from the *Euphorbia* genus are toxic due to the latex they contain, which can cause vomiting, purgation, and irritation. Species like *Euphorbia peplus* L. and *Euphorbia esula* L. have been traditionally used for cauterizing calluses and warts. Upon local application, the latex can produce blisters, while internal consumption can lead to inflammation, and accidental contact with the eyes can result in blindness. If accidentally consumed by animals along with hay, these species can cause hematuria and diarrhea.[3]

Euphorbia cyparissias L. (Cypress spurge)

Scientific name: *Euphorbia cyparissias* L.
Common name: Cypress spurge

DOI: 10.1201/9781003270515-16

Figure 15.1 *Euphorbia cyparissias* L. (Cypress spurge).

Source: 6.

Romanian name: Alfior, Ariu[5]
Taxonomic classification
Kingdom: *Plantae*
Subkingdom: *Cormobionta*
Phylum: *Spermatophyta*
Subphylum: *Magnoliophytina (Angiospermae)*
Class: *Magnoliatae (Magnoliopsida, Dicotyledonatae)*
Subclass: *Rosidae*
Order: *Euphorbiales*
Family: *Euphorbiaceae*
Genus: *Euphorbia*
Species: *cyparissias* L. (Figure 15.1)[1-2,4]

SPREAD
Frequently found in ruderalized areas, from plains to mountains, growing as a weed in grassy, dry zones, pastures, cultivated areas, rocky slopes, clearings, and gravelly places.[1,3-4]

DESCRIPTION
A perennial herbaceous plant with an oblique, lignified, branching rhizome, and creeping stolons. The stems are densely packed, glabrous, yellowish-green, reddish at the base, with sterile branches in the superior part, reaching up to 50 cm in height. The narrow-linear, soft leaves, not coriaceous (up to 3 mm in width at most), spiral around the stems, being patent and imbriate towards the superior part, with obtuse or slightly notched tips. The involucral leaflets resemble the leaves but those of the involucels are rhomboid or subtriangular-ovate, entire, shortly acuminate, yellow-green, turning reddish later on. The flowers are organized in cyathia, 3 mm in length, with yellowish or brownish bicornate glands, shorter than the width of the gland. The fruits are glabrous

170

capsules, deeply trisulcate, slightly scabrous-dotted, measuring 3 mm in length. The seeds are smooth, round-ovate, gray, 2 mm long, with a reniform caruncle. It flowers from April to July.[1,3]

Main phytoconstituents	Ref.
	7-11

coumarins (esculetin, euphorbetin, isoeuphorbetin)

alkaloids

flavonoids (quercetin, kaempherol, myricetin, vitexin, naringerin, steppogenin, robidanol rhamnetin, eriodictyol)

terpenoids (cyparissins A-B, α-amyrin, euphol, α-amyrinone, β -amyrin, betulin, β-fernenone, betulinic acid, corollatadiol, β-amyrinone, lupenone, glutinol, glutinone, cycloartenol, ψ-taraxasterol, sterone, ingol, euphorbinol, friedelin, cyparissins A-B, germanicol, jatrophane, gypsogenic acid)

saponins

sterols (β-sitosterol, stigmasterol, campesterol, cholesterol)

fatty acids (ricinoleic acid)

phenolic acids (chlorogenic acid, gallic acid, vanillic acid, hydroxybenzoic acid)

tannins (ellagic acid)

anthocyanins (cyanidin, delphinidin, pelargonidin)

alcohols (octacosanol, n-hexacosanol)

phenanthrenes (micrandrol)

misceleous (glut-5-(6)-3n-3-one, glut-5-(6)-en-3-ol), quinic acid, shikimic acid, acalyphamide

Medicinal Use

A highly toxic plant, causing dermal lesions, gastrointestinal inflammation, and severe purgation. Contact of the eyes with the latex can lead to blindness. In the past, the plant was empirically used as an emetic, anthelmintic, and antieczematous agent. Externally, fresh latex was applied to remove warts (verrucas, papillomas).[1]

Biological Activity

cytotoxic agent (ovarian cancer)[11]
antioxidant activity[12]
antimicrobial effect[13]
natural inhibitor of the digestive enzymes, useful in metabolic syndrome treatment[14]

Other Uses

Euphorbia cyparissias has also been documented for its historical application by mothers as a stimulant for lactation and as a means to induce labor, as well as for its use in aiding the expulsion of the placenta. Additional applications have been noted in social practices, environmental purposes, ornamental horticulture, fuelwood and timber production, woodcraft, hedge/fence establishment, and, despite its high toxicity, even as a potential source of nutrients in food.[15]

Euphorbia amygdaloides L. (Wood spurge)

Scientific name: *Euphorbia amygdaloides* L.
Common name: Wood spurge
Romanian name: Alior,[3] Alior de pădure[5]
Taxonomic classification
Kingdom: *Plantae*
Subkingdom: *Cormobionta*
Phylum: *Spermatophyta*
Subphylum: *Magnoliophytina (Angiospermae)*
Class: *Magnoliatae (Magnoliopsida, Dicotyledonatae)*
Subclass: *Rosidae*
Order: *Euphorbiales*
Family: *Euphorbiaceae*
Genus: *Euphorbia*
Species: *amygdaloides* L. (Figure 15.2)[1-2,4]

Figure 15.2 *Euphorbia amygdaloides* L. (Wood spurge).

Source: 16.

SPREAD

A common species throughout the country, often found from the plains to the spruce forest zone, in forests, clearings, and thickets.[2,4]

DESCRIPTION

A perennial herbaceous plant with branched and lignified rhizomes. The stems reach a height of 30–60 cm; the sterile stems are numerous, lignified, forming tufts, and the terminal rosette persists through the winter, while the fertile stems are laxly foliate, glabrous or brown-hairy, arising from the previous year's stems. Sterile stems bear large, obovate leaves, acute or obtuse, narrowed at the base, leathery in consistency, reddish, sometimes. Leaves on fertile stems are smaller, obovate, soft, pubescent on the inferior side, yellowish. Involucral leaflets are obovate, obtuse, weakly emarginate or slightly acute, while involucel leaflets are semiorbicular, fused in pairs, yellowish in color. The flowers are organized in cyathia, 3–4 mm in length, with bicornuate glands, yellow or purplish. The fruits are capsules, 4 mm in length, deeply trisulcate, glabrous, and rough. The seeds, 2.5 mm in length, are smooth, gray, and furnished with a lobed caruncle.[3]

Main phytoconstituents   Ref.
coumarins (esculetin, euphorbetin, isoeuphorbetin)   8,10,15
alkaloids
resins (podophyllin)
flavonoids (quercetin, kaempherol, rhamnetin, myricetin, vitexin, naringerin, steppogenin, robidanol eriodictyol)
terpenoids (amygdaloidin A-L, euphocharacin, camphor, amyrin, euphol, betulin, β-fernenone, betulinic acid, corollatadiol, β-amyrinone, lupenone, glutinol, glutinone, cycloartenol, ψ-taraxasterol, sterone, ingol, euphorbinol, friedelin, A-B, germanicol)
saponins

172

sterols (β-sitosterol, stigmasterol, campesterol, cholesterol)
fatty acids (ricinoleic acid)
phenolic acids (chlorogenic acid, gallic acid, vanillic acid, hydroxybenzoic acid)
tannins (ellagic acid)
anthocyanins (cyanidin, delphinidin, pelargonidin)
alcohols (octacosanol, n-hexacosanol)
phenanthrenes (micrandrol)
misceleous (quinic acid, shikimic acid, acalyphamide)

Medicinal Use

The species was employed in traditional medicine to address a variety of ailments, including respiratory infections, skin irritations, digestive issues, inflammatory infections, discomfort, microbial ailments, snake or scorpion bites, pregnancy-related concerns, as well as sensory disorders.[15]

Biological Activity

antibacterial activity[17-19]
antioxidant activity[20-21]
antiviral activity[22]

Other Uses

In the past, the fresh latex was used to hasten the process of coagulating milk, and to clean water for domestic uses.[15]

REFERENCES

1. Bejenaru C., Mogoșanu G.D., Bejenaru L.E., Popescu H. 2020. *Botanică farmaceutică. Cormobionta*, ediția a III-a. Ed. Sitech, Craiova.

2. Ciocârlan V. 2009. *Flora ilustrată a României. Pteridophyta et Spermatophyta*. Ediția a 3-a revizuită și adăugită. Ed. Ceres, București.

3. Săvulescu T. (ed). 1953. *Flora RPR*, vol. II. Ed. Academiei RPR, București.

4. Sârbu I., Ștefan N., Oprea A. 2013. *Plante vasculare din România, Determinator ilustrat de teren.* Ed. Victor B Victor, București.

5. Drăgulescu C. 2018. *Dicționar de Fitonime Românești*. Ediția a 5-a completată, Editura Universității "Lucian Blaga" Sibiu.

6. https://dryades.units.it/rosandra_en/index.php?procedure=taxon_page&id= 2965&num=1543

7. Hemmers H., Gülz P.G. 1989. Tetra- and pentacyclic triterpenoids from epicuticular wax of *Euphorbia cyparissias. Zeitschrift für Naturforschung A* 44c:563–567.

8. Benjamaa R., Moujanni A., Kaushik N., Choi E.H., Essamadi A.K., Kaushik N.K. 2022. *Euphorbia* species latex: A comprehensive review on phytochemistry and biological activities. *Frontiers in Plant Science* 13:1008881.

9. De P.T., Urones J.G., Marcos I.S., Basabe P., Cuadrado M.S., Fernandez Moro R. 1987. Triterpenes from *Euphorbia broteri. Phytochemistry* 26(6):1767–1776.

10. Rizk F.M. 1986. The chemical constituents and economic plants of the *Euphorbiaceae. Botanical Journal of the Linnean Society* 94(1–2):293–326.

11. Lanzotti V., Barile E., Scambia G., Ferlini C. 2015. Cyparissins A and B, jatrophane diterpenes from *Euphorbia cyparissias* as Pgp inhibitors and cytotoxic agents against ovarian cancer cell lines. *Fitoterapia* 104:75–79.

12. Stankovic M., Zlatić N. 2014. Antioxidant activity and concentration of secondary metabolites in the plant parts of *Euphorbia cyparissias* L. *Kragujevac Journal of Science* 36:121–128.

13. Semnani S.N., Rahnema M., Alizadeh H., Ghasempour H. 2013. Evaluation of antimicrobial effects of *Euphorbia cyparissias* extracts on intramacrophages *Salmonella typhi*. *Journal of Biologically Active Products from Nature* 3(1):64–71.

14. Alonazi M., Horchani H., Alwhibi M., Ben Bacha A. 2020. Cytotoxic, antioxidant, and metabolic enzyme inhibitory activities of *Euphorbia cyparissias* extracts. *Oxidative Medicine and Cellular Longevity* vol. 2020, Article ID 9835167, 10 pages.

15. Kemboi D., Peter X., Langat M., Tembu J. 2020. A review of the ethnomedicinal uses, biological activities, and triterpenoids of *Euphorbia* species. *Molecules* 25(17):4019.

16. https://dryades.units.it/floritaly/index.php?procedure=taxon_page&tipo=all&id=2968

17. Papp N. 2004. Antimicrobial activity of extracts of five Hungarian *Euphorbia* species and some plant metabolits. *Acta Botanica Hungarica* 46(3–4):363–371.

18. Barile E., Corea G., Lanzotti V. 2008. Diterpenes from *Euphorbia* as potential leads for drug design. *Natural Product Communications* 3(6):1003–1020.

19. Cicek S., Gungor A.A., Adiguzel A., Nadaroglu H. 2015. Biochemical evaluation and green synthesis of nano silver using peroxidase from *Euphorbia* (*Euphorbia amygdaloides*) and its antibacterial activity. *Journal of Chemistry* vol. 2015, Article ID 486948, 7 pages.

20. Nadaroglu H., Onem H., Alayli Gungor A. 2017. Green synthesis of Ce_2O_3 NPs and determination of its antioxidant activity. *IET Nanobiotechnology* 11:411–419.

21. Hosseinzadeh R., Mohadjerani M., Pakzad K. 2013. Antioxidant properties and metal chelating activity of *Euphorbia amygdaloides* roots. *Iranian Seminar of Organic Chemistry*. SID. Available from: https://sid.ir/paper/938399/en

22. Nothias-Scaglia L.F., Retailleau P., Paolini J., Pannecouque C., Neyts J., Dumontet V., Roussi F., Leyssen P., Costa J., Litaudon M. 2014. Jatrophane diterpenes as inhibitors of chikungunya virus replication: structure-activity relationship and discovery of a potent lead. *Journal of Natural Products* 77(6):1505–1512.

16 *Fabaceae* Lindl. (*Leguminosae* Adans.) Family

Cornelia Bejenaru[1], Antonia Radu[1], George Dan Mogoşanu[2],
Ludovic Everard Bejenaru[2], Andrei Biţă[2], and Adina-Elena Segneanu[3]
[1] Department of Pharmaceutical Botany, Faculty of Pharmacy, University of Medicine
and Pharmacy of Craiova, 2 Petru Rareş Street, Craiova, Dolj County, Romania
[2] Department of Pharmacognosy & Phytotherapy, Faculty of Pharmacy, University of
Medicine and Pharmacy of Craiova, 2 Petru Rareş Street, Craiova, Dolj County, Romania
[3] Institute for Advanced Environmental Research–West University of
Timisoara (ICAM–WUT), 4 Oituz Street, Timisoara, Romania

BOTANICAL DESCRIPTION

The plants of this family are herbaceous annuals, biennials, or perennials (rarely woody) and have nodules with nitrogen-fixing bacteria at the root level. The stem is erect, rarely voluble. The leaves are stipulate, with stipules sometimes large and serving the function of leaves, which have been transformed into tendrils. Stipules can be foliaceous, persistent or caducous, sometimes absent, or transformed into thorns. The leaves are arranged alternately, sessile or petiolate, rarely simple, typically compound (pinnate or palmate). The flowers are hermaphroditic, zygomorphic, pentamerous, solitary, or in axillary or terminal racemose inflorescences. The floral envelope consists of sepals and petals, which can be free or partially united. The corolla has the superior petal (vexillum or stindard) larger and covering the two lateral petals (wings or alae), which in turn cover the two equal and fused inferior petals (forming the keel). The androecium consists of 10 stamens: most commonly, 9 are united by their filaments, and 1 is free (diadelphous androecium), less commonly, all 10 stamens are united in a single bundle by their filaments (monadelphous androecium), and very rarely (in *Sophora*), all 10 stamens are free (dialistemonous androecium). The gynoecium consists of a single carpel with a superior, unilocular ovary containing campylotropous ovules arranged in two rows along the suture line of the carpel. The fruit is a pod or loment, and the seeds are exalbuminous, containing a fleshy, thick embryo with two cotyledons.[1-4]

The *Fabaceae* family includes over 10,000 species primarily distributed in temperate and cold regions, with fewer species in tropical and subtropical areas. In Romania, this family comprises 30 genera with 108 species.[5]

The following species were selected and presented as the most representative medicinal plants from this family.

Laburnum anagyroides Medik. (Golden chain tree)

Scientific name: *Laburnum anagyroides* Medik.
Common name: Golden chain tree
Romanian name: Salcâm galben
Taxonomic classification
Kingdom: *Plantae*
Subkingdom: *Cormobionta*
Phylum: *Spermatophyta*
Subphylum: *Magnoliophytina (Angiospermae)*
Class: *Magnoliatae (Magnoliopsida, Dicotyledonatae)*
Subclass: *Rosidae*
Order: *Fabales (Leguminosales)*
Family: *Fabaceae (Leguminosae)*
Genus: *Laburnum*
Species: *anagyroides* Medik. (Figure 16.1)[1-3]

SPREAD

The golden chain tree is found along the edges of forests or in shrubby areas in the sessile oak and beech forest zone.[1-3]

DESCRIPTION

It is a woody plant that grows in the form of a tree, occasionally a shrub, with a height of up to 7 meters, spontaneously found in Mehedinţi County. The shoots are green-gray, vigorous, and

DOI: 10.1201/9781003270515-17

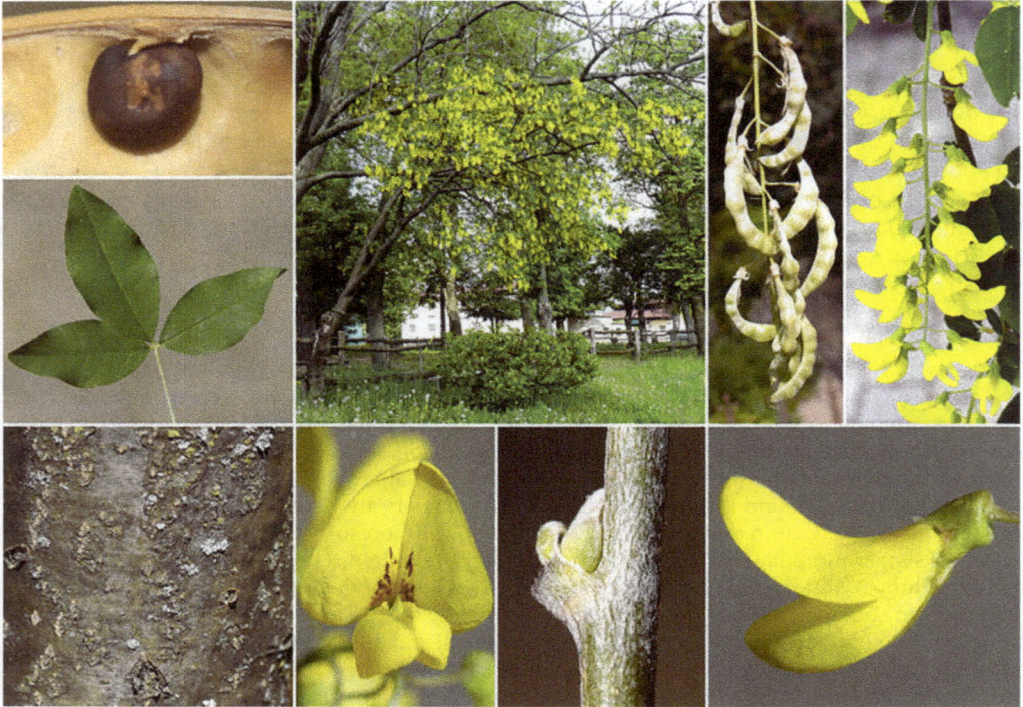

Figure 16.1 *Laburnum anagyroides* Medik. (Golden chain tree).

Source: 6.

adpressed pubescent. The leaves are arranged alternately, have long, hairy petioles and consist of 3 folioles (trifoliate leaves), lacking stipules. The folioles are elliptical to elongated elliptical, nearly sessile, obtuse, and shortly mucronate, with a gray-green inferior side, and are sericeous-pubescent when young. The pleasantly fragrant flowers are melliferous, yellow, about 2 cm long, forming large, pendulous, lax racemes, up to 35 cm in length. The pods are adpressed-sericeous pubescent, with a non-winged superior margin, containing 3–7 seeds, without a strophiole. It blooms from May to July.[1-4]

Main Phytoconstituents

Laburnum anagyroides contains quinolizidine or norlupin alkaloids, approximately 1.5–3% in the seeds, 0.3% in the leaves, and 0.2% in the flowers, represented by cytisine and methylcytisine.[7]

Main phytoconstituents	Ref.
amino acids & peptides (lectins, phenylalanine, lysine proline, methionine, glutamic acid aspartic acid, glutamic acid)	8-9

alkaloids (cystine, N-methylcytisine)

terpenoids (lupeol)

flavonoids (quercetin, kaempferol, isorhamnetin, rutin, genistein, derrubone, tricin)

saponin

tannins (catechin, epicatechin)

sterols (β-sitosterol)

fatty acids (linoleic acid, oleic acid, stearic acid, palmitic acid, caprylic acid, myristic acid)

phenolic acids (gallic acid, ferulic acid, vanillic acid, caffeic acid, benzoic acid, p-hydroxy acid, sinapinic acid, syringic acid)

antrachinones (chrysophanol, emodin, fistulic acid, sennoside A, sennoside B, physcion, rhein)

hydrocarbons (hentriacontanone, 5-nonatetracontanone, triacontane)

miscellaneous (β-carotene)

Medicinal Use

The seeds of golden chain tree (*Cytisi semen*) are used as a respiratory analeptic in asphyxia, shock, narcotic intoxication, and for combating tobacco addiction.[1]

Biological Activity

Around 400 years ago, empirically, golden chain tree seeds were used for their emmenagogue effect, as a respiratory stimulant, anti-inflammatory agent, being active in neuralgia and in conditions resembling asthma.

Citizine is a respiratory analeptic with effects similar to lobeline, recommended for asphyxia, shock, narcotic poisoning, and combating tobacco addiction (smoking cessation therapy).

In phytotherapy, golden chain tree leaves are used for their choleretic and cholagogue effects, but they should be administered with caution due to the toxicity of citizine and methylcytizine.[7]

Other Uses

- Ornamental in parks and gardens[4]

Melilotus officinalis (L.) Lam. (Yellow sweet clover)

Scientific name: *Melilotus officinalis* (L.) Lam.
Common name: Yellow sweet clover
Romanian name: Sulfină galbenă
Taxonomic classification
Kingdom: *Plantae*
Subkingdom: *Cormobionta*
Phylum: *Spermatophyta*
Subphylum: *Magnoliophytina (Angiospermae)*
Class: *Magnoliatae (Magnoliopsida, Dicotyledonatae)*
Subclass: *Rosidae*
Order: *Fabales (Leguminosales)*
Family: *Fabaceae (Leguminosae)*
Genus: *Melilotus*
Species: *officinalis* (L.) Lam. (Figure 16.2)[1-3]

SPREAD

Yellow sweet clover is frequently encountered in ruderal places or meadows, pastures, crop fields, along the roadside, from the lowlands to the subalpine zone.[1-4]

DESCRIPTION

A biennial herbaceous species, melliferous, 40–100 cm tall. The stem is erect or ascending, branched, with prismatic branches, and the superior ones are sparsely pubescent. The leaves are trifoliate, with serrated-dentate margins of the folioles, obtuse apex, and cuneate base, sparsely pubescent on the inferior surface. The stipules are entire, narrow, and subulate. The flowers are small, nutant, fragrant, arranged in axillary racemes, lax, with long peduncles. The glabrous calyx has 5 triangular teeth and is persistent on the pod, while the yellow corolla has the vexillum and wings equal, longer than the keel. The ovary is pedicellate, pubescent, and contains 5–8 ovules, rarely more. The pod, 5 mm long, is indehiscent, globular, and glabrous, transversely veined-reticulate, with 1–2 ovoid seeds, yellow-green in color. The flowering period is from June to September.[1-4]

Main phytoconstituents
amino acids
flavonoids (kaempferol, quercetin, luteolin, rutin)
coumarins: (hydrocoumarin)
terpenoids (eudesmol, lobulol, farnesol, camphor, terpinene-4-ol, aromadendrene, farnesane, isophytol, spathulenol, viriidiflorol, bisabolol, lupanone, lupeol, betulinic acid, oleanolic acid)
sterols
fatty acids (octadecatrienoic acid, palmitic acid)
phenolic acids (caffeic acid, alicylic acid, *p*-coumaric acid)

Ref.
7,10

177

Figure 16.2 *Melilotus officinalis* (L.) Lam. (Yellow sweet clover).

Source: 8.

Saponins (oleanene glucuronide, soyasaponin, dehydrosoyasaponin, acetyl-soyasaponin, oyasapogenols B and E)
carbohydrates
tannins
hydrocarbons (eicosane)
organic acids (fumaric acid)
esters
miscellaneous (ionone)

Medicinal Use

emollient, gout, gastritis, insomnia, hemorrhoids, carminative, gastrointestinal disorders, hepatoregenerative, colds, sedative, antispasmodics, hypertension, urinary disorders (cystitis, pyelitis, pyelonephritis, urethritis), oropharyngeal inflammation, eye infections, expectorant, bruising, boils, erysipelas, rheumatism, arthritis, neuralgia[1,7,11-12]

Biological Activity

■ antioxidant, antibacterial, antifungal, antimicrobial, anti-inflammatory, antispasmodic, hepatoprotective, antidiabetic, anticoagulant, hepatoprotective, neuroprotective, sedative, immunostimulatory, anxiolytic, antitumor, lymphokinetic, vasodilator, venotonic[1,7,10-11]

Other Uses

■ melliferous

■ aromatherapy

■ culinary: salad (young leaves, flowers), side dishes

178

- flavouring: sweet smell of vanilla and freshly cut hay (young leaves, flowers, seeds) pastry products, drinks

- tea

- repellent (moths)

- cosmetics (moisturizing creams, hand creams, tonic lotion, bath salt, make-up removers, hair conditioner, sunburn soothing lotion, etc.)

- eroded, degraded lands

- fixing sandy ground[10,13]

Sarothamnus scoparius (L.) W. D. J. Koch (*Cytisus scoparius* (L.) Link) (Scotch broom)

Scientific name: *Sarothamnus scoparius* (L.) W. D. J. Koch
Common name: Common broom, Scotch broom
Romanian name: Drob, Mături
Taxonomic classification
Kingdom: *Plantae*
Subkingdom: *Cormobionta*
Phylum: *Spermatophyta*
Subphylum: *Magnoliophytina (Angiospermae)*
Class: *Magnoliatae (Magnoliopsida, Dicotyledonatae)*
Subclass: *Rosidae*
Order: *Fabales (Leguminosales)*
Family: *Fabaceae (Leguminosae)*
Genus: *Sarothamnus*
Species: *scoparius* (L.) W. D. J. Koch (Figure 16.3)[1-3]

Figure 16.3 *Sarothamnus scoparius* (L.) W. D. J. Koch (*Cytisus scoparius* (L.) Link) (Scotch broom).
Source: 14.

SPREAD

It sporadically grows from the plains to the mountainous regions, in clearings and forest edges.[1]

Description

It is a 2-meter tall shrub, erect, with a principal taproot and lateral roots with digitate-lobate nodules. The stem is densely branched, with straight, thin, green branches with 5 edges. The leaves are trifoliate (those at the tip of the stem or flowering branches are simple) and early deciduous, arranged alternately, with obovate leaflets. The flowers are yellow, about 2(3) cm long, arranged either singly or in pairs in the leaf axils, and the androecium is monadelphous. The fruit is a flattened, dehiscent pod, black in color with dense vilous hairs on the margins, containing numerous round, shiny, dark brown to black seeds with a caruncle. The flowering period is from May to June.[1-4]

Main Phytoconstituents

Sarothamni herba contains: 0.5–1.5% alkaloids with a quinolizidine nucleus (sparteine, α-izosparteine or genistein alkaloid, cytisine, lupanine) and derivatives (ammodendrine); 0.2–0.6% flavonoids; isoflavones; coumarins; polyphenolcarboxylic acids (caffeic acid); carotenoids; biogenic amines (tyramine, oxytyramine, dioxiphenylalanine).

The term "genistein alkaloid" for (-)-α-izosparteine is used to avoid confusion with genistein, an isoflavone identified in the species *Genista tinctoria* L., dyer's greenweed (*Fabaceae*).

Flavonoids are represented by scoparoside or scoparine, C^8-flavonic heteroside, with diuretic and cardiotonic properties, identified in the flowering branches of the plant, but especially in the flower buds. Such flavonic heterosides, called vitexins (without the $–OCH_3$ group at $C^{3'}$), are also found in *Crataegus* species and exhibit cardiotonic effects with much lower toxicity.

Under the influence of phenoloxidases, the amines present in the flower buds are transformed into melanin, which is why the vegetable product browns during preservation or slow drying.[7]

Main phytoconstituents	Ref.
amino acids	7,15-16

alkaloids (sparteine, sarotamine, lupanine, α-isolupanine, anagyrine, angustifoline, tyramine, multiflorine hydroxytyramine, cytisine, N-methyl cytisine, N-formylcytisine, 17-oxosparteine, tetra hydro rhombifoline, genisteine, ammodendrine)

coumarins (aesculetin, scoparone)

terpenoids (verbenol, α-terpineol, verbenone, linalool)

flavonoids (rutin, quercetin, isorhamnetin, quercitrin, kaempferol, scoparin, genistein, orobol, spiraeoside)

carotenoids (chrysanthemaxanthin, xanthophyll, xanthophillepoxide, β- carotene, α-carotene, γ-carotene, lutein)

sterols (cholesterol, campesterol, β-sitosterol, stigmasterol)

alcohols& phenols (cis-3-hexen-1-ol, 1-octen-3-ol, benzyl alcohol, phenylethyl alcohol, phenol, cresols, guaiacol, eugenol)

organic acids (benzoic acid, quinic acid, shikimic acid)

fatty acids (palmitic acid, arachidic acid, capric acid, caproic acid, lauric acid, myristic acid, oenathic acid, pelargonic acid)

phenolic acids (gallic acid, protocatechuic acid, chlorogenic acid, caffeic acid, p-coumaric acid)

miscellaneous (tyramine, 3-hydroxy tyramine)

Medicinal Use

Since the Middle Ages, the flowers and flowering branches were used for their purgative and diuretic effects. In phytotherapy, the flowering branches are used for their content of scoparoside, which has a cardiotonic and diuretic action. From a cardiotonic perspective, the total extract acts synergistically by enhancing the effects of sparteine through scoparoside. Due to the presence of biogenic amines with vasoconstrictor effects, infusions or decoctions from the young branches are administered for epistaxis.[7]

Sarothamni herba is a source for the extraction of sparteine.

Sparteine has oxytocic action, antiarrhythmic action of cardiotonic type, useful for parturients with cardiac disorders by promoting the beginning of labor and providing antiarrhythmic protection. At the cardiac level, it initially causes a transient ganglionic excitation state and then decreases the excitability, conductivity, frequency, and amplitude of myocardial contractions. Compared to quinidine and ajmaline, sparteine is less toxic and is recommended for neurotonic-origin sinus tachycardia, atrial fibrillation, and extrasystoles.

Professor Léon René Binet (1891–1971) and his student Gabriel Billard (1873–1929) observed that viper venom is neutralized, through precipitation, by sparteine, the latter being an antidote in viper bites.[7]

Toxicity

Prolonged treatment with sparteine causes hyperglycemia.

In high doses, sparteine leads to tachycardia, ventricular fibrillation, vomiting, diarrhea, dizziness, headache, migraines, tetanic contractions of the myometrium, visual disturbances, sweating, and hemolysis. The toxic effects are not antagonized by atropine. Death occurs due to cardiac arrest, caused by central and ganglionic paralysis.[7]

Biological Activity

Sparteine has oxytocic, antiarrhythmic of cardiotonic type actions and is an antidote in cases of viper bites.[1]

Other Uses

■ ornamental in parks[4]

Galega officinalis L. (Goat's rue)

Scientific name: *Galega officinalis* L.
Common name: Goat's rue, Professor-weed
Romanian name: Ciumărea
Taxonomic classification
Kingdom: *Plantae*
Subkingdom: *Cormobionta*
Phylum: *Spermatophyta*
Subphylum: *Magnoliophytina (Angiospermae)*
Class: *Magnoliatae (Magnoliopsida, Dicotyledonatae)*
Subclass: *Rosidae*
Order: *Fabales (Leguminosales)*
Family: *Fabaceae (Leguminosae)*
Genus: *Galega*
Species: *officinalis* L. (Figure 16.4)[1-3]

SPREAD

The species is commonly distributed from the lowlands to the beech zone, in hedgerows, moist and flood-prone meadows, along watercourses and irrigation canals, being toxic to sheep during flowering.[1-4]

DESCRIPTION

Perennial species, 40–120 cm tall, with an erect, fistulous, glabrous stem and imparipinnate compound leaves, consisting of 11–17 uniform, linear-lanceolate folioles with obtuse, mucronate tips and entire margins. The stipules are free, large, and acuminate. The flowers are white or pale lilac and form lax, long-pedunculate racemes. The androecium is monadelphous. The fruit is a dehiscent, reddish-brown, glabrous pod, slightly constricted between seeds. The flowering period encompasses the months of July and August.[1-4]

Main Phytoconstituents

Galegae herba contains approximately 0.1–0.3% galegin (isoamylene-guanidine); 4-hydroxygalegin; guanidine; α-pipecoline; flavonoids, in flowers; tannin; quinazoline alkaloids (peganine); saponins; lectins; and chromium salts.[17]

Figure 16.4 *Galega officinalis* L. (Goat's rue).

Source: 14.

Main phytoconstituents   Ref.
amino acids (lysine, threonine, leucine, histidine, phenylanine, methionine, alanine, tyrosine,
    arginine, isoleucine, valine, proline, aspartic acid, glycine, serine, glutamic acid)   [17-19]
alkaloids (galegine, vasicine, vasicinol, vasicinone)
flavonoids (quercetin, taxifolin, kaempferol, astragalin, narcissin, quercitrin, hyperoside,
    clitorin, mauritianin, rutin, isorhamnetin)
saponin
fatty acids (palmitic acid, linoleic acid, α-linolenic acid)
phenolic acids (caffeic acid, p-coumaric acid, chlorogenic acid, ferulic acid, protocatechuic
    acid, hydroxycinnamic acid)
miscellaneous (galegine, hydroxygalegine)

Medicinal Use

The vegetable product *Galegae herba* has hypoglycemic (galegin and derivatives, chromium salts), galactagogic, diaphoretic, diuretic, anthelmintic, and stimulating effects on the adrenal function and pancreas.[1,17]

Biological Activity

Experiments *in vitro*, conducted on human epithelial cell cultures, have demonstrated the inhibition of glucose transport by extracts from *Galegae herba*. In rats with experimental alloxan-induced diabetes, galegin induced the hypergenesis of β-Langerhans cells and lowered glycemia. However, the hypoglycemic action of extracts from *Galegae herba* has not been fully confirmed, and their use is prohibited in severe forms of diabetes.

The medicinal product is recommended in the complementary treatment of incipient forms of diabetes, gastroduodenitis, diarrhea, dyspepsia, flatulence, and intestinal paralysis.[17]

Toxicity

Galega officinalis can cause lethal intoxications in herbivores through ingestion in fodder. Compared to cattle, where intoxication is rare, sheep are much more sensitive.[17]

Risks and Pharmacological Interactions

Due to the uncertain hypoglycemic action and the toxicity of galegin, the product *Galegae herba* is recommended only in incipient forms of diabetes, in moderate doses, for a limited time. It should not be associated with hypoglycemic medications and should not substitute them in the treatment of diabetes.[17]

Other Uses

- culinary – cooked leaves are similar to spinach

- cosmetic – hand and foot baths[20]

Glycyrrhiza glabra L. (Liquorice)

Scientific name: *Glycyrrhiza glabra* L.
Common name: Liquorice
Romanian name: Lemn dulce
Taxonomic classification
Kingdom: *Plantae*
Subkingdom: *Cormobionta*
Phylum: *Spermatophyta*
Subphylum: *Magnoliophytina (Angiospermae)*
Class: *Magnoliatae (Magnoliopsida, Dicotyledonatae)*
Subclass: *Rosidae*
Order: *Fabales (Leguminosales)*
Family: *Fabaceae (Leguminosae)*
Genus: *Glycyrrhiza*
Species: – *glabra* L.[1-3] presents two varieties: *glabra* with glabrous pods and *glandulifera* (Waldst. et Kit.) Regel et Herder (*G. glandulifera* Waldst. et Kit.) with glandular-setose pods (Figure 16.5)[2-3]

SPREAD

A rare species in the lowland to hilly areas, found in wet locations, especially along riverbanks, meadows, and shrubs.[1-4]

DESCRIPTION

Perennial herbaceous species, with an above-ground part of 0.5–1 m. It has a thick, woody root and thick rhizomes. The stem is erect, vigorous, sparsely branched, and rough in the superior part. The leaves are imparipinnate compound with 5–10 pairs of ovate folioles, glandular-punctate on the dorsal side. The stipules are small and deciduous. The flowers have a blue-violet corolla (8–12 mm long), lanceolate bracts, caducous, diadelphous androecium, and form an elongated and lax spiciform raceme (with rare flowers). Plants bloom in June-July.[1-4]

Main Phytoconstituents

Liquiritiae radix contains: 2–12% triterpene saponins; flavanone (liquiritoside) and chalcone (isoliquiritoside) heterosides; coumarins; polyholozides (glycyrrhizans); bitter principles (glycyrrhamarine); sterols; mannitol; resins; 20–30% starch; 5–15% sugars; proteins; mineral salts.

Triterpene saponins have glycyrrhizic acid as their main representative, with glycyrrhetic acid as the aglycone. The calcium (Ca^{2+}) and magnesium (Mg^{2+}) salts of glycyrrhizic acid are called glycyrrhizin. Compared to glycyrrhizic acid, which is not sweet-tasting, glycyrrhizin is 50 times sweeter than sucrose.

According to the Romanian Pharmacopoeia (F.R. X), the medicinal product must contain at least 5% glycyrrhizic acid and 20–25% hydrosoluble substances. The yellow color of the vegetable product is imparted by the flavononic and chalconic heterosides, which, during conservation, partially hydrolyze into aglycones (liquiritigenol and izoliquiritigenol, respectively).[17]

Figure 16.5 *Glycyrrhiza glabra* L. (Liquorice).

Source: 21.

Main phytoconstituents   Ref.

amino acids (tyrosine, leucine, alanine, serine, threonine, phenylalanine, aspartic, valine, histidine, glycine, glutamic acid, proline, isoleucine, lysine, tyrosine)   22-24

coumarins (licopyranocoumarin, coumarin-GU-12)

terpenoids (terpinen-4-ol, linalool, geraniol, α-terpineol)

flavonoids (luteolin, acacetin, apigenin, hesperetin, liquiritin, isoangustone A, isoliquiritin, glabridin, shinpterocarpin, hispaglabridins A and B, glabrene, rhamnoliquirilin, liquiritigenin, prenyllicoflavone A, glucoliquiritinapioside, 1-metho-xyphaseolin, shinflavanone, semilicoisoflavone B, licoriphenone, 1-methoxyficifolinol, kanzonol R)

saponin (glycyrrhizin)

carbohydrates (maltol)

fatty acids (palmitic acid, linoleic acid)

phenolic acids (p-coumaric acid, chicoric acid)

nucleosides (ribavirin)

alcohols (hexanol, pentanol, 2,3-butanediol)

aldehydes&ketones (furfuraldehyde, methyl ethyl ketone,

organic acids (propionic acid, acetic acid, fumaric acid, citric acid, butyric acid, malic acid, tartaric acid, benzoic acid)

miscellaneous (tetramethyl pyrazine, 1-methyl-2-formylpyrrole, ethyl linoleate, furfuryl formate, trimethylpyrazie)

Medicinal Use

The medicinal product *Liquiritiae radix* (underground organs with or without suber) is used as an expectorant, laxative, diuretic, sweetening agent, and flavor corrector.[1]

Biological Activity

Recently, it has been discovered that glycyrrhetic acid has a *hormone-like SR* (similar to the main corticosuprarenal hormone) action, even more pronounced when the active principle is isolated as such. For this reason, for the cortisone-like effect, *Liquiritiae radix* is successfully used in the treatment of Addison's disease. The toxicity of the medicinal product is lower than that of cortisone at the same intensity of pharmacological effect.

Licorice root is also used in the treatment of gastric ulcers, gastritis, and gastro-duodenitis because glycyrrhetic acid has anti-inflammatory, cicatrizing, and gastric secretion-inhibiting properties. Additionally, due to its cortisone-like effect, *Liquiritiae radix* is recommended for the treatment of dermatitis, neurodermatitis, psoriasis, and eczema.[17]

Other Uses

- sweetener and flavoring agent for lemonade, beer
- glycyrrhizin, when mixed with water, produces foam and is used in fire extinguishers[4]

Ononis spinosa L. (Spiny restharrow)

Scientific name: *Ononis spinosa* L.
Common name: Spiny restharrow
Romanian name: Osul iepurelui
Taxonomic classification
Kingdom: *Plantae*
Subkingdom: *Cormobionta*
Phylum: *Spermatophyta*
Subphylum: *Magnoliophytina (Angiospermae)*
Class: *Magnoliatae (Magnoliopsida, Dicotyledonatae)*
Subclass: *Rosidae*
Order: *Fabales (Leguminosales)*
Family: *Fabaceae (Leguminosae)*
Genus: *Ononis*
Species: *spinosa* L.[1-3]
Subspecies: – *spinosa* (*O. campestris* W.D.J. Koch et Ziz)[3] has robust thorns, a corolla of 10–20 mm, and the fruit is equal to or exceeds the calyx (Figure 16.6).[2-3]

- *austriaca* (Beck) Gams has weak thorns, a corolla of 15–20 mm, and the fruit is shorter than the calyx.[2-3]

SPREAD

Species is sporadic in the lowland zone up to the sessile oak forest level, growing in meadows, arid areas, sandy places, along watersides, pastures, steppes, and spiny shrublands.[1-4]

DESCRIPTION

A spiny subshrub, glandular-hairy, with erect stems (30–80 cm), without stolons and initially creeping branches, then ascending or erect, with rigid thorns, one terminal and 2–3 lateral thorns. The leaves are trifoliate (on the stem) or simple (unifoliate on the flowering branches), uniformly serrated-dentate. Stipules are elongated ovate and amplexicaul. The flowers are pink and found singly in the axils of the simple, bracteate leaves, forming lax racemes. Flowering occurs in June and July.[1-4]

Main Phytoconstituents

Ononidis radix contains tetracyclic triterpene saponins (α- and β-onocerin) with aglycones α- and β-onocerol; isoflavonoids (ononin or 7-O-glucoside of formononetin); lignans; sterols (β-sitosterol); volatile oil; glycosylated phenyl-benzyl ketones (onospin, with aglycone ononetin); mineral salts.[17]

Main phytoconstituents	Ref.
coumarins (aesculetin)	26-29
terpenoids (α-amyrin,β-amyrin, carvone, menthol, menthone, isomenthone, borneol linalool, clitorienolactone A)	

Figure 16.6 *Ononis spinosa* L. (Spiny restharrow).

Source: 25.

flavonoids (quercetin, kaempherol, apigenin, luteolin, chrysoeriol, formononetin, chrysoeriol, acacetin, nepetin, hispidulin, sativanone, velutin, hymenoxin, daidzein, calycosin D, medicagol, genkwanin, ononin)
saponin (azukisaponin V)
sterols (β-sitosterol, stigmasterol,campesterol, cholesterol, daucosterol, α-spinasterol)
fatty acid (linoleic acid)
phenolic acids (caffeic acid, ferrulic acid, chiconic acid, gallic acid, protocatechuic acid, caftaric acid, hydroxybenzoic acid, coutaric acid, fertaric acid, vanillic acid, syngic acid, p-coumaric acid, cinnamic acid, sinapin acid, homopipecolic acid, salicylic acid gentisin acid)
miscellaneous (clitorienolactone B, isoliquiritigenin, puerol A, trans-anethole, cis-anethole, estragole, maackiain, medicarpin)

Medicinal Use

The underground part of the plant (*Ononidis radix*) has a diuretic action, of saluretic type.[1]

Biological Activity

It is recommended for urinary lithiasis, in the form of an infusion, in courses of 4–5 days, alternating with breaks of 8–10 days, as the diuretic effect diminishes over time.[17]

Other Uses

- culinary
- nitrogen fixer[30]

Genista tinctoria L. (Dyer's greenweed)

Scientific name: *Genista tinctoria* L.
Common name: Dyer's greenweed, Common woadwaxen, Broom
Romanian name: Drobiță
Taxonomic classification
Kingdom: *Plantae*
Subkingdom: *Cormobionta*
Phylum: *Spermatophyta*
Subphylum: *Magnoliophytina (Angiospermae)*
Class: *Magnoliatae (Magnoliopsida, Dicotyledonatae)*
Subclass: *Rosidae*
Order: *Fabales (Leguminosales)*
Family: *Fabaceae (Leguminosae)*
Genus: *Genista*
Species: *tinctoria* L.[2,3]
Subspecies: – *oligosperma* (Andrae) Borza is endemic in our country, being present in the Vrancea Mountains and the Southern Carpathians, frequently found from the spruce floor to the subalpine floor. It has creeping stems, is 20 cm tall, and has a short raceme of 1–3 cm.[2-3]

- *tinctoria* is widespread at lower altitudes, with erect stems and racemes over 4 cm long. It has 2 varieties: *tinctoria* has glabrous ovary and fruit, and *hungarica* (A. Kern.) Nyman (var. *banatica* Simonk.) with hairy ovary and fruit (Figure 16.7).[2-3]

SPREAD
Widespread throughout the country from the plains to the mountainous area, found in meadows, pastures, forest edges, and sunny steep slopes.[4]

Figure 16.7 *Genista tinctoria* L. (Dyer's greenweed).
Source: 31.

DESCRIPTION

A subshrub with a height of 10–60 cm, with a taproot and erect stems, abundantly branched or unbranched. The branches are striped or with a broom-like aspect, with 5 edges or ridged, hairy or becoming adpressed hairy towards the top, and terminating with an elongated inflorescence. The leaves are lanceolate-elliptical, with an acute tip, nearly sessile, with few prominent lateral veins, dark green and glabrous on the superior surface, and light green and ciliate on the inferior surface. The leaf margin is ciliate, and the stipules are subulate. The yellow-golden flowers are short-pedicelled, with a glabrous corolla, and are grouped in a long raceme of (3) 4–10 (15) cm. The pods are linear, glabrous, or shortly pubescent. The flowering period is from June to July.[2-4]

Main phytoconstituents   Ref.
amino acids   [32-33]
lectines
flavonoids (hyperoside, isoquercitrin, rutin, quercitrin, quercetinluteolin, apigenin, luteolin, apigenin, daidzin, daidzein, genistein, prunetin, isoprunetin, biochanin)
coumarins: (iso-coumarin)
alkaloid (cytosine, anagyrine N-methylcytisine,
terpenoids (linalool, α-terpineol, nerol, safranal, geraniol, β-damascenone, β-bisabolene, nerolidol, farnesol, phytol, camphenilone)
sterols
fatty acids (stearic acid, lauric acid, myristic acid, caprylic acid, palmitic acid, margaric acid, pentadecanoic acid)
phenolic acids (caffeic acid, chlorogenic acid, ferulic acid, *p*-coumaric acid)
carbohydrates
saponins
tannins
hydrocarbons (docosane, hexacosane, heptacosane, pentacosane, tetracosane, tricosane)
organic acids
esters
miscellaneous (eugenol)

Medicinal Use

laxative, cardiotonic, stomachic, neuralgia, anorexia, sudorific, diuretic, depurative, antiemetic, vasoconstrictor, epithelizing, cicatrizing[34-36]

Biological Activity

antitumoral, antioxidant, antimicrobial, cardioprotective[33]

Other Uses

■ culinary (pods)

■ textile dye[34]

Vicia cracca L. (Tufted vetch)

Scientific name: *Vicia cracca* L.
Common name: Tufted vetch, Bird vetch, Cow vetch
Romanian name: Măzăriche
Taxonomic classification
Kingdom: *Plantae*
Subkingdom: *Cormobionta*
Phylum: *Spermatophyta*
Subphylum: *Magnoliophytina (Angiospermae)*
Class: *Magnoliatae (Magnoliopsida, Dicotyledonatae)*
Subclass: *Rosidae*
Order: *Fabales (Leguminosales)*
Family: *Fabaceae (Leguminosae)*
Genus: *Vicia*
Species: *cracca* L. (Figure 16.8)[2-3]

Figure 16.8 *Vicia cracca* L. (Tufted vetch).

Source: 26.

SPREAD

A common species in our country, widespread from the plains to the beech forest zone, found in meadows, orchards, shrubs, pastures, forest edges, and along fences.[2-4]

DESCRIPTION

Perennial herbaceous plant with a long, creeping rhizome, extensively branched, and a stem ranging from 30 to 150 cm in height, slender, simple or branched, edged, glabrous or hairy. The leaves are paripinnate compound, with (6)10–12 (20) pairs of folioles, adpressed sericeous-hairy on the inferior surface, narrowly elliptical, with an acute or rounded tip, and the rachis ends in a branched tendril, with small, non-dentate stipules and no extrafloral nectaries. The inflorescence is an elongated raceme, consisting of 20–40 flowers with a corolla 3–4 times longer than the calyx, measuring 12–18 mm, and the vexil longer than the ungula, violet-blue or rarely white. The pods are elongated-rhombic, glabrous, brown to black in color, and contain 4–8 globular seeds. The flowering period is from June to August.[2-4]

Main phytoconstituents   Ref.
                                                                                   [37-39]
amino acids (vicine, isoleucine, tyrosine)
flavonoids (catechin, quercetin, apigenin, kaempferol, diosmetin, naringenin, luteolin)
terpenoids (caryophilene, ocimene, elemene, phytol, linalool, farnesane, limonene, camphor, germacrene D, hexahydro-farnesylacetone)
sterols
fatty acids
phenolic acids (gallic acid, ellagic acid, syringic acid, salicylic acid)
carbohydrates
saponins
tannins (ellagic acid)

hydrocarbons
organic acids (malic acid, citric acid)
esters
miscellaneous

Medicinal Use

eczema, dermatitis, skin irritations[40]

Biological Activity

antioxidant, hepatoprotective[38]

Other Uses

- animal feed

- melliferous

- culinary: the fruits (pods) dishes similar to peas[40]

Astragalus glycyphyllos L. (Milk vetch)

Scientific name: *Astragalus glycyphyllos* L.
Common name: Milk vetch, Licorice milkvetch
Romanian name: Unghia găii
Taxonomic classification
Kingdom: *Plantae*
Subkingdom: *Cormobionta*
Phylum: *Spermatophyta*
Subphylum: *Magnoliophytina (Angiospermae)*
Class: *Magnoliatae (Magnoliopsida, Dicotyledonatae)*
Subclass: *Rosidae*
Order: *Fabales (Leguminosales)*
Family: *Fabaceae (Leguminosae)*
Genus: *Astragalus*
Species: *glycyphyllos* L. (Figure 16.9)[2-3]

SPREAD

The species is commonly found from lowlands to the beech zone, on sunny slopes, in forests, shrubs, and forest edges.[2-4]

DESCRIPTION

Perennial herbaceous plant, reaching heights of 30–100 cm, with a deep taproot, a branched basal axis in the superior part, and a prostrate, flexuous, edged stem, with rare, adpressed hairs or glabrous. The leaves consist of 4–7 pairs of large, oval to broadly elliptic folioles, rounded or shortly aristate-acute, glabrous on the superior side and hairy and slightly glaucous on the inferior side, with 5–7 pairs of veins. The stipules are free, lanceolate, and acute. The flowers are short-pedicelled, yellow-green, erect, patent, grouped in ovate elongated racemes, at the base of membranous, ciliate bracts longer than the floral pedicels. The pod is sessile, long, linear, curved, without a carpophore, about 7–8 times longer than wide, and contains at least 10 reniform, smooth, reddish-brown seeds in each locule. The flowering period is from May to July.[2-4]

Main phytoconstituents Ref.
amino acids [42-4 7]
flavonoids (apigenin, fisetin, rhamnetin, isorhametin, naringerin, genistein, myricetin,
   kaempherol, genistein, tamarixetin, daidezin, luteolin, sulfuretin, isoliquiritigenin, pendulone)
terpenoids (ursolic acid)
carbohydrates
sterols (β-sitosterol)
fatty acids

Figure 16.9 *Astragalus glycyphyllos* L. (Milk vetch).

Source: 41.

phenolic acids (p-coumaric, ferulic acid, chlorogenic acid)
saponins
tannins
anthocyans
hydrocarbons

Medicinal Use

urinary disorders, heart disease, diabetes, superior respiratory infections, hypertension, diuretic, laxative, expectorant[43,45,47]

Biological Activity

immunostimulant, anti-inflammatory, antiviral, antidiabetic, diuretic, antioxidant, antiallergic, antiproliferative, antitumoral, sedative, hypotensive, hepatoprotective, cardioprotective[46-47]

Other Uses

- ornamental

- culinary use (salads, pesto)[46-47]

Trifolium montanum L. (Mountain clover)

Scientific name: *Trifolium montanum* L.
Common name: Mountain clover
Romanian name: Trifoi de munte[48]
Taxonomic classification
Kingdom: *Plantae*
Subkingdom: *Cormobionta*

Figure 16.10 *Trifolium montanum* L. (Mountain clover).

Source: 49.

Phylum: *Spermatophyta*
Subphylum: *Magnoliophytina (Angiospermae)*
Class: *Magnoliatae (Magnoliopsida, Dicotyledonatae)*
Subclass: *Rosidae*
Order: *Fabales (Leguminosales)*
Family: *Fabaceae (Leguminosae)*
Genus: *Trifolium*
Species: *montanum* L. (Figure 16.10)[2-3]

SPREAD

This species is commonly distributed from the plains to the montane zone in sunny or semi-shaded areas, including meadows, forest edges, shrublands, pastures, and orchards.[2-4]

DESCRIPTION

This perennial herbaceous plant reaches a height of 15–40 cm and features a multicapitate, lignified taproot and a stem that is erect or procumbent, simple, and adpressed sericeous hairy. The trifoliate leaves have short petioles and consist of narrowly elliptic or lanceolate folioles with acute tips, rounded bases, and sharp setaceous serrated margins, densely hairy on the inferior surface. The stipules are entire, subulate, with veins. The flowers have a white corolla, measuring 7–9 mm in length, and are grouped in globular, multiflowered capitula. The pods contain 2 ovate, greenish seeds. The flowering period occurs from May to July.[2-4]

Main phytoconstituents   Ref.
amino acids   50-52
flavonoids (formononetin, daidzein, genistein, quercetin, myricetin, gallocatechin, epigallocatechin, gallocatechin, Biochanin A)

terpenoids (β-ocimenes, longifolene, β-phellandrene, β-menthane, thymol, carophyllene)
sterols
coumarins (coumarin)
fatty acids (linolenic acid, palmitic acid, caprylic acid, pelargonic acid, oleic acid)
phenolic acids (gallic acid, chlorogenic acid, gerrulic acid, caffeic acid, rosmarimic acid, *p*-coumaric acid, cinnamic acid)
saponins (soyasaponin)
carbohydrates
tannins
hydrocarbons
organic acids (succinic acid)
miscellaneous

Medicinal Use

estrogenic action, reduction in severity of menopausal symptoms, laxative, antioxidant, heart disorders, diuretic, expectorant, emmenagogue, rheumatism[50-52]

Biological Activity

antimicrobial, antidiabetic, antitumoral, antirheumatic, cardioprotective, hepatoprotective, analgesic[50-52]

Other Uses

- tea
- animal feed[50-52]

REFERENCES

1. Bejenaru C., Mogoșanu G.D., Bejenaru L.E., Popescu H. 2020. *Botanică farmaceutică. Cormobionta*, ediția a III-a. Ed. Sitech, Craiova.

2. Ciocârlan V. 2009. *Flora ilustrată a României. Pteridophyta et Spermatophyta.* Ediția a 3-a revizuită și adăugită. Ed. Ceres, București.

3. Sârbu I., Ștefan N., Oprea A. 2013. *Plante vasculare din România, Determinator ilustrat de teren.* Ed. Victor B Victor, București.

4. Săvulescu T. (ed). 1957. *Flora RPR*, vol. V. Ed. Academiei RPR, București.

5. Pop I., Hodișan I., Mititelu Gh.M., Lungu L., Cristurean I., Mihai Gh. 1983. *Botanică sistematică*. Ed. Didactică și Pedagogică, București.

6. https://dryades.units.it/scuole/index.php?procedure=taxon_page&id=2254&num=553

7. Bejenaru L.E., Mogoșanu G.D., Cornelia B., Popescu H. 2015. *Farmacognozie-Fitoterapie.* Vol. I, ediția a III-a. Ed. Sitech, Craiova.

8. Usman M., Khan W.R., Yousaf N., Akram S., Murtaza G., Kudus K.A., et. al. 2022. Exploring the phytochemicals and anti-cancer potential of the members of *Fabaceae* family: A comprehensive review. *Molecules* 27(12):3863.

9. Bahorun T., Neergheen V.S, Aruoma O.I. 2005. Phytochemical constituents of *Cassia fistula*, *African Journal of Biotechnology* 4 (13):1530–1540.

10. Al-Snafi A.E. 2020. Chemical constituents and pharmacological effects *of Melilotus officinalis*-A Review. *OSR Journal of Pharmacy* 10(1):26–36.

11. https://pfaf.org/user/plant.aspx?LatinName=Melilotus+officinalis

12. https://bionutris.ro/sulfina-beneficii-proprietati-terapeutice-si-mod-de-utilizare/

13. https://gradinahobby.blogspot.com/2015/12/sulfina-melilotus-officinalis.html

14. https://dryades.units.it/floritaly/index.php?procedure=taxon_page&tipo=all&id=2269

15. Caramelo D., Barroca C., Guiné R., Gallardo E., Anjos O., Gominho J. 2022. Potential applications of the *Cytisus* shrub species: *Cytisus multiflorus, Cytisus scoparius,* and *Cytisus striatus. Processes* 10(7): 1287.

16. Raja S., Koduru R. 2014. *Cytisus scoparius*: A review of ethnomedical, phytochemical and pharmacological information. *Indo American Journal of Pharmaceutical Research* 4(04):2151–2169.

17. Mogoșanu G.D., Bejenaru L.E., Bejenaru C., Popescu H. 2015. *Farmacognozie-Fitoterapie*, Vol. II, ediția a III-a. Ed. Sitech, Craiova.

18. Bednarska K., Kuś P., Fecka I. 2020. Investigation of the phytochemical composition, antioxidant activity, and methylglyoxal trapping effect of *Galega officinalis* L. herb *in vitro. Molecules* 25(24):5810.

19. Peiretti P.G., Gai F. 2006. Chemical composition, nutritive value, fatty acid and amino acid contents of *Galega officinalis* L. during its growth stage and in regrowth. *Animal Feed Science and Technology* 130(3-4):257–267.

20. https://pfaf.org/user/Plant.aspx?LatinName=Galega+officinalis

21. https://dryades.units.it/lamone/index.php?procedure=taxon_page&id=2395&num=4786

22. El-Saber Batiha G., Magdy Beshbishy A., El-Mleeh A.M., Abdel-Daim M., Prasad Devkota H. 2020. Traditional uses, bioactive chemical constituents, and pharmacological and toxicological activities of *Glycyrrhiza glabra* L. *(Fabaceae). Biomolecules* 10(3):352.

23. Sharma V., Katiyar A., Agrawal R.C. 2018. *Glycyrrhiza glabra*: Chemistry and pharmacological activity. In: Mérillon JM., Ramawat K. (eds). *Sweeteners: Pharmacology, Biotechnology, and Applications*. Reference Series in Phytochemistry, Springer, Cham, Switzerland, pp. 87–100.

24. Babich O., Ivanova S., Ulrikh E., Popov A., Larina V., Frolov A., Prosekov A. 2022. Study of the chemical composition and biologically active properties of *Glycyrrhiza glabra* extracts. *Life* 12(11):1772.

25. https://dryades.units.it/Roma/index.php?procedure=taxon_page&id=2536&num=4818

26. Stojković D., Dias M.I., Drakulić D., Barros L., Stevanović M. 2020. Methanolic extract of the herb *Ononis spinosa* L. is an antifungal agent with no cytotoxicity to primary human cells. *Pharmaceuticals* 13(4):78.

27. Stojkovic D., Drakulic D., Gašić U., Zengin G., Stevanovic M., et.al. 2020. *Ononis spinosa* L. an edible and medicinal plant: UHPLC-LTQ-Orbitrap/MS chemical profiling and biological activities of the herbal extract. *Food & Function* 11:7138–7151.

28. Besbas S., Mouffouk S., Haba H., Marcourt L., Wolfender J.-L., Benkhaled M. 2020. Chemical composition, antioxidant, antihemolytic and anti-inflammatory activities of *Ononis mitissima* L. *Phytochemistry Letters* 37:63–69.

29. Spiegler V., Gierlikowska B., Saenger T., Addotey J.N., Sendker J., Jose J., Kiss A.K., Hensel A. 2020. Root extracts from *Ononis spinosa* inhibit IL-8 release via interactions with toll-like receptor 4 and lipopolysaccharide. *Frontiers in Pharmacology* 11:519045.

30. https://pfaf.org/user/Plant.aspx?LatinName=Ononis+spinosa

31. https://dryades.units.it/floritaly/index.php?procedure=taxon_page&tipo=all&id=2279

32. Hanganu D., Olah N.K., Benedec D., Mocan A., Crisan G., Vlase L., Popica I., Oniga I. 2016. Comparative polyphenolic content and antioxidant activities of *Genista tinctoria* L. and *Genistella sagittalis* (L.). *Pakistan Journal of Pharmaceutical Sciences* 1(Suppl):301–307.

33. Rigano D., Russo A., Formisano C., Cardile V., Senatore F. 2010. Antiproliferative and cytotoxic effects on malignant melanoma cells of essential oils from the aerial parts of *Genista sessilifolia* and *G. tinctoria*. *Natural Product Communications* 5(7):1127–1132.

34. www.remediilenaturii.ro/remediu_fitoterapie.php?n=Drobi%C8%9B%C4%83&ID=217

35. www.wikiwand.com/ro/Drobi%C8%9B%C4%83

36. www.sfatulmedicului.ro/plante-medicinale/drobita-genista-tinctoria_14594

37. Salehi, B., Abu-Reidah, I. M., Sharopov, F., Karazhan, N., Sharifi-Rad, J., Akram, M., Pezzani, R. 2020. *Vicia* plants – A comprehensive review on chemical composition and phytopharmacology. *Phytotherapy Research* 35(2):790–809.

38. Shokrzadeh M., Rahimi F., Ziar A., Ebrahimzadeh M.A. 2018. Antioxidant and hepatoprotective properties of *Vicia cracca* against carbon tetrachloride induced oxidative stress in mice. *Journal of Mazandaran University of Medical Sciences* 27(156):50–65.

39. Altundag E., Ozturk M. 2011. Ethnomedicinal studies on the plant resources of east Anatolia, Turkey. *Procedia Social and Behavioral Sciences* 19:756–777.

40. www.plantlife.org.uk/uk/discover-wild-plants-nature/plant-fungi-species/common-vetch

41. https://dryades.units.it/lamone/index.php?procedure=taxon_page&id=2354&num=565

42. Li X., Qu L., Dong Y., Han L., Liu E., Fang S., Zhang Y., Wang T.A. 2014. A review of recent research progress on the *Astragalus* genus. *Molecules* 19(11):18850–18880.

43. Amiri M.S., Joharchi M.R., Nadaf M., Nasseh Y. 2020. Ethnobotanical knowledge of *Astragalus* spp.: The world's largest genus of vascular plants. *Avicenna Journal of Phytomedicine* 10(2):128–142.

44. Bartók A., Hurdu B.I., Szatmari P.M., Ronikier M., Puşcaş M., Novikoff A., Bartha L., Vonica G. 2016. New records for the high-mountain flora of the Făgăraş Mts. (Southern Carpathians) with discussion on ecological preferences and distribution of studied taxa in the Carpathians. *Contribuţii Botanice Grădina Botanică "Alexandru Borza" Cluj-Napoca* LI:77–153.

45. Bratkov V.M., Shkondrov A.M., Zdraveva P. K., Krasteva I.N. 2016. Flavonoids from the genus *Astragalus*: phytochemistry and biological activity. *Pharmacognosy Reviews* 10(19):11–32.

46. Georgieva A., Popov G., Shkondrov A., Toshkova R., Krasteva I., Kondeva-Burdina M., Manov V. 2021. Antiproliferative and antitumour activity of saponins from *Astragalus glycyphyllos* on myeloid Graffi tumour. *Journal of Ethnopharmacology* 267:113519.

47. Lysiuk R., Darmohray R. 2016. Pharmacology and ethnomedicine of the genus *Astragalus*. *International Journal of Pharmacology, Phytochemistry and Ethnomedicine* 3:46–53.

48. www.afaceriardelene.ro/altapagina.php?url=trifoiul-rosu-efecte-terapeutice-trifoiul-rosu-efecte-terapeutice

49. https://dryades.units.it/floritaly/index.php?procedure=taxon_page&tipo=all&id=2602

50. Sultana S., Foster K., Lim L.Y., Hammer K., Locher C.A. 2022. Review of the phytochemistry and bioactivity of clover honeys (*Trifolium* spp.). *Foods* 11(13):1901.

51. Kolodziejczyk-Czepas J. 2016. *Trifolium* species – the latest findings on chemical profile, ethnomedicinal use and pharmacological properties. *Journal of Pharmacy and Pharmacology* 68(7):845–861.

52. Sabudak T., Guler N. 2009. *Trifolium* L.–A review on its phytochemical and pharmacological profile. *Phytotherapy Research* 23(3):439–446.

17 Fungi

Cornelia Bejenaru[1], Antonia Radu[1], George Dan Mogoşanu[2],
Ludovic Everard Bejenaru[2], Andrei Biţă[2], and Adina-Elena Segneanu[3]
[1] Department of Pharmaceutical Botany, Faculty of Pharmacy, University of Medicine
and Pharmacy of Craiova, 2 Petru Rareş Street, Craiova, Dolj County, Romania
[2] Department of Pharmacognosy & Phytotherapy, Faculty of Pharmacy, University of
Medicine and Pharmacy of Craiova, 2 Petru Rareş Street, Craiova, Dolj County, Romania
[3] Institute for Advanced Environmental Research–West University of
Timisoara (ICAM–WUT), 4 Oituz Street, Timisoara, Romania

BOTANICAL DESCRIPTION

The estimated number of mushroom species on Earth is 140,000, of which approximately 22,000 are identified. In the spontaneous flora of Romania, there are 1,600 species of mushrooms, of which only 400 are edible.[1,2]

Mushrooms are inferior plants, without chlorophyll, with various shapes and sizes, unicellular or multicellular, found in different environments worldwide. Their vegetative body consists of thin, branching hyphae that anastomose and intertwine with each other, forming the mycelium. Mushrooms have heterotrophic nutrition and can be saprophytic, parasitic, or symbiotic.[3,4]

The distribution range varies from one species to another, depending on climatic conditions, altitude, and the superior plants present in that specific area.[3]

Reproduction can occur vegetatively, asexually, and sexually.[4]

Edible mushrooms have drawn the attention of consumers because, in addition to being tasty, they also have remarkable nutritional value due to their high content of proteins, carbohydrates, vitamins, and minerals, as well as their low fat content.[5-7]

MAIN PHYTOCONSTITUENTS

Mushrooms have a high protein content and low fat (sterols, fatty acids) content, along with significant concentrations of phenolic compounds, vitamins B, C, D, and K, minerals, such as potassium and phosphorus, as well as selenium. Additionally, they contain important proportions of dietary fibers, chitin, and beta-glucans, as functional elements.[8,9]

BIOLOGICAL ACTIVITY

Edible mushrooms exhibit numerous pharmacological properties, such as antifungal, anti-inflammatory, antiviral, antibacterial, hepatoprotective, antidiabetic, hypolipidemic, hypotensive, and cytotoxic activities.[10-12]

The primary medicinal effects attributed to mushrooms include antibiotic, antitumor, antiviral, immunostimulatory, and hypolipidemic properties. Numerous studies have documented the presence of antielastase, anticollagenase, antihyaluronidase, and antityrosinase activities in fungal extracts and the individual compounds isolated from them. At the same time, wound healing stimulation has been one of the most researched positive abilities of mushrooms, described in various scientific studies, through the stimulation of immune epithelial cells and the release of cytokines and growth factors.[13-19]

The medicinal features of mushrooms are closely related to their content of bioactive phytoconstituents, which primarily involve polysaccharides, terpenoids, glucans, phenolic compounds, statins, and lectins. In addition to that, melanin pigments, chitin, and chitosan present in their cell walls, along with extracellular enzymes are also highly efficient molecules.[20-22]

The **phylum *Ascomycota*** comprises approximately 150,000 species of fungi characterized by a fertile structure called an ascus, where ascospores are formed. It is divided into 4 classes:

■ Class *Laboulbeniomycetes* includes fungi that are parasitic on insects and terrestrial or aquatic arthropods, rarely on algae.

■ Class *Acarpoascomycetes* (*Hemiascomycetes*) includes primitive fungi in the Order *Taphrinales* and the Order *Endomycetales*.

■ Class *Plectomycetes* includes fungi that passively release ascospores. Fungi in this class are grouped into 4 orders: *Ophiostomatales*, *Eurotiales*, *Onygenales* (*Gymnoascales*), and *Elaphomycetales*.

DOI: 10.1201/9781003270515-18

■ Class *Hymenoascomycetes* presents the asci arranged in a hymenium within the ascocarp. This class is divided into 5 subclasses: *Erysiphomycetideae, Pyrenomycetideae, Loculoascomycetideae, Lecanoromycetideae,* and *Pezizomycetidae.*[2]

MORCHELLACEAE FAMILY

Includes terrestrial saprophytic fungi, characterized by a relatively large, fistulous, fleshy ascocarp, differentiated into a sterile stipe and a fertile part resembling a cap, with a sinuous-pleated, wrinkled, honeycomb-like surface. The stipe is smooth or furrowed, fragile, hollow, lighter in color compared to the ascocarp.[3]

The asci are pedunculate, cylindrical, and contain 2–8 ascospores. The family includes 4 genera: *Morchella, Mitrophora, Ptychoverpa,* and *Verpa.*[2]

The genus *Morchella* in the flora of Romania includes the following species: *Morchella esculenta* (L.) Pers. ex Fr. with the varieties *rotunda* Pers., *crassipes* Vent ex Pers., and *alba* Boud., *Morchella elata* Fr., *Morchella conica* Pers. with 2 varieties *deliciosa* Fr. and *rigida* Krbh., *Morchella hybrida* Sow. Pers., *Morchella rotunda* Pers., and *Morchella crassipes* Vent. ex. Pers.[2,3]

Morchella esculenta (L.) Pers. ex Fr. (Common morel)

Scientific name: *Morchella esculenta* (L.) Pers. ex Fr.
Common name: Common morel, Morel, Yellow morel, True morel, Morel mushroom, Sponge morel
Romanian name: Zbârciog
Taxonomic classification
Kingdom: *Fungi* (*Mycophyta*)
Subkingdom: *Dikarya*
Phylum: *Ascomycota*
Subphylum: *Pezizomycotina*
Class: *Hymenoascomycetes*
Subclass: *Pezizomycetideae*
Order: *Pezizales*
Suborder: *Pezizineae*
Family: *Morchellaceae*
Genus: *Morchella*
Species: *esculenta* (L.) Pers. ex Fr. (Figure 17.1)[2,3]

SPREAD

The distribution range extends from lowland areas to mid-altitude hills, and it grows both in isolation and in groups. It prefers the ecosystem of deciduous forests but can also be found on roadside verges, meadows, parks, orchards, gardens, and wetlands. It thrives particularly well in the habitat of arboreal species such as poplars, elms, ash trees, and acacias.[3,4,24-25]

DESCRIPTION

The apothecia are pedunculate and differentiated into a fertile part considered the cap and a sterile part represented by the stipe.

The cap has a diameter of 2–6 cm and a height of 8–15 cm, with an ovoid, globular, or conical shape and an ochre-yellow color. On the surface of the cap, there are numerous irregular, angular, sinuous alveoli separated by thick, darker-colored crests that are sterile. The inner surface of these alveoli is lined with a hymenial layer composed of asci containing ascospores. The cap is hollow inside and connects to the stipe. The stipe is 3–6 cm tall and 2–3 cm in diameter, with a whitish or ochre color. It is cylindrical, smooth, cracked, thickened at the base, and also hollow inside. The flesh is thin, fragile, white, with a pleasant taste and odor. The spores are ellipsoidal, smooth, and yellowish. This species is edible but should be consumed with caution as it can cause gastric disorders. The Common morel is a precocious species, appearing in early spring from March to June.[3,25]

Morchella esculenta (L.) Pers. ex Fr. presents the following varieties:

■ – var. *rotunda* Pers.,

■ – var. *crassipes* Vent ex Pers.

■ – var. *alba* Boud.[2]

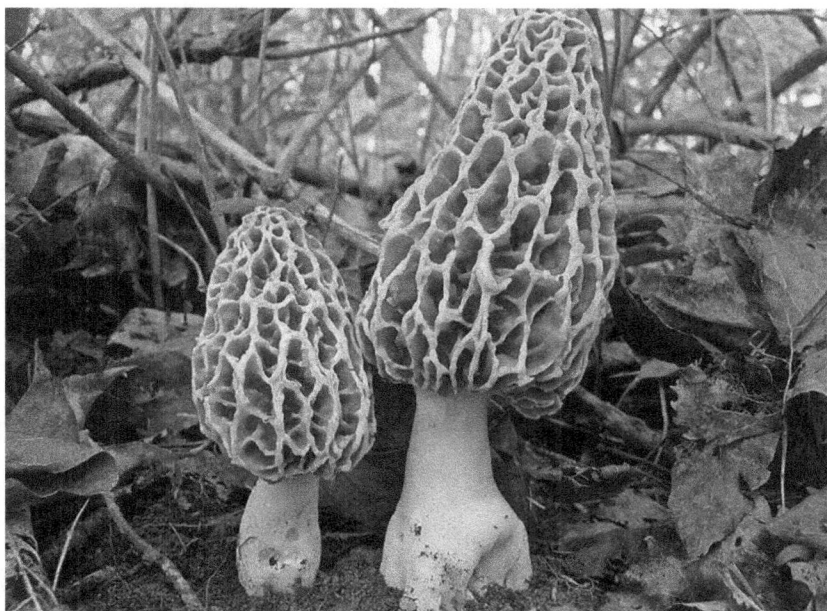

Figure 17.1 *Morchella esculenta* (L.) Pers. ex Fr. (Common morel).

Source: 23.

Main phytoconstituent	Ref.
amino acids (glutamic acid, isoleucine, leucine, tyrosine, tryptophan, valine, cysteine, methionine, aspartic acid)	26-29
peptide & proteins	
alkaloids	
flavonoids	
terpenoids	
nucleosides (adenosine, uridine)	
sterols (campesterol, ergosterol, fungisterol, lanosterol, neoergosterol, brassicasterol)	
fatty acids (linoleic acid, oleic acid, stearic acid, palmitic acid)	
phenolic acids (coumaric acid, caffeic acid, syringic acid, cinnamic acid, gallic acid, tannic acid)	
glycosides	
carbohydrates (glucose, mannose, galactose, rhamnose, xylos, arabinose)	
organic acids (oxalic acid, citric acid, malic acid, quinic acid, fumaric acid, ascorbic acid)	
tocopherols	
aldehyde & ketone	
hydrocarbons	

Medicinal Use

It is frequently employed to alleviate indigestion, address excessive phlegm, and manage asthma. The powdered form of *Morchella esculenta* functions as an antiseptic, aiding in wound healing and providing relief for stomachaches. It also possesses purgative properties and can be utilized as an emollient.[30]

Biological Activity

antioxidant activity; demonstrates strong inhibitory activity against acetylcholinesterase (AChE) and butyryl cholinesterase (BChE), suggesting potential value in the treatment of Alzheimer's and Parkinson's diseases;[26] antitumor, antimicrobial, immunostimulatory and anti-inflammatory properties;[30] hepatoprotective effect;[3] hypocholesterolemic, immunosuppressive[29,31]

Other Uses

■ Besides its great nutritional importance, it includes utilization as animal fodder, as well as in the crafting of thatched roofs and baskets by the indigenous population.[30,32]

TUBERACEAE FAMILY

Includes species of mycorrhizal fungi, equipped with a tuberculiform, globular, irregular, fleshy, occasionally woody ascocarp, with variable size and hypogeous development. The peridium is glabrous, furfuraceous, fuzzy, granular, or verrucous. The gleba is compact, fleshy, traversed by external veins that are oriented towards the apex or base of the ascocarp and open through one or more apertures. Internal veins are darker in color, fertile, and bear hymenium.[3,25]

In Romania, the family includes 2 genera: *Tuber* and *Balsamia*, with a total of 11 species. The genus *Balsamia* in the Romanian flora comprises 2 species: *Balsamia polysperma* Vitt. and *Balsamia vulgaris* Vitt.[3]

The genus *Tuber* includes the following species: *Tuber aestivum* Vitt., *Tuber melanosporum* Vitt., *Tuber rapaeodorum* Tul., inedible *Tuber dryophilum* Tul., *Tuber excavatum* Vitt., *Tuber fulgens* Quél., *Tuber maculatum* Vitt., *Tuber rufum* Pico ex Fr., and *Tuber puberulum* Bk. et Br.[2,3]

Tuber aestivum Vitt. (Summer truffle)

Scientific name: *Tuber aestivum* Vitt.
Common name: Summer truffle
Romanian name: Trufă de vară
Taxonomic classification
Kingdom: *Fungi (Mycophyta)*
Subkingdom: *Dikarya*
Phylum: *Ascomycota*
Subphylum: *Pezizomycotina*
Class: *Hymenoascomycetes*
Subclass: *Pezizomycetideae*
Ordinul: *Tuberales*
Family: *Tuberaceae*
Genus: *Tuber*
Species: *aestivum* Vitt. (Figure 17.2)[2,3]

SPREAD

It grows in oak forests in the warmer regions of Romania, as well as in humus-rich soils or sandy and clayey soils in hilly areas, but it prefers calcareous soils. In deciduous forests, it lives in symbiosis with the roots of hazelnuts, hornbeams, oaks, beeches, and lime trees. It is rarely found in coniferous forests. It develops in soil at a depth of 3–15 cm.[3]

DESCRIPTION

The appearance is tuberculiform, irregular, with sharp pyramid-shaped verrucae, brownish-black. The ascocarp has a diameter of 4–10 cm, is black-brown in color, has a pleasant odor, and typically with hypogeous development, rarely on the soil surface. The peridium is blackish-blue or blackish-brown, with large polygonal verrucae. The gleba is compact, whitish at first, then turning yellow with a fine marbled texture of brownish veins, and eventually becomes brown at maturity, with a fragrant odor and a nutty taste. The asci are pedunculate, spherical or oval, and contain 1–6 ellipsoidal spores with a reticulate membrane. Initially, the spores are colorless, then they become brownish-yellow. It is considered a culinary delicacy.[3,24,25]

Main phytoconstituents	Ref.
amino acids (glutamic acid, isoleucine, leucine, tyrosine, tryptophan, valine, cysteine, methionine, aspartic acid)	31,34
peptide & proteins	
flavonoids (vanillin, catechin, quercetin, myricetin, rutin, kaempferol, luteolin)	
terpenoids	
nucleosides	

Figure 17.2 *Tuber aestivum* Vitt. (Summer truffle).

Source: 33.

sterols (brassicasterol, ergosterol)
fatty acids (linoleic acid, palmitic acid, oleic acid, myristic acid, lauric acid, caproic acid)
phenolic acids (gallic acid, quinic acid, ferulic acid)
glycosides
carbohydrates
organic acids
tocopherols
aldehyde & ketone
hydrocarbons

Medicinal Use

Locals have traditionally employed *Tuber aestivum* to enhance fertility and to alleviate eye disorders and fatigue. This truffle was also used as an aphrodisiac, antidepressant, hypocholesterolemic, and immunostimulant in folk medicine.[35-36]

Biological Activity

anti-inflammatory, antioxidant, antitumor, antihyperglycemic, antimutagenic, antimicrobial, hepatoprotective and antiangiogenic activities[31,34-38]

Other Uses

■ Nutritional value – *Tuber aestivum* is among the most renowned and delicious truffle varieties in the global economic market.[38]

The **phylum *Basidiomycota*** comprises 30,000 species grouped into 1,500 genera. They are characterized by a sporiferous organ called a basidium, which produces basidiospores.

Class *Teliomycetes* (*Hemibasidiomycetes*)
Class *Phragmobasidiomycetes*
Class *Homobasidiomycetes*[2]

CORIOLACEAE FAMILY

In this family, there are saprophytic fungi characterized by semi-circular, sessile fruiting bodies with variable consistency. The family includes the following genera: *Trametes*, *Poria*, *Antrodia*, *Lenzites*, *Pycnoporus*, and *Gloeophyllum*.[2]

Trametes versicolor (L. ex Fr.) Quél. (Turkey tail)

Scientific name: *Trametes versicolor* (L. ex Fr.) Quél. syn. *Coriolus versicolor* (L. ex Fr.) Quél.
Common name: Turkey tail, Many-zoned polypore, Conifer wood
Romanian name: Iasca de cioată a foioaselor[24]
Taxonomic classification
Kingdom: *Fungi*
Subkingdom: *Dikarya*
Phylum: *Basidiomycota* (*Basidiomycetes*)
Subphylum: *Agaricomycotina*
Class: *Homobasidiomycetes*
Subclass: *Aphyllophoromycetideae*
Order: *Polyporales*
Family: *Coriolaceae*
Genus: *Trametes*
Species: *versicolor* (L. ex Fr.) Quél. (Figure 17.3)[2,3]

SPREAD

This mushroom grows in layered colonies on dry hardwood of deciduous trees, occasionally on conifers, and even on wooden structures.[3]

DESCRIPTION

The basidiocarp has a diameter of 4–10 cm and is very thin, sessile, elastic, corky, semicircular, fan-shaped, or reniform. It can be fused at the base with the basidiocarp of other specimens, forming

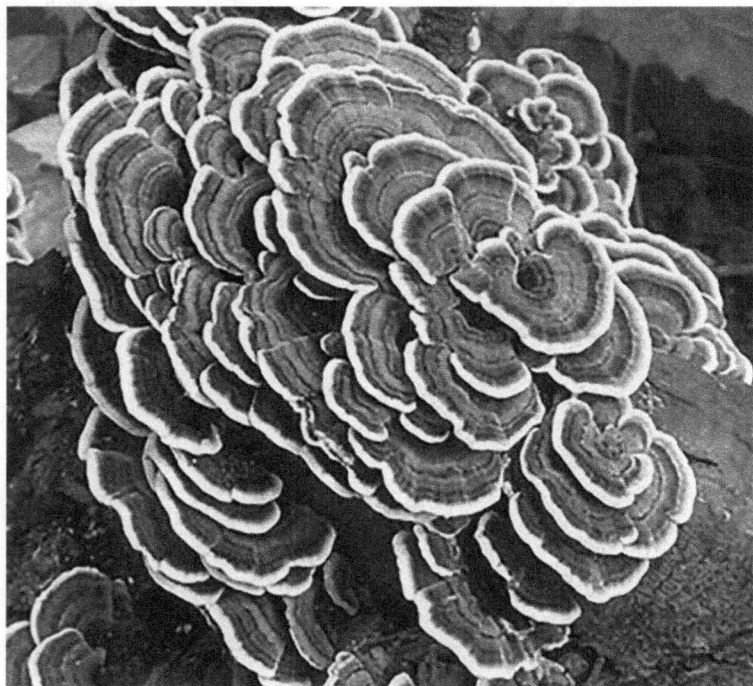

Figure 17.3 *Trametes versicolor* (L. ex Fr.) Quél. (Turkey tail).
Source: 39.

a rosette shape, with a wavy and lobed margin. It exhibits concentric zones that are smooth and shiny alternated with zones that are finely velvety and matte. The basidiocarp displays a range of colors including greenish, brown, grayish-yellow, black, brown-reddish, or brownish-blue. The sporiferous tubes do not exceed a length of 1 mm and have very small pores, which can be slightly angular or rounded, and they are whitish, later becoming pale yellowish. The flesh is thin, tough, corky, and becomes woody at maturity, without a characteristic odor or taste. It is an inedible species and can be found throughout the year.[2,24,25]

Main phytoconstituents   Ref.

amino acids   [31, 40-43]

peptide & proteins

coumarins

flavonoids (quercetin, rutin, kaempferol, apigenin, bailalin, vitexin, daidzein, amentoflavone,
    α-nigerose)

terpenoids (ursolic acid, oleanolic acid, lupenone, friedelin, esculentic acid, corosolic acid,
    lupeol, glutinol, betulinic acid, caryophyllene, copaene, elemene, selinene, bisabolene)

nucleosides

sterols (brassicasterol, ergosterol, cerevisterol)

fatty acids (linoleic acid, palmitic acid, oleic acid, myristic acid, lauric acid, caproic acid,

phenolic acids (p-coumaric acid, syringic acid, gallic acid, quinic acid, vanilic acid, ferulic acid)

glycosides

carbohydrates (glucose, galactose, mannose, arabinose, xylose, rhamnose, arabinose)

organic acids

tocopherols

aldehyde & ketone

hydrocarbons

Medicinal use

For thousands of years, it has enjoyed a prestigious reputation in Japan and China as a medicinal remedy, frequently employed in the management of cancer and hepatitis.[44]

Biological Activity

antioxidant, anti-inflammatory, immunostimulatory, antitumor, antiviral, antimicrobial, antihepatopathy, prebiotic, antihyperlipidemia, antidiabetic, AChE inhibitory activities[41,45]

Other uses

■ This mushroom has been a main choice for florists in commercial floral design. Moreover, it plays a crucial role in forest ecosystems as a recycler of dead or decaying trees around the world.[44]

PLEUROTACEAE FAMILY

It includes saprophytic fungi that grow on the trunks of trees, at a considerable height above the ground. The fruiting body is suspended in the console and attaches to the substrate through a thick, short stipe. The basidiocarp has a fleshy or coriaceous consistency, rarely suberose, is resistant to breaking, and can be either conchoidal or differentiated into a cap and stipe. The gills are very fleshy and very rare. In young specimens, the cap is velvety, while in mature ones, it is smooth and shiny. The stipe has an eccentric insertion, very rarely central, or it may be absent. The flesh is compact, and in some cases, it can be tough. This family is represented in the flora of our country by the following genera: *Panellus, Pleurotus, Panus, Lentinus*.[2,3,25]

The genus *Pleurotus* includes the following species that are found in the spontaneous flora of Romania: *Pleurotus eryngii* (D.C. ex Fr.) Quél., *Pleurotus ostreatus* (Jacq. ex Fr.) Kumm. syn. *P. columbinus* Quél., *Pleurotus cornucopiae* Paul ex Fr., and *Pleurotus dryinus* (Pers. ex Fr.) Kumm.[2]

Pleurotus ostreatus (Jacq. ex Fr.) Kumm. (Oyster mushroom)

Scientific name: *Pleurotus ostreatus* (Jacq. ex Fr.) Kumm. syn. *P. columbinus* Quél.
Common name: Oyster mushroom, Oyster fungus, Hiratake, Pearl oyster mushroom

Figure 17.4 *Pleurotus ostreatus* (Jacq. Ex Fr.) Kumm. (Oyster mushroom).

Source: 46.

Romanian name: Păstrăv de fag
Taxonomic classification
Kingdom: *Fungi*
Subkingdom: *Dikarya*
Phylum: *Basidiomycota* (*Basidiomycetes*)
Subphylum: *Agaricomycotina*
Class: *Homobasidiomycetes*
Subclass: *Agaricomycetideae*
Order: *Tricholomatales*
Family: *Pleurotaceae*
Genus: *Pleurotus*
Species: *ostreatus* (Jacq. ex Fr.) Kumm. (Figure 17.4)[2,3]

SPREAD

The species is widespread in all regions of the world with a tropical or temperate climate. It grows in numerous groups of 1–15 specimens on the trunks of deciduous trees.[2,3]

DESCRIPTION

The cap has a diameter of 5–15 cm, is asymmetrical, subspherical-irregular, initially convex, and then becomes flat and at maturity, it is thick, deeply depressed in the shape of a shell (conchoidal), tongue, or spade. It is fleshy, thick, with a smooth and glossy surface, undulating, and with a slightly upturned margin. The cap can be gray, brown-gray, violet, olive, or blackish, turning yellowish at maturity, sometimes covered with a white spore layer. The gills are whitish or gray, anastomosed at the base of the stipe, occasionally spaced towards the edge. The stipe is 1–4 cm to 10–15 cm in height, when there are more fungi developed in a group, and 1–3 cm in thickness, compact, solid, thickened at the top and narrowed in the inferior part. It is white, sometimes covered with fine white fuzz, with an eccentric, lateral insertion. The flesh is white, elastic, soft, tender, with a floury smell and a sweet taste. The spores are smooth, ellipsoidal, initially pink-lilac, and then hyaline. It is an edible species, highly prized, and develops almost throughout the year, from spring to late autumn.[3,4,25]

Main phytoconstituents	Ref.
amino acids (phenylalanine, threonine, lysine, valine, methionine, isoleucine, tryptophane, leucine)	
	31,47-49

peptide & proteins (pleurostrin, lectine)
alkaloids
coumarins
flavonoids (rutin, myrictin, chrysin, naringin)
terpenoids (oleanolic acid, ursolic acid)
nucleosides
sterols (stigmaterol, ergosterol, sitosterol)
fatty acids (palmitic acid, stearic acid, oleic acid, linoleic acid, behemic acid)
phenolic acids (gallic acid, protocatechuic acid, homogentisic acid)
glycosides
carbohydrates (glucose, manose, arabinose, rhamnose, xylose, glucan)
organic acids (ascorbic acid)
tocopherols
aldehyde & ketone
hydrocarbons

Medicinal Use

Pleurotus ostreatus enjoys worldwide popularity as a culinary delight because of its delicious taste, distinctive flavor, impressive nutritional content, and medicinal properties. Due to its rich nutritional composition and multiple active ingredients, it has been used in traditional medicine for its diverse range of benefits, including its role in managing diabetes, combating bacterial infections, regulating cholesterol levels, alleviating arthritis symptoms, providing antioxidant protection, potential anticancer properties, supporting eye health, and demonstrating antiviral effects.[50]

Biological Activity

antioxidant, anti-inflammatory, anticarcinogenic, antiviral, antifungal, antibacterial, antidiabetic, antiangiogenic, immunomodulatory, hypoglycemic, and hepatoprotective pharmacological properties.[51]

Other Uses

■ It receives recognition as a nutraceutical functional food, due to its significant nutritional content and therapeutic properties;

■ Aroma compounds hold significant importance in both the perfume industry and the production of cosmetics. Creams incorporating *Pleurotus ostreatus* exhibited beneficial effects as a supportive therapy for atopic dermatitis.[51]

CANTHARELLACEAE FAMILY

Comprises species of terrestrial saprophytic fungi with differentiated basidiocarp, consisting of cap and stipe, sometimes with a conical or trumpet-like appearance and fleshiness or slight coriaceous consistency. The fertile surface has thick, dichotomously branched, decurrent folds, resembling radially arranged gills. Occasionally, they may have the shape of veins, wrinkles, or a smooth aspect. The stipe can be either fistulous or full, and the spores are smooth, oval, and white in color.[2,3,25]

In the flora of Romania, the following species grow: *Cantharellus cibarius* Fr., *Cantharellus lutescens* (Pers.) Kűhn-Romagn., *Cantharellus friesii* Quél., *Cantharellus tubaeformis* Bull. ex Fr.[2]

Cantharellus cibarius Fr. (Golden chanterelle)

Scientific name: *Cantharellus cibarius* Fr.
Common name: Golden chanterelle, Girolle
Romanian name: Bureți galbeni
Taxonomic classification
Kingdom: *Fungi*

Figure 17.5 *Cantharellus cibarius* Fr. (Golden chanterelle).

Source: 52.

Subkingdom: *Dikarya*
Phylum: *Basidiomycota (Basidiomycetes)*
Subphylum: *Agaricomycotina*
Class: *Homobasidiomycetes*
Order: *Cantharellales*
Family: *Cantharellaceae*
Genus: *Cantharellus*
Species: *cibarius* Fr. (Figure 17.5)[2,3]

SPREAD

It grows in deciduous forests, forming symbiotic relationships with oak and beech roots, but it prefers damp and shaded coniferous forests in the mountainous region. It grows on the ground in clusters, rarely solitary. It appears in large numbers after warm rains.[3,4]

DESCRIPTION

The cap has a diameter of 3–10 cm, is fleshy, tough, and compact. It is initially spherical and convex, then slightly depressed, and at maturity, it becomes flat with a central depression like a hat. The cap surface is matte and yellow, becoming yellowish-white at maturity, with an irregular, wavy edge, sometimes cracked and downturned. The fertile layer on the inferior side of the cap is in the form of folds (pseudogills) that resemble gills. These folds are thick, decurrent on the stipe, and bifurcate toward the edge of the cap. The stipe measures 3–6 cm in height and 0.8–2 cm in thickness, with the same color as the cap. It is robust, fleshy, full, smooth, fibrous, tapered at the base, and sometimes curved. In the superior part, it continues into the cap. The flesh is compact, fibrous, tough, and elastic, with a light yellow color, a sweet taste, and a fruity smell. Those growing in coniferous forests have a peppery and bitter taste. The spores are ovoid, smooth, and hyaline. This species grows from May to September and is edible, well-known, and apprecia ted.[3,4,24,25]

The species has 2 varieties:

■ var. *pallidus* R. Sch. has a distribution area limited to oak and oak-hornbeam forests.

■ var. *amethysteus* Quél. grows in beech forests in the mountainous region and has a cap covered with a fluffy layer of lilac color.[2]

Main phytoconstituents	Ref.
	32, 53-54

amino acids
peptides & proteins
alkaloids
coumarins
flavonoids (quercetin, rutin, myrictin)
terpenoids (phenandrene, limonene, pinene)
nucleosides
sterols (stigmaterol, ergosterol, sitosterol)
fatty acids (myristic, linoleic acid, oleic acid, pentadecanoic acid, palmitic acid, stearic acid)
phenolic acids (caffeic acid, gallic acid, p-hydroxybenzoic acid, protocatechuic acid)
glycosides
carbohydrates (glucose, manose, arabinose, rhamnose, xylose, glucan)
organic acids (malic acid, citric acid)
tocopherols
aldehyde & ketone
hydrocarbons

Medicinal Use

Cantharellus cibarius has a history of use in traditional medicine due to its beneficial properties, which encompass antibacterial and anticancer effects, antioxidant capabilities, antihypoxic properties, antihyperglycemic potential, wound-healing attributes, and anti-inflammatory qualities.[55]

Biological Activity

immunomodulatory, anti-inflammatory, antioxidant, neuroprotective, antiviral, antihypertensive, antimicrobial, wound-healing, cytotoxic and antigenotoxic properties[56-60]

Other Uses

- High nutritional value (culinary use); it is considered one of the most valued and currently commonly harvested types of edible mushrooms in Europe.[60]

AGARICACEAE FAMILY

Comprises saprophytic mushrooms, sometimes coprophilous, consisting of a stipe and cap that may be easily detachable or inseparable, and they have a fleshy consistency. The species typically have a fleshy, soft cap, usually furfuraceous, squamous, flocculus, non-viscous, often umbonate. The centrally located stipe is fleshy with a fibrous structure, it has a membranous ring, sometimes with a curtain, without a volva. The flesh is white but changes color with age or upon exposure to air, turning yellowish or reddish.[3,25]

The spontaneous flora of Romania includes the following genera: *Agaricus, Melanophyllum, Lepiota, Macrolepiota, Leucocoprinus,* and *Leucoagaricus.*[2]

The genus *Agaricus* (*Psalliota*) in the spontaneous flora of Romania includes the following species: *Agaricus arvensis* Sch. ex Fr., *Agaricus campestris* (L.) Fr., *Agaricus bitorquis* Quel., *Agaricus hortensis* (Cke) Pil., *Agaricus bisporus* (Lge) Sing., *Agaricus silvaticus* Schff. ex Secr., *Agaricus silvicola* (Vitt.) Sacc., *Agaricus vaporarius* (Pers. ex Vitt.) Mos, *Agaricus augustus* Fr., *Agaricus haemorrhoidarius* Kalchbr. et Schulz,[3] *Agaricus comtulus* Fr. var. *rusiophyllus* Lasch., *Agaricus semotus* Fr. var. *purpurellus* Moel.,
Agaricus macrosporus (Moel. ex Schff.) Pil.[2]

Inedible: *Agaricus abruptibulus* Peck., *Agaricus bernardii* (Quel.) Sacc., *Agaricus langei* Moell., *Agaricus maskae* Pil., *Agaricus meleagris* J. Schff., and *Agaricus perrarus* Schulz. *Agaricus xanthodermus* Genecier cause gastrointestinal disturbances.[2]

Agaricus arvensis Sch. (Horse mushroom)

Scientific name: *Agaricus arvensis* Sch.
Common name: Horse mushroom
Romanian name: Ciuperca de câmp

Figure 17.6 *Agaricus arvensis* Sch. (Horse mushroom).

Source: 61.

Taxonomic classification
Kingdom: *Fungi*
Subkingdom: *Dikarya*
Phylum: *Basidiomycota* (*Basidiomycetes*)
Subphylum: *Agaricomycotina*
Class: *Homobasidiomycetes*
Subclass: *Agaricomycetideae*
Order: *Agaricales*
Family: *Agaricaceae*
Genus: *Agaricus*
Species: *arvensis* Sch. (Figure 17.6)[2,3]

SPREAD

It is a terrestrial species that is distributed from the hill zone to the mountain zone through meadows, clearings, deciduous and coniferous forests, pastures, gardens, forest edges, parks, orchards, grassy fields. The species grows either isolated or in numerous groups, called "circles of witches."[3,4,25]

DESCRIPTION

This species has a cap with a diameter of 8–20 cm, initially ovoid, campanulate, hemispherical and later becoming convex or slightly flattened, sometimes slightly mamillate, and it is white, velvety, smooth, and fleshy. The color changes to yellowish-white or dirty-yellow-ochraceous as it ages. The cap of mature specimens has fibrous brown or yellowish scales, with small, uneven, fragile fringes and remnants of veil on the curved-down edges. The gills are dense, thin, free, initially white, then pink, red-gray, and at maturity brown or brownish-black, with striated, light-colored edges. The stipe is 6–15 cm tall and 1–2 cm thick, cylindrical, straight, thickened at the base, without a volva, with a central insertion on the cap and a slightly visible bulb at the base. In the superior half, it has a double ring with fringed edges, turned downwards, membranous, initially white and later turning yellow. The stipe starts smooth and full but becomes hollow with age. The flesh is compact, tender, soft, white, and turns ochre when broken. This mushroom is edible, has an anise-like odor, and a mild taste. The spores are ovoid, elliptical, smooth, and purplish-brown in color. The fruiting bodies of the mushroom appear from May to November.[3,4,25]

Main phytoconstituents
Ref.
amino acids
[31,62]
peptide & proteins (lectins)
alkaloids
coumarins
flavonoids (myricetin)
terpenoids
nucleosides
sterols (stigmaterol, ergosterol, sitosterol)
fatty acids (palmitic acid, stearic acid, oleic acid, linoleic acid, behemic acid)
phenolic acids (chlorogenic acid, gallic acid, caffeic acid, p-coumaric acid, ferrulic acid,
    trans-cinnamic acid, protocatechuic acid)
glycosides
carbohydrates (glucan)
organic acids (oxalic acid, fumaric acid, lactic acid, acetic acid, succinic acd)
tocopherols
aldehyde & ketone
hydrocarbons

Medicinal Use

In traditional medicine, it has been employed for its antioxidant properties. In China, it is used because of its anticancer attributes and has been utilized for alleviating lower back pain as well as pain in tendons and veins.[63]

Biological Activity

antioxidant, antimicrobial, interferon-like, cytotoxic, and apoptotic activities.[62-66]

Other Uses

■ *Agaricus arvensis* is regarded as one of the most delicious edible fungi, also being used in the nutritional supplements industry.[63]

Agaricus campestris (L.) Fr. (Field mushroom)

Scientific name: *Agaricus campestris* (L.) Fr.
Common name: Field mushroom, Meadow mushroom
Romanian name: ciuperca de bălegar
Taxonomic classification
Kingdom: *Fungi*
Subkingdom: *Dikarya*
Phylum: *Basidiomycota* (*Basidiomycetes*)
Subphylum: *Agaricomycotina*
Class: *Homobasidiomycetes*
Subclass: *Agaricomycetideae*
Order: *Agaricales*
Family: *Agaricaceae*
Genus: *Agaricus*
Species: *campestris* (L.) Fr. (Figure 17.7)[2,3]

SPREAD

It grows on humus-rich, sufficiently moist, and sunny soils, preferably in places where decomposed manure is present. It is found in meadows, orchards, parks, gardens, and occasionally in forests, typically in isolation or in small to larger groups, forming "circles of witches."[3]

DESCRIPTION

The cap has a diameter of 5–10 cm, with some specimens reaching up to 15 cm. It is initially globular, convex, then semispherical, and towards the end of the development, it becomes somewhat flattened, plain. The surface is soft, smooth, silky-fibrous or with brown scales,

Figure 17.7 *Agaricus campestris* (L.) Fr. (Field mushroom).

Source: 67.

sometimes it has cracked edges and colors ranging from white to dirty-white, yellowish-white and easily detaches from the stipe. The gills are broad, free, distanced and initially white, turning light-pink, dark-pink and then brownish-black as they reach the end of the development period. The stipe is 3–8 cm tall and 1–2 cm thick, cylindrical or fusiform, straight, firm, full, smooth, slightly squamose in the superior part, and rather thickened at the base, white or the same color as the cap. There is a thin, narrow, caducous, soft, fragile, downward-facing ring on the stipe. The flesh is soft and thick, white in the cap and superior stipe, turning yellowish and then rusty brown in the inferior stipe. When sectioned, it turns pink. The mushroom has an anise-like smell and a good taste, making it an appreciated edible species. The spores are ovoid, smooth, and brown to dark purple-brown in color. This species grows from spring to autumn.[2,3,24]

Main phytoconstituents

Ref.
[31,62]

amino acids
peptides & proteins
alkaloids
coumarins
flavonoids (myrictin)
terpenoids
nucleosides
sterols (stigmaterol, ergosterol, sitosterol)
fatty acids (palmitic acid, stearic acid, oleic acid, linoleic acid, behemic acid)
phenolic acids (gallic acid, caffeic acid, *p*-hydroxybenzoic acid, p-coumaric acid, ferrulic acid)
glycosides
carbohydrates (glucan)
organic acids (citric acid, formic acid, fumaric acid, lactic acid, malonic acid, oxalic acid, succinic acid)
tocopherols
aldehyde&ketone
hydrocarbons

Medicinal Use

Agaricus campestris has primarily been used in the treatment of ulcers, as material for fungal dressings, and for addressing bed sores. In various regions of Scotland, *Agaricus* slices have been applied to scalds and burns, serving as traditional remedies.[68]

Biological Activity

antioxidant, antimicrobial, antidiabetic, haem-agglutination and anticancer properties;[68-70] natural larvicidal agent against *Aedes aegypti* (vector mosquito for dengue, chikungunya, and yellow fever).[71]

Other Uses

- culinary use: in rural areas of Romania, is harvested from spontaneous mycoflora and consumed because of its nutritional value.[72]

Agaricus bisporus (Lge) Sing. (Button mushroom)

Scientific name: *Agaricus bisporus* (Lge) Sing. (J. E. Lange) Imbach.
Common name: Button mushroom, Table mushroom, Cultivated mushroom
Romanian name: Ciuperca de cultură
Taxonomic classification
Kingdom: *Fungi*
Subkingdom: *Dikarya*
Phylum: *Basidiomycota* (*Basidiomycetes*)
Subphylum: *Agaricomycotina*
Class: *Homobasidiomycetes*
Subclass: *Agaricomycetideae*
Order: *Agaricales*
Family: *Agaricaceae*
Genus: *Agaricus*
Species: *bisporus* (Lge) Sing. (Figure 17.8)[2,3]

SPREAD

Its distribution range spans from lowland areas to medium altitudes in mountainous regions, commonly found in parks, meadows, pastures, open fields, forest edges, orchards, and in places rich in humus, on manure, and along road edges. It grows both in isolation and in groups.[3,25]

Figure 17.8 *Agaricus bisporus* (Lge) Sing. (Button mushroom).

Source: 73.

DESCRIPTION

The cap has a diameter of 5–10 cm and is initially white-brown, globose, later becoming convex, and at the end of the development stage completely flat, compact, fleshy, silky, with a membranous edge, often with fuzz, slightly striate, and covered with small brown scales, or folded, or with spots or cracks. The stipe is 3–8 cm in height and 1–2 cm thick, cylindrical, slightly domed at the base, straight, smooth, full, and in older specimens, it is hollow in the middle. It is white, with a tendency to become red under the cap. In the proximity of the cap, the stipe has a persistent, thick, white ring. The basidial gills are pink, thin, and very tall. The flesh is consistent, white, with well-defined odor and taste. The mushroom typically grows from June to September.[3,25]

Main phytoconstituents   Ref.

amino acids (valine, isoleucine, methionine, leucine, lysine, threonine, phenylalanine,
    tryptophane)   [31, 49, 63, 74]
peptide & proteins (lectine)
alkaloids
coumarins
flavonoids (myrictin)
terpenoids (pinene, limonene, phenandrene)
nucleosides
sterols (stigmaterol, ergosterol, sitosterol)
fatty acids (palmitic acid, stearic acid, oleic acid, linoleic acid, behemic acid)
phenolic acids (gallic acid, caffeic acid, *p*-hydroxybenzoic acid, p-coumaric acid, chlorogenic
    acid, feruulic acid, *trans*-cinnamic acid, protocatechuic acid)
glycosides
carbohydrates (glucose, manose, arabinose, rhamnose, xylose, glucan)
organic acids (acetic acid, formic acid, fumaric acid, malonic acid, lactic acid, malic acid,
    succinic acid, oxalid acid)
tocopherols
aldehyde & ketone
hydrocarbons

Medicinal Use

It has been used in traditional medicine to treat heart diseases, cancer, diabetes, conditions related to the immune system, viral, bacterial, and fungal infections.[75]

Biological Activity

■ antitumor, antioxidant, antiobesity, antidiabetic, antimicrobial, immunomodulatory, cardiovascular, hepatoprotective, antinociceptive, hypoglycemic, hypolipidemic and anti-inflammatory activities;

■ dermatological properties: bioactive compounds extracted from *A. bisporus* have beauty-enhancing benefits and serve as important agents in combatting skin inflammation, signs of aging, and hyper-pigmentation. They also exhibit antioxidant properties and possess antibacterial actions. Furthermore, shampoos can be formulated to address various hair issues such as dandruff, oily hair, and hair loss;[75-79]

■ natural larvicidal agent against *Aedes aegypti* (vector mosquito for dengue, chikungunya, and yellow fever).[71]

Other Uses

■ functional food: in Romania, 90% of mushroom production is covered by *Agaricus bisporus* species (champignon), being consumed primarily in urban areas because of its nutritional value.[72]

The genus *Macrolepiota* is represented in the Romanian flora by *Macrolepiota procera* (Scop. ex Fr.) Sing. and *Macrolepiota rhacodes* (Vitt.) Sing.[3]

Macrolepiota procera (Scop. ex Fr.) Sing. (Parasol mushroom)

Scientific name: *Macrolepiota procera* (Scop. ex Fr.) Sing.
Common name: Parasol mushroom
Romanian name: Burete șerpesc[3,4]
Taxonomic classification
Kingdom: *Fungi*
Subkingdom: *Dikarya*
Phylum: *Basidiomycota* (*Basidiomycetes*)
Subphylum: *Agaricomycotina*
Class: *Homobasidiomycetes*
Subclass: *Agaricomycetideae*
Order: Agaricales
Family: *Agaricaceae*
Genus: *Macrolepiota*
Species: *procera* (Scop. ex Fr.) Sing. (Figure 17.9)[2,3]

SPREAD

This species is terrestrial and grows from lowland to mountainous regions, primarily in deciduous or coniferous forests, thickets, meadows, along roadsides and in acacia plantations. It can be found both isolated and in "circles of witches."[3,4,25]

DESCRIPTION

This species has a large cap, with a diameter ranging from 10 to 30 cm, sometimes even up to 50 cm. Initially, it is ovoid, then becomes campanulate, semispherical, and at its maximum development, it is flat with a central, smooth, uniformly colored in brown mamelon. The cap has a light brown-reddish, gray, or dark brown color with numerous thick scales that are darker than the cap itself, which are arranged in concentric circles, denser towards the center and sparser towards the periphery. The margin is fringed and whitish-gray. The gills are wide, free, fleshy, and almost folded, originally white, gray, or yellowish and turning brown as they mature. The stipe is tall, ranging from 15 to 40 cm, and thick, measuring 2–4 cm in diameter. It is straight, cylindrical, fistulous, compact, fragile, bulbous, and white under the cap. The stipe has a hard, fibrous consistency, whitish-gray color, and it is covered in a zigzag pattern of brown, fibrous scales, resembling the skin of a snake. The stipe can detach from the cap, and in its superior half,

Figure 17.9 *Macrolepiota procera* (Scop. ex Fr.) Sing. (Parasol mushroom).

Source: 80.

it features a thick, double, lacy, mobile ring that can slide towards the inferior part. This ring is whitish towards the cap and brownish-gray towards the base. The flesh inside the cap is thin, soft, elastic, white, sometimes pinkish, with a sweet taste of nuts or hazelnuts and a pleasant smell, while the flesh in the stipe is fibrous and woody. The spores are ellipsoidal, hyaline, and smooth. The mushroom is edible and typically appears from July to October.[3,4,24,25]

Main phytoconstituents	Ref.
amino acids	[80,81]

amino acids
peptide & proteins
alkaloids
coumarins
flavonoids (quercetin)
terpenoids (lepiotaprocerins)
nucleosides
sterols
fatty acids (linoleic acid, oleic acid, palmitic acid, stearic acid, palmitoleic acid, pentadecanoic acid)
phenolic acids
glycosides
carbohydrates (glucan)
organic acids
tocopherols
aldehyde & ketone
hydrocarbons
tannins

Medicinal Use

Macrolepiota procera has been used in traditional medicine in the treatment of diabetes, hypertension, and inflammation, also being able to enhance the immune system.[82]

Biological Activity

antibacterial, antifungal, antiviral, antioxidant, anti-inflammatory, regulatory, antidepressant, and anticancer activities[83-88]

Other Uses

- culinary use (nutraceutical)
  - Green synthesis of silver nanoparticles using *Macrolepiota procera* have applications as antitumor agents[89]

The genus *Lepiota* is represented in the flora of our country by the following species: *Lepiota excoriata* (Schaeff. ex Fr.) Sing., *Lepiota mastoidea* (Fr.) Sing., *Lepiota clypeolaria* (Bull. ex Fr.) Kumm., *Lepiota alba* (Bres.) Sacc., and *Lepiota ventriosospora* Reid.[3]

Lepiota mastoidea (Fr.) Sing. (Slender parasol)

Scientific name: *Lepiota mastoidea* (Fr.) Sing.
Common name: Slender parasol
Romanian name: Nană de pădure
Taxonomic classification
Kingdom: *Fungi*
Subkingdom: *Dikarya*
Phylum: *Basidiomycota* (*Basidiomycetes*)
Subphylum: *Agaricomycotina*
Class: *Homobasidiomycetes*
Subclass: *Agaricomycetideae*
Order: Agaricales
Family: *Agaricaceae*
Genus: *Lepiota*
Species: *mastoidea* (Fr.) Sing. (Figure 17.10)[3]

Figure 17.10 *Lepiota mastoidea* (Fr.) Sing. (Slender parasol).

Source: 90.

SPREAD

It grows in small groups, in forest edges, and meadows.[3]

DESCRIPTION

The cap has a diameter of 8–14 cm, is granulose-furfuraceous, initially globose, becoming convex at maturity. It is white or cream-ocher, covered with numerous granulose scales of ochre color, rarer towards the edge of the cap, and with a prominent, sharp mamelon at the center. The gills are dense, soft to the touch, and range from white-cream to pale yellow. The stipe measures 8–11 cm in height and 0.8–1.5 cm in thickness, it is white, cylindrical, slightly bulbous, covered with yellowish-ochre scales, and it has a fleshy, free ring that slides towards the inferior side. The flesh is soft, white, compact, with a taste of hazelnuts and a pleasant odor. The spores are elliptical. This species is edible.[3,25]

Main phytoconstituents	Ref.
amino acids	[91]
peptide & proteins	
alkaloids	
coumarins	
flavonoids (rutin, myrictin, chrysin, naringin)	
terpenoids	
nucleosides	
sterols (stigmaterol, ergosterol, sitosterol)	
fatty acids (palmitic acid, stearic acid, oleic acid, linoleic acid)	
phenolic acids (cinnanic acid, p-coumaric acid, p-hydroxybenzoic acid)	
glycosides	
carbohydrates	
organic acids (malic acid, oxalic acid)	
tocopherols	
aldehyde & ketone	
hydrocarbons	

Medicinal Use

In the area where it is collected, *Lepiota mastoidea* is employed as a medicinal treatment for stomach and heart conditions by the local population.[92]

Biological Activity

antioxidant, antimicrobial and cytotoxic activities[91,93]
inhibitory activity against enzymes involved in diabetes and Alzheimer's disease[94]

Other Uses

Lepiota mastoidea is a frequently consumed wild mushroom in the collection area, prized for its nutritional content (comprising 33% protein, 55% total carbohydrates, and 4.5% crude fat) whenever it is abundantly present. To ensure long-term availability, some locals opt to dehydrate fresh mushrooms. These aspects collectively enhance its economic significance.[92]

RUSSULACEAE FAMILY

It includes saprophytic, terrestrial fungi that form ectomycorrhizae and do not have a volva, ring, or veil. The basidiocarp consists of an inseparable stipe and cap. The cap is convex, then flat, and sometimes centrally depressed, brightly colored. The surface of the cap is smooth, often viscous. The gills are nearly attached and highly curved on the inferior side. The stipe is centrally attached to the cap, cylindrical, non-fibrous, usually short and thick, full, often spongy or hollow. The flesh is almost always white, granular, crumbly, with latex and a taste that can be sweet, bitter, or peppery. The spores are elongated, white, or yellowish. The family comprises 2 genera in the Romanian flora: *Russula* and *Lactarius*.[2,3,25]

The genus *Lactarius* has the following representatives in the spontaneous flora of Romania: *Lactarius piperatus* (L. ex Fr.) S.F. Gray, *Lactarius semisanguifluus* Heim et Leclair, *Lactarius deliciosus* (L. ex Fr.) S. F. Gray, *Lactarius glyciosmus* Fr., *Lactarius lilacinus* Lasch. ex Fr., *Lactarius lignyotus* Fr., *Lactarius mammosus* Fr., *Lactarius salmonicolor* Heim et Leclair, *Lactarius subdulcis* Bull. ex Fr., *Lactarius volemus* Fr.; inedible: *Lactarius acerrinus* Britz., *Lactarius aspideus* Fr., *Lactarius badiosanguineus* Kűhn. et Romagn., *Lactarius blennius* Fr., *Lactarius bresadolianus* Sing., *Lactarius camphoratus* (Bull.) Fr., *Lactarius circellatus* Fr., *Lactarius controversus* Pers. ex Fr., *Lactarius cyathula* Fr., *Lactarius decipiens* Quel., *Lactarius flexuosus* Fr., *Lactarius fuliginosus* Fr., *Lactarius hysginus* Fr., *Lactarius ichoratus* Batsch. ex Fr., *Lactarius insulsus* Fr., *Lactarius lacunarum* Romagn. ex Hora, *Lactarius mitissimus* Fr., *Lactarius pallidus* Pers. ex Fr., *Lactarius pergamenus* (Swartz ex Fr.) Fr., *Lactarius picinus* Fr., *Lactarius pterosporus* Romagn., *Lactarius pyrogalus* Bull. ex Fr., *Lactarius quietus* Fr., *Lactarius rufus* (Scop.) Fr., *Lactarius sphagneti* Fr., *Lactarius serifluus* DC ex Fr., *Lactarius theiogalus* (Bull.) Fr., *Lactarius trivialis* Fr., *Lactarius turpis* (Weinm.) Fr., *Lactarius uvidus* Fr., *Lactarius vellereus* (Fr.) Fr., *Lactarius vietus* Fr., *Lactarius violascens* (Otto) Fr., and *Lactarius zonarius* Bull. ex Fr.

Some *Lactarius* spp. cause gastrointestinal disorders: *Lactarius acris* Bolt. ex Fr., *Lactarius azonites* Bull. ex Fr., *Lactarius chrysorrheus* Fr., *Lactarius helvus* Fr., *Lactarius pubescens* Fr., *Lactarius repraesentaneus* Britz., *Lactarius scrobiculatus* (Scop. ex Fr.) Fr., and *Lactarius torminosus* (Schff. ex Fr.) S. F. Gray.[2]

Lactarius piperatus (L. ex Fr.) S.F. Gray (Peppery milkcap)

Scientific name: *Lactarius piperatus* (L. ex Fr.) S.F. Gray
Common name: Peppery milkcap
Romanian name: Burete piperat,[3] Iuțari[2]
Taxonomic classification
Kingdom: *Fungi*
Subkingdom: *Dikarya*
Phylum: *Basidiomycota* (*Basidiomycetes*)
Subphylum: *Agaricomycotina*
Class: *Homobasidiomycetes*
Subclass: *Agaricomycetideae*
Order: *Russulales*
Family: *Russulaceae*
Genus: *Lactarius*
Species: *piperatus* (L. ex Fr.) S.F. Gray (Figure 17.11)[2,3]

Figure 17.11 *Lactarius piperatus* (L. ex Fr.) S.F. Gray (Peppery milkcap).

Source: 95.

SPREAD

It is widespread in mountainous areas, rarely in lowland regions, in oak, oak-hornbeam, and coniferous or mixed forests. It forms numerous groups on the ground.[3]

DESCRIPTION

The cap has a diameter of 8–20 cm, initially hemispherical, then flattened or broadly infundibuliform, brittle, fleshy, smooth, and glabrous, with a rolled margin. As it matures, the margin becomes flat, wavy, crested, forming lobes. Initially, the cap is white, matte, and towards maturation, it turns yellow or ochraceous. During drought periods, it cracks into small plates. The gills are white, thin, close together, decurrent down the stipe, often bifurcating at the base, and they turn yellow with age. The stipe is slender, 4–8 cm tall and 2–4 cm thick, cylindrical, white, tough, full, slightly narrowed at the base, with a smooth surface, and sometimes slightly lumpy. The flesh is white, later turning yellowish, brittle, tough, and when broken, it snaps and secrets a white latex with a peppery taste. Upon drying, it turns greenish, and with potassium hydroxide, it becomes golden-yellow. The spores are white, verrucose, and ovoid. This species is edible and develops from July to September.[3,4,24,25]

Main phytoconstituents	Ref.
amino acids (leucine, valine, threonine, phenylalanine, isoleucine, lysine, tryptophane)	49,74

alkaloids
coumarins
flavonoids (rutin, myrictin, chrysin, naringin)
terpenoids (pinene, limonene)
nucleosides
sterols (stigmaterol, ergosterol, sitosterol)
fatty acids (palmitic acid, stearic acid, oleic acid, linoleic acid)
phenolic acids (gallic acid, protocatechuic acid, homogentisic acid)
glycosides
carbohydrates (glucose, manose, arabinose, rhamnose, xylose)
organic acids (ascorbic acid)
tocopherols
aldehyde & ketone
hydrocarbons

Medicinal Use

Lactarius piperatus has been used in traditonal medicine as protective agent to help human body reduce oxidative damage, boost the immune system, prevent cancer, and fight infections.[96]

Biological Activity

antioxidant, antimicrobial, genoprotective, anticancer, and neuroprotective potential[97-98]

Other Uses

- The mushroom is dried and grinded, then used as a substitute for pepper.[4,25]
- An efficient biosorbent, used to remove dyes from aqueous solutions.[99]

BOLETACEAE FAMILY

It includes terrestrial fungi, symbiotic, and occasionally saprophytic on tree trunks. The basidiocarp consists of fleshy, inseparable, often massive and robust cap and stipe. The cap has smooth, glabrous, hairy, squamose, velvety, or viscous cuticle, and is sometimes fissured. The stipe, typically with central insertion, but occasionally eccentric, can be smooth or reticulate, and sometimes has a membranous or viscous ring. The soft hymenophore is composed of tubes that are fused together, slightly detachable from the cap, rarely inseparable. It opens in the superior part through round or polygonal pores. The flesh is soft and putrescible. The spores are fusiform or ellipsoidal, and smooth.[2,3,25]

In the Romanian flora, there are found the following species from the genus *Boletus*: *Boletus aereus* Bull. ex Fr., *Boletus appendiculatus* Schff., *Boletus edulis* Bull. ex Fr., *Boletus frechtneri* Vel., *Boletus pinicola* (Vitt.) Venturi, *Boletus regius* Krbh., *Boletus reticulatus* Schff. ex Boud., *Boletus duriusculus* Schulz., *Boletus erythropus* (Fr.) Kbch., *Boletus fragrans* Vitt., *Boletus aeruginascens* Secr., *Boletus crocipodius* (Let.) Kűhn., *Boletus griseus* Quel., *Boletus impolitus* Fr. syn. *Boletus obsonium* Fr., *Boletus luridus* Schff. ex Fr., *Boletus piperatus* Fr. syn. *Chalciporus piperatus* (Bull. ex Fr.) Bat., *Boletus pseudorubinus* Thirring., *Boletus purpureus* Fr., *Boletus quéletii* Schulz, *Boletus rhodoxanthus* Kbch., *Boletus sulphureus* Fr., *Boletus testaceo-scabrus* Secr.; inedible: *Boletus calopus* Fr., *Boletus flavidus* Fr., *Boletus oxydabilis* Sing., *Boletus porphyrosporus* (Fr. et Hők) Bat. syn. *Porphyrellus porphyrosporus* (Fr.) Gilb., *Boletus radicans* Pers. And ex Fr., *Boletus vulpinus* Wat.; causing gastrointestinal disorders: *Boletus lupinus* Fr., and causing severe intoxications, sometimes lethal: *Boletus satanas* Lenz.[2]

Boletus edulis Bull. ex Fr. (Porcini)

Scientific name: *Boletus edulis* Bull. ex Fr.
Common name: Porcini, King mushroom, Cep
Romanian name: Hrib, Mânătarcă
Taxonomic classification
Kingdom: *Fungi*
Subkingdom: *Dikarya*
Phylum: *Basidiomycota (Basidiomycetes)*
Subphylum: *Agaricomycotina*
Class: *Homobasidiomycetes*
Subclass: *Agaricomycetideae*
Order: *Boletales*
Family: *Boletaceae*
Genus: *Boletus*
Species: *edulis* Bull. ex Fr. (Figure 17.12)[2,3]

SPREAD

This species is found on the ground in deciduous and coniferous forests, either isolated or in groups. It prefers warm and sunny locations, especially in forest edges and along paths. It is present in hilly and mountainous areas, but rarely in lowland regions.[3,25]

Figure 17.12 *Boletus edulis* Bull. ex Fr. (Porcini).

Source: 100.

DESCRIPTION

The cap has a diameter of 7–15 cm, it is fleshy, initially hemispherical, then stretched, and in wet weather, it is viscous and matte. The cap is glabrous, glossy, and light to dark brown. The stipe is 10–15 cm tall and 4–6 cm thick, thick, robust, cylindrical, compact, and light brown. Under the cap, it has a reticulate, dense, whitish appearance. The tubes are thin, long, free, whitish, then yellow-green, and detach easily from the cap. The pores are small and circular. The spore dust is brown-violet. The flesh is compact in young specimens and spongy in mature ones, white, red under the cuticle, with a nutty taste, and a pleasant odor. The spores are fusiform or ovoid, yellow or ochraceous. and develops from August to October.[3]

Main phytoconstituents Ref.

amino acids (aspartic acid, serine, phenylalanine, threonine, lysine, arginine, glutamic acid,
    proline, glycine, alanine, cysteine, valine, methionine, isoleucine, tyrosine, histidine)   [74]
alkaloids
coumarins
flavonoids (rutin, myrictin, chrysin, naringin)
terpenoids (myrcene)
nucleosides
sterols (stigmaterol, ergosterol, sitosterol)
fatty acids (palmitic acid, stearic acid, oleic acid, linoleic acid)
phenolic acids (gallic acid, protocatechuic acid, homogentisic acid)
glycosides
carbohydrates (glucose, manose, arabinose, rhamnose, xylose)
organic acids (ascorbic acid)
tocopherols
aldehyde & ketone
hydrocarbons

Medicinal Use

Boletus edulis has a history of use in traditional medicine for addressing numerous chronic and degenerative ailments, including Alzheimer's, Parkinson's, atherosclerosis, diabetes mellitus, chronic inflammation, and cancer. This species is known to produce a range of nutraceuticals, each with a diverse array of medicinal properties, including antioxidant, anticancer, and antimicrobial effects.[101]

Biological Activity

Bioactive compounds and chemical extracts derived from *Boletus edulis* have been scientifically confirmed to exhibit a range of beneficial properties, including the prevention of constipation, antioxidant effects, antineoplastic activity, anti-inflammatory properties, hepato-protective benefits, antibacterial actions, and antiviral activity.[102-103]

Other Uses

Among the native edible species, *Boletus edulis* takes a prominent position due to its extensive consumption and economic significance. This mushroom is characterized by its low fat and easily digestible carbohydrates content, while it boasts high levels of protein, vitamins, minerals, and dietary fiber. It is primarily harvested from the wild, often undergoing drying, packaging, and global distribution. Even after drying, it retains its flavor, making it suitable for conversion into powdered form, which can be conveniently used in various culinary applications.[101]

REFERENCES

1. Kim S.E., Hwang B.S., Song J.G., Lee S.W., Lee I.K., Yun B.S. 2013. New bioactive compounds from Korean native mushrooms. *Mycobiology* 41(4):171–176.

2. Popescu M.-L. 2006. *Ciuperci – potențial terapeutic și toxicologic.* Ed. Tehnoplast Company S.R.L., București.

3. Tudor I. 2010. *Ciupercile din flora spontană a României. Manualul culegătorului și consumatorului autohton. Ghid de identificare a peste 300 de specii de ciuperci comestibile și otrăvitoare de pe teritoriul României.* Ed. Gramen, București.

4. Begu A., Manic Șt., Salaru V., Simonov G. 2005. *Lumea vegetală a Moldovei.* Vol. 1, Ed. Știința, Chișinău.

5. Islam T., Yu X., Xu B. 2016. Phenolic profiles, antioxidant capacities and metal chelating ability of edible mushrooms commonly consumed in China. *LWT – Food Science and Technology* 72:423–431.

6. Fogarasi M., Socaci S.A., Dulf F.V., Diaconeasa Z.M., Fărcaş A.C., Tofană M., Semeniuc C.A. 2018. Bioactive compounds and volatile profiles of five Transylvanian wild edible mushrooms. *Molecules* 23(12):3272.

7. Bekiaris G., Tagkouli D., Koutrotsios G., Kalogeropoulos N., Zervakis G.I. 2020. *Pleurotus* mushrooms content in glucans and ergosterol assessed by ATR-FTIR spectroscopy and multivariate analysis. *Foods* 9:535.

8. Badalyan S.M. 2014. *Potential of mushroom bioactive molecules to develop healthcare biotech products.* In: Proceedings of the 8th International Conference on Mushroom Biology and Mushroom Products (ICMBMP8). Yugantar Prakashan Pvt. Ltd., New Delhi, India, pp. 373–378.

9. Manzi P., Aguzzi A., Pizzoferrato L. 2001. Nutritional value of mushrooms widely consumed in Italy. *Food Chemistry* 73:321–325.

10. Aprotosoaie A.C., Zavastin D.E., Mihai C.T., Voichita G., Gherghel D., Silion M., Trifan A., Miron A. 2017. Antioxidant and antigenotoxic potential of *Ramaria largentii* Marr & D.E. Stuntz, a wild edible mushroom collected from Northeast Romania. *Food and Chemical Toxicology* 108:429–437.

11. Piskov S., Timchenko L., Grimm W.D., Rzhepakovsky I., Avanesyan S., Sizonenko M., Kurchenko V. 2020. Effects of various drying methods on some physico-chemical properties and the antioxidant profile and ACE inhibition activity of oyster mushrooms (*Pleurotus ostreatus*). *Foods* 9:160.

12. Ragucci S., Pacifico S., Ruocco M.R., Crescente G., Nasso R., Simonetti M., Masullo M., Piccolella, S., Pedone P.V., Landi N., et al. 2019. Ageritin from poplar mushrooms: Scale-up purification and cytotoxicity towards undifferentiated and differentiated SH-SY5Y cells. *Food & Function* 10:6342–6350.

13. Chowdhury M.M.H., Kubra K., Ahmed S.R. 2015. Screening of antimicrobial, antioxidant properties and bioactive compounds of some edible mushrooms cultivated in Bangladesh. *Annals of Clinical Microbiology and Antimicrobials* 14:8.

14. Taofiq O., González-Paramás A.M., Martins A., Barreiro M.F., Ferreira I.C. 2016. Mushrooms extracts and compounds in cosmetics, cosmeceuticals and nutricosmetics – A review. *Industrial Crops and Products* 90:38–48.

15. Zengin G., Uren M.C., Kocak M.S., Gungor H., Locatelli M., Aktumsek A., et al. 2017. Antioxidant and enzyme inhibitory activities of extracts from wild mushroom species from Turkey. *International Journal of Medicinal Mushrooms* 19:327–336.

16. Schepetkin I.A., Quinn M.T. 2006. Botanical polysaccharides: macrophage immunomodulation and therapeutic potential. *International Immunopharmacology* 6:317–333.

17. Cheng P.G., Phan C.-W., Sabaratnam V., Abdullah N., Abdulla M. A., Kuppusamy U.R. 2013. Polysaccharides-rich extract of *Ganoderma lucidum* (MA Curtis: Fr.) P. Karst accelerates wound healing in streptozotocin-induced diabetic rats. *Evidence-Based Complementary and Alternative Medicine* 2013:671252.

18. Amin Z.A., Ali H.M., Alshawsh M.A., Darvish P.H., Abdulla M.A. 2015. Application of *Antrodia camphorata* promotes rat's wound healing *in vivo* and facilitates fibroblast cell proliferation *in vitro*. *Evidence-Based Complementary and Alternative Medicine* 2015:317693.

19. Krupodorova T.A., Klymenko P.P., Barshteyn V.Y., Leonov Y. I., Shytikov D.W., Orlova T.N. 2015. Effects of *Ganoderma lucidum* (Curtis) P. Karst and *Crinipellis schevczenkovi* Buchalo aqueous extracts on skin wound healing. *Journal of Phytopharmacology* 4(4):197–201.

20. Lindequist U., Niedermeyer T.H., Jülich W.D. 2005. The pharmacological potential of mushrooms. *Evidence-Based Complementary and Alternative Medicine* 2:285–299.

21. Badalyan S.M. 2014. *Potential of mushroom bioactive molecules to develop healthcare biotech products*. In: Proceedings of the 8th International Conference on Mushroom Biology and Mushroom Products (ICMBMP8). Yugantar Prakashan Pvt., Ltd., New Delhi, India, pp. 373–378.

22. Ruthes A.C., Smiderle F.R., Iacomini M. 2016. Mushroom heteropolysaccharides: A review on their sources, structure and biological effects. *Carbohydrate Polymers* 136:358–375.

23. https://alchetron.com/Morchella

24. Locsmándi C., Vasas G. 2013. *Ghidul culegătorului de ciuperci. Ciuperci comestibile și otrăvitoare.* Ed. Casa, Oradea.

25. Bielli E., Gâdei R. 1999. *Ciuperci. Ghid complet All.* Ed. All, București.

26. Badshah S.L., Riaz A., Muhammad A., Çayan G.T., Çayan F., Duru M.E., Ahmad N., Emwas H., Jaremko M. 2021. Isolation, characterization, and medicinal potential of polysaccharides of *Morchella esculenta*. *Molecules* 26(5):1459.

27. Heleno S.A., Stojković D., Barros L., Glamočlija J., Soković M., Martins A., Ferreira I.C.F.R. 2013. A comparative study of chemical composition, antioxidant and antimicrobial properties of *Morchella esculenta* (L.) Pers. from Portugal and Serbia. *Food Research International* 51(1):236–243.

28. Wu H., Chen J., Li, J. et al. 2021. Recent advances on bioactive ingredients of *Morchella esculenta*. *Applied Biochemistry and Biotechnology* 193:4197–4213.

29. Tietel Z., Masaphy S. 2017. True morels (*Morchella*) – nutritional and phytochemical composition, health benefits and flavor: A review. *Critical Reviews in Food Science and Nutrition* 58(11):1888–1901.

30. Ajmal M., Akram A., Ara A., Akhund S., Nayyar B.G. 2015. *Morchella esculenta*: An edible and health beneficial mushroom. *Pakistan Journal of Food Sciences* 25(2):71–78.

31. Cateni F., Gargano M.L., Procida G., Venturella G., Cirlincione F., Ferraro V. 2022. Mycochemicals in wild and cultivated mushrooms: nutrition and health. *Phytochemistry Reviews* 21:339–383.

32. Nitha B, Fijesh P.V., Janardhanan K.K. 2013. Hepatoprotective activity of cultured mycelium of Morel mushroom, *Morchella esculenta. Experimental and Toxicologic Pathology* 65(1–2):105–112.

33. www.trufe-negre.ro/trufe%20in%20romania.html

34. Shah N., Usvalampi A., Chaudhary S., et al. 2020. An investigation on changes in composition and antioxidant potential of mature and immature summer truffle (*Tuber aestivum*). *European Food Research and Technology* 246:723–731.

35. Marathe S.J., Hamzi W., Bashein A.M., Deska J., Seppänen-Laakso T., Singhal R.S., Shamekh S. 2020. Anti-angiogenic and anti-inflammatory activity of the summer truffle (*Tuber aestivum* Vittad.) extracts and a correlation with the chemical constituents identified therein. *Food Research International* 137:109699.

36. Lee H., Nam K., Zahra Z., Farooqi M.Q.U. 2020. Potentials of truffles in nutritional and medicinal applications: a review. *Fungal Biology and Biotechnology* 7:9.

37. Bhotmange D.U., Wallenius J.H., Singhal R.S., Shamekh S.S. 2017. Enzymatic extraction and characterization of polysaccharide from *Tuber aestivum. Bioactive Carbohydrates and Dietary Fibre* 10:1–9.

38. Mudliyar D., Wallenius J., Bedade D., Singhal R., Madi N., Shamekh S. 2019. Ultrasound assisted extraction of the polysaccharide from *Tuber aestivum* and its *in vitro* antihyperglycemic activity. *Bioactive Carbohydrates and Dietary Fibre* 20:100198.

39. www.sciencedirect.com/topics/immunology-and-microbiology/trametes-versicolor

40. Habtemariam S. 2020. *Trametes versicolor* (synn. *Coriolus versicolor*) polysaccharides in cancer therapy: targets and efficacy. *Biomedicines* 8(5):135.

41. Pop R.M., Puia I.C., Puia A., Chedea V.S., Leopold N., Bocsan I.C., Buzoianu A.D. 2018. Characterization of *Trametes versicolor*: medicinal mushroom with important health benefits. *Notulae Botanicae Horti Agrobotanici Cluj-Napoca* 46(2):343–349.

42. Jin M., Zhou W., Jin C., Jiang Z., Diao S., Jin Z., Li G. 2018. Anti-inflammatory activities of the chemical constituents isolated from *Trametes versicolor. Natural Product Research* 33(16):2422–2425.

43. Hosseinihashemi S.K., Salem M.Z.M., HosseinAshrafi S.K., Latibari A.J. 2016. Chemical composition and antioxidant activity of extract from the wood of *Fagus orientalis:* water resistance and decay resistance against *Trametes versicolor. BioResources* 11(2):3890–3903.

44. Hobbs C. 2004. Medicinal value of Turkey tail fungus *Trametes versicolor* (L.:Fr.) Pilat (*Aphyllophoromycetideae*). A literature review. *International Journal of Medicinal Mushrooms* 6(3):195–218.

45. Elkhateeb W.A., Elnahas M.O., Thomas P.W., Daba G.M. 2020. *Trametes versicolor* and *Dictyophora indusiata* champions of medicinal mushrooms. *Open Access Journal of Pharmaceutical Research* 4(1):000192.

46. https://en.wikipedia.org/wiki/Pleurotus

47. Hadar Y., Cohen-Arazi E. 1986. Chemical composition of the edible mushroom *Pleurotus ostreatus* produced by fermentation. *Applied and Environmental Microbiology* 51(6):1352–1354.

48. Mishra V., Tomar S., Yadav P., Vishwakarma S., Singh M.P. 2022. Elemental analysis, phytochemical screening and evaluation of antioxidant, antibacterial and anticancer activity of *Pleurotus ostreatus* through *in vitro* and *in silico* approaches. *Metabolites* 12(9):821.

49. Nagy M., Socaci S., Tofană M., Biris-Dorhoi E.S., Țibulcă D., Petruț G., Salanta C.L. 2017. Chemical composition and bioactive compounds of some wild edible mushrooms. *Bulletin of University of Agricultural Sciences and Veterinary Medicine Cluj-Napoca* 74(1):1–8.

50. Galappaththi M.C.A., Dauner L., Madawala S., Karunarathna S.C. 2021. Nutritional and medicinal benefits of Oyster (*Pleurotus*) mushrooms: a review. *Fungal Biotec* 1(2):65–87.

51. Kaisun N.L., Mayeen U.K., Faruque Mohammad R.I., Rohit S., Fahadul I., Saikat M., Talha B.E. 2022. Nutritional value, medicinal importance, and health-promoting effects of dietary mushroom (*Pleurotus ostreatus*). *Journal of Food Quality* 2022, Article ID 2454180, pp. 9.

52. www.first-nature.com/fungi/cantharellus-cibarius.php

53. Režić Mužinić N., Veršić Bratinčević M., Grubić M., Frleta Matas R., Čagalj M., Visković T., Popović M. 2023. Golden chanterelle or a gold mine? Metabolites from aqueous extracts of golden chanterelle (*Cantharellus cibarius*) and their antioxidant and cytotoxic activities. *Molecules* 28(5):2110.

54. Daniewski W.M., Danikiewicz W., Gołębiewski W.M., Gucma M., Łysik A., Grodner J., Przybysz E. 2012. Search for bioactive compounds from *Cantharellus cibarius. Natural Product Communications* 7(7):917–918.

55. Vlasenko V., Turmunkh D., Ochirbat E., Dondov B., Nyamsuren K., Samiya J., Burenbaatar C., Vlasenko A. 2019. Medicinal potential of extracts from the chanterelle mushroom, *Cantharellus cibarius* (Review) and prospects for studying its strains from differs plant communities of ultracontinental regions of the Asia. *BIO Web of Conferences* 16:00039.

56. Sevindik M. 2019. Wild edible mushroom *Cantharellus cibarius* as a natural antioxidant food. *Turkish Journal of Agriculture – Food Science and Technology* 7(9):1377–1381.

57. Nasiry D., Khalatbary A.R., Ebrahimzadeh M.A. 2017. Anti-inflammatory and wound-healing potential of golden chanterelle mushroom, *Cantharellus cibarius* (*Agaricomycetes*). *International Journal of Medicinal Mushrooms* 19(10):893–903.

58. Lemieszek M.K., Nunes F.M., Cardoso C., Marques G., Rzeski W. 2018. Neuroprotective properties of *Cantharellus cibarius* polysaccharide fractions in different *in vitro* models of neurodegeneration. *Carbohydrate Polymers* 197:598–607.

59. Nowakowski P., Markiewicz-Żukowska R., Gromkowska-Kępka K., Naliwajko S., Moskwa J., Bielecka J., Grabia M., Borawska M., Socha K. 2021. Mushrooms as potential therapeutic agents in the treatment of cancer: Evaluation of anti-glioma effects of *Coprinus comatus*, *Cantharellus cibarius*, *Lycoperdon perlatum* and *Lactarius deliciosus* extracts. *Biomedicine & Pharmacotherapy* 133:111090.

60. Elkhateeb W., Daba G. 2021. Highlights on the golden mushroom *Cantharellus cibarius* and unique shaggy ink cap mushroom *Coprinus comatus* and smoky bracket mushroom *Bjerkandera adusta* ecology and biological activities. *Open Access Journal of Mycology & Mycological Sciences* 4:1–8.

61. https://verovita.ro/portfolio-item/ciuperca-de-camp/

62. Gąsecka, M., Magdziak, Z., Siwulski, M. et al. 2018. Profile of phenolic and organic acids, antioxidant properties and ergosterol content in cultivated and wild growing species of *Agaricus. European Food Research and Technology* 244:259–268.

63. Lodonjav M., Namjil E., Tselmuungarav B., Amartuvshin B., Bolor Tsolmon, Regdel D., Odonmajig P. 2014. Chemical composition and biological activities of the *Agaricus* mushrooms. *Mongolian Journal of Chemistry* 14(40):41–45.

64. Korkmaz A.I., Eraslan E.C., Uysal I., Akgül, H. 2021. Functional food *Agaricus arvensis*: antioxidant and antimicrobial potentials. *Eurasian Journal of Medical and Biological Sciences* 1(2):86–91.

65. Dogan A., Dalar A., Sadullahoglu C. et al. 2018. Investigation of the protective effects of horse mushroom (*Agaricus arvensis* Schaeff.) against carbon tetrachloride-induced oxidative stress in rats. *Molecular Biology Reports* 45:787–797.

66. Özlem K., Nazan Gökşen T., Rizvan I., Ibrahim T.. Isa G., Aykut Ö. 2022. Biosynthesis and characterization of silver nanoparticles from *Tricholoma ustale* and *Agaricus arvensis* extracts and investigation of their antimicrobial, cytotoxic, and apoptotic potentials. *Journal of Drug Delivery Science and Technology* 69:103178.

67. https://healing-mushrooms.net/agaricus-campestris

68. Waheed I., Ahmad M., Rasool S. 2013. Investigation of haem-agglutination activity in *Agaricus campestris* and *Ballota limbata. International Journal of Advanced Pharmaceutical Sciences and Research* 5(1):51–56.

69. Elkhateeb W.A., Daba G.M. 2022. Bioactive potential of some fascinating edible mushrooms *Flammulina, Lyophyllum, Agaricus, Boletus, Letinula, Pleurotus* as a treasure of multipurpose therapeutic natural product. *Open Access Journal of Pharmaceutical Research* 6(1):1–10.

70. Kosanić M., Ranković B., Rancic A., Stanojkovic T. 2017. Evaluation of metal contents and bioactivity of two edible mushrooms *Agaricus campestris* and *Boletus edulis. Emirates Journal of Food and Agriculture* 29:98–103.

71. Arul S., Predheepan D., Dayalan H. 2017. Larvicidal activity of *Agaricus bisporus* and *Agaricus campestris* against the *Aedes aegypti* larvae. *International Journal of Materials and Product Technology* 55:272.

72. Bubueanu C., Popa G., Pirvu L.C. 2015. Comparative analysis of polyphenolic profiles and antioxidant activity of *Agaricus bisporus* and *Agaricus campestris. Scientific Bulletin. Series F. Biotechnologies* 19:29–33.

73. www.mykoweb.com/CAF/species/Agaricus_bisporus.html

74. Muszyńska B., Kała K., Firlej A., Sułkowska-Ziaja K. 2016. *Cantharellus cibarius:* culinary-medicinal mushroom content and biological activity. *Acta Poloniae Pharmaceutica* 73(3):589–598.

75. Usman M., Murtaza G., Ditta A. 2021. Nutritional, medicinal, and cosmetic value of bioactive compounds in button mushroom (*Agaricus bisporus*): a review. *Applied Sciences* 11:5943.

76. Muszyńska B., Kała K., Rojowski J., Grzywacz A., Opoka W. 2017. Composition and biological properties of *Agaricus bisporus* fruiting bodies – a review. *Polish Journal of Food and Nutrition Sciences* 67:173–182.

77. Dhamodharan G., Mirunalini S. 2010. A novel medicinal characterization of *Agaricus bisporus* (white button mushroom). *Pharmacology Online* 2:456–463.

78. Bhushan A., Kulshreshtha M. 2018. The medicinal mushroom *Agaricus bisporus*: review of phytopharmacology and potential role in the treatment of various diseases. *Journal of Natural Science, Biology, and Medicine* 1(1):4–9.

79. Yuqin F., Jixian Z., Chaoting W., Courage S.D., Igbokwe C.J., Yuqing D., Haihui Z. 2020. Recent advances in *Agaricus bisporus* polysaccharides: Extraction, purification, physicochemical characterization and bioactivities. *Process Biochemistry* 94:39–50.

80. https://ro.wikipedia.org/wiki/Macrolepiota_procera

81. Bengu A.S. 2020. The fatty acid composition in some economic and wild edible mushrooms in Turkey. *Progress in Nutrition* 22(1):185–192.

82. Georgiev Y.N., Vasicek O., Dzhambazov B., Batsalova T.G., Denev P.N., Dobreva L.I., Danova S.T., Simova S.D., Wold C.W, Ognyanov M.H., Paulsen B.S., Krastanov A.I. 2022. Structural features and immunomodulatory effects of water-extractable polysaccharides from *Macrolepiota procera* (Scop.) Singer. *Journal of Fungi (Basel)* 8(8):848.

83. Secme M., Kaygusuz O., Eroglu C., Dodurga Y., Colak O.F., Atmaca P. 2018. Potential anticancer activity of the parasol mushroom, *Macrolepiota procera* (*Agaricomycetes*), against the A549 human lung cancer cell line. *International Journal of Medicinal Mushrooms* 20(11):1075–1086.

84. Erbiai E.H., da Silva L.P., Saidi R., Lamrani Z., Esteves da Silva J.C.G., Maouni A. 2021. Chemical composition, bioactive compounds, and antioxidant activity of two wild edible mushrooms *Armillaria mellea* and *Macrolepiota procera* from two countries (Morocco and Portugal). *Biomolecules* 11:575.

85. Adamska I., Tokarczyk G. 2022. Possibilities of using *Macrolepiota procera* in the production of prohealth food and in medicine. *International Journal of Food Science* 2022, Article ID 5773275, 27 pp.

86. Aytar E.C., Akata I., Açık L. 2020. Antioxidant and antimicrobial activities of *Armillaria mellea* and *Macrolepiota procera* extracts. *Mantar Dergisi* 11(2):121–128.

87. Chen H.P., Zhao Z.Z., Li Z.H., Huang Y., Zhang S.B., Tang Y., Yao J.N., Chen L., Isaka M., Feng T., Liu J.K. 2018. Anti-proliferative and anti-inflammatory lanostane triterpenoids from the Polish edible mushroom *Macrolepiota procera*. *Journal of Agricultural and Food Chemistry* 66(12):3146–3154.

88. Kosanić M, Ranković B, Rančić A, Stanojković T. 2016. Evaluation of metal concentration and antioxidant, antimicrobial, and anticancer potentials of two edible mushrooms *Lactarius deliciosus* and *Macrolepiota procera*. *Journal of Food and Drug Analysis* 24(3):477–484.

89. Özgür A., Kaplan Ö., Tosun N.G., Türkekul I., Gökçe I. 2023. Green synthesis of silver nanoparticles using *Macrolepiota procera* extract and investigation of their HSP27, HSP70, and HSP90 inhibitory potentials in human cancer cells. *Particulate Science and Technology* 41(3):330–340.

90. www.hautesavoiephotos.com/champis/photos_macrolepiota_mastoidea.htm

91. Ciric A., Kruljevic I., Stojkovic D., Fernandes, A., Barros, L., Calhelha, R. C., et.al. 2019. Comparative investigation on edible mushrooms *Macrolepiota mastoidea*, *M. rhacodes* and *M. procera*: Functional food with diverse biological activities. *Food & Function* 10:7678–7686.

92. Kolcuoğlu Y., Colak A., Sesli E., Yildirim M., Saglam N. 2007. Comparative characterization of monophenolase and diphenolase activities from a wild edible mushroom (*Macrolepiota mastoidea*). *Food Chemistry* 101(2):778–785.

93. Barros L., Baptista P., Correia D.M., Sá Morais J., Ferreira Isabel C.F.R. 2007. Effects of conservation treatment and cooking on the chemical composition and antioxidant activity of Portuguese wild edible mushrooms. *Journal of Agricultural and Food Chemistry* 55(12):4781–4788.

94. Akata I., Zengin G., Picot C.M.N., Mahomoodally M.F. 2018. Enzyme inhibitory and antioxidant properties of six mushroom species from the *Agaricaceae* family. *South African Journal of Botany* 120:95–99.

95. https://wpamushroomclub.org/product/lactarius-piperatus/

96. Bebu A., Andronie L., Matieş A., Micle S., Darjan S., Culcear O. 2020. Comparative analysis between mushrooms *Lactarius piperatus* and *Agaricus bisporus* (champignon) using FT-IR spectroscopy. *Scientific Papers Series B Horticulture* 64(1):633–637.

97. Barros L, Baptista P, Ferreira IC. 2007. Effect of *Lactarius piperatus* fruiting body maturity stage on antioxidant activity measured by several biochemical assays. *Food and Chemical Toxicology* 45(9):1731–1737.

98. Kosanic M., Petrović N., Milosevic-Djordjevic O., Grujičić D., Tubic J., Marković A., Stanojkovic T.P. 2020. The health promoting effects of the fruiting bodies extract of the peppery milk cap mushroom *Lactarius piperatus* (*Agaricomycetes*) from Serbia. *International Journal of Medicinal Mushrooms* 22(4):347–357.

99. Birol I., Ugraskan V., Cankurtaran O. 2022. Effective biosorption of methylene blue dye from aqueous solution using wild macrofungus (*Lactarius piperatus*). *Separation Science and Technology* 57(6):854–871.

100. www.gastronomiavasca.net/en/gastro/glossary/boletus-edulis

101. Novakovic A., Karaman M., Kaisarevic S., Radusin T., Llic N. 2017. Antioxidant and antiproliferative potential of fruiting bodies of the wild-growing king bolete mushroom, *Boletus edulis* (*Agaricomycetes*), from Western Serbia. *International Journal of Medicinal Mushrooms* 19(1):27–34.

102. Tan Y., Zeng N.K., Xu B. 2022. Chemical profiles and health-promoting effects of porcini mushroom (*Boletus edulis*): A narrative review. *Food Chemistry* 390:133199.

103. Rosa G.B., Sganzerla W.G., Ferreira A.L.A., Xavier L.O., Veloso N.C., da Silva J., de Oliveira G.P., Amaral N.C., Veeck A.P.L., Ferrareze J.P. 2020. Investigation of nutritional composition, antioxidant compounds, and antimicrobial activity of wild culinary-medicinal mushrooms *Boletus edulis* and *Lactarius deliciosus* (*Agaricomycetes*) from Brazil. *International Journal of Medicinal Mushrooms* 22(10):931–942.

18 *Gentianaceae* Juss. Family

Cornelia Bejenaru[1], Antonia Radu[1], George Dan Mogoşanu[2],
Ludovic Everard Bejenaru[2], Andrei Biţă[2], and Adina-Elena Segneanu[3]

[1] Department of Pharmaceutical Botany, Faculty of Pharmacy, University of Medicine and Pharmacy of Craiova, 2 Petru Rareş Street, Craiova, Dolj County, Romania
[2] Department of Pharmacognosy & Phytotherapy, Faculty of Pharmacy, University of Medicine and Pharmacy of Craiova, 2 Petru Rareş Street, Craiova, Dolj County, Romania
[3] Institute for Advanced Environmental Research–West University of Timisoara (ICAM–WUT), 4 Oituz Street, Timisoara, Romania

BOTANICAL DESCRIPTION

Herbaceous, terrestrial, rarely aquatic plants, glabrous, with leaves that are sometimes slightly fused at the base, sessile, entire, without stipules, and arranged opposite each other (very rarely alternate). The flowers are hermaphroditic, actinomorphic, of 4 or 5 type, gamosepalous and gamopetalous, arranged solitarily or forming cymose or racemose inflorescences, either axillary or terminal. The flowers occasionally have developed nectaries. The persistent calyx is composed of 4–8 fused sepals, rarely free. The corolla is 4–8-lobed, with twisted perfloration. The androecium consists of 4 or 5 stamens attached to the corolla tube, alternately positioned with its incisions, each with bilocular introrse anthers, and the gynoecium has a superior ovary, unilocular, with 2 parietal placentas and a glandular disc at the base. The style is simple, very short, terminated by two free or united stigmas, and the ovules are anatropous. The fruit is a septicidal capsule containing numerous albuminous seeds, globular, and edged.[1-4]

The family comprises 65 genera, with approximately 800 species, distributed worldwide.

In the spontaneous flora of Romania, there are 8 genera: *Blackstonia* Huds., *Centaurium* Hill, *Gentiana* L., *Gentianopsis* Ma, *Gentianella* Moench, *Comastoma* (Wettst.) Toyok., *Lomatogonium* A. Braun, and *Swertia* L.[2-3]

Gentiana L. (Gentian)

The genus comprises 13 species in the Romanian flora, which grow in mountainous regions: *Gentiana lutea* L., *Gentiana cruciata* L., *Gentiana phlogifolia* Schott et Kpotschy (*G. cruciata* L. subsp. *phlogifolia* (Schott et Kotshy) Tutin), *Gentiana punctata* L., *Gentiana nivalis* L., *Gentiana utriculosa* L., *Gentiana asclepiadea* L., *Gentiana pneumonanthe* L., *Gentiana frigida* Haenke, *Gentiana acaulis* L., *Gentiana clusii* J. O. E. Perrier et Songeon, *Gentiana verna* L., and *Gentiana orbicularis* Schur (*G. verna* var. *favratii* Rittener).[2-3]

The species are annual, biennial, or perennial, with thick rhizomes, erect or cespitose stems, large elliptical or broadly ovate leaves, sessile, without stipules, while the flowers have a corolla with a slightly inconspicuous tube, with a glabrous throat, lacking appendages at the base of the laciniae, but with nectaries at the base of the ovary.[1,4]

Gentiana lutea L. (Great yellow gentian)

Scientific name: *Gentiana lutea* L.
Common name: Great yellow gentian
Romanian name: Ghinţură galbenă
Taxonomic classification
Kingdom: *Plantae*
Subkingdom: *Cormobionta*
Phylum: *Spermatophyta*
Subphylum: *Magnoliophytina (Angiospermae)*
Class: *Magnoliatae (Magnoliopsida, Dicotyledonatae)*
Subclass: *Asteridae*
Order: *Gentianales*
Family: *Gentianaceae*
Genus: *Gentiana*
Species: *lutea* L. (Figure 18.1)[1-3]

DOI: 10.1201/9781003270515-19

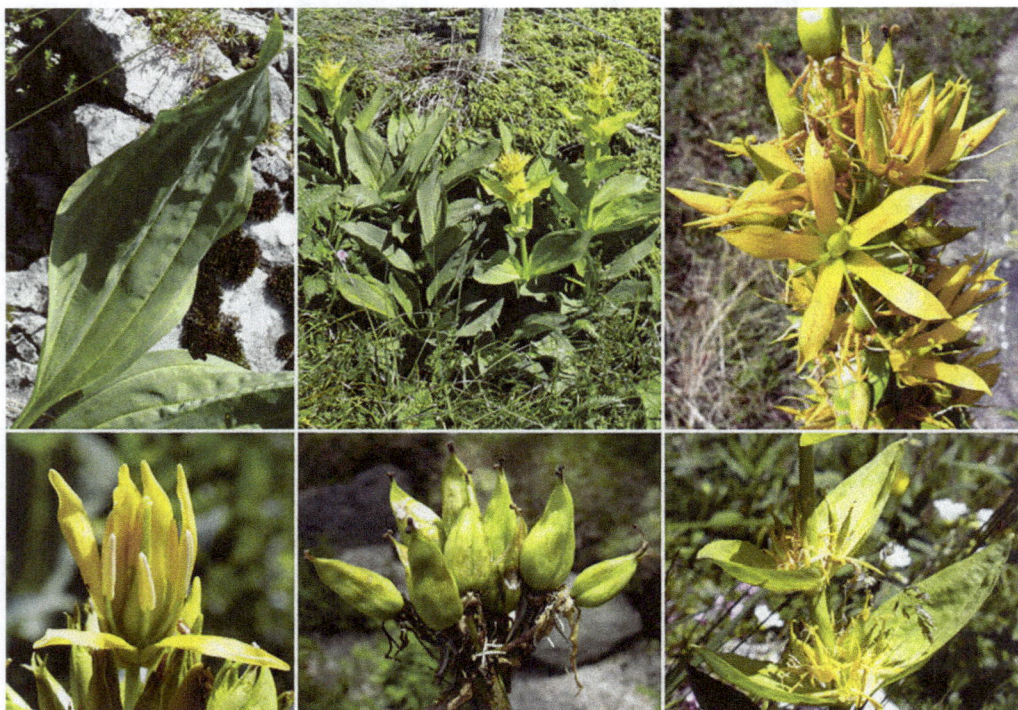

Figure 18.1 *Gentiana lutea* L. (Great yellow gentian).

Source: 5.

SPREAD

The plant is sporadically encountered from the spruce forest zone up to the subalpine region, found in mountain meadows or on sunny slopes, on rocky soils rich in humus.[1,2,4]

DESCRIPTION

A robust perennial herbaceous plant, glabrous, with a height ranging from 40 to 120 cm. The rhizome is thick and short, and the taproots are thick and long. The stem is cylindrical, fistulous, erect, unbranched, with a thickness of up to 1 cm. The leaves are large, arranged opposite, the inferior ones have petioles, broadly ovate elliptical or elongated, with 5–7 prominent, almost parallel veins, while the superior ones are sessile, ovate or ovate-elliptical, pointed, united at the base in a short sheath. The plant has large yellow flowers, long-pedicellate, grouped in the axils of the superior leaves, in compact corymbiform multifloral cymes. The calyx is membranous and white, shorter than the corolla, longitudinally cleft down to the base. The corolla is rotate and yellow, with 5–6 (9) lobes, with narrow lanceolate, acute laciniae, much longer than the corolla tube, measuring 2–3 cm. The androecium has stamens with free anthers, rarely closely united, while the gynoecium has a long stigma that twists into a spiral after flowering. The fruit is a conical, elongated, unilocular capsule. The seeds are numerous, up to 100, rounded, elongated, compressed, with a broadly winged margin. Designated as a natural monument, the Great Yellow Gentian blooms from June to August.[1-4]

Main Phytoconstituents

Gentianae radix contains approximately 1–3% bitter principles, atypical monoterpenoid secoiridoid compounds such as gentiopicroside, amarogentin, and swertiamarin (sweroside); sweroside was first identified in the species *Swertia japonica* (Roem. et Schult.) Makino (*Gentianaceae*); an alkaloid with a secoiridoid structure (gentiannine) is found, which some authors speculate could be an artifact resulting from ammonia treatment of the root powder; xanthones, substituted dibenzopyrones (gentisine, isogentisine, gentioside) are present, imparting the yellow color to the root; other components include sterols (gentiosterol), oligosides – gentiobiose (α-glucosyl–β-glucosyl), gentianose, or gentiotriose (fructosyl–gentiobiosyl), tannin (gentiotanic acid), gentisic acid (2,5-dihydroxybenzoic acid), pectin, enzymes, and mineral salts.

During fermentation, enzymes hydrolyze gentianose into sucrose, gentioside into isogentisin (strongly colored), and partially gentiopicroside into gentiogenol, leading to a reduction in the bitter taste.

For the medicinal product, the Romanian Pharmacopoeia (F.R. X) specifies a bitterness index (BI) of at least 1/10,000 and a minimum of 33% soluble substances.[6-7]

Main phytoconstituents   Ref.
[8-15]

alkaloids (gentianine)
flavonoids (luteolin, isoorientin, apigenin, chrysoeriol, isoscoparin, acacetin, isovitexin)
terpenoids (oleanolic acid, betulin, β-amirin, lupeol, oleanane, ursane)
sterols (stigmasterol)
fatty acids
phenolic acids (ferulic acid, caffeic acid)
secoiridoids (swertiamarin, gentiopicroside, amarogentin sweroside, eustoside, gentiopicrin, amarogentin (B), amaroswerin (D))
xanthonoids (gentisine, isogentisine, mangiferin, methylgentisine, gentiseine, 1-hydroxy-3,7-dimethoxyxanthone, 1,3,7-trimethoxyxanthone, dihydroxy-1,3-dimethoxy-2,7-xanthone and gentisine-1-O-primveroside and gentioside-7-O-primveroside)
irinoids (loganic acid, swertiamarin, gentiopicrin)

Medicinal Use

The vegetable product is a bitter tonic, eupeptic and possesses mild choleretic-cholagogue properties. It is recommended for conditions like anorexia, anemia, hypochlorhydria (low acid gastric ulcer), and convalescence. It also has fever-reducing and antimalarial effects similar to quinine, and has been used since ancient times as an antimalarial remedy. After the discovery of quinine-rich trees, it was referred to as "indigenous quinquina" (*quinquina indigène*). *Gentianae radix* is often combined with antimalarials due to its febrifugal effect.

Gentianine has antipsychotic (antidepressant) properties and low toxicity.[7]

Biological Activity

Various studies have highlighted the therapeutical activities of extracts from *Gentiana lutea* in combating oxidative stress, microbial infections, inflammation, obesity, and atherosclerosis. Furthermore, it has been shown to possess properties such as analgesic, antifungal, stomachic, appetizer, and immunomodulatory.[12,16-18]

Other Uses

■ From the rhizome and roots, liqueurs, brandies, and bitter tonic wines are produced, which have digestive and appetizing qualities.[4]

Gentiana asclepiadea L. (Willow gentian)

Scientific name: *Gentiana asclepiadea* L.
Common name: Willow gentian, Milkweed gentian
Romanian name: Lumânărica pământului
Taxonomic classification
Kingdom: *Plantae*
Subkingdom: *Cormobionta*
Phylum: *Spermatophyta*
Subphylum: *Magnoliophytina (Angiospermae)*
Class: *Magnoliatae (Magnoliopsida, Dicotyledonatae)*
Subclass: *Asteridae*
Order: *Gentianales*
Family: *Gentianaceae*
Genus: *Gentiana* L.
Species: *asclepiadea* L. (Figure 18.2)[1-3]

Figure 18.2 *Gentiana asclepiadea* L. (Willow gentian).

Source: 19.

SPREAD

A common species found in forests, forest clearings, and shrubby areas, ranging from the Turkey oak zone to the spruce zone.[2,3]

DESCRIPTION

A perennial herbaceous plant with a cylindrical, thick, knotty rhizome, and a glabrous, fistulous, simple stem that is densely foliate and grows to a height of 20–60 cm. The leaves are ovate-lanceolate, 5-nerved, acuminate, sessile, arranged opposite. The flowers are large, short-pedicellate, with a cuneate campanulate corolla consisting of 5 triangular, sharp-pointed laciniae, each with a small tooth between the folds, colored blue with dark spots inside, grouped in pairs of 2–4 in the axils of the superior leaves. The calyx is gamosepalous, campanulate, membranous, with teeth shorter than the tube. The androecium has stamens with fused anthers, while the gynoecium has a short style and twisted stigma. The fruit is a distinctly pedicellate capsule, elongated, narrowed at the base. The seeds are fusiform with broad wings. Flowering occurs from June to September.[1,4]

Main phytoconstituents   Ref.
amino acids   20-23
flavonoids (quercetin, apigenin, isoorientin, isovitexin, luteolin, naringenin, rutin, chrysin)
terpenoids (ursolic acid, spathulenol, τ-muurolol, β-damascenone, β-ionone, α-terpineol caryophyllene, linalool, geraniol, myrtenal, safranal)
sterols (β-sitosterol)
fatty acids (lauric acid, oleic acid, palmitic acid, perlargonic acid)
phenolic acids (chlorogenic acid, caffeic acid, p-coumaric acid)
tannins
organic acids
furanes (2-furfural, 2,4-dimethylfuran, 2-pentylfurane, 2-(2-pentenyl)furan)

xanthones (magniferin, isogentisin)
glycosides
esters
irinoids (gentiopicrine, swertiamarine, sweroside, amarogentin)
hydrocarbons (eicosane, tetramethylheptane, ethylbenzene, tridecane, tetradecane,
   pentadecane, tricosane, toluene)
aldehydes&ketones
miscellaneous (eugenol, *β*-asarone)

Medicinal Use

The underground part exhibits the same properties as those of the species *G. lutea*.[1]

Biological Activity

Gentiana asclepiadea exhibits hepatoprotective, antimicrobial, antioxidant, and cytotoxic actvities.[21,23-25]

Other Uses

■ ornamental

The genus *Centaurium* is represented in the spontaneous flora of Romania by 4 species: *Centaurium erythraea* Raf., *Centaurium spicatum* (L.) Fritsch, *Centaurium pulchellum* (Sw.) Druce, and *Centaurium littorale* (Turner) Gilmour.[2,3]

Centaurium erythraea Raf. (*C. umbellatum* Gilib.) (Common centaury)

Scientific name: *Centaurium erythraea* Raf.
Common name: Common centaury, European centaury
Romanian name: Fierea pământului
Taxonomic classification
Kingdom: *Plantae*
Subkingdom: *Cormobionta*
Phylum: *Spermatophyta*
Subphylum: *Magnoliophytina (Angiospermae)*
Class: *Magnoliatae (Magnoliopsida, Dicotyledonatae)*
Subclass: *Asteridae*
Order: *Gentianales*
Family: *Gentianaceae*
Genus: *Centaurium*
Species: *erythraea* Raf. (Figure 18.3)[1-3]

SPREAD

These plants are commonly found from the plains to the mountainous regions, occurring in meadows, wet grasslands, shrubs, and forest edges.[1-3]

DESCRIPTION

Annual or biennial plants, reaching a height of 10–30 cm. The stem is glabrous, four-angled, erect, simple, and rarely branched. At ground level, a basal rosette of leaves forms, with 3–5 prominent veins, which are obovate-elliptic, obtuse, and taper into petioles. Stem leaves are ovate or elliptic, three-veined, sessile, and arranged opposite each other. The nearly sessile flowers are arranged in umbelliform cymes. The bracts are broadly linear, and the calyx length is half that of the corolla tube after flowering. The corolla is pinkish-red, about twice as long as the calyx, with anthers that twist into a spiral after opening, and a filiform style. The fruit is a narrow cylindrical capsule, and the seeds are small and reticulate. It blooms from July to September.[1-3-4]

The species has 3 subspecies:

■ subsp. *turcicum* (Velen.) Melderis is characterized by the stem edges, leaf margins, sepal laciniae, and bracts being densely and shortly scabrous-hairy, and stem leaves being lanceolate-elliptic. It is rarely found in the counties of Brașov, Ialomița, Teleorman, Giurgiu, Tulcea, and Constanța.[1-3]

Figure 18.3 *Centaurium erythraea* Raf. (*C. umbellatum* Gilib.) (Common centaury).
Source: 26.

- subsp. *erythraea* has glabrous leaf margins, sepal laciniae, and bracts, with ovate-elliptic leaves. The calyx is half the length of the corolla tube, and the anthers are 1–1.5 mm long. This subspecies is frequently encountered.

- subsp. *austriaca* (Ronn.) Holub features glabrous leaf margins, sepal laciniae, and bracts, with ovate-elliptic leaves. The calyx is two-thirds the length of the corolla tube, and the anthers are 1.8–2.5 mm long. It is rarely distributed in Transylvania, Banat, Crișana, and Maramureș.[2,3]

Main phytoconstituents

Ref.
27-33

amino acids (L-glutamic acid, L-arginine, L-aspartic acid, and L-cystine)

alkaloids (gentianine, eritricine)

flavonoids (quercetin, sakuranin, kaempherol, rutin, isovitexin)

terpenoids (carvacrol, α-pinene, sabinene, α-phellandrene, m-cymene, α-3-carene, α-terpinene, p-camphorene, thymol, menthol, linalool, borneol, methone, β-thujone, camphor, γ-terpinene, p-cymene, neophytadiene, tricosane, caryophyllene oxide, δ-cadinene, lupane, oleanane, ursane)

sterols (stigmast-5-en-3-ol, 3-keto-urs-12-ene, and stigmasta-3,5-dien-7-one)

fatty acids (palmitic acid, linolenic acid, myristic acid, 2-heptadecanone, methyl palmitate, ethyl palmitate, methyl stearate, ethyl linoleate, ethyl oleate)

phenolic acids (ferulic acid, p-coumaric acid, gallic acid, cinnamic acid, chlorogenic acid, caffeic acid, cichoric acid, sinapic acid, rosmarinic acid, hydroxy terephthalic acid, 2,5-dihydroxyterephthalic acid)

lactones (eritaurone)

xanthonoids (eustomin, demethyleustomin, decussatin, methylbellidifolin, dihydroxy-tetrametoxy-O-pentosyl-hexosylxanthone, trihydroxymonometoxy-xanthone, dihydroxy-dimethoxy-xanthone, dihydroxy-tetramethoxy-xanthone, mono-hydroxy-trimetoxy-xanthone, trihydroxy-dimetoxy-xanthone, 1-hydroxy-3-methoxy-xanthone-3 rhamnose, 1-hydroxy-3-methoxy-xanthone, 3-hydroxy-1-methoxy-xanthone, 3-hydroxy-3-methoxy-xanthone, 1,3,8-trihydroxy-5,6-dimethoxy-xanthone)

irinoids (loganic acid, eritaurin, amarogentin, gentiopicroside, sweroside, secologanoside, swertiamarin, gentiopicroside, sweroside, gentiopicrin, centapicrin)

Medicinal Use

The vegetable product exhibits the following pharmacological actions: bitter tonic, febrifuge, excites the sympathetic vegetative nervous system and nerve endings at the capillary level (increases capillary tonus, enhancing blood circulation). The product also has mild laxative, anti-inflammatory, and antibacterial properties, and is used for conditions such as anorexia, hypochlorhydria (low acid gastric ulcer), abdominal bloating, and feverish states.[27]

Biological Activity

The essential oils and extracts derived from *Centaurium erythraea* have displayed a wide range of biological activities, including antibacterial, antioxidant, antifungal, antileishmanial, anticancer, antidiabetic, anti-inflammatory, insecticidal, diuretic, gastroprotective, hepatoprotective, dermatoprotective, neuroprotective properties, and the ability to inhibit larval development.[34]

Other Uses

- The plant is employed as a flavoring agent in bitter herbal liqueurs and is included as an ingredient in vermouth.[35]

The genus *Swertia* comprises 2 species in the spontaneous flora of our country: *Swertia perennis* L. and *Swertia punctata* Baumg.[2,3]

Swertia perennis L. (Felwort)

Scientific name: *Swertia perennis* L.
Common name: Felwort, Star swertia
Romanian name: Siminic vânăt[5]
Taxonomic classification
Kingdom: *Plantae*
Subkingdom: *Cormobionta*
Phylum: *Spermatophyta*
Subphylum: *Magnoliophytina (Angiospermae)*
Class: *Magnoliatae (Magnoliopsida, Dicotyledonatae)*
Subclass: *Asteridae*
Order: *Gentianales*
Family: *Gentianaceae*
Genus: *Swertia*
Species: *perennis* L. (Figure 18.4)[2-3]

SPREAD

The species is sporadically distributed from the Turkey oak zone to the subalpine zone through swampy meadows, peat bogs, and small woods on the riverbanks.[2,4]

DESCRIPTION

A perennial plant with a short, cylindrical rhizome from which numerous roots emerge. The stem is simple, edged, violet or reddish-brown, with a height ranging from 15 to 50 cm. The inferior leaves are elliptical or ovate, narrowed into petioles, with a strong central vein, while the superior leaves are sessile, arranged opposite each other (rarely alternate), semi-amplexicaul, lanceolate or lanceolate-elongate, and multi-veined. The flowers have angular pedicels with four wings and are grouped in cymose racemes. The calyx is gamosepalous, 5-lobed, with violet-greenish linear, acute laciniae, while the corolla is gamopetalous, rotate, divided nearly to the base, with long, acute, linear-lanceolate laciniae of dark violet color with darker lines or dots. At the base of the corolla, there is a pair of small nectar pits. The androecium consists of stamens with twice the length of the ovary, with sagittate-shaped anthers. The fruit is a polyspermous, unilocular capsule. It blooms from June to August.[2,4]

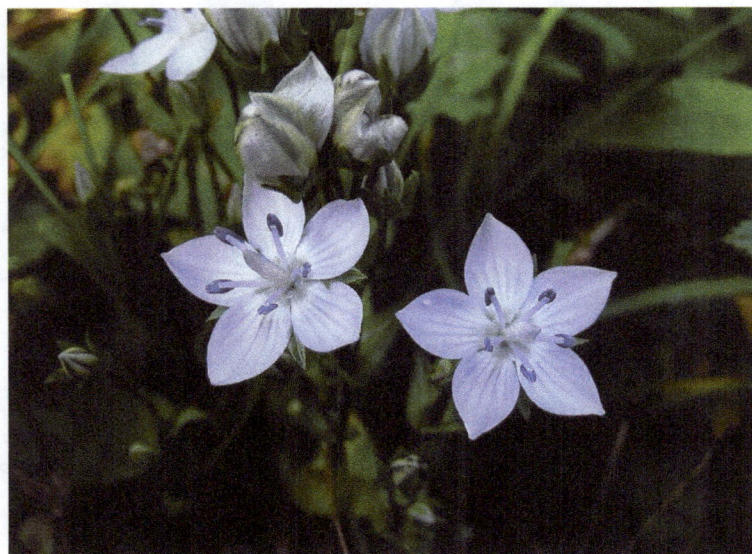

Figure 18.4 *Swertia perennis* L.

Source: 36.

The species has 2 subspecies:

- subsp. *perennis*, found in lower regions, has stem leaves arranged opposite to each other, strongly branched inflorescences, and a height of 20–50 cm.[1]

- subsp. *alpestris* (Baumg. ex Fuss) Simonk., found in higher regions, has alternate arrangement of inferior stem leaves and opposite arrangement of superior stem leaves, weakly branched inflorescences, large flowers, and a height of 15–30 cm.[1]

Main phytoconstituents   Ref.
amino acids   [37-38]
flavonoids (apigenin, quercetin, isoorientin, isovitexin, kaempherol, luteolin)
terpenoids (oleanolic acid, ursolic acid, β-amyrin, lupeol, betulinic acid, hederagenin)
sterols (β-sitosterol, sigmasterol)
fatty acids
phenolic acids
tannins (catechin, epicatechin)
xanthones (mangiferin, norswertianin)
glycosides
esters
irinoids (swertiamarin, amarogentin, amaroswerin, sweroside, gentiopicroside)
hydrocarbons
aldehydes & ketones
miscellaneous

Medicinal Use

Swertia perennis has been used in traditional medicine to treat gastritis.[39]

Biological Activity

Studies report the antioxidant and enzyme inhibitory activities of *Swertia perennis*.[39]

    The genus *Lomatogonium* is represented in the spontaneous flora of Romania by a single species.[2,3]

Lomatogonium carinthiacum (Wulfen) Rchb.

Scientific name: *Lomatogonium carinthiacum* (Wulfen) Rchb.
Common name: Blue feltwort
Romanian name:
Taxonomic classification
Kingdom: *Plantae*
Subkingdom: *Cormobionta*
Phylum: *Spermatophyta*
Subphylum: *Magnoliophytina (Angiospermae)*
Class: *Magnoliatae (Magnoliopsida, Dicotyledonatae)*
Subclass: *Asteridae*
Order: *Gentianales*
Family: *Gentianaceae*
Genus: *Lomatogonium*
Species: *carinthiacum* (Wulfen) Rchb. (Figure 18.5)[2-3]

SPREAD

The species is rarely present in the alpine zone, on rocky soils, stony places, and meadows.[2,4]

DESCRIPTION

It is an annual plant, reaching a height of 3–10 cm, with branching stems from the base, rarely simple, glabrous, four-edged, and erect. The leaves are arranged opposite to each other; the basal ones are shortly petiolate, obtuse, elongate-ovate, while the stem leaves are sessile, elliptic, ovate, or elongated, multi-veined, and pointed. The flowers are of 5 (rarely 4) type and are borne on long, slender pedicels. The gamosepalous calyx is 5 (4)-divided, with lanceolate-ovate, three-veined laciniae, while the rotate corolla is deeply 5 (4)-lobed, with pale blue or white laciniae that are pointed and elliptical, with two basal nectar pockets. The stamens have white filaments and blue anthers, shorter than the elongated, unilocular ovary. The fruit is a polyspermous capsule with smooth, elongated seeds. It blooms from June to September.[1,4]

Figure 18.5 *Lomatogonium carinthiacum* (Wulfen) Rchb.
Source: 40.

Main phytoconstituents	Ref.
	41-43

amino acids
flavonoids (apigenin, quercetin, isoorientin, isovitexin, swertisin, chrysin, luteolin)
terpenoids (oleanolic acid, ursolic acid, Swertiamarin, amarogentin)
coumarins (erythrocentaurin)
sterols (β-sitosterol, daucosterol)
fatty acids
phenolic acids (ferrulic acid, caffeic acid)
tannins (catechin, epicatechin)
xanthones (1-hydroxyl-4,6,8-trimethoxyxanthone, 1,8-dihydroxy-3,5-dimethoxy xanthone, 1-hydroxy-3,7-dimethoxyxanthone, mangiferin, norswertianin)
glycosides
esters
irinoids (swertiamarin, mangiferin, gentiopicroside)
hydrocarbons
aldehydes & ketones
miscellaneous

Medicinal Use

Lomatogonium carinthiacum was utilized in traditional medicine for addressing hepatic and biliary conditions.[41]

Biological Activity

Lomatogonium carinthiacum demonstrates efficacy in various conditions, such as icteric hepatitis, cholecystitis, gastritis, and other digestive tract diseases. Recent medical investigations reveal that its extracts possess hypotensive, hypoglycemic, antioxidant, antitumor, and hypolipidemic effects. They also exhibit an inhibitory impact on the Hepatitis B virus.[42, 44]

REFERENCES

1. Bejenaru C., Mogoșanu G.D., Bejenaru L.E., Popescu H. 2020. *Botanică farmaceutică. Cormobionta*, ediția a III-a. Ed. Sitech, Craiova.

2. Ciocârlan V. 2009. *Flora ilustrată a României. Pteridophyta et Spermatophyta*. Ediția a 3-a revizuită și adăugită. Ed. Ceres, București.

3. Sârbu I., Ștefan N., Oprea A. 2013. *Plante vasculare din România, Determinator ilustrat de teren*. Ed. Victor B Victor, București.

4. Săvulescu T. (ed). 1961. *Flora RPR*, vol. VIII. Ed. Academiei RPR, București.

5. https://dryades.units.it/floritaly/index.php?procedure=taxon_page&tipo=all&id=4012

6. Mogoșanu G.D., Bejenaru L.E., Bejenaru C., Popescu H. 2015. *Farmacognozie-Fitoterapie*. Vol. II, ediția a III-a. Ed. Sitech, Craiova.

7. Drăgulescu C. 2018. *Dicționar de Fitonime Românești*. Ediția a 5-a completată, Editura Universității "Lucian Blaga" Sibiu.

8. Olennikov D.N., Gadimli A.I., Isaev J.I., Kashchenko N.I., Prokopyev A.S., Kataeva T.N., Chirikova N.K., Vennos C. 2019. Caucasian *Gentiana* species: untargeted LC-MS metabolic profiling, antioxidant and digestive enzyme inhibiting activity of six plants. *Metabolites* 9(11):271.

9. Mirzaee F., Hosseini A., Jouybari H.B., Davoodi A., Azadbakht M. 2017. Medicinal, biological and phytochemical properties of *Gentiana species*. *Journal of Traditional and Complementary Medicine* 7(4):400–408.

10. Joksic G., Radak D., Sudar-Milovanovic E., Obradovic M., Radovanovic J., Isenovic E. 2021. Effects of *Gentiana lutea* root on vascular diseases. *Current Vascular Pharmacology* 19(4):359–369.

11. European Medicines Agency-Science Medicine Health. 2018. Assessment Report on *Gentiana lutea* L., *radix*, Based on Article 16d(1), Article 16f and Article 16h of Directive 2001/83/EC (traditional use), EMA/607863/2017 Committee on Herbal Medicinal Products (HMPC).

12. Prakash O., Singh R., Kumar S., Srivastava S., Ved A. 2017. *Gentiana lutea* Linn. (Yellow Gentian): A comprehensive review. *Journal of Ayurvedic and Herbal Medicine* 3(3):175–181.

13. Xu Y., Li Y., Maffucci K.G., Huang L., Zeng R. 2017. Analytical methods of phytochemicals from the genus *Gentiana*. *Molecules* 22(12):2080.

14. Jiang M., Cui B.W., Wu Y.L., Nan J.X., Lian L.H. 2020. Genus *Gentiana*: A review on phytochemistry, pharmacology and molecular mechanism. *Journal of Ethnopharmacology* 264:113391.

15. Menković N., Šavikin-Fodulović K., Savin K. 2000. Chemical composition and seasonal variations in the amount of secondary compounds in *Gentiana lutea* leaves and flowers. *Planta Medica* 66(02):178–180.

16. Šavikin K., Menković N., Zdunić G., Stević T., Radanović D., Janković T. 2009. Antimicrobial activity of *Gentiana lutea* L. *extracts*. *Zeitschrift für Naturforschung C* 64(5-6):339–342.

17. Öztürk N., Can Başer K.H., Aydin S., Öztürk Y., Çaliş I. 2002. Effects of *Gentiana lutea* ssp. *symphyandra* on the central nervous system in mice. *Phytotherapy Research* 16:627–631.

18. Ponticelli M., Lela L., Moles M., Mangieri C., Bisaccia D., Faraone I., Falabella R., Milella L. 2023. The healing bitterness of *Gentiana lutea* L., phytochemistry and biological activities: A systematic review. *Phytochemistry* 206:113518.

19. https://dryades.units.it/ampezzosauris/index.php?procedure=taxon_page&id=4018&num=2010

20. Kitanov G.M., Van D.T. Asenov I. 1991. Chemical composition of the roots of *Gentiana asclepiadea*. *Chemistry of Natural Compounds* 27:369–370.

21. Mihailovic V., Matic S., Mišic D., Solujic S., Stanic S., Katanic J., Mladenovic M., Stankovic N. 2013. Chemical composition, antioxidant and antigenotoxic activities of different fractions of *Gentiana asclepiadea* L. Roots extract. *EXCLI Journal* 12:807–823.

22. Mihailović V., Vuković N., Stojanović J., Nićiforović N. 2009. Chemical composition and antimicrobial activity of the essential oil of the underground parts of *Gentiana asclepiadea* L. *Planta Medica* 75:PJ122.

23. Buza V., Niculae M., Hanganu D., Pall E., Burtescu R.F., Olah K., Matei-Lațiu C., Vlasiuc I., Iozon I., Szakacs A.R., Ielciu I., Ștefănuț L.C. 2022. Biological activities and chemical profile of *Gentiana asclepiadea* and *Inula helenium* ethanolic extracts. *Molecules* 27(11):3560.

24. Mihailović V., Mihailović M., Uskoković A., Arambašić J., Mišić D., Stanković V.., Katanić J., Mladenović M., Solujić S., Matić S. 2013. Hepatoprotective effects of *Gentiana asclepiadea* L. extracts against carbon tetrachloride induced liver injury in rats. *Food and Chemical Toxicology* 52:83–90.

25. Mihailovic V., Vukovic N., Martinović Neda, Solujić Slavica, Mladenović M., Maskovic P., Stankovic M. 2011. Studies on the antimicrobial activity and chemical composition of the essential oils and alcoholic extracts of *Gentiana asclepiadea* L. *Journal of Medicinal Plant Research* 5(7):1164–1174.

26. https://dryades.units.it/trieste/index.php?procedure=taxon_page&id=3999&num=2006

27. Mogoșanu G.D., Bejenaru L.E., Bejenaru C., Popescu H. 2015. *Farmacognozie-Fitoterapie*, Vol. II, ediția a III-a. Ed. Sitech, Craiova.

28. Jovanovic O., Radulović N., Stojanović G., Gudzic B. 2009. Chemical composition of the essential oil of *Centaurium erythraea* Rafn (*Gentianaceae*) from Serbia. *Journal of Essential Oil Research* 21(4):317–322.

29. Mansar-Benhamza L., Djerrou Z., Hamdi Pacha Y. 2013. Evaluation of anti-hyperglycemic activity and side effects of *Erythraea centaurium* (L.) Pers. in rats. *African Journal of Biotechnology* 12(50):6980–6985.

30. Mihaylova D., Vrancheva R., Popova A. 2019. Phytochemical profile and in vitro antioxidant activity of *Centaurium erythraea* Rafn. *Bulgarian Chemical Communications* 51(SI A):95–100.

31. Šiler B., Mišić D., Nestorović J., Banjanac T., Glamočlija J., Soković M. A, Ćirić A. 2010. Antibacterial and antifungal screening of *Centaurium pulchellum* crude extracts and main secoiridoid compounds. *Natural Product Communications* 5(10):1525–1530.

32. El Menyiy, N., Guaouguaou, F.E., El Baaboua, A., El Omari, N., Taha, D., Salhi, N., et.al. 2021. Phytochemical properties, biological activities and medicinal use of *Centaurium erythraea* Rafn. *Journal of Ethnopharmacology* 276:114171.

33. Budniak L., et al. 2021. Determination of amino acids of some plants from *Gentianaceae* family. *Pharmacia* 2: 441.

34. el Menyiy N., Guaouguaou F-E., El Baaboua A., El Omari N., Taha D., Salhi N., Shariati M. A., Tarik A., Benali T., Zengin G., El-Shazly M., Bouyahya A. 2021. Phytochemical properties, biological activities and medicinal use of *Centaurium erythraea* Rafn. *Journal of Ethnopharmacology* 276:114171.

35. https://pfaf.org/user/Plant.aspx?LatinName=Centaurium+erythraea

36. www.infoflora.ch/en/flora/swertia-perennis.html

37. Kshirsagar P., Jagtap U., Gaikwad N., Bapat V. 2019. Ethnopharmacology, phytochemistry and pharmacology of medicinally potent genus *Swertia*: An update. *South African Journal of Botany* 124:444–483.

38. Li J., Zhao Y.L., Huang H.Y., Wang Y.Z. 2017. Phytochemistry and pharmacological activities of the genus *Swertia* (*Gentianaceae*): A review. *American Journal of Chinese Medicine* 45(04):667–736.

39. He X., Chen M., Sun H., Guo X., Sun Y., Li L., Zhu J., Xia G., Zang H. 2022. Multidirectional insights into the phytochemical, biological, and multivariate analysis of extracts from the aerial part of *Swertia perennis* Linnaeus. *Natural Product Research* 38(1):135–139.

40. https://powo.science.kew.org/taxon/urn:lsid:ipni.org:names:370047-1

41. Li L., Li M.H., Zhang N., Huang L.Q. 2011. Chemical constituents from *Lomatogonium carinthiacum* (*Gentianaceae*). *Biochemical Systematics and Ecology* 39(4–6):766–768.

42. Aodungerile X., Tunumula B.G, Sudabilige C., He C., Qirigeer B.L., Bai S. 2022. Chemical constituents of *Lomatogonium carinthiacum* and *Halenia corniculata*. *Chinese Herbal Medicine* 14(3):459–463.

43. Dai L.L., Eni R.G., Fu M.H., Ba G.N. 2022. Botanical, chemical, and pharmacological characteristics of *Lomatogonium rotatum*: A review. *World Journal of Pharmacology* 11(2):6–15.

44. Jia L., Guo H., Jia B., Sun Q. 2011. Anti-tumour activities and a high-performance liquid chromatography mass spectrometric method for analysis of the constituents of *Lomatogonium carinthiacum*. *Natural Product Research* 25(2):100–107.

19 *Hypericaceae* Juss. Family

Cornelia Bejenaru[1], Antonia Radu[1], George Dan Mogoşanu[2],
Ludovic Everard Bejenaru[2], Andrei Biţă[2], and Adina-Elena Segneanu[3]

[1] Department of Pharmaceutical Botany, Faculty of Pharmacy, University of Medicine and Pharmacy of Craiova, 2 Petru Rareş Street, Craiova, Dolj County, Romania
[2] Department of Pharmacognosy & Phytotherapy, Faculty of Pharmacy, University of Medicine and Pharmacy of Craiova, 2 Petru Rareş Street, Craiova, Dolj County, Romania
[3] Institute for Advanced Environmental Research–West University of Timisoara (ICAM–WUT), 4 Oituz Street, Timisoara, Romania

BOTANICAL DESCRIPTION

In the spontaneous flora of Romania, the family groups herbaceous plants from the genus *Hypericum* L: *Hypericum perforatum* L., *Hypericum hirsutum* L., *Hypericum humifusum* L., *Hypericum maculatum* Crantz, *Hypericum montanum* L., *Hypericum tetrapterum* Fr. (*H. quadrangulum* L.), *Hypericum rochelii* Griseb. et Schenk, *Hypericum rumeliacum* Boiss., *Hypericum elegans* Stephan, *Hypericum umbellatum* A. Kern., and *Hypericum richeri* Vill.[1-3]

The plants are perennial, with simple leaves arranged opposite to each other, transparently dotted or black-spotted. The flowers are usually yellow, of type 5, forming cymose terminal inflorescences. The flowers dialisepal and dialipetal, usually contorted or convolute in the bud, with numerous stamens grouped in 3–5 bundles (polyadelphous androecium), sometimes with intercalary hypogynous glands and a superior ovary with numerous ovules. The fruit is a capsule with 1–5 locules and numerous seeds.[1,4]

Hypericum perforatum L. (St. John's wort)

Scientific name: *Hypericum perforatum* L.
Common name: St. John's wort
Romanian name: Sunătoare, Pojarniţă,[1] Iarba Sf. Ion[4]
Taxonomic classification
Kingdom: *Plantae*
Subkingdom: *Cormobionta*
Phylum: *Spermatophyta*
Subphylum: *Magnoliophytina (Angiospermae)*
Class: *Magnoliatae (Magnoliopsida, Dicotyledonatae)*
Subclass: *Dilleniidae*
Order: *Theales*
Family: *Hypericaceae*
Genus: *Hypericum*
Species: *perforatum* L. (Figure 19.1)[1-3]

SPREAD

The species is frequently found in meadows, shrubs, roadside edges, dry, siliceous, and calcareous places, in neglected fields, and along watersides, ranging from the steppe zone up to the beech forest level.[1,4]

DESCRIPTION

The plant is glabrous, with a height of 10–100 cm, having a cylindrical stem, smooth or with two longitudinal edges, and branched in the superior part, while the inferior part is lignified. The leaves are sessile, glabrous, oval to elliptical, with the inferior ones having a rounded base, and the superior ones narrowing towards the base. The flowers have entire lanceolate sepals, acute, with the margin lacking black glands, yellow petals, and numerous stamens (around 50), shorter than the petals. When viewed in the light, the leaves and petals appear perforated due to the presence, at their level, of pockets containing volatile oil. The fruits are oval capsules, irregular-vesicular, and twice as long as the calyx. The seeds are cylindrical, finely alveolate, with both ends pointed. Flowering occurs from June to September.[1-4]

DOI: 10.1201/9781003270515-20

Figure 19.1 *Hypericum perforatum* L. (St. John's wort).

Source: 5.

Main phytoconstituents   Ref.
                                                                                6-8
amino acids (tryptophan)
flavonoids (quercetin, apigenin, kaempherol, luteolin, hyperoside, rutin, quercitrin, myricetin)
terpenoids (pimene, germacrene D, caryophyllene, limonene, cineole, myrcene, geraniol,
    farnesene, humulene)
sterols
phloroglucinols (hyperforin, adhyperforin)
tanins (proanthocyanidin)
fatty acids (palmitic acid, stearic acid, myristic acid)
phenolic acids (chlorogenic acid, vanillic acid, caffeic acid, p-coumaric acid)
hydrocarbons
organic acids (citric acid, malic acid, isovalerianic acid, nicotinic acid, ascorbic acid)
anthraquinones (hypericin, isohypericin, protohypericin, pseudohypericin, protopseudohypericin)
glycosides
carbohydrates (glucose, fructose, saccharose, lactose)
misceleous (xanthones, choline, carotenoids)

From a biogenetic perspective, hypericin, the active lipophilic component with a reddish-violet color, derives from emodol-anthrone.[6] The red color and fluorescence of the oily macerate obtained through hot maceration procedure from the floral tops are due to the content of naphthodianthrones, flavonoid aglycones, and xanthones.[6]

Medicinal Use

Hyperici herba presents a whole complex of pharmacological actions:

■ Psychotropic action, of antidepressant type, according to some authors, attributed to naphthodianthrones, while others correlate it to flavonoids and xanthones (still insufficiently researched), used in the treatment of mild or moderate psychogenic depressions, endo- or exogenous, pre- or post-climacteric, and in states of nervousness;[6]

■ Biocatalytic action, by stimulating, toning, and regulating vital functions: a cup of 2% infusion, or two tea bags in a cup of 200 ml tea, administered in the evening, treats frigidity or impotence;[6]

■ P-vitaminoid-like action, by correcting and improving circulation, reducing capillary permeability, and increasing capillary resistance;

- Bacteriostatic and bactericidal action (hyperforin), used in the treatment of cystitis;

- Cicatrizing and antieczematous action, attributed to both hyperforin and anthranolic monomers, through their antimicrobial effects, showing activity in dermal infections caused by *Staphylococcus aureus*;

- Other actions: anti-inflammatory (antiphlogistic), antihemorrhagic, antihemorrhoidal, cicatrizing in dermal ulcerations and burns (empirical uses), as well as in gastro-duodenal ulcers;

- Used in enuresis nocturna.[6]

The antidepressant effect does not appear immediately but rather after a longer period from the administration of *Hyperici herba* extracts. In depressive patients, clinical studies have shown improvement in the general condition and alleviation or disappearance of specific symptoms such as: anxiety, ataraxia, feelings of worthlessness, somatic disorders, sleep disturbances, and concentration difficulties.

In vitro, hypericin inhibits cerebral, pulmonary, and cutaneous tumors, and *in vitro* and *in vivo*, it has antiviral action against certain viruses (*Influenza*, *Herpes simplex*) or retroviruses (HIV-1).[6]

Toxicity. Contraindications

In prolonged treatment, albino individuals may suffer from an intoxication known as "hypericism," characterized by photodynamic reactions (photosensitization), psychomotor excitement, skin eruptions, dermal necrosis, hemolysis, and even epileptic seizures leading to death. However, intoxications are more common in herbivores than in humans.

Preparations made from *Hyperici herba* are contraindicated both during pregnancy, due to their uterotonic effect, and during breastfeeding, as they inhibit pituitary prolactin.[6]

Drug Interactions

Due to the hepatic enzyme induction effect, through the stimulation of cytochrome P450, St. John's wort extracts should not be administered concomitantly with anticoagulants, anticonvulsants, antivirals used in HIV therapy (indinavir, ritonavir), oral contraceptives, and digoxin. The effects of reserpine are antagonized by St. John's wort preparations.[6]

Biological Activity

- It is recommended for its anti-inflammatory, antihemorrhagic, antihemorrhoidal, and cicatrizing action in dermal ulcerations, burns, and gastro-duodenal ulcers.[1]

- immunomodulatory activity.[9]

- Antioxidant, antidiabetic, antifungal, cholagogic and choleretic, analgesic, anticholinesterase, antiulcerous, anticonvulsant, and cytotoxic activities.[7,10]

- Antidepressant, antiviral, and antibacterial effect.[11-12]

- *Hypericum perforatum* essential oils have demonstrated significant biological activities, such as antiviral, wound healing, antioxidant, antimicrobial, antifungal, anxiolytic, and anticonvulsant properties.[13]

Other Uses

- dermatocosmetics (tonic, moisturizer, etc.)[14]

Hypericum maculatum Crantz (Imperforate St. John's wort)

Scientific name: *Hypericum maculatum* Crantz
Common name: Imperforate St. John's wort
Romanian name: Drobşor, Pojarnică, Pojarniţă, Sunătoare neagră (Horezu)[15]
Taxonomic classification
Kingdom: *Plantae*
Subkingdom: *Cormobionta*
Phylum: *Spermatophyta*
Subphylum: *Magnoliophytina (Angiospermae)*

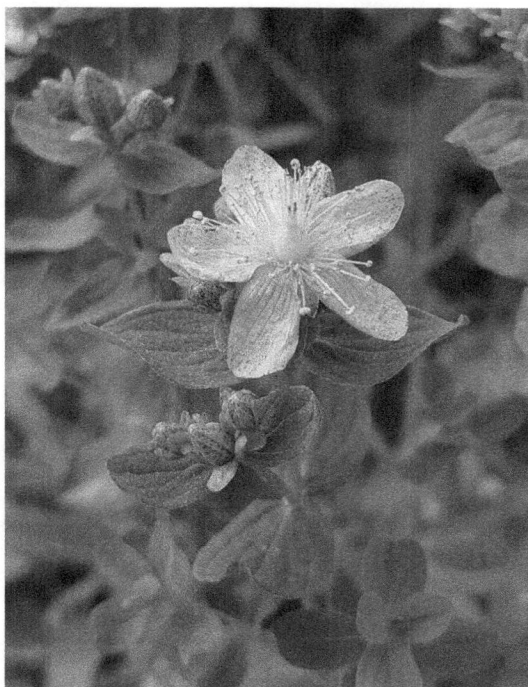

Figure 19.2 *Hypericum maculatum* Crantz (Imperforate St. John's wort).

Source: 16.

Class: *Magnoliatae (Magnoliopsida, Dicotyledonatae)*
Subclass: *Dilleniidae*
Order: *Theales*
Family: *Hypericaceae*
Genus: *Hypericum*
Species: *maculatum* Crantz (Figure 19.2)[1-3]

SPREAD

The species is widespread from the oak forest level up to the subalpine level, occurring in shrubs, meadows, forest edges, and clearings, in wet places, between juniper bushes, wet meadows, on calcareous, granitic, and schistose soils.[2-4]

DESCRIPTION

It is a perennial plant with a height of 15–70 cm, having 4 pronounced edges, not winged on the stem and upper branches. The leaves are sessile, with the inferior ones being broad-oval or elliptical, with prominent reticulate veins, without translucent dots, and the superior ones are semi-amplexicaul. Few, large flowers arranged in terminal dicasiums. The sepals are entire, oval to elliptical, and the petals are yellow, 3–4 times longer than the sepals. The fruits are ovoid capsules, 1–2 times longer than the calyx. The seeds are finely alveolate. It blooms from July to August.[2-4]

The species has two subspecies:

Hypericum maculatum Crantz subsp. *maculatum* has flowers with a diameter of 20–25 mm, entire, broad-elliptical sepals, and petals with black dots on both sides and short black lines.[2,3]
Hypericum maculatum Crantz subsp. *obtusiusculum* (Tourlet) Hayek has floral diameter of 25–35 mm, elliptical-lanceolate sepals, with weakly dentate tips, and petals with shorter lines and long black streaks.[2,3]

243

Main phytoconstituents Ref.

[14,17-19]

amino acids (tryptophan)

flavonoids (luteolin, kaempferol, quercitrin, quercetin, apigenin, rutin, hyperoside, isoquercitrin)

terpenoids (pinene, sabinene, camphene, cymene, sativene, caryophyllene, gurjunene, germacrene, acoradiene, calacorene, ocimene, verbenone, elemene, thujene, muurolene, myrcene, α-phellandrene, α-terpinene, limonene, borneol, α-cubebene, copaene, longipinene, ylangene, bourbonene, cadalene, β-cedrene, farnesene, humulene)

sterols

fatty acids (palmitic acid)

phenolic acids (chlorogenic acid, caftaric acid)

anthraquinones (hypericin, isohypericin, protohypericin, pseudohypericin, protopseudohypericin)

tannins

organic acids

aldehydes & ketones

miscellaneous (xanthones, junenol, carotenoids)

Medicinal Use

Hypericum maculatum is recognized for its medicinal properties. In traditional medicine, it is utilized both internally, as tea or oil extract, and externally, as oil extract, ointment, or cold maceration in ethanol. It has been administered to treat various conditions affecting the skin, locomotor system, nervous system, gastrointestinal tract, respiratory tract, kidneys, urinary tract, cardiovascular system, as well as infections, rheumatism, and gout.[14,20-21]

Biological Activity

Hypericum maculatum and its phytoconstituents demonstrate cytotoxic, antimicrobial, anti-inflammatory, antioxidant, antidepressant, anti-Alzheimer's, antifungal, astringent, antihyperglycemic, and hepatoprotective activities both *in vitro* and/or *in vivo*. Additionally, they exhibit acetylcholinesterase and monoamine oxidase inhibitory activities.[14,21-22]

Other Uses

- cosmetics[14]

Hypericum hirsutum L. (Hairy St. John's wort)

Scientific name: *Hypericum hirsutum* L.
Common name: Hairy St. John's wort
Romanian name: Sunătoare păroasă[15]
Taxonomic classification
Kingdom: *Plantae*
Subkingdom: *Cormobionta*
Phylum: *Spermatophyta*
Subphylum: *Magnoliophytina (Angiospermae)*
Class: *Magnoliatae (Magnoliopsida, Dicotyledonatae)*
Subclass: *Dilleniidae*
Order: *Theales*
Family: *Hypericaceae*
Genus: *Hypericum*
Species: *hirsutum* L. (Figure 19.3)[2-3]

SPREAD

Spread from the oak forest zone to the beech forest level, through shrubs, forest edges, deforested areas, clearings, and forest clearcuts.[2-4]

DESCRIPTION

Perennial plant with a short, woody rhizome, roots emerging from nodes, and one or more stems. The stems are horizontal or oblique at the base, suddenly ascending and erect, unbranched, cylindrical, with short branches. The stem reaches a height of 40–80 cm, covered with short hairs.

Figure 19.3 *Hypericum hirsutum* L. (Hairy St. John's wort).

Source: 23.

The leaves are pubescent, without black glands on the margin, sessile or very shortly petiolate, oval or oval-elongated, obtuse, with translucent dots. The flowers, with short linear-lanceolate bracts with slightly glandular edges, are grouped in a terminal dichasium. The calyx consists of narrowly lanceolate sepals, marginally provided with black, short stipitate, or sessile globular glands. The corolla has oval, narrowly ligulate, yellow petals, slightly longer than the stamens, with a few black glands. The fruits are long capsules with longitudinal resinous stripes, containing cylindrical, finely papillose seeds. The flowering period is from June to August.[2-4]

Main phytoconstituents Ref.
amino acids [14,24-32]
flavonoids (quercetin, apigenin, rutin, myricetin, hyperoside, isoquercitrin, quercitrin, luteolin)
terpenoids (longipinene, farnesene, α-himachalene, caryophyllene, cedrol, α-longipinene, germacrene D, α-selinene)
sterols
fatty acids (palmitic acid)
phenolic acids (ferulic acid, p-coumaric acid)
tannins
organic acids
anthraquinones
glycosides
hydrocarbons (patchoulene, n-nonane, n-undecane)
miscellaneous

Medicinal Use

Despite containing lower concentrations of active principles compared to *Hypericum perforatum*, the species *Hypericum hirsutum* still has potential for producing the composite medicine *Hyperici herba* and standardized products suitable for treating biliary disorders, anxiety, depression, insomnia and promoting wound healing, because they meet the quality requirements outlined in the European Pharmacopoeia concerning the total hypericins content.[26]

Biological Activity

■ anti-inflammatory, antimicrobial and antioxidant activities.[27-31]

Other Uses

Apart from its therapeutic attributes, the Hairy St. John's Wort has been employed for various alternative uses. Historically, people would ignite this plant as incense to repel malevolent spirits and shield against adverse energies. Additionally, it served as a natural dye, yielding a yellow pigment that embellished fabrics. It holds significance in the realm of wildlife as well. Its blossoms offer nectar to bees, butterflies, and other pollinators, whereas the leaves and stems serve as sustenance for caterpillars. Moreover, the plant offers a habitat and shelter for small mammals and avian species.[32]

REFERENCES

1. Bejenaru C., Mogoșanu G.D., Bejenaru L.E., Popescu H. 2020. *Botanică farmaceutică. Cormobionta*, ediția a III-a. Ed. Sitech, Craiova.

2. Ciocârlan V. 2009. *Flora ilustrată a României. Pteridophyta et Spermatophyta*. Ediția a 3-a revizuită și adăugită. Ed. Ceres, București.

3. Sârbu I., Ștefan N., Oprea A. 2013. *Plante vasculare din România, Determinator ilustrat de teren*. Ed. Victor B Victor, București.

4. Săvulescu T. (ed). 1956. *Flora RPR*, vol. IV. Ed. Academiei RPR, București.

5. https://gradinabotanica.umfst.ro/project/hypericum-perforatum/

6. Mogoșanu G.D., Bejenaru L.E., Bejenaru C., Popescu H. 2015. *Farmacognozie-Fitoterapie*. Vol. II, ediția a III-a. Ed. Sitech, Craiova.

7. Alahmad A., Alghoraibi I., Zein R., Kraft S., Dräger G., Walter J.G., Scheper T. 2022. Identification of major constituents of *Hypericum perforatum* L. extracts in Syria by development of a rapid, simple, and reproducible HPLC-ESI-Q-TOF MS analysis and their antioxidant activities. *ACS Omega* 7(16):13475–13493.

8. Nahrstedt A., Butterweck V. 1997. Biologically active and other chemical constituents of the herb of *Hypericum perforatum* L. *Pharmacopsychiatry* 30(Suppl 2):129–134.

9. Schepetkin I.A., Özek G., Özek T., Kirpotina L.N., Khlebnikov A.I., Quinn M.T. 2020. Chemical composition and immunomodulatory activity of *Hypericum perforatum* essential oils. *Biomolecules* 10(6):916.

10. Babotă M., Frumuzachi O., Mocan A., Tămaș M., Dias M.I., Pinela J., Stojković, D., Soković M., Bădărău A.S., Crișan G., et al. 2022. Unravelling phytochemical and bioactive potential of three *Hypericum* species from Romanian spontaneous flora: *H. alpigenum, H. perforatum* and *H. rochelii. Plants* 11:2773.

11. Barnes J., Anderson L.A., Phillipson J.D. 2001. St John's wort (*Hypericum perforatum* L.): a review of its chemistry, pharmacology and clinical properties. *Journal of Pharmacy and Pharmacology* 53(5):583–600.

12. Nahrstedt A, Butterweck V. 1997. Biologically active and other chemical constituents of the herb of *Hypericum perforatum* L. *Pharmacopsychiatry* 30(Suppl 2):129–134.

13. Sharopov F.S., Gulmurodov I.S., Setzer W.N. 2010. Essential oil composition of *Hypericum perforatum* L. and *Hypericum scabrum* L. growing wild in Tajikistan. *Journal of Chemical and Pharmaceutical Research* 2(6):284–290.

14. Silva A.R., Taofiq O., Ferreira I.C.F. R., Barros L. 2021. Hypericum genus cosmeceutical application – A decade comprehensive review on its multifunctional biological properties. *Industrial Crops and Products* 159:113053.

15. Drăgulescu C. 2018. *Dicţionar de Fitonime Româneşti*. Ediţia a 5-a completată, Editura Universităţii "Lucian Blaga" Sibiu.

16. https://identify.plantnet.org/ro/k-world-flora/species/Hypericum%20maculatum%20Crantz/data

17. Rusalepp L., Raal A., Püssa T., Mäeorg U. 2017. Comparison of chemical composition of *Hypericum perforatum* and *H. maculatum* in Estonia. *Biochemical Systematics and Ecology* 73:41–46.

18. Bagdonaite E, Janulis V, Ivanauskas L, Labokas J. 2012. Between species diversity of *Hypericum perforatum* and *H. maculatum* by the content of bioactive compounds. *Natural Product Communications* 7(2):199–200.

19. Oniga I., Toiu A., Benedec D., Tomuţă I., Vlase L. 2016. Phytochemical analysis of *Hypericum maculatum* in order to obtain standardized extracts. *Farmacia* 64:171–174.

20. Băcilă I., Coste A., Halmagyi Deliu A. 2010. Micropropagation of *Hypericum maculatum* Cranz an important medicinal plant. *Romanian Biotechnological Letters* 15:86–91.

21. Caldeira G.I., Gouveia L.P., Serrano R., Silva O.D. 2022. *Hypericum* genus as a natural source for biologically active compounds. *Plants* 11:2509.

22. Ethorđević A.S., Lazarević J.S., Petrović G.M., Zlatković B.K., Solujić S.R. 2014. Chemical and biological evaluation of *Hypericum maculatum* Crantz essential oil. *Chemistry & Biodiversity* 11(1):140–149.

23. https://dryades.units.it/floritaly/index.php?procedure=taxon_page&tipo=all&id=1236

24. Semerdjieva I., Zheljazkov V. D., Dincheva I., Piperkova N., Maneva V., Cantrell C. L., Astatkie T., Stoyanova A., Ivanova T. 2023. Essential oil composition of seven Bulgarian *Hypericum* species and its potential as a biopesticide. *Plants* 12(4):923.

25. Oniga I., Toiu A., Benedec D., Vlase L. 2022. Comparative phytochemical profile of *Hypericum perforatum* and *Hypericum hirsutum* (*Hypericaceae*). *Farmacia* 70(6):1046–1049.

26. https://gradinabotanica.umfst.ro/project/hypericum-perforatum/

27. Radulović N., Stankov-Jovanovic V., Stojanović G., Šmelcerović A., Spiteller M., Asakawa Y. 2007. Screening of in vitro antimicrobial and antioxidant activity of nine *Hypericum* species from the Balkans. *Food Chemistry* 10:15–21.

28. Saroglou V., Marin P., Rancic A., Veljic M., Skaltsa H. 2007. Composition and antimicrobial activity of the essential oil of six *Hypericum* species from Serbia. *Biochemical Systematics and Ecology* 35:146–152.

29. Šavikin K., Dobrić S., Tadić V., Zdunić G. 2007. Antiinflammatory activity of ethanol extracts of *Hypericum perforatum* L., *H. barbatum* Jacq., *H. hirsutum* L., *H. richeri* Vill. and *H. androsaemum* L. in rats. *Phytotherapy Research* 21:176–180.

30. Ilieva Y., Marinov T., Trayanov I., Kaleva M., Zaharieva M.M., Yocheva L., Kokanova-Nedialkova Z., Najdenski H., Nedialkov P. 2023. Outstanding antibacterial activity of *Hypericum rochelii* – comparison of the antimicrobial effects of extracts and fractions from four *Hypericum* species growing in Bulgaria with a focus on prenylated phloroglucinols. *Life* 13:274.

31. Zdunic G., Godjevac D., Savikin K., Petrovic S. 2017. Comparative analysis of phenolic compounds in seven *Hypericum* species and their antioxidant properties. *Natural Product Communications* 12(11):1805–1811.

32. www.wildflowerweb.co.uk/plant/1802/hairy-st-johns-wort

20 *Lamiaceae* Martinov (*Labiatae* Adans.) Family

Cornelia Bejenaru[1], Antonia Radu[1], George Dan Mogoşanu[2],
Ludovic Everard Bejenaru[2], Andrei Biţă[2], and Adina-Elena Segneanu[3]
[1] Department of Pharmaceutical Botany, Faculty of Pharmacy, University of Medicine
and Pharmacy of Craiova, 2 Petru Rareş Street, Craiova, Dolj County, Romania
[2] Department of Pharmacognosy & Phytotherapy, Faculty of Pharmacy, University of
Medicine and Pharmacy of Craiova, 2 Petru Rareş Street, Craiova, Dolj County, Romania
[3] Institute for Advanced Environmental Research–West University of
Timisoara (ICAM–WUT), 4 Oituz Street, Timisoara, Romania

BOTANICAL DESCRIPTION

It includes herbaceous plants, as well as some species of subshrubs or shrubs, with quadrilateral, erect, oblique, or prostrate stems and simple, non-stipulate, opposite leaves. The flowers are hermaphroditic, zygomorphic, pentamerous, grouped in dichasial cymes joined in verticils, in the axil of bracts. The floral envelope is gamosepalous and gamopetalous, with the calyx often bilabiate (unequal teeth), campanulate to long tubular, and the corolla bilabiate (the superior labium may sometimes be reduced or even absent), tubular, or infundibuliform. The superior labium of the corolla consists of 2 petals (with 2 teeth, sometimes fused into one), and the inferior labium is formed by 3 petals (with 2 teeth). The androecium is didynamous, consisting of 4 stamens fused to the corolla, sometimes reduced to only 2 stamens or the stamens may have only half of a fertile anther. Pollen grains are ovoid and exhibit 3 or 6 longitudinal ridges. The gynoecium consists of 2 fused carpels, with a tetralocular ovary (due to the early formation of a false wall) and a superior arrangement. Each ovary locus has a basal, apotropous, and ascending ovule. The style is gynobasic, and the stigma is bilobed. The nectariferous disc is bi- or tetralobate. The fruit is a tetra-achene, with independent detachment of achenes from the flat or convex receptacle. The fruits have a hard and smooth pericarp, rarely with verrucae or hairs. The seed is usually erect, rarely transverse and curved, with a slightly present endosperm that is absent in mature seeds, and an embryo with an inferior radicle and thick, flat cotyledons.[1-4]

The plants of this family are widespread in the warm regions of the globe, with many species characteristic of the Mediterranean region.[1] This family encompasses 150 genera with approximately 3000 species.[4] In the flora of our country, numerous species grow both spontaneously and cultivated for medicinal, aromatic, or ornamental purposes.[1]

In the spontaneous flora of Romania, the following genera are found: *Ajuga* L., *Teucrium* L., *Scutellaria* L., *Marrubium* L., *Dracocephalum* L., *Nepeta* L., *Glechoma* L., *Prunella* L., *Melittis* L., *Phlomis* L., *Galeobdolon* Adans., *Lamium* L., *Galeopsis* L., *Leonurus* L., *Ballota* L., *Stachys* L., *Salvia* L., *Ziziphora* L., *Melissa* L., *Clinopodium* L., *Calamintha* Mill., *Acinos* Mill., *Satureja* L., *Micromeria* Benth., *Origanum* L., *Thymus* L., *Lycopus* L., *Mentha* L., and *Elsholtzia* Willd.[2-3]

Lamium album L. (White dead nettle)

Scientific name: *Lamium album* L.
Common name: White dead nettle
Romanian name: Urzică moartă
Taxonomic classification
Kingdom: *Plantae*
Subkingdom: *Cormobionta*
Phylum: *Spermatophyta*
Subphylum: *Magnoliophytina (Angiospermae)*
Class: *Magnoliatae (Magnoliopsida, Dicotyledonatae)*
Subclass: *Asteridae*
Order: *Lamiales*
Family: *Lamiaceae (Labiatae)*
Genus: *Lamium*
Species: *album* L. (Figure 20.1)[1-3]

DOI: 10.1201/9781003270515-21

Figure 20.1 *Lamium album* L. (White dead nettle).

Source: 5.

SPREAD

White dead nettle is a plant that grows in dense clusters, in thickets, riverside coppices, forest edges, meadows, and fertile places. It is a frequent species, widespread from the plains to the mountainous areas.[1-4]

DESCRIPTION

Perennial herbaceous species with underground stolons and a long rhizome. The erect or ascending stems reach a height of 30–50 cm, are unbranched, tetragonous, violet at the base, dispersed hairy in the middle, and have medium-sized internodes. The petiolate leaves are cordate-ovate and acuminate, dispersed hairy and serrate on both sides. The flowers, measuring 2–2.5 cm in length, are grouped in clusters of 3–6 at the base of the superior leaves. The campanulate calyx has equal teeth that are long and subulate; after the corolla falls, they become divergent. The tubular corolla is white or pale yellow, pubescent, with an oblique hairy ring inside. It has a clearly vaulted superior labium, like a helmet, and a bilobed inferior labium with two small, dentiform lobes at the base and olive-green spots at the base. The anthers are dark brown and covered by the superior labium. The elongated tetra-achenes are trigonous, olive-green, with finely tuberculate surfaces and a voluminous white caruncle. It blooms from April to June.[1-4]

Main phytoconstituents   Ref.

terpenoids (α-thujene, β-amyrin, linalool, oleanoic acid, elemol, myrcene, camphene, β-phellandrene, γ-terpinene, isogeranial, α-terpineol, borneol, cedrol, elemene, longicyclene, α-copaene, germacrene D, humulene, caryophyllene, piperitone, limonene, β-citronellol, β-damascenone, carvone, γ-eudesmol, α-bergamotene, cymene, β-bisabolene, cadinene)   6-9

iridoids (aucubin, caryoptoside, lamalbid, alboside, shanzhiside, barlerin, lamiusides)

flavonoids (quercetin, trifolin, hyperoside, nicotiflorin**,** isoscutellarein, isoquercetin, rutin) ecdysteroids

250

glycosides (verbascoside (acteoside), isoverbascoside, lamalboside)
sterols (β-sitosterol, stigmasterol, daucosterol)
alcohols (octen-3-ol, 3-hexene-1-ol)
phenolic acids (vanillic acid, caffeic acid, ferrulic acid, gallic acid, chlorogenic acid, protocatechuic acid, gentisic acid, syringic acid, *p*-coumaric acid)
carbohydrates (mannitol)
hydrocarbons (n-tetradecane)
aldehydes & ketones (benzeneacetaldehyde, geranial, neral, isoneral)
miscellaneous (allantoin)

Medicinal Use

The flowering aerial parts (*Lamii herba*), and in some cases, just the flowers (*Lamii flos*), are used as anti-inflammatory, sedative, depurative, expectorant, vasoconstrictor, and cicatrizing agents.[1]

Biological Activity

- antioxidant, cytotoxic, haemostatic, antiviral, antimicrobial, anticancer, cytoprotective, wound healing activities[10-15]

Other Uses

- Edible leaves, tea, attracts wildlife, food forest, ground cover[16]

Leonurus cardiaca L. (Motherwort)

Scientific name: *Leonurus cardiaca* L.
Common name: Motherwort
Romanian name: Talpa gâştei
Taxonomic classification
Kingdom: *Plantae*
Subkingdom: *Cormobionta*
Phylum: *Spermatophyta*
Subphylum: *Magnoliophytina (Angiospermae)*
Class: *Magnoliatae (Magnoliopsida, Dicotyledonatae)*
Subclass: *Asteridae*
Order: *Lamiales*
Family: *Lamiaceae (Labiatae)*
Genus: *Leonurus*
Species: *cardiaca* L.[1-3]
Subspecies: – *cardiaca* are glabrescent to glabrous plants, provided with a corolla measuring 9 mm in length (Figure 20.2).[2]

- – *villosus* (Desf. ex Spreng.) Hyl. (*L. quinquelobatus* Gilib.) are densely patent-hairy and have a corolla that is 12 mm in length.[2-3]

SPREAD

Very common in ruderal areas, fallow lands, plantations, and forest clearings throughout the country, from the plains to the sessile oak forest zone.[1-4]

DESCRIPTION

Perennial herbaceous plant with short, lignified rhizomes and erect, branched stems that are reddish or brown, with basal lignified parts, tetragonous, with short, uncinate hairs along the edges, and reaching a height of 1.5 meters. The petiolate leaves are elliptical-suborbicular, subcordate at the base, divided, with palmate venation (palmate-lobate to palmate-partite), and 3–7 acute lobes. The flowers have cuneiform bracteal leaves at the base, entire or irregularly dissected, slightly glandular on the inferior surface. The flowers are arranged in spaced terminal verticils, forming a foliate inflorescence, 10–30 cm long. The floral envelope consists of a campanulate actinomorphic calyx, 5-nerved, slightly bilabiate, glabrous or glabrescent, with 5 prickly teeth,

Figure 20.2 *Leonurus cardiaca* L. (Motherwort).

Source: 17.

triangular-subulate, 2 anterior teeth reflexed, and a pink or rarely white corolla, 9–12 mm in length (with the length twice that of the calyx), with the superior labium arched, villous on the inferior surface, the inferior labium trilobate, shorter, maculate, and the tube white, swollen, with an oblique hairy ring. The stamens have brown anthers and are arranged below the superior labium, and the style is filiform. The fruits are ovoid, truncated, and hairy at the top. The plant blooms from June to August.[1-4]

Main Phytoconstituents

Leonuri herba contains approximately 0.5–1% incomplete iridoids, represented by ajugol (leonuride) and galiridoside; small amounts of volatile oil rich in monoterpenoids; bitter principles with a bicyclic diterpenoid structure; cardiotonic heterosides of the sciladienolide type; sterols; saponins; flavonoids; tannins; polyphenol carboxylic acids; vitamins; organic acids; and mineral salts.[18]

Main phytoconstituents   Ref.

terpenoids (leocardin, marubin, ursolic acid, caryophyllene, α-humulene, α-pinene, β-pinene, linalool, limonene, ursolic acid)   18-20
alkaloids (stachydrine, leonurine)
flavonoids (rutin, hyperoside, isoquercitrin, apigenin, genkwanin)
iridoids (ajugol, leosibiricin, galiridoside, harpagide acetate, reptosid, 7-chloro-6-desoxy-harpagide)
glycosides (lavandulifolioside, verbascoside)
tannins (ellagic acid)
tannins (catechine)
phenolic acids (ferulic acid, chlorogenic acid, caffeic acid, cichoric acid, rosmarinic acid)

Medicinal Use

The aerial part of the plant forms the vegetable product *Leonuri herba* and has central nervous system sedative, cardiac sedative, and antispasmodic actions.[1]

Biological Activity

The herbal product *Leonuri herba* is appreciated in phytotherapy as an effective central nervous system sedative, cardiac sedative, and antispasmodic agent, and it is virtually non-toxic. It is recommended for palpitations and purely nervous heart disorders (cardiac neuroses). It also exhibits other actions, including astringent, antidiarrheal, expectorant, bitter tonic, cicatrizing, and emmenagogue properties.[14]

Other Uses

- condiment

- tea

- dye[21]

Melissa officinalis L. (Lemon balm)

Scientific name: *Melissa officinalis* L.
Common name: Lemon balm
Romanian name: Roiniță, Iarba stupului
Taxonomic classification
Kingdom: *Plantae*
Subkingdom: *Cormobionta*
Phylum: *Spermatophyta*
Subphylum: *Magnoliophytina (Angiospermae)*
Class: *Magnoliatae (Magnoliopsida, Dicotyledonatae)*
Subclass: *Asteridae*
Order: *Lamiales*
Family: *Lamiaceae (Labiatae)*
Genus: *Melissa*
Species: *officinalis* L. (Figure 20.3)[1-3]

SPREAD

The species is found from lowland to mountainous regions in dry, rocky, shady, ruderal places, shrublands, forest edges, clearings, and glades.[1-4]

DESCRIPTION

The genus name is derived from the Greek word "melissa," which means bee, referring to the melliferous properties of the plant.[4]

It is a perennial herbaceous species with a sporadic, tufted growth habit. The plant is characterized by a specific lemony scent. The rhizome is horizontally positioned, with short, underground stolons. The stems are erect or ascending, tetragonal, becoming glabrescent or only pubescent at the top, with mixed articulate hairs and glandular hairs, short, and branched, reaching a height of 30 to 80 cm. The leaves are petiolate, ovate, with a cordate base and crenate margins. The superior surface is glabrescent, with short hairs mixed with glandular hairs, and the inferior surface is glabrous, with 3–4 pairs of secondary veins. The flowers are grouped in clusters of 5–6 in verticils (which have longer bracteal leaves at the base), arranged in a loose inflorescence. The flowers are short-pedicelled, approximately 1 cm in length, with a bilabiate calyx that is slightly curved upward, 13-nerved, with articulate patule hairs mixed with glandular hairs. The corolla is bilabiate, infundibuliform, with the superior labium bilobate and the inferior labium trilobate, pale yellowish or pale lilac, and a stigma with equal lobes. The fruits are ovoid, smooth, and chestnut-colored. Lemon balm blooms from June to August.[1-4]

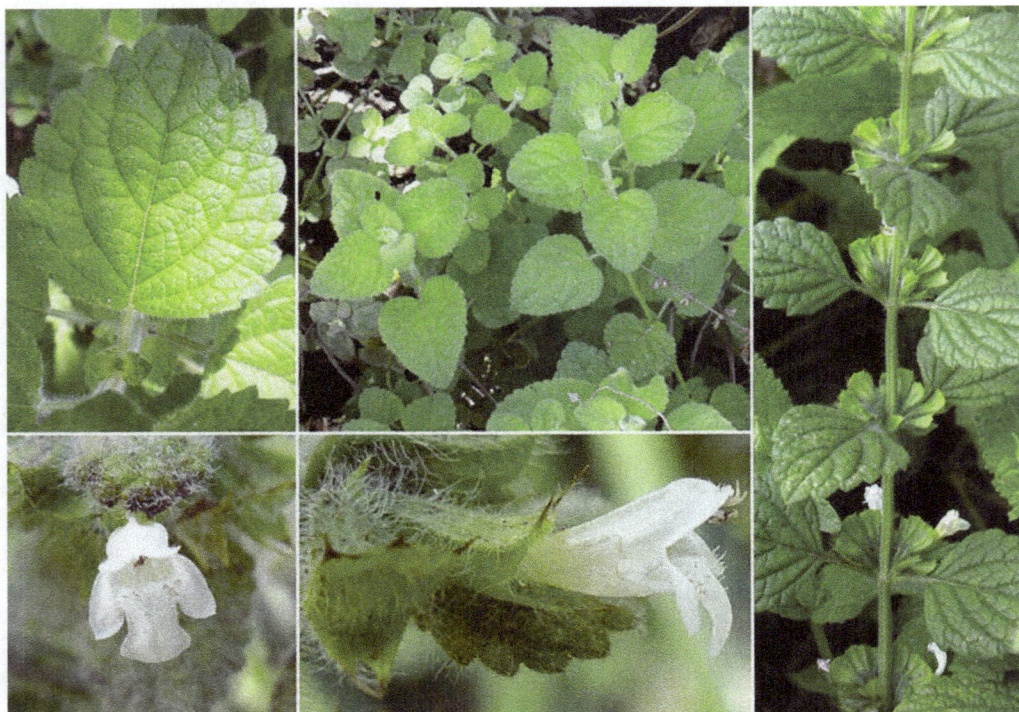

Figure 20.3 *Melissa officinalis* L. (Lemon balm).

Source: 16.

Main Phytoconstituents

Melissae folium contains approximately 0.05–0.15% volatile oil, gallic tannin, flavonoids (derivatives of quercetol, luteol, and kaempferol), coumarins (aesculetin), bitter principles, pentacyclic triterpenoid acids (ursolic acid, oleanolic acid), polyphenolcarboxylic acids, mucilages, and resins.

The volatile oil mainly consists of oxygenated derivatives of acyclic monoterpenes, including alcohols (approximately 25–45% linalool, geraniol) and aldehydes (25–50% citronellal, geranial, or citral). These components contribute to its very pleasant aroma.[18]

Main phytoconstituents   Ref.

terpenoids (α-pinene, betulinic acid camphene, oleanolic acid, ursolic acid, menthol, nerol, camphor, carvacrol, citronellal, citral, geranial, p-cymene, γ-muurolol ocimene, linalool, verbenol, patchoulene, isogeraniol, carane, β-caryophyllene, phytol, linalool, tymol, sabinene, aromadendrene, longifolene, cubenole, himachalene, cis-chrysanthenol, Melissioside A, B, and C)   18,22-24

flavonoids (luteolin, rhamnocitrin, naringin, hesperidin, quercetin, rhamnocitrin, rutin, apigenin, daidzein, myricetin)

phenolic acids (rosmarinic acid, caffeic acid, chlorogenic acid, ferrulic acid, *p*-coumaric acid, gentisic acid, gallic acid, chicoric acid, caftaric acid)

organic acids (tartaric acid, quinic acid, malic acid, citric acid, succinic acid)

aldehydes & ketones (nonanal, benzene acetaldehyde, nonen-1-al)

Medicinal Use

The leaves (*Melissae folium*) have carminative, stomachic, nervine sedative, and antipyretic properties, while the volatile oil obtained by hydrodistillation of lemon balm leaves (*Melissae aetheroleum*) is mildly sedative and antispasmodic.[1]

Biological Activity

Lemon balm leaves are used to prepare infusions with a beneficial effect on indigestion, dyspepsia, diarrhea, and flatulence due to their carminative, stomachic, nervine sedative, and diaphoretic actions.

The volatile oil is mildly sedative and antispasmodic. Most commonly, it is used in the preparation of Melissa water (*Aqua Melissae*) containing 4‰ volatile oil, which is administered internally for digestive ulcers. Lemon balm oil is also used to prepare *Spiritus Melissae*, which is applied externally in lotions or aqueous preparations for vaginal washes.

The volatile oil can be administered directly in its pure form because it is relatively non-toxic at therapeutic doses. The low proportion of volatile oil in the leaves has led to the use of alternative plant materials to produce *Melissae aetheroleum*, resulting in a substitute product known as *Citronellae aetheroleum*.[18]

Toxicity

In large doses, lemon balm volatile oil can cause dizziness, bradycardia, and even a stupefying effect.[18]

Other Uses

- condiment

- tea

- insect repellent[25]

Salvia glutinosa L. (Jupiter's distaff)

Scientific name: *Salvia glutinosa* L.
Common name: Jupiter's distaff, Sticky sage
Romanian name: Cinsteț
Taxonomic classification
Kingdom: *Plantae*
Subkingdom: *Cormobionta*
Phylum: *Spermatophyta*
Subphylum: *Magnoliophytina (Angiospermae)*
Class: *Magnoliatae (Magnoliopsida, Dicotyledonatae)*
Subclass: *Asteridae*
Order: *Lamiales*
Family: *Lamiaceae (Labiatae)*
Genus: *Salvia*
Species: *glutinosa* L. (Figure 20.4)[2-3]

SPREAD

Widespread throughout the country from the sessile oak zone to the spruce zone, through groves, shady forests, along streams, and forest clearings.[2-4]

DESCRIPTION

Perennial herbaceous plant with a thick, oblique rhizome and simple or branched stems, with articulate hairs, predominantly glandular-viscid in the superior part. The leaves are long-petiolate, ovate, with a cordate-hastate-sagittate base, with acute basal lobes, a long pointed apex, reticulate venation, glabrous, green on the superior surface, and grayish-green on the inferior surface. Lanceolate bracteal leaves with an extended apex, entire margins, glandular-hairy. The inflorescence is simple, with 10–12 distant verticils, each with 2–6 flowers. The flowers have pedicels densely glandular-hairy, with a length equal to half the length of the calyx. The calyx is tubular-campanulate, with 14 prominent veins, long glandular hairs, and reddish or yellow punctiform glands on the exterior. The corolla is yellow, sometimes with brown spots, glandular-pubescent on the exterior, longer than the calyx, with the superior labium curved like a sickle, with an emarginate apex, and the inferior labium slightly longer and marked with brown or reddish patterns in the middle. The fruits are ellipsoidal, smooth, brown, with a darker network. The flowering period is from June to August.[2-4]

Figure 20.4 *Salvia glutinosa* L. (Jupiter's distaff).

Source: 26.

Main phytoconstituents   Ref.
flavonoids (apigenin, quercetin, naringin, myricetin, carvacrol, naringerin, luteolin, rutin,
    cathechin, kaempferol, chrysin)   27-32
terpenoids (myrcene, camphor, carene, copaene, limonene, pinene, sabinene, terpinene,
    ocimene, terpinolene, linalool, elemene, valencene, germacrene, humulene, spathulenol,
    farnesene, cadinene, cubebene, phytol, nerolidol, cariophyllene, bourbonene, terpineol)
sterols (sitosterol)
fatty acids (palmitic acid, linoleic acid, oleic acid)
aldehyde & ketone (nonanal, decanal, benzenacetaldehyde)
phenolic acids (Chlorogenic acid, caffeic acid, rosmarinic acid, p-coumaric)
hydrocarbons (tridecane)

Medicinal Use

- wound healing, antidiabetic, analgesic[27-31]

Biological Activity

- ϖ antioxidant, wound healing, dermatological disorders, analgesic, antimicrobial, anbacterial, antifungal activity, neuroprotective, antidiabetic, anticholinesterase, cytotoxic activity[27-32]

Other Uses

- flavouring wines
- ornamental plant[28-29]

Figure 20.5 *Salvia pratensis* L. (Meadow clary).

Source: 33.

Salvia pratensis L. (Meadow clary)

Scientific name: *Salvia pratensis* L.
Common name: Meadow clary, Introduced sage
Romanian name: Salvie de câmp
Taxonomic classification
Kingdom: *Plantae*
Subkingdom: *Cormobionta*
Phylum: *Spermatophyta*
Subphylum: *Magnoliophytina (Angiospermae)*
Class: *Magnoliatae (Magnoliopsida, Dicotyledonatae)*
Subclass: *Asteridae*
Order: *Lamiales*
Family: *Lamiaceae (Labiatae)*
Genus: *Salvia*
Species: *pratensis* L. (Figure 20.5)[2-3]
Subspecies: – *pratensis* has ovate or ovate-oblong leaves and hermaphrodite flowers, which are long, measuring about 20–25 mm in length, with a curved superior labium.[2-3]

 – *dumetorum* (Andrz. ex Besser) Ciocârlan[2] has narrower, oblong leaves, and the corolla of its hermaphrodite flowers is about 12–15 mm long, with the superior labium nearly straight.[2-3]

SPREAD

The species is commonly found from the lowlands up to the beech forest level, in meadows, forest edges, dry hayfields and pastures, shrubs, and grassy rocky areas.[2-4]

DESCRIPTION

Perennial herbaceous plant, with a stem measuring 30–80 cm, erect, simple or branched in the superior part and covered with stipitate glandular hairs, and in the inferior part with tangled multicellular hairs. The stem has 1–3 pairs of leaves. Basal leaves are ovate, cordate at the base, with a sharp apex, irregularly doubly crenate margin, reticulate veined, green on the inferior surface, pubescent, long-petiolate, the stem leaves are smaller, short-petiolate or sessile, and those at the base of the inflorescence branches are amplexicaul, ovate-lanceolate, long-acuminate, with dentate and tangled lanate margin on the inferior surface. Bracts are rounded, long-acuminate, with entire margin, amplexicaul, hairy on the superior surface and with dense and tangled hairs on the inferior surface. The inflorescence is simple, with 1–3 pairs of branches, with 5–12 distant verticils consisting of 4–6 flowers, some specimens have large hermaphrodite flowers, others have smaller female flowers. Floral pedicels are long and densely hairy, the calyx is campanulate, with 13 prominent nerves and covered with long multicellular hairs and stipitate glandular hairs, and between the nerves with sessile punctiform glands. The corolla is long, blue-violet, rarely white or pink, with a tube longer than the calyx, with the superior labium curved, emarginate at the apex, with stipitate glandular hairs on the exterior midline, and otherwise with orange sessile glands, while the inferior labium has multicellular hairs and sessile glands on the exterior surface. The style is longer than the corolla, and the stigma has unequal lobes. The fruits are ovoid, three-edged, brown-blackish. The flowering period is from May to July.[2-4]

Main phytoconstituents

Ref.
[34-39]

flavonoids (apigenin, quercetin, myricetin, luteolin, rutin, hyperoside)
terpenoids (myrcene, camphor, carene, thujene, copaene, limonene, pinene, sabinene, terpinene, camphene, borneol, ocimene, terpinolene, linalool, elemene, valencene, germacrene, humulene, spathulenol, muurolene, eugenol, cadinene, cubebene, cariophyllene, terpineol)
sterols (sitosterol)
fatty acids (palmitic acid, linoleic acid, oleic acid)
aldehyde & ketone (2,4-decadienaI, phenylacetaldehyde, 4-octen-3-one)
phenolic acids (gallic acid, caffeic acid, ferulic acid, ellagic acid, rosmarinic acid, p-coumaric acid)
other (2,3-dihydrobenzofuran, 2-phenylethyl alcohol, indole, resveratrol)

Medicinal Use

- bactericidal, stomachic, antiasthmatical[37-39]

Biological Activity

- antimicrobial, antidiabetic, antioxidant, anti-inflamatory, anticancer, spasmolytic, antiseptic, astringent, antiviral, cardiovascular[37-39]

Other Uses

- flavouring wines
- cooking spice
- salads
- food decoration
- ornamental plant
- hair lotion[39]

Stachys officinalis (L.) Trevis. (Wood betony)

Scientific name: *Stachys officinalis* (L.) Trevis. (*Betonica officinalis* L.)
Common name: Wood betony, Common hedgenettle, Betony, Woundwort
Romanian name: Crețișor, Vindecea,[2] Cuișor,[3] Cuișoriță[4]
Taxonomic classification
Kingdom: *Plantae*
Subkingdom: *Cormobionta*
Phylum: *Spermatophyta*
Subphylum: *Magnoliophytina (Angiospermae)*

Figure 20.6 *Stachys officinalis* (L.) Trevis. (Wood betony).

Source: 40.

Class: *Magnoliatae (Magnoliopsida, Dicotyledonatae)*
Subclass: *Asteridae*
Order: *Lamiales*
Family: *Lamiaceae (Labiatae)*
Genus: *Stachys*
Species: *officinalis* (L.) Trevis. (Figure 20.6)[2-3]

SPREAD

The species is commonly distributed from the plains to the spruce zone, in wet meadows, shrubs, forest edges, ruderal areas, orchards, and hayfields.[2-4]

DESCRIPTION

It is a perennial herbaceous plant with erect stems, slightly ascending at the base, simple or with one pair of branches towards the top. It can be glabrescent or slightly hairy, with hairs pointing downwards, reaching a height of 30–100 cm. The basal rosette leaves are long-petiolated, elongated ovate, cordate at the base, prominently crenate at the margins, clearly reticulate-veined, and disperse and attached hairy on both sides. Stem leaves are short-petiolate, arranged in pairs of 2–3 per stem. The floral leaves are sessile, lanceolate, and either entire or dentate. The inflorescence is 3–6 cm long and consists of compact multi-flowered verticils. The calyx is tubular-campanulate, sparsely hairy, not reticulate-veined, and measures 5–9 mm in length. The corolla is purple, attached-hairy on the exterior, with a tube longer than the calyx, an ovate superior labium with entire margins, an inferior labium with three lobes, and a length of 12–14 mm. The stamens are shorter than half the length of the superior labium, and the fruits are smooth, ovoid, and brown. It blooms from June to August.[2-4]

Main phytoconstituents Ref.

flavonoids (catechin, apigenin, tricin, tricetin, orientin, scutellarein, quercetin, casticin, luteolin, chryseoriol, eupatorin, viscosine, kaempferol, hesperidin, naringin) 41-49

terpenoids (farnesene, selinene, bourbonene, roseotetrol, oleanolic acid, ursolic acid, amyrin, thujene, copaene, pinene, terpinene, ocimene, terpinolene, stachysolone, elemene, valencene, germacrene, humulene, spathulenol, cadinene, cubebene, cariophyllene, terpineol, bergamotene)

sterols (sitosterol, stigmasterol)

fatty acids (palmitic acid, linoleic acid)

phenolic acids (chlorogenic acid, gallic acid, vanillic acid, caffeic acid, ferulic acid, ellagic acid, hydroxycinnamic acid, coumaric acid, sinapic acid)

glycoside (lavandulifolioside, betonyoside, sampneoside)

other (sesamin, coniferin, sulphur compounds)

Medicinal Use

■ respiratory disorders (throat inflammations, cough, cold), migraines, astringent, antidiarrheal, diuretic, antiseptic, febrifuge, tonic, antirheumatic.[41-48]

Biological Activity

■ astringent, antidiarrheal, cholagogue, diuretic, antiseptic, antipyretic, cholagogue, antimicrobial, antibacterial, antinephritic, citotoxic, antioxidant, anti-inflammatory, antihemorrhagic[41-48]

Other Uses

■ herb tea

■ seasonings

■ flavourings[41-48]

Thymus balcanus Borbás (Balkan thyme)

Scientific name: *Thymus balcanus* Borbás (*T. praecox* Opiz subsp. *polytrichus* (A. Kerner ex Borás) Jalas)[2]
Common name: Balkan thyme
Romanian name: Cimbrişor de Balcani[4]
Taxonomic classification
Kingdom: *Plantae*
Subkingdom: *Cormobionta*
Phylum: *Spermatophyta*
Subphylum: *Magnoliophytina (Angiospermae)*
Class: *Magnoliatae (Magnoliopsida, Dicotyledonatae)*
Subclass: *Asteridae*
Order: *Lamiales*
Family: *Lamiaceae (Labiatae)*
Genus: *Thymus*
Species: *balcanus* Borbás (*T. praecox* Opiz subsp. *polytrichus* (A. Kerner ex Borás) Jalas) (Figure 20.7).[2]

SPREAD

This species is commonly distributed from the beech zone to the subalpine zone, growing on rocky, gravelly soils, scree slopes, and crystalline rocks.[2-4]

DESCRIPTION

This perennial herbaceous plant reaches a height of 10 cm and has a thick, woody rhizome. The stems are prostrate, and the flowering branches are short, with hairs only on two opposite sides, alternatively, occasionally having hairs only along the edges. The leaves are densely arranged on the inferior part of the stem, small and elliptical, with prominent pseudo-marginate veins. The middle leaves are rounded spatulate, while those on the sterile shoots are larger. The inflorescence

Figure 20.7 *Thymus balcanus* Borbás (Balkan thyme).

Source: 50.

is globular, and the flowers have a red corolla, rarely white. The flowering period is from June to August.[2-4]

Main phytoconstituents   Ref.
flavonoids (apigenin, eriodictyol, thymusin, thymonin, isorhamnetin, luteolin, isosakuranetin, chrysoeriol, kaempferol, naringin, cirsimaritin, hesperetin, pectolinarigenin) [51-54]
terpenoids (tymol, carvacrol, camphor, linalool, eucalyptol, cymene, geraniol, carene, camphene, myrcene, bisabolene, nerol, farnesene, sabinene, pinene, limonene, linalool, thujene, terpinene, ocimene, elemene, valencene, germacrene, humulene, spathulenol, cadinene, cariophyllene, phellandrene, muurolene, ursolic *acid*, oleanolic acid)
sterols (sitosterol, stigmasterol)
fatty acids (palmitic acid, linoleic acid)
phenolic acids (rosmarinic acid, caffeic acid,)

Medicinal Use

■ cold, tonic, antitussive, antiseptic, carminative[51-54]

Biological Activity

■ antimicrobial, antifungal, antioxidant, anti-inflammatory, antiproliferative[51-54]

Other Uses

■ herbal tea
■ culinary uses[51-54]

REFERENCES

1. Bejenaru C., Mogoșanu G.D., Bejenaru L.E., Popescu H. 2020. *Botanică farmaceutică. Cormobionta*, ediția a III-a. Ed. Sitech, Craiova.

2. Ciocârlan V. 2009. *Flora ilustrată a României. Pteridophyta et Spermatophyta*. Ediția a 3-a revizuită și adăugită. Ed. Ceres, București.

3. Sârbu I., Ștefan N., Oprea A. 2013. *Plante vasculare din România, Determinator ilustrat de teren.* Ed. Victor B Victor, București.

4. Săvulescu T. (ed). 1961. *Flora RPR*, vol. VIII. Ed. Academiei RPR, București.

5. https://dryades.units.it/casentinesi/index.php?procedure=taxon_page&id=4492&num=4679

6. Sulborska A., Konarska A., Matysik-Woźniak A., Dmitruk M., Weryszko-Chmielewska E., et al. 2020. Phenolic constituents of *Lamium album* L. subsp. *album* flowers: anatomical, histochemical, and phytochemical study. *Molecules* 25(24):6025.

7. Konarska A., Sulborska A., Polak B., Dmitruk M., Stefańczyk B., Rejdak, R. 2021, Histochemical and phytochemical analysis of *Lamium album* subsp. *album* L. corolla: essential oil, triterpenes, and iridoids. *Molecules* 26(14):4166.

8. Botirov E.K. 2019. Flavonoids and phenolcarboxylic acids from *Lamium album*. *Chemistry of Natural Compounds* 55:1159–1160.

9. Salehi B., Armstrong L., Rescigno A., Yeskaliyeva B., Seitimova G., Beyatli A., et.al., 2019, *Lamium* plants – a comprehensive review on health benefits and biological activities. *Molecules* 24(10):1913.

10. Paduch R., Matysik G., Wójciak-Kosior M., Kandefer-Szerszen M., Skalska-Kaminska A., Nowak-Kryska M., Niedziela P. 2008. *Lamium album* extracts express free radical scavenging and cytotoxic activities. *Polish Journal of Environmental Studies* 17(4):569–580.

11. Bubueanu C., Iuksel R., Panteli M. 2019. Haemostatic activity of butanolic extracts of *Lamium album* and *Lamium purpureum* aerial parts. *Acta Pharmaceutica* 69(3):443–449.

12. Uwineza P.A., Gramza-Michałowska A., Bryła M., Waśkiewicz A. 2021. Antioxidant activity and bioactive compounds of *Lamium album* flower extracts obtained by supercritical fluid extraction. *Applied Sciences* 11(16):7419.

13. Yordanova Z.P., Zhiponova M.K., Iakimova E.T., Dimitrova M.A., Kapchina-Toteva V.M. 2014. Revealing the reviving secret of the white dead nettle (*Lamium album* L.). *Phytochemistry Reviews* 13:375–389.

14. Bubueanu C., et al. 2013. Antioxidant activity of butanolic extracts of Romanian native species–*Lamium album* and *Lamium purpureum*. *Romanian Biotechnological Letters* 18(6):8855–8862.

15. Chipeva V.A., Petrova D.C., Geneva M.E., Dimitrova M.A., Moncheva P.A., Kapchina-Toteva V.M. 2013. Antimicrobial activity of extracts from *in vivo* and *in vitro* propagated *Lamium album* L. plants. *African Journal of Traditional, Complementary and Alternative Medicines* 10(6):559–562.

16. https://pfaf.org/user/Plant.aspx?LatinName=Lamium+album

17. https://dryades.units.it/floritaly/index.php?procedure=taxon_page&tipo=all&id=4503

18. Mogoșanu G.D., Bejenaru L.E., Bejenaru C., Popescu H. 2015. *Farmacognozie-Fitoterapie*. Vol. II, ediția a III-a. Ed. Sitech, Craiova.

19. Fierascu R.C., Fierascu I., Ortan A., Fierascu I.C., Anuta V., Velescu B.S., Pituru S.M., Dinu-Pirvu C.E. 2019. *Leonurus cardiaca* L. as a source of bioactive compounds: an update of the European Medicines Agency Assessment Report (2010). *BioMed Research International* 2019:4303215.

20. Flemmig J., Noetzel, I., Arnhold J., Rauwald H.W. 2015. *Leonurus cardiaca* L. herb extracts and their constituents promote lactoperoxidase activity. *Journal of Functional Foods* 17:328–339.

21. https://pfaf.org/User/Plant.aspx?LatinName=Leonurus+cardiaca

22. Miraj S., Kiani S. 2017. *Melissa officinalis* L.: a review study with an antioxidant prospective. *Journal of Evidence-Based Complementary & Alternative Medicine* 22(3):385–394.

23. Petrisor G., Motelica L., Craciun L.N., Oprea O.C., Ficai D., Ficai A. 2022. *Melissa officinalis*: composition, pharmacological effects and derived release systems – A review. *International Journal of Molecular Sciences* 23(7):3591.

24. Sharifi-Rad J., Quispe C., Herrera-Bravo J., Akram M., et.al. 2021. Phytochemical constituents, biological activities, and health-promoting effects of the *Melissa officinalis*. *Oxidative Medicine and Cellular Longevity* 2021, Article ID 6584693, 20 pp.

25. https://pfaf.org/User/plant.aspx?latinname=Melissa+officinalis

26. https://dryades.units.it/ampezzosauris/index.php?procedure=taxon_page&id=4672&num=978

27. Mocan A., Babotă M., Pop A., Fizeșan I., Diuzheva A., Locatelli M., et.al. 2020. Chemical constituents and biologic activities of Sage species: a comparison between *Salvia officinalis* L., *S. glutinosa* L. and *S. transsylvanica* (Schur ex Griseb&Schenk) Schur. *Antioxidants* 9:480.

28. Senatore F., De Fusco R., De Feo V., 1997. Essential oils from *Salvia* spp. (*Lamiaceae*). I. Chemical composition of the essential oils from *Salvia glutinosa* L. growing wild in Southern Italy. *Journal of Essential Oil Research* 9(2):151–157.

29. Tavassoli A., Esmaeili A., Ebrahimzadeh M.A., Safaeyan S., Akbarzade M., Rustaiyan A. 2009. Chemical composition of essential oil and antibacterial activity of *Salvia glutinosa* L. growing wild in Iran. *Journal of Applied Chemical Researchers (JACR)* 3(10):5–10.

30. Velickovic D., Ristic M., Velickovic A. 2003. Chemical composition of the essential oils obtained from the flower, leaf and stem of *Salvia aethiopis* L. and *Salvia glutinosa* L. originating from the southeast region of Serbia. *Journal of Essential Oil Research* 15(5):346–349.

31. Mervi M., Bival Stefan M., Kindl M., Blazekovi B., Marijan M., Vladimir-Knezevi S. 2022. Comparative antioxidant, anti-acetylcholinesterase and anti-glucosidase activities of mediterranean *Salvia* species. *Plants* 11: 625.

32. Orhan I.E., Senol F.S., Ercetin T., Kahraman A., Celep F., Akaydin G., Sener B., Dogan M. 2013. Assessment of anticholinesterase and antioxidant properties of selected sage (*Salvia*) species with their total phenol and flavonoid contents. *Industrial Crops and Products* 41:21–30.

33. https://dryades.units.it/lamone/index.php?procedure=taxon_page&id=4674&num=976

34. Anačkov G., Božin B., Zorić L., Vukov D., Mimica-Dukić N., Merkulov L., et.al. 2009. Chemical composition of essential oil and leaf anatomy of *Salvia bertolonii* Vis. and *Salvia pratensis* L. *(Sect. Plethiosphace, Lamiaceae). Molecules* 14:1–9.

35. Carović-Stanko K., Petek M., Grdiša M., Pintar J., Bedeković D., Herak Ćustić M., Satovic Z. 2016. Medicinal plants of the family *Lamiaceae* as functional foods – A review. *Czech Journal of Food Sciences* 34(5):377–390.

36. Coisin M., Necula R., Grigoraş V., Gille E., Rosenhech E., Zamfirache M.M. 2012. Phytochemical evaluation of some *Salvia* species from Romanian flora. *Analele Ştiinţifice ale Universităţii "Al. I. Cuza" Iaşi s. II a. Biologie vegetală* 58(1):35–44.

37. Velickovic D.T., Randjelovic N.V., Ristic M.S., Velickovic A.S. 2002. Chemical composition and antimicrobial action of the ethanol extracts of *Salvia pratensis* L., *Salvia glutinosa* L. and *Salvia aethiopis* L. *Journal of the Serbian Chemical Society* 67(10):639–646.

38. Srećković N., Katanić Stanković J., Mihailovic V. 2020. *Salvia pratensis* L. as a valuable source of phenolic compounds with promising antimicrobial activity. 6th International Electronic Conference on Medicinal Chemistry, 1–30 Nov, 7292.

39. Kucekova Z., Mlcek J., Humpolicek P., Rop O. 2013. Edible flowers – antioxidant activity and impact on cell viability. *Central European Journal of Biology* 8:1023–1031.

40. https://dryades.units.it/euganei/index.php?procedure=taxon_page&id=4518&num=994

41. Imbrea I.M., Nicolin A.L., Imbrea F. 2007. Identifying the main medicinal and aromatic plants in the Almăj depression (Caraş-Severin County, România). *Research Journal of Agricultural Science* 39(1):179–184.

42. Jerkovic I., Gugic M., Males Z., Hazler Pilepic K. 2012. Chemical composition of the essential oil from *Stachys serotina. Chemistry of Natural Compounds* 48(3):508–509.

43. Lazarević J.S., Devic A.S., Kitic D. 2013. Chemical composition and antimicrobial activity of the essential oil of *Stachys officinalis* (L.) Trevis. *(Lamiaceae). Chemistry & Biodiversity* 10(7):1335–1349.

44. Šliumpaitė, I., Venskutonis, P.R., Murkovic, M., Ragažinskienė, O. 2013. Antioxidant properties and phenolic composition of wood betony (*Betonica officinalis* L., syn. *Stachys officinalis* L.). *Industrial Crops and Products* 50:715–722.

45. Avni H., Behxhet M., Chlodwig F., Novak J. 2011. Variability of essential oils of *Betonica officinalis* (*Lamiaceae*) from different wild populations in Kosovo. *Natural Product Communications* 6(9):1345.

46. Bączek K., Kosakowska O., Przybył J.L., Węglarz Z. 2016. Accumulation of phenolic compounds in the purple betony herb (*Stachys officinalis* L.) originated from cultivation. *Herba Polonica* 62(2):7–16.

47. Tobyn G., Denham A., Whitelegg M. 2011. *Stachys officinalis*, wood betony. *Medical Herbs* 29:307–316.

48. Tomou E.M., Barda C., Skaltsa H. 2020. *Genus Stachys*: A review of traditional uses, phytochemistry and bioactivity. *Medicines (Basel, Switzerland)* 7(10):63.

49. Conforti F., Menichini F., Formisano C., Rigano D., Senatore F., Arnold N.A., Piozzi F. 2009. Comparative chemical composition, free radical-scavenging and cytotoxic properties of essential oils of six *Stachys* species from different regions of the Mediterranean Area. *Food Chemistry* 116:898–905.

50. https://dryades.units.it/ampezzosauris/index.php?procedure=taxon_page&id= 4625&num=4552

51. Amiri H. 2012. Essential oils composition and antioxidant properties of three thymus species. *Evidence-Based Complementary and Alternative Medicine: eCAM* 2012:728065.

52. Vidic D., Cavar S., Solić M.E., Maksimović M. 2010. Volatile constituents of two rare subspecies of *Thymus praecox*. *Natural Product Communications* 5(7):1123–1126.

53. Maksimović Z., Stojanović D., Šoštarić I., Dajić Z., Ristić M. 2008. Composition and radical-scavenging activity of *Thymus glabrescens* Willd. (*Lamiaceae*) essential oil. *Journal of the Science of Food and Agriculture* 88(11):2036–2041.

54. Marin P., Grayer R., Kite G., Matevski V. 2003. External leaf flavonoids of *Thymus species* from Macedonia. *Biochemical Systematics and Ecology* 31(11):1291–1307.

21 *Liliaceae* Adans. Family

Cornelia Bejenaru¹, Antonia Radu¹, George Dan Mogoşanu²,
Ludovic Everard Bejenaru², Andrei Biţă², and Adina-Elena Segneanu³

¹ Department of Pharmaceutical Botany, Faculty of Pharmacy, University of Medicine and Pharmacy of Craiova, 2 Petru Rareş Street, Craiova, Dolj County, Romania
² Department of Pharmacognosy & Phytotherapy, Faculty of Pharmacy, University of Medicine and Pharmacy of Craiova, 2 Petru Rareş Street, Craiova, Dolj County, Romania
³ Institute for Advanced Environmental Research–West University of Timisoara (ICAM–WUT), 4 Oituz Street, Timisoara, Romania

BOTANICAL DESCRIPTION

Includes perennial herbaceous plants, rarely shrubs or trees. The roots are fasciculate, filiform, and often contractile. In the plants of this family, underground stems like rhizomes, bulbs, or bulbotubers are frequently present. The aerial stem is unbranched, erect, rarely voluble, and exceptionally strongly branched. The leaves are sessile, of varying shapes (from ovate to linear), with an arcuate venation and entire margins. Occasionally, reticulate venation or dentate margins are encountered. The arrangement of the leaves is alternate, both on stem leaves and basal leaves, sometimes giving the impression of whorled leaves. The flowers are hermaphroditic (exceptionally unisexual), actinomorphic (rarely zygomorphic), pentacyclic trimers (exceptionally dimers or tetramers), arranged solitarily or in diverse inflorescences. The floral envelope is a petaloid perigonium, with free or united tepals, sometimes very well-developed. The stamens, arranged in two verticils of three, have cylindrical or dilated filaments. The tricarpellate gynoecium typically has a superior ovary, rarely inferior or semi-inferior, trilocular, and multiovulate. The ovules can be anatropous, sometimes orthotropous, and have two integuments. The fruit is a loculicidal or septicidal capsule, occasionally a multispermous berry. The seeds have diverse coats and contain a horny or fleshy endosperm.[1-4]

Many species from this family are appreciated for their medicinal and ornamental qualities, which is why they are cultivated. The family is divided into six subfamilies: *Melanthioideae, Colchicoideae, Asphodeloideae, Lilioideae, Scilloideae,* and *Asparagoideae.*

The subfamily *Melanthioideae* includes 2 genera with representatives in the spontaneous flora of Romania: *Veratrum* L. and *Tofieldia* Huds. The subfamily *Colchicoideae* comprises 3 genera: *Colchicum* L., *Bulbocodium* L., and *Merendera* Ramond. The subfamily *Asphodeloideae* includes the genera *Asphodeline* Rchb. and Anthericum L., with representatives in the spontaneous flora, as well as the genus *Hemerocallis* L., with species cultivated for ornamental purposes.

The subfamily *Lilioideae* includes 7 genera with species that are both native to Romania's flora and also cultivated: *Lilium* L., *Fritillaria* L., *Tulipa* L., *Lloydia* Salisb. ex Rchb., *Gagea* Salisb., *Erythronium* L., and *Bellevalia* Lapeyr.

The subfamily *Scilloideae* includes 5 genera: *Scilla* L., *Hyacinthus* L., *Hyacinthella* Schur, *Muscari* Mill., and *Ornithogalum* L.

The subfamily *Asparagoideae* includes 6 genera: *Asparagus* L., *Ruscus* L., *Convallaria* L., *Polygonatum* Mill., *Streptopus* Michx., and *Maianthemum* Weber.[2-3]

The following medicinal plants from the spontaneous flora of Romania are presented, belonging to this family, on which chemical or biological screening were carried out.

Veratrum album L. (False helleborine)

Scientific name: *Veratrum album* L.
Common name: False helleborine, White hellebore, European white hellebore, White veratrum
Romanian name: Ştirigoaie
Taxonomic classification
Kingdom: *Plantae*
Subkingdom: *Cormobionta*
Phylum: *Spermatophyta*
Subphylum: *Magnoliophytina (Angiospermae)*
Class: *Liliatae (Liliopsida, Monocotyledonatae)*
Subclass: *Liliidae*
Order: *Liliales*

DOI: 10.1201/9781003270515-22

Figure 21.1 *Veratrum album* L. (False helleborine).

Source: 5.

Family: *Liliaceae*
Subfamily: *Melanthioideae*
Genus: *Veratrum*
Species: *album* L.[1-3]
Subspecies: – *album* — frequently found in subalpine meadows and marshes. It is characterized by tepals that are green on the exterior and white on the interior. The branches of the inflorescence are patent and reflexed (Figure 21.1);[2-3]

- *lobelianum* (Bernh.) Arcang. — is sporadically found in coniferous forests from the montane to subalpine zones. It is characterized by its greenish-yellow color on both sides and the erect-patent branches of the inflorescence.[2-3]

SPREAD
It grows sporadically from the oak forest zone to the sessile oak zone. It can be found in wet meadows, forest clearings, shrubs, forest edges, pastures, orchards, fallow lands, or weedy areas. However, it is important to note that it is considered toxic and is actively managed by livestock farmers in the mountainous areas.[1-4]

Description
It is a perennial herbaceous species with a felshy, thickened, short, conical rhizome and long, fleshy, thick, whitish or yellowish roots. It has a tall (50–150 cm), erect, thick, fistulous, puberulent aerial pseudostem formed from the leaf sheaths. The leaves are entire, elliptical, and large, over 10 cm in length, with a finely pubescent inferior surface, especially on the veins, and arcuate venation. The flowers have an unpleasant, intoxicating odor and feature tepals that are white, pale green, or yellowish, measuring 7–8 mm in length. The floral pedicels are hairy and shorter than the bracts, arranged in a compound raceme with a terminal disposition. The stamens are shorter than the perigonium, with subulate, white, or pale green filaments and reniform anthers that dehisce

apically through 2 valves. The fruit is a septicidal capsule. The flowering period is from June to August.[1-4]

The name *Veratrum* derives from the Greek word "veratrum," which means "to cause sneezing."[6]

Main phytoconstituents	Ref.
	6-9

amino acids

alkaloids (jervine, pseudojervine, rubijervine, cevadine, germitrine, germidine, veratralbine, veratroidine, neogermitrine, neoprotoveratrine, protoveratrine A and B, veratridine)

flavonoids

terpenoids

sterols

fatty acids

phenolic acids

tannins

Medicinal Uses

- heart diseases (hypertension, heart failure, palpitations in Basedow's disease), skin conditions (eczema, scabies, itching, psoriasis, zoster), stomach issues (cramps, dyspepsia), respiratory problems (whooping cough, pneumonia, fever), mental disorders (anger), rheumatic pains, gout[9-11]

Biological Activity

- anti-inflammatory, vasodilator, hypotensive, anthelmintic, antirheumatic, insecticide, parasiticide, sedative[6,10-11]

At present time, the medicinal product from this plant serves as a raw material for the semisynthesis of both male and female sex hormones, as well as for steroid derivatives with anabolic effects (Decanofort) and steroid anti-inflammatories of adrenal cortical hormones type (cortisone, hydrocortisone) and their derivatives (prednisone and prednisolone). For these purposes, rubijervine and isorubijervine are extracted and processed from the plant.[6]

Protoveratrines A and B, in the form of maleate (Hipoveral), have a very potent antihypertensive effect, but therapeutic doses are very close to toxic doses. At the first signs of exceeding the doses, vomiting occurs. Protoveratrines A and B also act as local anesthetics and cutaneous revulsants. Due to the risk of intoxication with these alkaloids and their inconsistent action (resulting from easily hydrolyzable esters, with the resulting alkamines being pharmacologically inactive), Romanian medications containing protoveratrines for antihypertensive purposes were withdrawn from use with the introduction of Hiposerpil (containing reserpine).[6]

Toxicity

Species of *Veratrum* are highly toxic to both humans and animals. Intoxication with *Veratrum* species is similar to that of *Aconitum* species. It often occurs accidentally, typically due to the plant being mistaken for *Gentiana lutea*, which is used as a bitter tonic. Symptoms include numbness and cooling of the extremities, nausea, vomiting, abdominal colic, bradycardia, severe arterial hypotension, and disruptions in the electrocardiogram (parasympathetic stimulation).[6]

Other Uses

- insecticide
- parasiticide[6]

Colchicum autumnale L. (Autumn crocus)

Scientific name: *Colchicum autumnale* L.
Common name: Autumn crocus, Meadow saffron
Romanian name: Brânduşă de toamnă
Taxonomic classification
Kingdom: *Plantae*
Subkingdom: *Cormobionta*

Figure 21.2 *Colchicum autumnale* L. (Autumn crocus).

Source: 12.

Phylum: *Spermatophyta*
Subphylum: *Magnoliophytina (Angiospermae)*
Class: *Liliatae (Liliopsida, Monocotyledonatae)*
Subclass: *Liliidae*
Order: *Liliales*
Family: *Liliaceae*
Subfamily: *Colchicoideae*
Genus: *Colchicum*
Species: *autumnale* L. (Figure 21.2)[1-3]

SPREAD

A common species, especially in hilly and mountainous meadows, but less frequently found in lowland areas.[1]

Description

It is a toxic perennial herbaceous plant, with a bulbo-tuber that is ovoid and voluminous, covered in brown tunics on the exterior, and may or may not have roots during the flowering period. The leaves appear in the spring following the emergence of flowers and are lanceolate, assimilatory, glabrous, smooth, with a length of 15–35 cm and a width of 2–5 (7) cm. The flowers have tepals of light pink-lilac color, uniform or with a checkerboard-like pattern, and they appear in the autumn months of September to October, emerging from the underground part of the plant. The tepals are fused into a long tube that originates from the bulbo-tuber. The ovary is superior, tricarpellate, with 3 free, long styles that exceed the length of the perigonium. The fruit is a voluminous capsule with soft walls, measuring (30) 45–60 mm in length, initially formed in the soil and emerging above the surface in the spring, along with the young leaves. The fruit reaches maturity in May. The seeds are brown, rugous, with a pitted alveolate appearance, short-appendaged, and brown to nearly black.[1-4]

Main phytoconstituents   Ref.

alkaloids (colchicine, demecolcine, y-lumicolchicine, antitubullin, colchifoline, cornigerine, regelinone, schelhammeridine)   6,13-15

terpenes (carnosol, carnosic acid, rosmadial)

flavonoids (apigenin, luteolin, kaempferol, quercetin, eriodictyol, naringenin, hesperetin, eriocitrin, baicalein, scutellarein)

sterols (β-sitosterol)

fatty acids (palmitic acid, linoleic acid, pentadecanoic acid

phenolic acids (coumaric acid, ferulic acid, caffeic acid, vanillic acid, 2-hydroxybenzoic acid, gentisic acid)

anthocyans (petunidin, cyanidin, pelargonidin, malvinidin, vitisin)

chalcones (butein, phloretin, phloridzin, phlorin)

organic acids (benzoic acid, salicylic acid)

tannins

miscellaneous (arbutin)

Colchici semen contains alkaloid amides with a tropolonic structure, including colchicine (present throughout the entire plant), colchicoside (found only in bulbs and seeds), demecolcine, and N-desacetyl-N-formyl-colchicine (found only in bulbs and flowers). Additionally, it contains 17% lipids, sterols, catechic tannin, organic acids and starch. Protoalkaloids are present in concentrations of 0.3–1.2% in seeds and 0.2–0.6% in bulbs.[6]

Medicinal Use

The medicinal products *Colchici semen* and *Colchici bulbus* have preventive and therapeutic uses due to their antigout, anti-inflammatory, antitumor, mitoclastic (related to cell division), and applications in treating leukemia and cutaneous carcinomas.[1,6,13-15]

Biological Activity

Colchicine has a specific anti-inflammatory action, particularly effective in the treatment of gouty arthritis attacks. It inhibits the phagocytosis of uric acid crystals by polymorphonuclear leukocytes. As a result, the crystals are unable to penetrate the cells, and uric acid is not released into circulation, preventing the onset of gout attacks (inflammation and pain). It is important to note that colchicine does not have analgesic effects and does not alter the values of uricemia or uricosuria.

Colchicine tablets are administered to stimulate intestinal peristalsis after extensive surgical procedures, especially in abdominal surgeries. This assists patients who are under the influence of a muscle relaxant (curarized) in regaining their intestinal function and eliminating gases accumulated in the intestinal tract.

Colchicine, at a dose of 1–2 mg per day, for a duration of 4–11 years, is also used in the treatment of Familial Mediterranean Fever (FMF). FMF is a disease with an unknown etiology characterized by periodic painful episodes.

Colchicine also exhibits antineuralgic and diuretic actions, particularly in the treatment of cardiac insufficiency.

When the tropolonic ring is opened, the pharmacological action of colchicine disappears. Compared to colchicine, colchicoside and demecolcine are less toxic. Colchicoside is approximately 100 times less toxic than colchicine. However, it is important to note that the entire plant remains highly toxic despite these differences in toxicity between its constituents.[6]

Toxicity Contraindications

In toxic doses, colchicine can cause: a burning sensation in the oral mucosa, thirst, difficulty in deglutition, nausea, vomiting, severe gastric pain, diarrhea, biliary colic, hematuria, arterial hypotension, and progressive paralysis. Death can occur due to organ exhaustion, asphyxia, and circulatory collapse. For an adult, the lethal dose is 5 grams of seeds or 10 milligrams of colchicine.

The contraindications for colchicine include pregnancy (due to mutagenic and teratogenic effects), breastfeeding period (as it passes into breast milk and can cause diarrhea in infants), advanced renal insufficiency, and severe hepatic conditions.[6]

Other Uses

The toxic alkaloid colchicine is employed to modify the genetic characteristics of plants, aiming to discover novel and enhanced varieties. This is achieved through a process that involves doubling the chromosome count.[16]

Lilium martagon L. (Turkscap lily)

Scientific name: *Lilium martagon* L.
Common name: Turkscap lily
Romanian name: Crin de pădure
Taxonomic classification
Kingdom: *Plantae*
Subkingdom: *Cormobionta*
Phylum: *Spermatophyta*
Subphylum: *Magnoliophytina (Angiospermae)*
Class: *Liliatae (Liliopsida, Monocotyledonatae)*
Subclass: *Liliidae*
Order: *Liliales*
Family: *Liliaceae*
Subfamily: *Lilioideae*
Genus: *Lilium*
Species: *martagon* L. (Figure 21.3)[2-3]

SPREAD

This species is commonly distributed from lowland areas to the inferior alpine zone, occurring in meadows, on rocky hillsides, at forest edges, in pastures, and ruderal areas.[2-4]

Figure 21.3 *Lilium martagon* L. (Turkscap lily).

Source: 17.

DESCRIPTION

It is a perennial herbaceous species with an ovoid bulb composed of numerous narrow ovoid fleshy scales, arranged laxly, without appendages at the tip, and with numerous roots, some of which contractile. The stem is erect, glabrous, green with reddish spots, and finely powdery scabrous. The inferior and middle stem leaves are arranged in verticils of 5–6, while the superior leaves are alternate. The leaves are glabrous on both surfaces, narrowly to broadly elliptic-lanceolate. The nutant flowers are pink, reddish, or violet, with purple dots, arranged in a lax terminal raceme and having a specific scent. The tepals curve outward, with an elongated nectariferous foveole. The stamens have filaments shorter than the perigonium, and the anthers are red and linear. The fruit is an ovoid capsule with triangular, thin, light brown seeds with a rugose surface. The flowering period is from May to June.[2-4]

Main phytoconstituents
Ref.
[18]

amino acids
flavonoids (rutin, quercetin, kaempferol, myricetin, eriodictyol, phloridzin, isorhamnetin, hyperoside, lancifolide A, isoquercitrin, hesperidin, baicalin)
alkaloids
tanins (catechin, epi-catechin)
sterols
fatty acids
phenolic acids
steroidal saponins (spirostanols, isospirostanols, pseudo-spirostanols, furostanols)
carbohydrates
anthocyans (cyanidin)

Medicinal Uses

- diuretic, emmenagogue, emollient, cicatrizing, expectorant, cardiotonic, heart diseases (angina pectoris), hepatitis, abscesses, toothache[19-21]

Biological Activity

- antioxidant, antifungal, hypoglycemic, anti-inflammatory, sedative, analgesic, hemostatic, anti-hemorrhoidal, antitumor[18,21]

Other Uses

- ornamental

- culinary (bulb)

- cosmetic (emollient, skin conditioner, antiaging, scar fading products, skin lightening)[18-19,21]

Fritillaria meleagris L. (Snake's head fritillary)

Scientific name: *Fritillaria meleagris* L.
Common name: Snake's head fritillary, Chess flower, Frog-cup, Guinea-hen flower, Leper lily
Romanian name: Bibilică,[2] Lalea pestriță[3]
Taxonomic classification
Kingdom: *Plantae*
Subkingdom: *Cormobionta*
Phylum: *Spermatophyta*
Subphylum: *Magnoliophytina (Angiospermae)*
Class: *Liliatae (Liliopsida, Monocotyledonatae)*
Subclass: *Liliidae*
Order: *Liliales*
Family: *Liliaceae*
Subfamily: *Lilioideae*
Genus: *Fritillaria*
Species: *meleagris* L.[2-3]

Figure 21.4 *Fritillaria meleagris* L. (Snake's head fritillary).

Source: 22.

Subspecies: *meleagris* (Figure 21.4)[3]

SPREAD

A protected plant recognized as a natural monument, it is sporadically found from the sylvosteppe zone to the spruce forest zone, occurring in floodplain forests, forest clearings, wetlands, swamps, peat bogs, damp meadows, and shrublands.[2-4]

DESCRIPTION

It is a perennial herbaceous plant with a flattened bulb, provided with two scales in each generation. The stem measures 20–30 cm and lacks leaves in the inferior part, while the superior part typically has 3–5 (6) alternate, linear-lanceolate leaves. The solitary, occasionally paired, flowers are odorless, with a wide campanulate perigonium that is brown-purple with square brown patches or checkerboard-like white markings. The linear nectariferous foveole is located about 10 mm above the base of the tepals. The stamens have purple or whitish filaments, shorter than the perigonium, and elongated, linear anthers. The capsule is erect, and the seeds are flat and compressed. The flowering period is from April to May.[2-4]

Main phytoconstituents   Ref.
alkaloids (cevanine, veratramine, jervine, solanidine, verazine)   23
flavonoids
terpenoids (cholestane)
steroidal saponins (spirostane, spirosolene, furostene)

273

sterols
fatty acids
phenolic acids
tannins
lignans
coumarins
anthocyans

Medicinal Uses

■ asthma, bronchitis, tracheitis, menometrorrhagia, impetigo[23]

Biological Activity

■ antioxidant, antibacterial, antiviral, expectorant, antitussive, antiasthmatic, anti-inflammatory, neuroprotective, antidiabetic, antineoplastic, antihypertensive, anticholinesterase, antiallergic, antinociceptive[23]

Other Uses

■ ornamental[23]

Erythronium dens-canis L. (Dogtooth violet)

Scientific name: *Erythronium dens-canis* L.
Common name: Dogtooth violet
Romanian name: Măseaua ciutei, Cocoșei
Taxonomic classification
Kingdom: *Plantae*
Subkingdom: *Cormobionta*
Phylum: *Spermatophyta*
Subphylum: *Magnoliophytina (Angiospermae)*
Class: *Liliatae (Liliopsida, Monocotyledonatae)*
Subclass: *Liliidae*
Order: *Liliales*
Family: *Liliaceae*
Subfamily: *Lilioideae*
Genus: *Erythronium*
Species: *dens-canis* L.[2-3]
Subspecies: – *dens-canis* has the leaf lamina with reddish spots and a pink-violet perigonium, measuring 2–3 cm in length (Figure 21.5);[2]

– *niveum* (Baumg.) Buia et Păun features the leaf lamina with yellowish spots and a white perigonium, measuring 4(5) cm in length.[2-3]

SPREAD

This species is frequently distributed from lowland areas to the beech forest zone, occurring in meadows, forest clearings, glades, and forests.[2-4]

DESCRIPTION

It is a perennial herbaceous plant, glabrous, with a height of 10–30 cm. It has a narrow, elongated bulb with sessile bulbils and a thin tunic. The stem is a scape, straight and bent at the top, under the flower, during flowering, becoming erect at fruiting. The inferior leaf is long and broad, while the superior one is shorter and narrower, ovate-lanceolate, glabrous, and brown-maculate. The flower is solitary, rarely paired, and can be white, pale yellowish, pink, or violet, unscented. It is nutant during flowering and gradually becomes erect at fruiting. The anthers are linear, violet-blackish, with greenish or light violet pollen. The ovoid capsule contains yellow-brownish seeds that are ovoid with an acuminate appendage. The flowering period is from March to April.[2-4]

Figure 21.5 *Erythronium dens-canis* L. (Dogtooth violet).
Source: 24.

Main phytoconstituents	Ref.
amino acids	25

amino acids
flavonoids (quercetin)
terpenoids
sterols
phenolic acids (chlorogenic acid, ferrulic acid, caffeic acid, p-coumaric acid)
fatty acids
alcaloids
tannins (catechin)
aldehydes & ketones (pentanal, benzaldehyde, hexanal, heptanal, ionone)
anthocyans

Medicinal Uses

- vermifuge, emollient, emetic[26]

Biological Activity

antioxidant, antibacterial, antiparasitic[25]

Other Uses

- culinary (salads, pasta, pastry)
- ornamental[27-28]

Figure 21.6 *Scilla bifolia* L. (Alpine squill).
Source: 29.

Scilla bifolia L. (Alpine squill)

Scientific name: *Scilla bifolia* L.
Common name: Alpine squill
Romanian name: Viorele
Taxonomic classification
Kingdom: *Plantae*
Subkingdom: *Cormobionta*
Phylum: *Spermatophyta*
Subphylum: *Magnoliophytina (Angiospermae)*
Class: *Liliatae (Liliopsida, Monocotyledonatae)*
Subclass: *Liliidae*
Order: *Liliales*
Family: *Liliaceae*
Subfamily: *Scilloideae*
Genus: *Scilla*
Species: *bifolia* L.[2-3]
Subspecies: – *drunensis* has tepals that are 9–10 mm in length and is rare in the counties of Mehedinți and Vâlcea;[2-3]

- *bifolia* is common and has tepals measuring 5–8 mm in length (Figure 21.6);[2-3]

- *subtriphylla* (Schur) Domin reaches a height of 1—30 cm.[3]

SPREAD

This species is distributed from the sylvosteppe zone to the spruce forest zone, occurring in forests, clearings, meadows, and shrublands.[2-4]

DESCRIPTION

It is a perennial herbaceous plant, 10–20 cm in height, with an almost spherical ovoid bulb, covered by dark brown tunics and without bulbils. It has 2 (5) flat or canaliculate, widely linear leaves that envelop the scape halfway. The stem is cylindrical scape, and the inflorescence is an erect raceme during flowering, with 2–6 (12) flowers, arranged in two rows, unilaterally. The flowers are azure blue or violet, occasionally pink or white, with tepals measuring 6–10 mm in length. The erect stamens have blue filaments that do not exceed the perigonium, and the anthers are violet, turning black when mature. The globose capsule contains seeds with a voluminous aril-shaped appendage, wrinkled, and whitish or yellowish. The flowering period is from March to April.[2-4]

Main phytoconstituents

Ref.

amino acids (arginine, leucine, alanine, glycine, serine, valine, methionine, tyrosine, phenylalanine) [30]

alkaloids (hyacinthacines A1, A2, A3, and B3)

flavonoids (4-chromanone, scillascillin, fistein, myricetin, isoquercitin, scillavone, fisetin, myricetin, apigenin, luteolin, hyperoside, isoquercitrin, quercetin, kaempherol, rutin, patuletin)

terpenoids (camphor, scillascillone)

sterols (sitosterol, stigmasterol, cholesterol, campesterol, ergosterol.)

fatty acids (linoleic acid, lauric acid, lauric acid)

phenolic acids (caffeic acid, gentisic acid, chlorogenic acid, p-coumaric acid, ferulic acid, caftaric acid, sinapic acid)

glyosides (scillaren A, proscillaridin A)

saponins (ruscogenin, neoruscogenin, furostanol, spirostanol)

anthocyanins (petunidin, malvidin)

miscellaneous (pinoresinol, mucilages)

Medicinal Use

- emetic, purgative, expectorant, diuretic, chronic bronchitis, chronic kidney disease, heart disease[30-31]

Biological Activity

anti-inflammatory, antitumor[30-31]

Other Uses

- ornamental

- dying natural fibers

- melliferous[30-31]

Muscari neglectum Guss. ex Ten. (*M. racemosum* (L.) Lam. et DC.) (Grape hyacinth)

Scientific name: *Muscari neglectum* Guss. ex Ten.
Common name: Grape hyacinth, Starch hyacinth, Blue bottle
Romanian name: Porumbei,[3] Zambile, Struguraș, Nazarinean, Ceapa ciorii[32]
Taxonomic classification
Kingdom: *Plantae*
Subkingdom: *Cormobionta*
Phylum: *Spermatophyta*
Subphylum: *Magnoliophytina (Angiospermae)*
Class: *Liliatae (Liliopsida, Monocotyledonatae)*
Subclass: *Liliidae*
Order: *Liliales*
Family: *Liliaceae*
Subfamily: *Scilloideae*
Genus: *Muscari*
Species: *neglectum* Guss. ex Ten. (Figure 21.7)[2-3]

Figure 21.7 *Muscari neglectum* Guss. ex Ten. (Grape hyacinth).

Source: 33.

SPREAD

The plant is distributed from lowland areas to the sessile oak forest zone, occurring on rocky slopes, stony hillsides, sandy areas, in fields, vineyards, shrublands, and forest clearings.[2-4]

DESCRIPTION

It is a perennial herbaceous plant, 10–25 cm in height, with a long brown bulb that has numerous bulbils forming a cluster around the main bulb. The stem is a scape with a reddish base, as are the leaves. The leaves appear in autumn, are canaliculate, long, and smooth, with 10–12 pronounced ribs on the inferior surface. The numerous flowers (10) 15–20 (30) form a long inflorescence and have a discolored perigonium, with a dark bluish tube and white, recurved teeth. The flowers are fragrant, with a scent resembling plums. The ovoid capsule measures 8–10 mm in length and contains smooth, black seeds. The flowering period is from March to April.[2-4]

Main phytoconstituents	Ref.
amino acids	34-36
flavonoids (4-chromanone, apigenin, kaempherol)	
alkaloids (hyacinthacines A1, A2, A3, and B3)	
terpenoids	
sterols	
fatty acids (octadecadienoic, octadecatrienoic, octadecanoic)	
phenolic acids (caffeic acid, gentisic acid, p-coumaric acid, quinic acid, fumaric acid)	
glyosides (muscarosides)	
anthocyanins (delphinidin, cyanidin, pelargonidin)	

Medicinal Use

- emollient, diuretic, wound healer, digestive and renal disorders[35]

Biological Activity

- anticancer, antiviral, antidiabetic, antiobesity, anti-inflamatory[35]

Other Uses

- ornamental
- culinary use (pickles from roots and flower buds, wine from flowers)[37]

Asparagus tenuifolius Lam. (Narrow-leaved asparagus)

Scientific name: *Asparagus tenuifolius* Lam.
Common name: Narrow-leaved asparagus
Romanian name: Sparanghel sălbatic,[2] Umbra iepurelui[3]
Taxonomic classification
Kingdom: *Plantae*
Subkingdom: *Cormobionta*
Phylum: *Spermatophyta*
Subphylum: *Magnoliophytina (Angiospermae)*
Class: *Liliatae (Liliopsida, Monocotyledonatae)*
Subclass: *Liliidae*
Order: *Liliales*
Family: *Liliaceae*
Subfamily: *Asparagoideae*
Genus: *Asparagus*
Species: *tenuifolius* Lam. (Figure 21.8)[2-3]

SPREAD

This species is distributed from the sylvosteppe zone to the sessile oak forest zone, occurring in clearings, forests, shrublands, and sandy areas.[2-4]

DESCRIPTION

It is a perennial herbaceous plant with rounded stems, 30–100 cm in height, initially simple at the base and then abundantly branched. The phylloclades are very thin, filiform, with a diameter of 0.1–0.2 mm, smooth, glabrous, erect, grouped in fascicles of 10–20 (40). The stamen filaments are shorter than the greenish-white perigonium, and the anthers are globose or rounded-cordate, 4 times shorter than the filament. The fruit is a round berry about the size of a cherry, and the seeds are black. The flowering period is from May to June.[2-4]

Main phytoconstituents   Ref.
amino acids   [39-43]
flavonoids (quercetin, myricetin, apigenin, isorhamnetin, naringin, hesperidin, astragaloside, kaempherol, odoratin, pratensein, astragalin, trifolin, vitexin, clodrin, hyperoside, nicotiflorin, ascaside, robinin, isoquercetin, quercitrin, luteolin, clovin, rutin)
terpenoids (pinene, carene, limonene, cubebene, caryophyllene, muurolene, aristolene, eudesmene, spathulenol, sabinene, thujene, elemene, linalool, selinene)
sterols (sitosterol, daucosterol, sigmasterol)
fatty acids (palmitic acid, stearic acid, oleic acid, linolenic acid)
phenolic acids (chlorogenic acid, caffeic acid, gallic acid, p-coumaric acid, ferulic acid, syringic acid, vanillic acid)
saponins
tanins
miscellaneous (resveratrol, carotenoids)

Figure 21.8 *Asparagus tenuifolius* Lam. (Narrow-leaved asparagus).

Source: 38.

Medicinal Use

■ laxative, jaundice, expectorant, hypoglycemic, hypotensive, sedative, anemia, rheumatism, gout, osteoarthritis.[39-43]

Biological Activity

■ anti-inflammatory, antipyretic, diuretic, antiseptic, antibacterial, neuroprotective, hepatoprotective, antidiabetic.[39-43]

Other Uses

■ ornamental

■ culinary use (salads, soups)[39]

Ruscus aculeatus L. (Butcher's broom)

Scientific name: *Ruscus aculeatus* L.
Common name: Butcher's broom, Knee holly, Piaranthus
Romanian name: Ghimpe
Taxonomic classification
Kingdom: *Plantae*
Subkingdom: *Cormobionta*
Phylum: *Spermatophyta*
Subphylum: *Magnoliophytina (Angiospermae)*
Class: *Liliatae (Liliopsida, Monocotyledonatae)*
Subclass: *Liliidae*

Figure 21.9 *Ruscus aculeatus* L. (Butcher's broom).

Source: 44.

Order: *Liliales*
Family: *Liliaceae*
Subfamily: *Asparagoideae*
Genus: *Ruscus*
Species: *aculeatus* L. (Figure 21.9)[1-3]

SPREAD

This semi-shrub is of Pontic-Mediterranean origin and grows spontaneously in a few places in Romania, where it is protected as a natural monument. It grows sporadically from the sylvosteppe zone to the beech forest zone, occurring in forests, shrublands, forest clearings, and rocky areas. It is present in the following counties: Mehedinți, Dolj, Caraș-Severin, Bacău, Bihor, Tulcea, Constanța, and Giurgiu.[1-4]

Description

It is a perennial suffrutescent plant with trailing rhizomes from which the aerial part, which is strongly branched, emerges, with the average height of about 20–50 cm, and can even reach up to 1 m, with a pyramidal ovate tuft-like appearance. The plant is sempervirent and has on its stem dark green, rigid, sclerified, distinctly veined, elliptical and leathery phylloclades with a sharp, 1–4 cm long, and very spiky spine at the tip. On the phylloclades, there are reduced leaves in the form of membranous scales and small flowers, 1–2 in the middle area, initially campanulate cylindrical, later expanding into a star shape, with greenish tepals with violet dots, and stamen filaments that are green or violet. The fruit is a red berry. The flowering period is from February to April.[1-4]

Main phytoconstituents

Ref.
39-40,45-50

amino acids

flavonoids (apigenin, quercetin, kaempherol, rutin, hesperidin, schaftosid, orientin, vitexin, rutin, isoquercitrin, nicotiflorin, narcissin)

coumarins (esculin, esculetin, sparteine)

terpenoids (camphor)

sterols (sitosterol)

fatty acids (lignoceric acid)

phenolic acids (caffeic acid, p-coumaric)

saponins (ruscogenin, neoruscogenin, furostanol, spirostanol)

glycosides

miscellaneous (glycolic acid, benzofuranes, linalyl acetate, anethole)

Medicinal Use

The rhizomes with roots (*Rusci rhizoma*) have anti-inflammatory, antihemorrhoidal, diuretic, venotonic, and vasoprotective actions.[1,43]

The medicinal product is used for circulatory disorders (atherosclerosis, varicose veins, leg cramps, leg inflammation, hemorrhoids), as laxative, cytotoxic agent, depurative, diaphoretic, diuretic.[47-50]

Biological Activity

■ antioxidant, antimicrobial, anti-inflammatory, diuretic, vasoconstrictive and venotonic properties (antithrombotic activities).[39-40,49-50]

It is used in venous and venolymphatic insufficiencies, as well as hemorrhoidal crises (with pain, itching, and anal congestion).[45]

Other Uses

■ ornamental

■ cosmetic (eye contour creams, eye concealer)

■ food supplements (against alopecia)[39-40,49-50]

Convallaria majalis L. (Lily-of-the-valley)

Scientific name: *Convallaria majalis* L.
Common name: Lily-of-the-valley
Romanian name: Lăcrămioare, Mărgăritărele
Taxonomic classification
Kingdom: *Plantae*
Subkingdom: *Cormobionta*
Phylum: *Spermatophyta*
Subphylum: *Magnoliophytina (Angiospermae)*
Class: *Liliatae (Liliopsida, Monocotyledonatae)*
Subclass: *Liliidae*
Order: *Liliales*
Family: *Liliaceae*
Subfamily: *Asparagoideae*
Genus: *Convallaria*
Species: *majalis* L. (Figure 21.10)[1-3]

SPREAD

The species grows spontaneously in mountain forests and meadows, shrublands, and groves, from oak forests to the beech forest zone, and is also cultivated as an ornamental plant.[1-4]

Figure 21.10 *Convallaria majalis* L. (Lily-of-the-valley).

Source: 51.

DESCRIPTION

Perennial herbaceous plant with a horizontal, elongated rhizome, branching, with the extremity bent upwards and the base of the stem surrounded by fibrous remains of the sheaths from the previous years. The aerial part, 15–25 cm in height, consists of 2(3) large lanceolate-elliptical leaves, glabrous, attenuated at both ends, long-petiolate, with arcuate, convergent, anastomosed veins, and a terminal inflorescence, in the form of a unilateral raceme, with white flowers, clearly gamotepalous (urceolate), nutant, and beautifully fragrant. The stamens have short and thick filaments, attached at the base of the perigonium, and elongated yellow anthers. The fruit is a red berry. The flowering period is from May to June.[1-4]

Main phytoconstituents	Ref.
	45,52-55

amino acids (asparagine, proline, aspartic acid, tyrosine)
flavonoids (kampferol, quercetin, myrcetin, xanthophylls, isorhamnetin)
alkaloids
terpenoids (citronellol, geraniol, linalool, farnesol, myrcene, ocimene)
sterols
fatty acids (stearic acid, oleic acid)
phenolic acids (caffeic acid, ferulic acid, chlorogenic acid)
glyosides (convallamarin, convallatoxin, convalloside, lokunjoside, neoconvalloside strofantigenol, strofantidol, glucoconvalatoxolozide, convalatoxolozide, convalatoxol) bipindogenol, lokundiozide, strophanthidin, cannogenol, sarmentogenin, dipindogenin, hydroxysarmentogenin)
sterolic saponides (convalaroside (convalamarogenol heteroside), convalamaroside, covalagenol A, covalagenol B)
organic acids (citric acid, malic acid)
miscellaneous

Medicinal Use

The aerial part of the plant (*Convallariae herba*) is used for its cardiotonic and coronary vasodilatory effects, particularly in interdigital pauses.[1,45]

- would healing, cardiac tonic, coronary dilator, cardiac insufficiency, arrhythmia, hypertension, kidney diseases, emphysema, asthma, pulmonary edema, diuretic, laxative, conjunctivitis, joint pain, headaches, neurotonic.[55]

    Convallariae herba is the raw material for the industrial extraction of convallatoxin, which has a similar action to strophanthides (rapid, transient, without accumulation), but unlike them, it is better absorbed in the digestive system.[45]

Biological Activity

- neurotic cardiac conditions, coronary atherosclerosis, antispasmodic, cardiotonic, angiotensin, vasoconstrictive, inotropic, antitumor, anti-inflammatory, sedative, diuretic, emetic, febrifuge, laxative[45,54-57]

Toxicity

The plant is toxic, especially for children who may ingest the red berries or white flowers.[43]

Other Uses

- melliferous
- dye (textile fibres)
- perfumery
- cosmetics (shampoo, soaps, hand creams, depigmentation creams, etc.)
- aromatherapy[56,58]

POLYGONATUM MILL. (SOLOMON'S SEAL)

The genus comprises species of perennial herbaceous plants, characterized by the presence of a horizontal rhizome, on which specific scars remain, in the areas where the stems developed in previous years, leading to the popular name of these species. The stems with elliptical, entire leaves bear solitary, white flowers, arranged axillary, or pauciflorus inflorescences. The fruit is a berry, either black or red in color.[1]

In the Romanian flora, from the approximately 30 species of the genus *Polygonatum*, only four are spontaneous: *P. verticillatum* (L.) All., *P. latifolium* (Jacq.) Desf., *P. multiflorum* (L.) All., *P. odoratum* (Miller) Druce (*P. officinale* All.). These four species frequently grow in forests, woodland clearings, and shaded areas, ranging from the lowlands to the mountains.[1]

Polygonatum odoratum (Mill.) Druce (*P. officinale* All.) (Fragrant Solomon's seal)

Scientific name: *Polygonatum odoratum* (Mill.) Druce
Common name: Fragrant Solomon's seal
Romanian name: Pecetea lui Solomon
Taxonomic classification
Kingdom: *Plantae*
Subkingdom: *Cormobionta*
Phylum: *Spermatophyta*
Subphylum: *Magnoliophytina (Angiospermae)*
Class: *Liliatae (Liliopsida, Monocotyledonatae)*
Subclass: *Liliidae*
Order: *Liliales*
Family: *Liliaceae*
Subfamily: *Asparagoideae*
Genus: *Polygonatum*
Species: *odoratum* (Mill.) Druce[2-3]

Figure 21.11 *Polygonatum odoratum* (Mill.) Druce (*P. officinale* All.) (Fragrant Solomon's seal).

Source: 59.

Subspecies: – *odoratum* has 1–3 membranous scales at the base of the stem and glabrous leaves on the inferior surface. It is found in Oltenia, Banat, Transylvania, Maramureș, and Moldova counties (Figure 21.11);[2]

– *pruinosum* (Boiss.) Ciocârlan has an edged stem, without membranous scales at the base and minutely punctate scabrous leaves, with a pruinose appearance on the inferior surface. It grows in Northern and Central Dobrogea.[2]

SPREAD
The species is frequently distributed from the plain zone to the spruce fir forest level, through clearings, forest edges, shrublands, forests, and rocky areas.[2-4]

DESCRIPTION
Perennial herbaceous plant, with an elongated rhizome, composed of annual segments. The stem reaches a height of 15–50 cm, is glabrous, with 1–3 membranous scales at the base, not foliated in the inferior part, and edged when alive, with 1 (2) membranous scales, deciduous, in the middle section, leaving inelar scars after falling. The leaves are elongated, elliptic-ovoid, short-petiolate, coriaceous, glabrous, with uneven, closely spaced veins, and weakly anastomosed. The glabrous peduncles bear 1–2 flowers with a scent of bitter almonds, and the perigonium is 18–22 mm long, with a cylindrical tube, white, and pubescent-papillose on the internal surface towards the extremities. The stamens have flat, glabrous filaments and linear anthers. The berries are black, blue-pruinose, with yellow seeds, featuring shiny spots. The flowering period is from May to June.[2-4]

Main phytoconstituents
Ref.
60-61
amino acids
alkaloids (soyacerebroside II)
flavonoids (polygonatone A, polygonatone B, polygonatone C, polygonatone D, chroman-4-one, isorhamnetin, apigenin, disporopsin)
terpenoids (spirostene, cycloaltin)
sterols
fatty acids
phenolic acids
carbohydrates (mannose, glucosamine, rhamnose, glucose, galactose, arabinose)
quinones (polygonaquinone A, polygonaquinone B, emodin-8-O--D-glucopyranoside)
tannins

Medicinal Uses

wound healing, antitussive, cardiotonic, bruises, demulcent, diuretic, energizing, hypoglycemic, ophthalmic, resolvent, sedative, tonic, purgative, emetic, revulsive[62-64]

Biological Activity

■ antioxidant, anti-inflammatory, antifatigue, immunomodulating, diuretic, antirheumatic, antidiabetic, cardioprotective, antidote to vegetable poisons or heavy metal poisoning, hemolytic, astringent, antieczematous[61,64]

Other Uses

■ ornamental

■ emergency foods[65-66]

REFERENCES

1. Bejenaru C., Mogoşanu G.D., Bejenaru L.E., Popescu H. 2020. *Botanică farmaceutică. Cormobionta*, ediția a III-a. Ed. Sitech, Craiova.

2. Ciocârlan V. 2009. *Flora ilustrată a României. Pteridophyta et Spermatophyta*. Ediția a 3-a revizuită și adăugită. Ed. Ceres, București.

3. Sârbu I., Ştefan N., Oprea A. 2013. *Plante vasculare din România. Determinator ilustrat de teren.* Ed. Victor B Victor, București.

4. Săvulescu T. (ed). 1966. *Flora RPR*, vol. XI. Ed. Academiei RPR, București.

5. https://dryades.units.it/ampezzosauris/index.php?procedure=taxon_page&id=6744&num=2278

6. Bejenaru L.E., Mogoşanu G.D., Bejenaru C., Popescu H. 2015. *Farmacognozie-Fitoterapie*. Vol. I, ediția a III-a. Ed. Sitech, Craiova.

7. Chandler C.M., McDougal O.M. 2014. Medicinal history of North American *Veratrum*. *Phytochemistry Reviews* 13(3):671–694.

8. Gaillard Y., Pepin G. 2001. LC-EI-MS determination of veratridine and cevadine in two fatal cases of *Veratrum album* poisoning. *Journal of Analytical Toxicology* 25(6):481–485.

9. https://gd.eppo.int/taxon/VEAAL

10. www.sfatulmedicului.ro/plante-medicinale/steregoaie-veratrum-album_14579

11. www.pharmacy180.com/article/veratrum-151/

12. https://dryades.units.it/scuole/index.php?procedure=taxon_page&id=6764&num=320

13. Senizza B., Rocchetti G., Okur M.A., Zengin G., Yıldıztugay E., Ak G., Montesano D., Lucini L. 2020. Phytochemical profile and biological properties of *Colchicum triphyllum* (Meadow Saffron). *Foods* 9(4):457.

14. Davoodi A., Azadbakht M., Hosseinimehr S.J., Emami S., Azadbakht M. 2021. Phytochemical profiles, physicochemical analysis, and biological activity of three *Colchicum* species. *Jundishapur Journal of Natural Pharmaceutical Products* 16(2):e98868.

15. Hailu T., Sharma R., Mann S., Gupta P., Gupta R.K., Rani A. 2021. Determination of bioactive phytochemicals, antioxidant and anti-inflammatory activity of *Colchicum autumnale* L. (Suranjanshireen). *Indian Journal of Natural Products and Resources* 12(1):52–60.

16. https://pfaf.org/User/Plant.aspx?LatinName=Colchicum+autumnale

17. https://dryades.units.it/ampezzosauris/index.php?procedure=taxon_page&id=6806&num=323

18. Olegovich Bokov D., Nikolaevich Luferov A., Ivanovich Krasnyuk (Jnr) I., Vladimirovich Bessonov V. 2019. Ethno-pharmacological review on the wild edible medicinal plant, *Lilium martagon* L. *Tropical Journal of Pharmaceutical Research* 18(7):1559–1564.

19. Temperate Plants Database, Ken Fern. temperate.theferns.info. 2022-09-27. temperate.theferns.info/plant/Lilium+martagon

20. http://mediplantepirus.med.uoi.gr/pharmacology_en/plant_details.php?id=159

21. www.sfatulmedicului.ro/plante-medicinale/crinul-de-padure-lilium-martagon_14531

22. www.dreamstime.com/fritillaria-meleagris-flower-heads-two-beautiful-flower-heads-checkered-lily-fritillaria-meleagris-displayed-image142145975

23. Wang Y., Hou H., Ren Q., et al. 2021. Natural drug sources for respiratory diseases from *Fritillaria*: chemical and biological analyses. *Chinese Medicine* 16:40.

24. https://dryades.units.it/euganei/index.php?procedure=taxon_page&id=6797

25. Demasi S., Caser M., Donno D., Ravetto Enri S., Lonati M., Scariot V. 2021. Exploring wild edible flowers as a source of bioactive compounds: New perspectives in horticulture. *Folia Horticulturae* 33(1):1–22.

26. https://herbaria.plants.ox.ac.uk/bol/plants400/Profiles/EF/Eryth

27. https://koaha.org/wiki/Erythronium_dens-canis

28. www.rhs.org.uk/plants/6814/erythronium-dens-canis-dog-s-tooth-violet-dogs-tooth/details

29. https://dryades.units.it/FVG/index.php?procedure=taxon_page&id=6817&num=328

30. Mulholland D.A., Schwikkard S.L., Crouch N.R. 2013. The chemistry and biological activity of the *Hyacinthaceae*. *Natural Product Reports* 30(9):1165.

31. www.rhs.org.uk/plants/16835/scilla-bifolia/details

32. www.itis.gov/servlet/SingleRpt/SingleRpt#null

33. https://dryades.units.it/floritaly/index.php?procedure=taxon_page&tipo=all&id=6864

34. https://flw-ro.imadeself.com/33/cvety/muskari-57-foto-posadka-i-uhod-za-cvetami-v-otkrytom-grunte-opisanie-travanistogo-rastenia-mysinyj-giacint-ili-gaducij-luk-kogda-nuzno-vykapyvat-i-peresazivat

35. Olegovich Bokov D. 2019. *Muscari armeniacum* Leichtlin (Grape Hyacinth): phytochemistry and biological activities review. *Asian Journal of Pharmaceutical and Clinical Research* 12(1):68–72.

36. Eroglu A., Dogan A. 2023. Investigation of the phytochemical composition and remedial effects of southern grape hyacinth (*Muscari neglectum* Guss. ex Ten.) plant extract against carbon tetrachloride-induced oxidative stress in rats. *Drug and Chemical Toxicology* 46(3):491–502.

37. https://apnd.ro/biodiversitate/plante/muscari-neglectum-porumbei-zambila-motata/

38. https://dryades.units.it/floritaly/index.php?procedure=taxon_page&tipo=all&id=6968

39. Goreanu G. 2009. Studiul farmacognostic al unor specii saponifere din familiile *Asparagaceae* şi *Liliaceae*. *Curierul Medical* 4(310):42–49.

40. Vázquez-Castilla S., Jaramillo-Carmona S., Fuentes-Alventosa J.M., Jiménez-Araujo A., Rodríguez-Arcos R., Cermeño-Sacristán P., Espejo-Calvo J.A., Guillén-Bejarano R. 2013. Saponin profile of green *Asparagus* genotypes. *Journal of Agricultural and Food Chemistry* 61(46):11098–11108.

41. Bratkov V.M., Shkondrov A.M., Zdraveva P.K., Krasteva I.N. 2016. Flavonoids from the genus *Astragalus:* phytochemistry and biological activity. *Pharmacognosy Reviews* 10(19):11–32.

42. Bagci E. 2006. Fatty acid composition of some *Astragalus* species from Turkey. *Chemistry of Natural Compounds* 42:645.

43. Yoon J., Eun Hye K., Ju L., Sung K., Seung K., Jung L., Dae C., Yu C., Chang Y., Ill C. 2012. Variation of phenolic compounds contents in cultivated *Astragalus* membranaceus. *Korean Journal of Medicinal Crop Science* 20(6):447–453.

44. https://dryades.units.it/floritaly/index.php?procedure=taxon_page&tipo=all&id=6970

45. Mogoşanu G.D., Bejenaru L.E., Bejenaru C., Popescu H. 2015. *Farmacognozie-Fitoterapie*, Vol. II, ediţia a III-a. Ed. Sitech, Craiova.

46. De Marino S., Festa C., Zollo F, Iorizzi M. 2012. Novel steroidal components from the underground parts of *Ruscus aculeatus* L. *Molecules* 17(12):14002–14014.

47. Rodrigues J.P.B., Fernandes Â., Dias M.I., Pereira, C., Pires T.C.S.P., et.al. 2021. Phenolic compounds and bioactive properties of *Ruscus aculeatus* L. (*Asparagaceae*): The pharmacological potential of an underexploited subshrub. *Molecules* 26:1882.

48. Urbanek T. 2017. The clinical efficacy of *Ruscus aculeatus* extract: is there enough evidence to update the pharmacotherapy guidelines for chronic venous disease? *Phlebological Review* 25(1):75–80.

49. Hadžifejzović N., Kukić-Markovic J., Petrović, S., Soković M., Glamočlija J., Stojković D., Nahrstedt A. 2013. Bioactivity of the extracts and compounds of *Ruscus aculeatus* L. and *Ruscus hypoglossum* L. *Industrial Crops and Products* 49:407–411.

50. Masullo M., Pizza, C., Piacente S. 2016. *Ruscus* genus: A rich source of bioactive steroidal saponins. *Planta Medica* 82(18):1513–1524.

51. https://dryades.units.it/ampezzosauris/index.php?procedure=taxon_page&id=6953&num=351

52. Ramachanderan R., Schaefer B. 2019. Lily-of-the-valley fragrances. *ChemTexts* 5(2):11.

53. Paciana I., Butnariu M. 2021. Highlighting the compounds with pharmacological activity from some medicinal plants from the area of Romania. *Medicinal and Aromatic Plants (Los Angeles)* 10(1):370.

54. Stansbury J., Saunders P., Winston D., Zampieron E.R. 2012. The use of *Convallaria* and *Crataegus* in the treatment of cardiac dysfunction. *Journal of Restorative Medicine* 1(1):107–111.

55. Matsuo Y., Shinoda D., Nakamaru A., Kamohara K., Sakagami H., Mimaki Y. 2017. Steroidal glycosides from *Convallaria majalis* whole plants and their cytotoxic activity. *International Journal of Molecular Sciences* 18(11):2358.

56. https://plantemedicinale.site/plante-medicinale/lacramioara-convallaria-majalis-l/

57. https://pfaf.org/user/Plant.aspx?LatinName=Convallaria+majalis

58. https://plantemedicinale.site/plante-medicinale/lacramioara-convallaria-majalis-l/

59. https://dryades.units.it/ampezzosauris/index.php?procedure=taxon_page&id=6958&num=353

60. Quan L.T., Wang S.C., Zhang J. 2015. Chemical constituents from *Polygonatum odoratum*. *Biochemical Systematics and Ecology* 58:281–284.

61. Luo L., Qiu Y., Gong L., Wang W., Wen R.A. 2022. Review of *Polygonatum* Mill. genus: its taxonomy, chemical constituents, and pharmacological effect due to processing changes. *Molecules* 27:4821.

62. www.missouribotanicalgarden.org/PlantFinder/PlantFinderDetails.aspx?kempercode=l820

63. https://pfaf.org/user/Plant.aspx?LatinName=Polygonatum+odoratum

64. www.sfatulmedicului.ro/plante-medicinale/pecetea-lui-solomon-polygonatum-odoratum_14439

65. https://luontoportti.com/en/t/1252/angular-solomons-seal

66. https://pfaf.org/user/Plant.aspx?LatinName=Polygonatum+odoratum

22 *Lycopodiaceae* Pal. ex Mirb. Family

Cornelia Bejenaru[1], Antonia Radu[1], George Dan Mogoşanu[2],
Ludovic Everard Bejenaru[2], Andrei Biţă[2], and Adina-Elena Segneanu[3]
[1] Department of Pharmaceutical Botany, Faculty of Pharmacy, University of Medicine
and Pharmacy of Craiova, 2 Petru Rareş Street, Craiova, Dolj County, Romania
[2] Department of Pharmacognosy & Phytotherapy, Faculty of Pharmacy, University of
Medicine and Pharmacy of Craiova, 2 Petru Rareş Street, Craiova, Dolj County, Romania
[3] Institute for Advanced Environmental Research–West University of
Timisoara (ICAM–WUT), 4 Oituz Street, Timisoara, Romania

BOTANICAL DESCRIPTION

The family *Lycopodiaceae* comprises plants called isosporous (characterized by the similarity of sporangia and spores), perennial herbaceous plants, with dichotomously branched roots, creeping and ascending stems that grow longitudinally. The leaves are rigid, entire, without ligules, and small, exceeding 3 mm in length. The reniform sporangia are positioned axillary and isolated. At maturity, the sporangia have a yellow color and open transversely. From the spores, small prothalli with the appearance of tubers or nodules are formed, on which antheridia and archegonia develop. The underground prothallus often forms symbiotic relationships with fungal hyphae.[1-4]

The family includes approximately 200 species distributed in warm and temperate regions; seven of these species are found in our country's spontaneous flora[1], categorized into two genera: *Lycopodium* and *Huperzia*. The genus *Lycopodium* includes the following species: *Lycopodium clavatum* L., *Lycopodium annotinum* L., *Lycopodium inundatum* L. (*Lycopodiella inundata* (L.) Holub; *Lepidotis inundata* (L.) C. Bőrner), *Lycopodium alpinum* L. (*Diphasiastrum complanatum* (L.) Holub. *Diphasium alpinum* (L.) Rothm.), *Lycopodium complanatum* L. (*Diphasiastrum complanatum* (L.) Holub; *Diphasium complanatum* (L.) Rothm.), and *Lycopodium tristachyum* Pursh (*Diphasiastrum tristachyum* (Pursh) Holub; *Diphasium tristachyum* (Pursh) Rothm.). The genus *Huperzia* is represented by a single species, *Huperzia selago* (L.) Schrank et Mart., syn. *Lycopodium selago* L.[2-3]

Lycopodium clavatum L. (Common club moss)

Scientific name: *Lycopodium clavatum* L.
Common name: Common club moss, stag's-horn clubmoss, Running clubmoss, Ground pine
Romanian name: Pedicuţă
Taxonomic classification
Kingdom: *Plantae*
Subkingdom: *Cormobionta*
Phylum: *Pteridophyta*
Class: *Lycopodiatae* (*Lycopodiopsida*)
Order: *Lycopodiales*
Family: *Lycopodiaceae*
Genus: *Lycopodium*
Species: *clavatum* L. (Figure 22.1)[1-3]

SPREAD

The common club moss grows in damp and shady forests, in hilly and mountainous areas, as well as in meadows and shrubby areas.[1-3]

DESCRIPTION

It is a herbaceous perennial plant with creeping stems that can reach 1 meter in length. It has short, ascending lateral branches, measuring about 5–15 cm, which point upward. The leaves are abundant, on unilaterally curved branches, small, entire, linear, and pointed, with a prolonged, colorless aristate apex (extended midrib). This plant exhibits differentiated trophophylls and sporophylls. The trophophylls are normal leaves found at the base of the branches, terminating in long, cylindrical spikes. The spikes bear pointed and widened at the base sporophylls, each carrying a reniform sporangium. Usually, there are 1–3 sporiferous spikes, most commonly 2, on

DOI: 10.1201/9781003270515-23

Figure 22.1 *Lycopodium clavatum* L. (Common club moss).

Source: 5.

each fertile branch, arranged on a long pedicel covered by infrequent bracts. From the spores, a prothallus is formed, on which antheridia and archegonia develop. Following fertilization, an embryo will develop, giving rise to a new plant.[1-4]

Main Phytoconstituents

The *Lycopodium clavatum* alkaloids play a crucial role due to their unique chemical structures and biological activity. Recent research has identified lycopodine as a major alkaloid, alongside clavatine and clavatoxine, as well as polyphenolic acids such as dihydrocaffeic and triterpenes.[6]

Main phytoconstituents	Ref
alkaloids (palhinine A, clavatine, clavatoxine; obscurinene, lycojaponicumin C, huperzine, lycojaponicumin A, lycoflexine, lycodoline, lycojaponicumin E, lycoflexine, α-obscurine)	6-7
terpenoids (lycoclavanin, lycoclaninol, lycojaponicuminol A, lycojaponicuminol D, 3-epilycoclavanol, α-onocerin)	
flavonoids (apigenin)	
phenolic acids (vanillic acid, ferulic acid, p-coumaric acid, salicylic acid)	
miscellaneous (sporopollenin)	

Medicinal Use

The plant is toxic due to the presence of certain alkaloids (clavatotoxin, lycopodine).

The spores of the common club moss (*Lycopodii sporae*) were used in pharmaceutical technology; currently, they are only used in homeopathy. The aerial part of the plant (*Lycopodii herba*) is indicated for rheumatic pains, kidney or liver conditions, as well as in the treatment of smoking and alcohol cessation. Caution is advised when using this plant.[1]

Biological Activity

■ anti-inflammatory potential;[8]

■ antioxidant and antimicrobial actions;[9]

■ anticancer properties, hepatoprotective activity, analgesic and behavioral activity, effects on reproductive systems, central nervous system activity, immunomodulatory properties.[6]

Other Uses

■ Fresh plants are used for dyeing wool in green.[4]

Lycopodium annotinum L. (Stiff club moss)

Scientific name: *Lycopodium annotinum* L.
Common name: Stiff club moss
Romanian name: Cornișor
Taxonomic classification
Kingdom: *Plantae*
Subkingdom: *Cormobionta*
Phylum: *Pteridophyta*
Class: *Lycopodiatae* (*Lycopodiopsida*)
Order: *Lycopodiales*
Family: *Lycopodiaceae*
Genus: *Lycopodium*
Species: *annotinum* L. (Figure 22.2)[2-3]

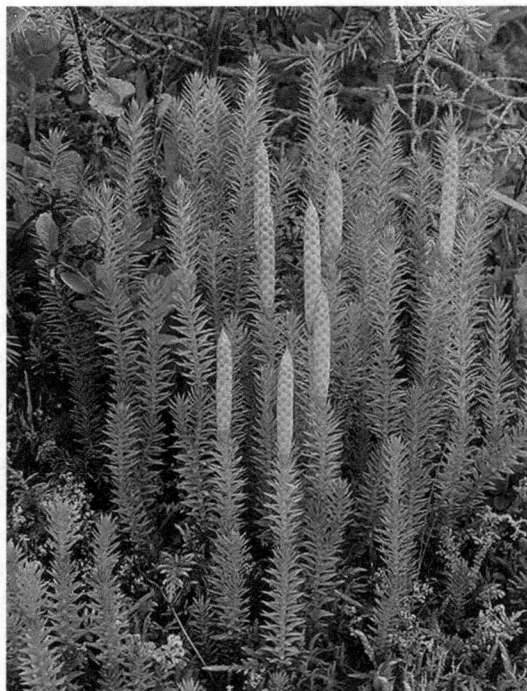

Figure 22.2 *Lycopodium annotinum* L. (Stiff club moss).
Source: 10.

SPREAD

The stiff club-moss is sporadically spread from the beech forest level up to the spruce forest level, in peats, wet areas, marshes, forests, on rocky places, on the edge of shady cliffs, up to an altitude of 1800 m.[2-4]

DESCRIPTION

Herbaceous, perennial plant with creeping stems up to 1 meter in length and ascending branches reaching 10–30 cm in height. The leaves are sparsely arranged, horizontally spreading or reflexed, lanceolate-linear, flat, acuminate, finely dentate, prominently veined on the lower surface, measuring 7 mm in length, and lacking an awn. The sporophylls form cylindrical, sessile spikes. The ascending branches bear a single sporiferous spike with a length of 4 cm.[2-4]

Main phytoconstituents	Ref
	[11-12]

alkaloids (acrifoline, lycopodin)
terpenoids (lycoclavanin, lycoclaninol, lycojaponicuminol A, lycojaponicuminol D, 3-epilycoclavanol, α-onocerin)
flavonoids (apigenin)

Medicinal Use

The species contains lycopodine, which is poisonous by paralysing the motor nerves and also clavatine, which is toxic to many mammals. The spores, however, are not toxic. There are no medicinal uses listed for *Lycopodium annotinum*.[13]

Biological Activity

- acetylcholinesterase inhibitory activity; acetylcholinesterase (AChE) inhibitors are used to treat Alzheimer's patients because they enhance cholinergic neurotransmission.[14]

Other Uses

The plant has been incorporated into clay to fill the gaps between logs in log cabins. It has also been mixed with potting compost to serve as a fertilizer and promote healthier growth for plants cultivated in it.[15]

Huperzia selago (L.) Schrank et Mart. syn. *Lycopodium selago* L.
(Northern firmoss, Fir clubmoss)

Scientific name: *Huperzia selago* (L.) Schrank et Mart. syn. *Lycopodium selago* L.
Common name: Northern firmoss, Fir clubmoss
Romanian name: Brădişor
Taxonomic classification
Kingdom: *Plantae*
Subkingdom: *Cormobionta*
Phylum: *Pteridophyta*
Class: *Lycopodiatae* (*Lycopodiopsida*)
Order: *Lycopodiales*
Family: *Lycopodiaceae*
Genus: *Huperzia*
Species: *selago* (L.) Schrank et Mart. (Figure 22.3)[1-3]

SPREAD

The northern firmoss is sporadically distributed from the beech forest zone up to the subalpine zone. It can be found in shady, slightly damp forests, on rocky terrain, and in grassy areas.[2,4]

DESCRIPTION

Herbaceous, perennial plant, with a height of 5-20 cm, having ascending stems that are dichotomously branched[2-3] and curved or straight branches, parallel, grouped in rigid, cylindrical fascicles. The leaves are entire, rigid, linear-lanceolate, acuminate, and completely cover the stem, positioned patent, erect, or appressed to the stem. The leaves are not differentiated into

Figure 22.3 *Huperzia selago* (L.) Schrank et Mart. (Northern firmoss, Fir clubmoss).

Source: 15.

trophophylls and sporophylls; they are similar. The reniform sporangia are arranged at the base of the leaves, axillary towards the tip.[2-4]

Main phytoconstituents   Ref
alkaloids (huperzine A, lycopodin, selagoline, serratidine, cermizine B,
   α-hydroxylycopodine, lycodoline, isolycodoline, acrifoline, huperzine B,
   β-hydroxyhuperzine A, α-obscurine, flabelline, lycoposerramine-L)   [16-19]
terpenoids (serratenediol, 21-episerratenediol)
flavonoids (selgin, tricin)
fatty acids (palmitic acid, stearic acid, oleic acid)

Medicinal Use

In traditional medicine was used the whole plant in a poultice applied to the head for headaches.[20]

In folk medicine, the aerial part of *Huperzia selago* was utilized for medicinal purposes, primarily to aid in alcohol withdrawal. Additionally, this plant was employed in complex therapy for conditions like pulmonary tuberculosis, urinary system disorders, neurasthenia, glaucoma, convulsions, metabolic disorders, also serving as a laxative, diuretic, and anti-inflammatory agent for cystitis. Moreover, it was also utilized as an anti-helminthic and anti-tumor agent. Externally, *Huperzia selago* was applied to treat conjunctivitis, baldness, and various skin conditions such as dermatitis, eczema, psoriasis, and others.

Potions are highly toxic. Incorrect usage of a decoction or concentrated form can lead to severe consequences, including vomiting, a decrease in blood pressure, arrhythmia, excessive sweating, diarrhea, speech impairments, breathing difficulties, and various other serious effects.[21]

Biological Activity

Studies have shown that alkaloids derived from *Huperzia selago* exhibit antioxidative properties, effectively scavenging free radicals and preventing lipid and protein oxidation.[22]

These properties make them a desirable mechanism of action for addressing neurodegenerative disorders. These alkaloids are potential active compounds for developing novel treatments for Alzheimer's and Parkinson's diseases. The most important one is huperzine A, which has garnered major therapeutic interest as a treatment for acetylcholine-deficit dementia, including Alzheimer's disease. Notably, *H. selago* is the only European and North American species known to contain huperzine A.[19,23]

REFERENCES

1. Bejenaru C., Mogoşanu G.D., Bejenaru L.E., Popescu H. 2020. *Botanică farmaceutică. Cormobionta*, ediția a III-a. Ed. Sitech, Craiova.

2. Ciocârlan V. 2009. *Flora ilustrată a României. Pteridophyta et Spermatophyta*. Ediția a 3-a revizuită și adăugită. Ed. Ceres, București.

3. Sârbu I., Ștefan N., Oprea A. 2013. *Plante vasculare din România, Determinator ilustrat de teren*. Ed. Victor B Victor, București.

4. Săvulescu T. (ed). 1952. *Flora RPR*, vol. I. Ed. Academiei RPR, București.

5. www.centralcoastbiodiversity.org/running-clubmoss-bull-lycopodium-clavatum.html

6. Banerjee J., Biswas S., Madhu N.R., Karmakar S.R., Biswas S.J. 2014. A better understanding of pharmacological activities and uses of phytochemicals of *Lycopodium clavatum*: A review. *Journal of Pharmacognosy and Phytochemistry* 3(1):207–210.

7. Li X., Kang M., Ma N., Pang T., et.al. 2019. Identification and analysis of chemical constituents and rat serum metabolites in Lycopodium clavatum using UPLC-Q-TOF/MS combined with multiple data-processing approaches. Evidence-Based Complementary and Alternative Medicine 2019, Article ID 5165029, 8 pp.

8. Orhan I., Küpeli E., Sener B., Yesilada E. 2007. Appraisal of anti-inflammatory potential of the clubmoss, *Lycopodium clavatum* L. *Journal of Ethnopharmacology* 109(1):146–50.

9. Orhan I., Özçelik B., Aslan S. et al. 2007. Antioxidant and antimicrobial actions of the clubmoss *Lycopodium clavatum* L. *Phytochemistry Reviews* 6:189–196.

10. www.rolv.no/bilder/galleri/fjellplanter/lyco_ann_ann.htm

11. Manske R.H.F., Marion L. 1947. The alkaloids of *Lycopodium* species. IX. *Lycopodium annotinum* var. *acrifolium*, fern. and the structure of annotinine. *Journal of the American Chemical Society* 69(9):2126–2129.

12. Descallar A.L., Nuñez M.P.S., Cabrera M.L.N., Martin T.T.B., Obemi, C.D.G., Lañojan R.S. 2017. Phytochemical analysis and antioxidant capacity of *Lycopodium clavatum* Linn. from Lake Sebu, South Cotabato, Philippines. *AIP Conference Proceedings* 1803(1):020021.

13. https://practicalplants.org/wiki/lycopodium_annotinum/#Medicinal_uses.28Warning.21.29

14. Halldórsdóttir E., Olafsdottir E., Jaroszewski J.W. 2008. Alkaloid content of the Icelandic club moss *Lycopodium annotinum*—acetylcholinesterase inhibitory activity *in vitro*. *Planta Medica* 74(09):PB53.

15. https://gobotany.nativeplanttrust.org/species/huperzia/selago/

16. Lenkiewicz A.M., Czapski G.A., Jęsko H., Wilkaniec A., Szypuła W., et al. 2016. Potent effects of alkaloid-rich extract from *Huperzia selago* against sodium nitroprusside-evoked PC12 cells damage via attenuation of oxidative stress and apoptosis. *Folia Neuropathologica* 2:156–166.

17. Xu M., Eiriksson F.F., Thorsteinsdottir M., Heidmarsson S., Omarsdottir S., Olafsdottir E.S. 2019. Alkaloid fingerprinting resolves *Huperzia selago* genotypes in Iceland. *Biochemical Systematics and Ecology* 83:77–82.

18. Armijos C., Gilardoni G., Ama, L., Lozano A., Bracco F., Ramirez J., et al. 2016. Phytochemical and ethnomedicinal study of *Huperzia* species used in the traditional medicine of Saraguros in Southern Ecuador; AChE and MAO inhibitory activity. *Journal of Ethnopharmacology* 193:546–554.

19. Kim Thu D., Vui D.T., Ngoc Huyen N.T., Duyen D.K., Thanh Tung B. 2019. The use of *Huperzia* species for the treatment of Alzheimer's disease. *Journal of Basic and Clinical Physiology and Pharmacology* 31(3):20190159.

20. Wagner W.H., Beitel J.M. 1993. *Huperzia selago*. In: Flora of North America Editorial Committee (ed.). Flora of North America North of Mexico (FNA). Vol. 2.

21. www.agfonds.lv/herbs/herbs-e-f/fir-clubmoss-huperzia-selago-l/

22. Czapski G., Szypuła W., Kudlik M., et al. 2014. Original article assessment of antioxidative activity of alkaloids from *Huperzia selago* and *Diphasiastrum complanatum* using *in vitro* systems. *Folia Neuropathologica* 52(4):394–406.

23. Szypuła W.J., Wileńska B., Misicka A., Pietrosiuk A. 2020. Huperzine A and huperzine B production by prothallus cultures of *Huperzia selago* (L.) Bernh. ex Schrank et Mart. *Molecules* 25:3262.

23 *Papaveraceae* Adans. Family

Cornelia Bejenaru[1], Antonia Radu[1], George Dan Mogoşanu[2],
Ludovic Everard Bejenaru[2], Andrei Biţă[2], and Adina-Elena Segneanu[3]
[1] Department of Pharmaceutical Botany, Faculty of Pharmacy, University of Medicine
and Pharmacy of Craiova, 2 Petru Rareş Street, Craiova, Dolj County, Romania
[2] Department of Pharmacognosy & Phytotherapy, Faculty of Pharmacy, University of
Medicine and Pharmacy of Craiova, 2 Petru Rareş Street, Craiova, Dolj County, Romania
[3] Institute for Advanced Environmental Research–West University of
Timisoara (ICAM–WUT), 4 Oituz Street, Timisoara, Romania

BOTANICAL DESCRIPTION

The *Papaveraceae* family is represented in the spontaneous flora of Romania by three
genera: *Papaver*, *Glaucium*, and *Chelidonium*. The genus *Papaver* includes the following 6 species
spread throughout the spontaneous flora: *Papaver alpinum* L. subsp. *corona-sancti-stephani* (Zapal.)
Borza, *Papaver rhoeas* L., *Papaver laevigatum* Bieb. (*P. maeoticum* Klokov), *Papaver dubium* L., *Papaver
hybridum* L., and *Papaver argemone* L. The genus *Glaucium* is represented in the Romanian flora by
the species *Glaucium flavum* Crantz and *Glaucium corniculatum* (L.) Rudolph.[1-3]

The *Papaveraceae* family includes herbaceous plants, annual, biennial, or perennial, glabrous
or hairy, which can have in their structures vessels with white or colored latex. The root can be
either taproot or extensively branched, and the stem is usually branched and leafy. The leaves are
simple, without stipules, and arranged alternately, with an entire or divided blade. The flowers are
hermaphroditic and actinomorphic, following a type 4 floral pattern. They are arranged terminally
or axillary, either solitary or grouped in umbel racemose inflorescences. The double floral envelope
consists of a deciduous calyx, formed by 2 free or united sepals, and a dialypetalous corolla,
composed of 4 petals arranged in two alternating opposite circles. The androecium is made up of
numerous stamens, and the gynoecium consists of 2–20 united carpels. The ovary is positioned
superiorly and contains anatropous or campylotropous ovules, each with 2 integuments. The fruit
is a capsule that dehisces through pores or valves, containing small, globular, oval, or reniform
seeds. The seeds can have a reticulate or smooth surface and contain oily endosperm.[1-4]

Papaver L. (Common poppy)

The species belonging to this genus mainly grow in the temperate zone of the Northern
Hemisphere. These herbaceous plants are annual or perennial, with milky sap, usually white, and
often have setose-hairy leaves that are pinnately divided. The flowers are pedunculate, solitary,
terminal, and pendulous before flowering. They can be white or brightly colored, large in size,
with an early deciduous floral envelope. The stigma is disc-shaped and sessile. The fruit is a
poricidal capsule, globular or subrounded, containing numerous small, reniform, reticulate seeds
with oily endosperm and a small embryo. This genus is represented in the flora of Romania by six
spontaneous species and two cultivated ones.[1-4]

Papaver alpinum L. subsp. *corona-sancti-stephani* (Zapal.) Borza

Scientific name: *Papaver alpinum* L. subsp. *corona-sancti-stephani* (Zapal.) Borza
Common name: Alpine poppy
Romanian name: Mac de munte, Mac galben de munte[5]
Taxonomic classification
Kingdom: *Plantae*
Subkingdom: *Cormobionta*
Phylum: *Spermatophyta*
Subphylum: *Magnoliophytina (Angiospermae)*
Class: *Magnoliatae (Magnoliopsida, Dicotyledonatae)*
Subclass: *Magnoliidae*
Order: *Papaverales*
Family: *Papaveraceae*
Genus: *Papaver*
Species: *alpinum* L. (Figure 23.1)[1-3]

DOI: 10.1201/9781003270515-24

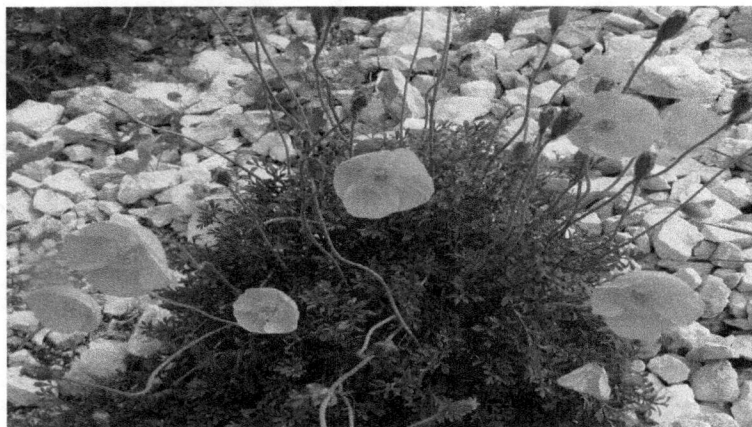

Figure 23.1 *Papaver alpinum* L. subsp. *corona-sancti-stephani* (Zapal.) Borza.

Source: 6.

SPREAD

The plant is widespread in the subalpine and alpine zones on rocky limestone scree, either mobile or partially fixed. It is found in the Southern and Eastern Carpathians, specifically in the Rodna, Ţibleş, Bucegi, Bîrsei, Făgăraş, and Retezat Mountains. It is an endemic species.[2-4]

DESCRIPTION

The perennial mountain plant has a height ranging from 5 to 20 cm. Its stem is scapiform, erect, and covered in setose hairs, typically bearing a single yellow flower. The leaves are simple pinnate, occasionally bipinnate at the base, with broad segments that are 0.7–4 mm wide and slightly glabrescent or glabrous. It blooms from July to August.[2-4]

Main phytoconstituents   Ref.

alkaloids (alpigenine, nudicaulins, papaverine, and the morphinanes morphine, codeine, thebaine**)**   [7]

flavonoids

terpenoids (phytol)

fatty acids (palmitic acid, caproic acid)

phenolic acids (syringic acid)

tannins (epicatechin)

alcohols

hydrocarbons (tricosane, pentacosane)

Medicinal Use

Papaver alpinum, commonly known as Alpine poppy, has some traditional medicinal uses, such as analgesic, sedative, and antispasmodic.[7]

Biological Activity

Some preliminary studies suggest potential biological activities for *Papaver alpinum*, including anti-inflammatory, antioxidant, cytotoxic and antimicrobial properties, due to their main phytocompounds, such as alpinigenine, alpinine, and epialpinine and certain flavonoids.[7-8]

Papaver rhoeas L. (Red poppy)

Scientific name: *Papaver rhoeas* L.
Common name: Red poppy, Flanders poppy, Corn poppy, Field poppy, Common poppy
Romanian name: Mac roşu de câmp

Figure 23.2 *Papaver rhoeas* L. var. *rhoeas*.

Source: 9.

Taxonomic classification
Kingdom: *Plantae*
Subkingdom: *Cormobionta*
Phylum: *Spermatophyta*
Subphylum: *Magnoliophytina (Angiospermae)*
Class: *Magnoliatae (Magnoliopsida, Dicotyledonatae)*
Subclass: *Magnoliidae*
Order: *Papaverales*
Family: *Papaveraceae*
Genus: *Papaver*
Species: *rhoeas* L. (Figure 23.2)[1-3]

SPREAD
The plant frequently grows from the steppe zone up to the oak forest zone, through ruderalized areas and cereal crops, ranging from plains to hilly regions. It is a calcicolous species.[1-4]

DESCRIPTION
The herbaceous species is annual, rarely biennial, with patent hairs and a height of 25–90 cm. The stems are erect, numerous, and can be simple or branched, with abundant leaves. The leaves are deeply divided (not amplexicaul), setose-hairy. The flowers are solitary and terminal, with long pedicels that can be patent or occasionally appressed hairy. They have two green sepals, setose-hairy and deciduous, and four red petals, sometimes with a black basal spot, along with numerous stamens with slender black or dark red filaments and short greenish-blue anthers. The capsule is obovate or globose, glabrous, with a persistent flat stigma, (6)8–12(18) radiated, and 7–9 septa inside. The seeds are reniform, reticulate, and dark brown in color. The Red poppy blooms from May to July.[1-4]

Papaver rhoeas L. has two varieties:

- var. *rhoeas*: It has floral pedicels covered with dense, patent setae and the flowers have red petals with a dark-colored spot at the base.[2-3]

- var. *strigosum* Boenningh.: This variety has pedicels with few, appressed setae and the flowers also have red petals, but they lack the dark-colored spot at the base.[2-3]

Main phytoconstituents   Ref.

alkaloids (rheagenine, rhoeadine, rhoeadic acid, papaveric acid, allotropine, protopine, chelerythrine, canadine, coulteropine, berberine, coptisine, sinactine, cryptopine isocorhydine, roemerine, epiberberine, epiglaucamine, glaucamine, glaudine, protopine, mecambrine, papaverrubine A-E, salutaridine, sanguinarine)   7,10

amino acids (leucine, proline, alanine, glutamic acid, thyrosine, isoleucine, valine, phenylalanine)

coumarins (coumarin)

flavonoids (apigenin, astragalin, hyperoside, isoquercitrin, hypolaetin, isorhamnetin, kaempferol, luteolin, myricetin, quercetin, rutin, vitexin)

terpenoids (phytol, a-pinene, a-terpineol, caryophyllene, β- ionene, germacrene D, farnesene)

fatty acids (palmitic acid, myristic acid, lauric acid, palmitoleic acid, margaric acid, stearic acid, linoleic acid, arachidic acid, lignoceric acid, behenic acid)

organic acid (citric acid, formic acid, fumaric acid, malic acid, malonic acid, succinic acid, quinic acid)

carbohydrates (inulin, rhamnose, sucrose, glucose)

phenolic acids (caffeic acid, chlorogenic acid, cichoric acid, protocatechuic acid, tyrosol)

anthocyanidins (malvidin)

hydrocarbons (heneicosane, nonadecane, tetracosane, tricosane)

miscellaneous (ascorbic acid, β-carotene, tocopherols)

Medicinal Use

- The petals (*Rhoeados flos*) have sedative and expectorant properties.[1]

- The seeds were used in traditional medicine internally for gastrointestinal diseases and diabetes, and externally for dermatological conditions.[7]

- Infusions of the aerial parts were employed to treat rheumatism, while the fresh leaves were used as a tonic.[10]

Biological Activity

Scientific studies have demonstrated various pharmacological activities of *Papaver rhoeas*, including antimicrobial, antioxidant, antiulcerogenic, cytotoxic, sedative, and effects on morphine-induced conditioned place preference tests and morphine-dependence.[10-11]

Other Uses

The seeds of the Red poppy are edible (bread, cakes, salads, etc.).

Leaves can be used in preparing soups, omelettes, or pies. However, it is advisable to use leaves moderately to avoid potential intoxication from alkaloids.

Petals, on the other hand, are collected to prepare a red coloring or can be used externally in infusion as an emollient for irritated and reddened skin.[10]

Glaucium Mill. (Horned poppy)

This genus includes annual, biennial, or perennial herbaceous plants that contain laticifers with yellow latex. The stems are branched, and the leaves of these plants are glaucous and divided, with pinnate venation, and those located in the upper part of the plant are amplexicaul. The flowers have 2 sepals, which are deciduous, glabrous, or setose-hairy, and 4 well-developed petals that are yellow, golden-yellow, or reddish-golden, with a reddish base, also deciduous. The flowers are large and solitary, situated on a long peduncle. The androecium consists of numerous stamens, and the gynoecium has a cylindrical ovary positioned superiorly, a short or absent style, and a thick, bilobed stigma with 2 horns. The fruit is a cylindrical-linear capsule, long (10–25 cm), valvicide,

Figure 23.3 *Glaucium flavum* Crantz (Yellow horned poppy).

Source: 12.

and bilocular, opening through 2 valves from the top downwards. The seeds are black and reticulate. These plants flower during the summer months.[1-4]

Glaucium flavum Crantz (Yellow horned poppy)

Scientific name: *Glaucium flavum* Crantz
Common name: Yellow horned poppy, Sea poppy, Horned poppy
Romanian name: Mac cornut galben
Taxonomic classification
Kingdom: *Plantae*
Subkingdom: *Cormobionta*
Phylum: *Spermatophyta*
Subphylum: *Magnoliophytina (Angiospermae)*
Class: *Magnoliatae (Magnoliopsida, Dicotyledonatae)*
Subclass: *Magnoliidae*
Order: *Papaverales*
Family: *Papaveraceae*
Genus: *Glaucium*
Species: *flavum* Crantz (Figure 23.3)[1-3]

SPREAD
The species is rarely found from the steppe zone up to the oak forest zone, primarily on sandy and arid terrains in the plains and hills.[2,4]

DESCRIPTION
The species is an annual or biennial, rarely perennial, with an erect, branched stem, sparsely hairy or glabrous, reaching a height of 30–50 cm. The leaves are thick, with the inferior ones petiolate, and the superior ones ovate with a cordate base, amplexicaul, pinnately lobed, with triangular-ovate lobes, unequally dentate at the margins, glabrous or sparsely hairy. The flowers are solitary, pedunculate, and arranged axillary. There are 2 sepals, deciduous, rarely glabrous or softly hairy,

and 4 oval-round petals, yellow as lemon or golden-yellow. The numerous stamens are yellow, and the ovary is linear, green, smooth, or scabrous-tuberculate and tapers towards the top. The fruits are linear-cylindrical, curved, tuberculate-scabrous or smooth. The species blooms from June to August.[2-4]

Two subspecies are described:

- subsp. *flavum*: It is characterized by flowers with hairy sepals and fruits (15–30 cm) that are tuberculate-scabrous along their entire length;[1-3]

- subsp. *leiocarpum* (Boiss.) Ciocârlan has flowers with glabrous sepals, and its fruits (10 cm) are tuberculate only towards the top.[1-3]

Main phytoconstituents	Ref.
alkaloids (sinoacutine, glauvine, glaucine, magnoflorine, chelidonine, sanguinarine, chelerythrine, bocconoline, isocorydine, corydine, cataline, protopine, 1,2,9,10-tetramethoxyoxoaporphine, α-allocryptopine, corunnine, isoboldine)	13-16
amino acids	
coumarins	
flavonoids (quercetin, rutin)	
terpenoids	
tannins (catechin)	
fatty acids	
organic acid	
carbohydrates	
phenolic acids (caffeic acid, chlorogenic acid, cinnamic acid, syringic acid)	
anthocyanidins	
hydrocarbons	
miscellaneous	

Medicinal Use

The aerial part of the plant (*Glaucii herba*) contains glaucine and is used to obtain preparations with antitussive and antilithiasic properties. Additionally, it is used in the treatment of certain cutaneous carcinomas.[1]

Biological Activity

- antitumoral, antioxidant, antiviral, antiproliferative, and anti-inflammatory effects[13,15-17]

Glaucium corniculatum (L.) Rudolph (Red horned poppy)

Scientific name: *Glaucium corniculatum* (L.) Rudolph
Common name: Red horned poppy, Blackspot hornpoppy
Romanian name: Mac cornut roşu
Taxonomic classification
Kingdom: *Plantae*
Subkingdom: *Cormobionta*
Phylum: *Spermatophyta*
Subphylum: *Magnoliophytina (Angiospermae)*
Class: *Magnoliatae (Magnoliopsida, Dicotyledonatae)*
Subclass: *Magnoliidae*
Order: *Papaverales*
Family: *Papaveraceae*
Genus: *Glaucium*
Species: *corniculatum* (L.) Rudolph (Figure 23.4)[1-3]

SPREAD

The Red horned poppy is sporadically found in ruderal or cultivated areas, in the plains and hills.[1]

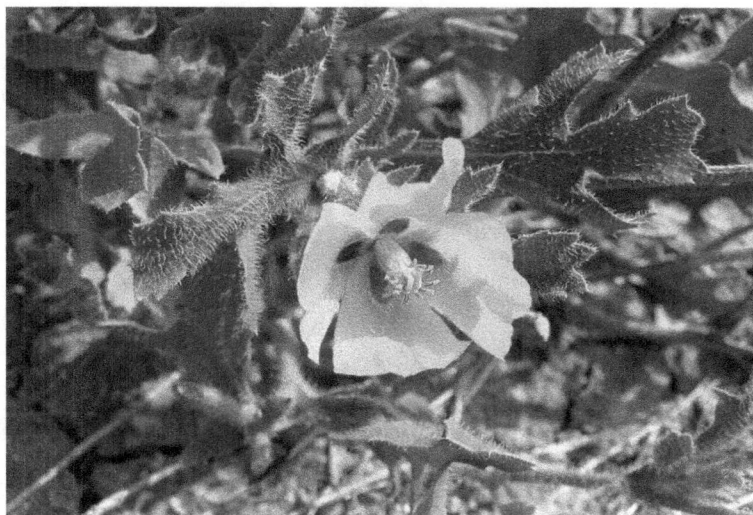

Figure 23.4 *Glaucium corniculatum* (L.) Rudolph (Red horned poppy).

Source: 18.

DESCRIPTION

The species is an annual or biennial plant with greenish-blue stems, erect, simple or branched, sparsely setose-hairy, and reaching a height of 15–50 cm. The inferior leaves have petioles, while the superior leaves are sessile and deeply pinnately divided, with lacerated, sharply dentate and hairy segments. The flowers are solitary, pedunculate, and arranged axillary. They have 2 hairy sepals, deciduous, and 4 broadly ovate petals of red or reddish-orange color, with a dark-lilac spot at the base, which are also deciduous. The numerous stamens are yellow, and the elongated cylindrical ovary is densely white-hairy, with a sessile bilobed stigma with 2 horns. The fruits are silicle-like, 10–25 cm long, shortly pedicellate, straight or slightly bent, and setose-hairy. They open into 2 valves from the top to the base. This species blooms from June to August.[1-4]

Main phytoconstituents	Ref.
alkaloids (glaucine, glauciumoline, *N*-methyl canadine, *trans*-protopinium, *cis*-protopinium)	[19-20]
amino acids	
coumarins	
flavonoids	
terpenoids	
tannins	
fatty acids (palmitic acid, linoleic acid)	
organic acid	
carbohydrates	
phenolic acids	
anthocyanidins	
hydrocarbons	
miscellaneous	

Medicinal Use

The flowering aerial part of the plant has sedative and antidiabetic properties.[1]

Biological Activity

■ – antibacterial, anticancer and neuroprotective activity (due to its antigenotoxic, acetylcholinesterase inhibitory, and antiproliferative effects)[19,21-22]

Other Uses

- ornamental species
- From the seeds, through cold pressing, an oil is extracted, used in soap-making and for illumination[4]

Chelidonium majus L. (Greater celandine)

Scientific name: *Chelidonium majus* L.
Common name: Greater celandine, Nipplewort, Tetterwort
Romanian name: Rostopască, Negelariţă, Iarbă de negi
Taxonomic classification
Kingdom: *Plantae*
Subkingdom: *Cormobionta*
Phylum: *Spermatophyta*
Subphylum: *Magnoliophytina (Angiospermae)*
Class: *Magnoliatae (Magnoliopsida, Dicotyledonatae)*
Subclass: *Magnoliidae*
Order: *Papaverales*
Family: *Papaveraceae*
Genus: *Chelidonium*
Species: *majus* L. (Figure 23.5)[1-3]

Figure 23.5 *Chelidonium majus* L. (Greater celandine).
Source: 23.

SPREAD

The plants are common from the plains to the mountainous areas, where they grow in ruderal places, grassy areas, shrubs, or forest edges.[1]

DESCRIPTION

The perennial species has a branched rhizome of dark brown color, and its stem is also branched, reaching a height of 30–60 cm. It contains yellow-orange latex. The leaves are large, arranged alternately, with the inferior and basal leaves having petioles and the superior ones being sessile. They are pinnately divided with 5–7 unequal lobes, glabrous or sparsely hairy on the inferior surface and golden-green in color. The pedunculate flowers are yellow, hermaphroditic, actinomorphic, of type 4, and arranged in groups of 2–8 in umbelliform inflorescences. There are two green sepals that are deciduous, and four broadly ovate petals arranged in two circles, which are also deciduous. The numerous stamens are yellow, with thickened filaments below the anthers, and the ovary is glabrous, bicarpellary, linear, containing anatropous ovules arranged in two rows. The style is short and thick, and the stigma is bilobed. The capsule is valvicide, linear (3–5 cm long), unilocular, and contains multiple seeds. It opens into two valves from the base to the top. The seeds are ovoid, black, with a reticulated surface and comb-like appendages. The species blooms from May to September.[1-4]

Main Phytoconstituents

Chelidonii herba et radix contains: 3–4% total alkaloids in the roots and 0.5–1.5% in the aerial part, 1–4% chelidonic acid, saponins, carotenoids, resins, flavonoids, catechic tannin, and trace amounts of volatile oil.

The alkaloids, approximately 20 in number, have diverse structures:

- Benzophenanthridine-type alkaloids: tertiary bases (chelidonine), quaternary bases (cheleritrine, chelirubine, sanguinarine);

- Protoberberines: berberine, coptisine, stilopine;

- Protopines: protopine, *α*- and *β-allo*criptopine;

- Aporphines: magnoflorine, only found in the roots.

Sanguinarine imparts a red color, while cheleritrine and berberine are yellow, together giving the fresh latex an orange hue. Chelidonine, in contrast, is colorless.

Chelidonic acid has a structure similar to meconic acid and solubilizes the alkaloids in greater celandine by transforming them into salts, balancing the basicity of quaternary alkaloids. The latex contains proteolytic enzymes, resinous substances, and tertiary and quaternary base alkaloids.

According to the Romanian Pharmacopeea (F.R. X), *Chelidonii herba* must contain at least 0.4% total alkaloids expressed as chelidonine.[24]

Main phytoconstituents   Ref.
amino acids (alanine, threonine)   25-28
alkaloids (stylopine, chelidonine, chelamidine, norchelidonine, isochelidonine, canadine, protopine, allocryptopine, sanguinarine, scoulerine, sanguilutine, coptisine, magnoflorine, berberine, corysamine, macarpine, sparteine, corytuberine, chelamine, dihydrochelilutine, turkiyenine, dihydrochelirubine, homochelidonine, tetrahydrocoptisine, oxysanguinarine)
flavonoids (rutin, quercetin, luteolin, hyperoside, luteolin, apigenin, fisetin, patuletin)
alcohols (hexacosanol, chelidoniol, nonacosanol)
sterols (α-spinasterol, ergosterol)
fatty acids (linoleic acid, oleic acid)
phenolic acids (gallic acid, chlorogenic acid, ferulic acid, hydroxycinnamic acid, *p*-coumaric acid, gentisic acid)
carotenoids (neoxanthin, violaxanthin, zeaxanthin, lutein, cryptoxanthin, β-carotene)
saponins
anthraquinones (physcion)
glycosides
organic acid (malic acid, citric acid, chelidonic acid, succinic acid)
hydrocarbons
aldehydes & ketones
miscellaneous (choline, histamine, tyramine, nicotinic acid)

Medicinal Use

The vegetable product *Chelidonii herba et radix* has several medicinal properties: choleretic-cholagogue, moderate central calming (analgesic) effects, and spasmolytic. Empirically, it is used as an anti-icteric agent to treat jaundice and as a cytostatic-proteolytic agent for treating warts (verrucae) or circumscribed cutaneous carcinomas.

The extract of *Chelidonium* increases bile secretion approximately 5 times, making it useful in conditions such as biliary disorders, cholelithiasis, gastrointestinal spasms, catarrhal jaundice, and indigestion.[24,28]

Biological Activity

Chelidonine has antitumoral properties, similar to colchicine and podophyllotoxin. It also has a weaker analgesic effect than morphine and a less intense spasmolytic action than papaverine. Due to the action of chelidonine, proteolytic enzymes and resins, the fresh latex applied locally can lead to the destruction of circumscribed skin tumors and warts, with comparable intensity to the use of liquid nitrogen, but with a longer latency period. With increased dosage, the cytostatic effect transitions into a cytotoxic action.[24]

Sanguinarine has antitumoral, anesthetic, narcotic, and sialagogue properties. It is used mainly for experimental glaucoma production in laboratory animals when testing drugs that reduce intraocular pressure (antiglaucomatous drugs). Berberine acts as a cholagogue and choleretic. Chelerithrine is highly toxic, with respiratory and cardiac paralyzing effects. Therefore, the main therapeutic actions and uses are attributed to chelidonine and berberine.[24,28]

Other Uses

- In the form of powder, the plant is used for spraying vegetable crops and animal feed that are attacked by harmful insects.[4]

REFERENCES

1. Bejenaru C., Mogoșanu G.D., Bejenaru L.E., Popescu H. 2020. *Botanică farmaceutică. Cormobionta.* Ediția a III-a. Ed. Sitech, Craiova.

2. Ciocârlan V. 2009. *Flora ilustrată a României. Pteridophyta et Spermatophyta.* Ediția a 3-a revizuită și adăugită. Ed. Ceres, București.

3. Sârbu I., Ștefan N., Oprea A. 2013. *Plante vasculare din România, Determinator ilustrat de teren.* Ed. Victor B Victor, București.

4. Săvulescu T. (ed). 1955. *Flora RPR*, vol. III. Ed. Academiei RPR, București.

5. Drăgulescu C. 2018. *Dicționar de Fitonime Românești.* Ediția a 5-a completată, Editura Universității "Lucian Blaga" Sibiu.

6. www.spontana.org/species.php?id=968

7. Butnariu M., Quispe C., Herrera-Bravo J., Pentea M., Sarac I., et.al. 2022. *Papaver* plants: current insights on phytochemical and nutritional composition along with biotechnological applications. *Oxidative Medicine and Cellular Longevity* 2022, Article ID 2041769, 23 pp.

8. Catană R. 2015. Direct somatic embryogenesis of the endemic taxon *Papaver alpinum* L. ssp. *Corona-sancti-stefani* (Zapal.) Borza for conservative purpose. *Oltenia. Studii și Comunicări. Științele Naturii* 31:47–51.

9. https://dryades.units.it/duino_en/index.php?procedure=taxon_page&id=1263&num=1243

10. Grauso L., de Falco B., Motti R., Lanzotti V. 2020. Corn poppy, *Papaver rhoeas* L.: a critical review of its botany, phytochemistry and pharmacology. *Phytochemistry Reviews* 20(1):227–248.

11. Hasplova K., Hudecova A., Miadokova E., et al. 2011. Biological activity of plant extract isolated from *Papaver rhoeas* on human lymfoblastoid cell line. *Neoplasma* 58(5):386–391.

12. https://cambriasalvatore.wixsite.com/flora-della-sicilia/glaucium-flavum-crantz

13. Boulaaba M., Kalai F.Z., Dakhlaoui S. et al. 2019. Antioxidant, antiproliferative and anti-inflammatory effects of *Glaucium flavum* fractions enriched in phenolic compounds. *Medicinal Chemistry Research* 28:1995–2001.

14. Yakhontova L.D., Sheichenko V.I., Tolkachev O.N. 1972. A study of the alkaloids of *Glaucium flavum*: the structure of glauvine. *Chemistry of Natural Compounds* 8:212–215.

15. Bournine L., Bensalem S., Wauters J., Bedjou F., Castronovo V., Bellahcène A., Tits M., Frédérich M. 2013. Identification and quantification of the main active anticancer alkaloids from the root of *Glaucium flavum*. *International Journal of Molecular Sciences* 14(12):23533–23544.

16. Bournine L., Bensalem S., Peixoto P., Gonzalez A., Maiza-Benabdesselam F., Bedjou F. et al. 2013. Revealing the anti-tumoral effect of Algerian *Glaucium flavum* roots against human cancer cells. *Phytomedicine* 20(13):1211–1218.

17. Dehghani F., Mosleh-Shirazi S., Shafiee M. et al. 2023. Antiviral and antioxidant properties of green synthesized gold nanoparticles using *Glaucium flavum* leaf extract. *Applied Nanoscience* 13:4395–4405.

18. https://ceb.wikipedia.org/wiki/Glaucium_corniculatum

19. Kocanci F.G., Aslim B. 2021. Chemical composition and neurotherapeutic potential of *Glaucium corniculatum* extracts. *Pharmacognosy Magazine* 17:67–75.

20. Saygi T.K., Tan N., Alim Toraman G.Ö., Gurer C.U., Tugay O., Topcu G. 2023. Isoquinoline alkaloids isolated from *Glaucium corniculatum* var. *corniculatum* and *Glaucium grandiflorum* subsp. *refractum* var. *torquatum* with bioactivity studies. *Pharmaceutical Biology* 61(1):907–917.

21. Allafchian A.R., Jalali S.A.H., Aghaei F., Farhang H.R. 2018. Green synthesis of silver nanoparticles using *Glaucium corniculatum* (L.) Curtis extract and evaluation of its antibacterial activity. *IET Nanobiotechnology* 12:574–578.

22. Kocanci F.G., Hamamcioglu B., Aslim B. 2022. The relationship between neuroprotective activity and antigenotoxic and acetylcholinesterase inhibitory effects of *Glaucium corniculatum* extracts. *Brazilian Journal of Pharmaceutical Sciences* 58:e19472.

23. www.etsy.com/pl/listing/784635811/greater-celandine-chelidonium-majus-20

24. Bejenaru L.E., Mogoșanu G.D., Bejenaru C., Popescu H. 2015. *Farmacognozie-Fitoterapie*. Vol. I. Ediția a III-a. EdiSitech, Craiova.

25. Arora D., Sharma A. 2013. A review on phytochemical and pharmacological potential of genus *Chelidonium*. *Pharmacognosy Journal* 5(4):184–190.

26. Jyoti B.S. 2013. *Chelidonium majus* L. – A review on pharmacological activities and clinical effects. *Global Journal of Research on Medicinal Plants & Indigenous Medicine* 2(4):238–245.

27. Parvu M., Vlase L., Fodorpataki L., Parvu O. et al. 2013. Chemical composition of celandine (*Chelidonium majus* L.) extract and its effects on *Botrytis tulipae* (Lib.) Lind fungus and the Tulip. *Notulae Botanicae Horti Agrobotanici Cluj-Napoca* 41(2):414–426.

28. Maji A.K., Banerji P. 2015. *Chelidonium majus* L. (Greater celandine)—A review on its phytochemical and therapeutic perspectives. *International Journal of Herbal Medicine* 3(1):10–27.

24 *Plantaginaceae* Juss. Family

Cornelia Bejenaru[1], Antonia Radu[1], George Dan Mogoşanu[2],
Ludovic Everard Bejenaru[2], Andrei Biţă[2], and Adina-Elena Segneanu[3]

[1] Department of Pharmaceutical Botany, Faculty of Pharmacy, University of Medicine
and Pharmacy of Craiova, 2 Petru Rareş Street, Craiova, Dolj County, Romania
[2] Department of Pharmacognosy & Phytotherapy, Faculty of Pharmacy, University of
Medicine and Pharmacy of Craiova, 2 Petru Rareş Street, Craiova, Dolj County, Romania
[3] Institute for Advanced Environmental Research–West University of
Timisoara (ICAM–WUT), 4 Oituz Street, Timisoara, Romania

BOTANICAL DESCRIPTION

It comprises herbaceous or semi-frutescent plants that have basal rosettes of leaves, rarely with alternate arrangement. The flowering stem has a terminal spiciform inflorescence, lacks stem leaves, rarely can be branched, and has leaves arranged in a decussate-opposite manner. The leaves are petiolate, without stipules, entire (occasionally divided), with arcuate venation. The flowers are hermaphroditic, actinomorphic, of type 4, grouped in spiciform inflorescences or, occasionally, capituliform, and very rarely solitary. The floral envelope consists of a gamosepalous, persistent calyx and a gamopetalous, tubular corolla with 4 membranous laciniae. The androecium consists of 4 stamens that protrude from the corolla tube to which they are attached. The ovary is superior, with central placentation and apotropous ovules. The fruit can be a polispermous capsule that dehisces through a superior cap (pyxidium) or an achene.[1-4]

In the flora of Romania, the family *Plantaginaceae* is represented by two genera: *Plantago* L. (with 17 species) and *Littorella* Bergius (with a single species).[1]

The most representative medicinal plants from the traditional Romanian medicine were selected in order to describe in detail.

Plantago major L. (Common plantain)

Scientific name: *Plantago major* L.
Common name: Common plantain
Romanian name: Pătlagină mare, Limba oii[4]
Taxonomic classification
Kingdom: *Plantae*
Subkingdom: *Cormobionta*
Phylum: *Spermatophyta*
Subphylum: *Magnoliophytina (Angiospermae)*
Class: *Magnoliatae (Magnoliopsida, Dicotyledonatae)*
Subclass: *Asteridae*
Order: *Plantaginales*
Family: *Plantaginaceae*
Genus: *Plantago*
Species: *major* L.[1-3]
Subspecies: – *major* common on non-saline soils, meadows, characterized by leaves with a truncate-cordate base, whose blade has 5–9 veins (Figure 24.1);[2]

- *winteri* (Wirtg.) W. Ludwig — spontaneously found only in certain areas (on the sands of the Danube Delta, in saline places), characterized by leaves with a tapered petiole, whose blade has only 3–5 veins.[2]

SPREAD

The species is frequent from the plains to the subalpine zone, occurring in cultivated fields, riversides, fallow lands, meadows, ruderal areas, wet pastures, and road edges.[1-4]

DESCRIPTION

Perennial herbaceous plant with a basal rosette of leaves and an unbranched, scapiform stem, cylindrical, smooth or slightly grooved, terminating in an inflorescence, with a height of 10–40 cm.

Figure 24.1 *Plantago major* L. (Common plantain).

Source: 5.

The leaves have an ovate-elliptic blade and a broad-winged petiole, at least half the length of the blade, with 3–7 prominent veins, entire or sparsely and distantly dentate or serrate, abruptly attenuated at the base into a sulcate and base-dilated petiole. The flowers form dense, continuous inflorescences, the type of cylindrical spikes. The floral envelope consists of a calyx with free sepals (with a median vein to the tip) or only slightly united at the base, broadly elliptical, carinated, green with a membranous white margin, and a corolla with united petals, yellowish-white, with a glabrous tube that remains dry around the fruit. The stamens are long-exserted, with stamen filaments twice as long as the corolla, and the anthers are pale violet at first, then dirty yellow. The fruit is an ovoid capsule, bilocular with 6–10 (14) dark brown seeds. Flowering occurs from June to August.[1-4]

Main phytoconstituents   Ref.

alkaloids (indicain, plantagonin)   [6-8]
iridoids (aucubin, majoroside, catapol, gardoside, geniposidic acid, melittoside)
flavonoids (baicalein, hispidulin, plantaginin, scutallarein, luteolin, hispidulin, apigenin, nepetin, aucubin)
terpenoids (loliolid, ursolic acid, oleanolic acid, sitosterol acid, α-linolenic acid)
fatty acids (lignoceric acid, myristic acid, palmitic acid, oleic acid, stearic acid, linoleic acid, arachidic acid, behenic acid, elaidic acid)
glycosides (verbascoside)
sterols (β-sitosterol)
phenolic acids (caffeic acid)
organic acids (citric acid, fumaric acid)
tannins (catechin)
carbohydrates (xylose, rhamnose, galactose, arabinose, galacturonic acid, glucuronic acid)
carotenoids
miscellaneous (allantoin, sulforaphane)

Medicinal Use

- cough, wound healing, constipation, infections, fever, bleeding, inflammation, diuretic, diabetes[7-8]

Plantaginis folium has the following pharmacological actions: demulcent (due to mucilages), anti-inflammatory (iridoids, sterols), cicatrizing, antiulcer (allantoin, proteolytic enzymes, flavonoids, carotenoids), antispasmodic (iridoids), antihemorrhagic (tannin, vitamin K), antibacterial (aucubigenol), and antitussive.[6]

Plantain leaf is a topical remedy used in local treatments. The leaves or the fresh juice obtained by pressing the leaves are cicatrizing and antibacterial, used in furuncles, acne, burns, open wounds, leg ulcers, hemorrhoids, and for antihemorrhagic purposes, in epistaxis.[6]

Biologic Activity

- antiulcerative, antidiarrhoeal, anti-inflammatory, anticancer, antinociceptive, antioxidant, anti-fatigue, antibacterial, antiviral, hepatoprotective, hypoglycemic effect.[7-8]

Internally, the product *Plantaginis folium* has an astringent and intestinal antimicrobial effect, used in dysentery and enteritis (tannin, aucuboside). Aucuboside hydrolyzes in the organism to aucubigenol, which has a strong antibacterial action, active against *Staphylococcus aureus*, similar to penicillins.

Aucuboside also has uricosuric, insecticidal, and potent antihepatotoxic actions, with beneficial effects in mushroom poisonings, such as *Amanita muscaria* (L.) Lam. and *A. phalloides* (Vaill. ex Fr.) Link, which contain toxic substances like amanitin and phalloidin.

Plantain seeds have a mechanical laxative action.[6]

Other Uses

- culinary

- insecticides

- fertilisers[7-8]

Plantago media L. (Hoary plantain)

Scientific name: *Plantago media* L.
Common name: Hoary plantain
Romanian name: Pătlagină[3]
Taxonomic classification
Kingdom: *Plantae*
Subkingdom: *Cormobionta*
Phylum: *Spermatophyta*
Subphylum: *Magnoliophytina (Angiospermae)*
Class: *Magnoliatae (Magnoliopsida, Dicotyledonatae)*
Subclass: *Asteridae*
Order: *Plantaginales*
Family: *Plantaginaceae*
Genus: *Plantago*
Species: *media* L. (Figure 24.2)[1-3]

SPREAD

The plants are commonly found from the lowlands to the mountainous regions, in dry meadows, grassy areas, shrubs, cultivated fields, grassy rock formations, ruderal areas, and forest clearings.[1-4]

DESCRIPTION

Perennial herbaceous plant with a pivoting rhizome and numerous roots, basal rosette leaves, and an unbranched stem (scape), distinctly striated or smooth, becoming appressed hairy towards the top with bifurcate hairs and terminating in a spiciform inflorescence, reaching a height of 20–40 cm. The leaves are erect or decumbent, with petioles shorter than half the lamina, elliptical and hairy on both sides, narrowed into the petiole, entire or sinuate-denticulate, and with (5) 7–9 veins, the

Figure 24.2 *Plantago media* L. (Hoary plantain).

Source: 9.

lateral ones slightly arched and joining at the tip. The inflorescence is a dense, cylindrical spike composed of glabrous, glossy white flowers. The stamens are highly long exserted, with violet filaments, their length being 4–5 times greater than the calyx, and pale violet or white anthers. The fruit is a small capsule (3–4 mm) with 3–4 black seeds. Flowering begins in May and continues until August.[1-4]

Main phytoconstituents   Ref.
terpenoids (ursolic acid)   10
Iridoids (aucubin, melittoside, catalpol, monomelittoside, 10-acetylaucubin)
flavonoids (kaempferol, luteolin, apigenin)
carotenoids (β-carotene, lutein, violaxanthin, zeaxanthin, neoxanthin)
carbohydrates (xylose, glucose, rhamnose, mannose, fucose)
glycosides (verbascoside, plantamajoside, homoplantaginin, martynoside)
sterols (campesterol, stigmasterol, sitosterol)
fatty acids (linoleic acid, palmitic acid, myristic acid, palmitoleic acid, behemic acid, erucic acid)
phenolic acids (chlorogenic acid, caffeic acid, ferulic acid, gallic acid, isochlorogenic acid)
organic acid (oxalic acid)

Medicinal Use

The hoary plantain is not only edible, with fresh young leaves suitable for fresh salads or cooking as a leafy green vegetable, but its historical uses were also widespread. Archaeological excavations from the Roman period Britain and earlier have unearthed its seeds, indicating its presence in ancient diets. Additionally, it had various applications in folk medicine, serving as an antimicrobial, anti-inflammatory, antihistaminic, hemostatic, cicatrizing, expectorant, and diuretic agent. For medicinal purposes, the leaves were commonly employed in the preparation of infusions.[10-13]

Biological Activity

■ antioxidant, antitumor, antiatherogenic, mycostatic, anti-inflammatory, cytotoxic, tyrosine kinase inhibitor[10,14-18]

Other Uses

Plantago media has a broader range of applications, including industrial purposes like the development of food or cosmetic products, applications in aquaculture or zootechny, environmental protection such as phytoremediation, and even in emerging fields like nanotechnology, where phytosynthesis of certain materials could hold significant promise across various domains.[10,19-22]

Plantago lanceolata L. (Ribwort plantain)

Scientific name: *Plantago lanceolata* L.
Common name: Ribwort plantain, Narrowleaf plantain, Ribleaf, Lamb's tongue, Buckhorn
Romanian name: Pătlagină cu frunze înguste
Taxonomic classification
Kingdom: *Plantae*
Subkingdom: *Cormobionta*
Phylum: *Spermatophyta*
Subphylum: *Magnoliophytina (Angiospermae)*
Class: *Magnoliatae (Magnoliopsida, Dicotyledonatae)*
Subclass: *Asteridae*
Order: *Plantaginales*
Family: *Plantaginaceae*
Genus: *Plantago*
Species: *lanceolata* L. (Figure 24.3)[1-3]

Figure 24.3 *Plantago lanceolata* L. (Ribwort plantain).

Source: 23.

SPREAD

Frequently found in meadows, wetlands, marshes, riverbanks, shrubs, forest clearings, grassy dry areas, ruderal or clover-rich places, from lowlands to mountains.[1-4]

DESCRIPTION

Perennial herbaceous plant, 10–30 cm in height, with numerous roots originating from a rhizome. The leaves form a basal rosette, from the center of which arises an ascending or erect (with five stripes) scapiform stem, terminating in a dense, ovoidal, or short cylindrical spike, approximately 5–8 mm thick. The leaves have a lanceolate blade, hairy, narrowed into the petiole, with 3–7 (8) entire or remotely denticulate veins, hairy to glabrescent. The floral envelope consists of a calyx with sepals fused for more than half their length, and a corolla with a glabrous tube. The stamen filaments are whitish, 2–3 times longer than the corolla, and the anthers are yellowish. The fruit is an ovoid capsule (3–4 mm), containing two blackish seeds. The flowering period is from May to August.[1-4]

Main phytoconstituents	Ref.
iridoids (catalpol, aucubin, acteoside, verbascoside, asperuloside, globularin, gardoside, geniposidic acid, mayoroside, melittoside, desacetylasperuloside)	1,24

flavonoids (vanillin, luteolin, apigenin, taxifolin, quercetin, kaempferol, hesperidin, rutin, isorhamnetin, eriodictyol)

terpenoids (β-farnesene, α-bergamotene, β-caryophyllene, nonatriene (DMNT), β-ocimene, loliolide, ursolic acid, and oleanolic acid)

sterols (stigmasterol)

fatty acids (capric acid, palmitic acid, margaric acid, linolenic acid, myristic acid, pentadecanoic acid)

phenolic acids (chlorogenic acid, caffeic acid, gallic acid, rosmarinic acid, vanillic acid, cinamic acid, syringic acid, p-coumaric acid, vanillic acid)

tannins (catechin, epicatechin)

polysaccharides (mannose, D-glucuronic acid, D-glucose, D-galactose, D-galacturonic acid, L-arabinose, D-mannose, L-fructose, D-xylose, pectic, rhamnogalacturonan, arabinogalactan, galacturonic acid, arabinose, glucuronic)

antocyans (cyanidin glycoside, delphinidin glycoside, peonidin glycoside, petunidin)

Medical Uses

The product *Plantaginis lanceolatae folium* is officinal in the European Pharmacopoeia, 8th edition.[6]

■ cough, disorders of the respiratory system, wound-healing, skin infections, insects bites[12-13]

BiologicalActivity

■ antioxidant, antibacterial, antimicrobial, antispasmodic, anti-inflammatory, cytotoxic, antiulcerogenic, antiobesity[25-26]

Other Uses

■ cosmetic products

■ food (soups, salads)

■ additive in wine

■ insecticide

■ agriculture (polluted soil remediation, heavy metals removal; reduces reactive nitrogen (N) losses, in particular N leaching)

■ ability to remove metals from areas with air pollution[25,27]

Plantago scabra Moench (*P. indica* L. nom. illegit.; *P. arenaria* Waldst. et Kit.)
(Branched plantain)

Scientific name: *Plantago scabra* Moench
Common name: Branched plantain, Sand plantain, Black psyllium
Romanian name: Ochiul lupului
Taxonomic classification
Kingdom: *Plantae*
Subkingdom: *Cormobionta*
Phylum: *Spermatophyta*
Subphylum: *Magnoliophytina (Angiospermae)*
Class: *Magnoliatae (Magnoliopsida, Dicotyledonatae)*
Subclass: *Asteridae*
Order: *Plantaginales*
Family: *Plantaginaceae*
Genus: *Plantago*
Species: *scabra* Moench[1-2]
Subspecies: – *scabra* with hairy laciniae of the calyx (Figure 24.4);[2]

 – *orientalis* (Soó) Tzvelev with glabrous laciniae of the calyx.[2]

SPREAD

The plants of this species sporadically appear in dry and sandy places, especially in the southern part of the country, in ruderal areas, cultivated fields, along roadsides, and are distributed from the plains to the mountainous areas.[1,2,4]

Figure 24.4 *Plantago scabra* Moench (*P. indica* L. nom. illegit.; *P. arenaria* Waldst. et Kit.) (Branched plantain).

Source: 28.

DESCRIPTION

Annual plant with a small taproot and a branched, erect or ascending stem, 10–40 cm in length. The leaves are linear and entire, sharp, lanate-hairy toward the base, arranged oppositely on the stem, with sterile axillary shoots. The flowers are brownish-white and arranged axillary in capituliform inflorescences on long peduncles. The fruit is an ovoid capsule with 2 boat-shaped seeds. The flowering period spans from June to August.[1,2,4]

Main phytoconstituents

Ref.
[29-31]

iridoids (catalpol, aucubin)

flavonoids (luteolin, apigenin, kaempferol)

terpenoids (borneol, terpin-4-ol, myrtenal, α-terpineol, eugenol, piperitone, β-damascenone, carvacrol, phytol, anethole, squalene germacrene D, ionene, spathulenol, γ-eudesmol, α-cadinol)

glycosides (plantarenaloside, verbacoside)

sterols (stigmasterol, campesterol, γ-sitosterol)

fatty acids (capronic acid, staric acid, lauric acid, palmitic acid, linolenic acid, myristic acid)

phenolic acids (p-methylcinnamic acid)

tannins (catechin)

carbohydrates (galactose, galacturonic acid, arabinose, xylose)

hydrocarbons (myristicin, tetradecane, octadecane, nonadecane, n-hexadecane, n-octadecane, tricosane, eicosane, pentacosane, heptacosane)

organic acids (benzoic acid)

miscellaneous (alpha-tocopherol, phenol, sulfolane, tetradecyloxirane, 6-methoxy-3-methyl-1H-indole, 7-ethyl-1,4-dimethylazulene, 9-methylanthracene)

Medicinal Use

The product *Psyllium*, also known as *Psyllii semen* (Ph.Eur. 10), representing the seeds of the species, has emollient and demulcent properties, acts as a mild laxative, regulates intestinal transit, functions as a detoxifier in the colon, and normalizes blood cholesterol levels.[1]

Biological Activity

■ anti-inflammatory, cicatrizing[32-33]

Other Uses

Plantago scabra has applications in novel salt-tolerant crops and is employed to enhance the salt tolerance of current oilseed crops, offering alternative sources of unsaturated fatty acids.[34]

REFERENCES

1. Bejenaru C., Mogoșanu G.D., Bejenaru L.E., Popescu H. 2020. *Botanică farmaceutică. Cormobionta*, ediția a III-a. Ed. Sitech, Craiova.

2. Ciocârlan V. 2009. *Flora ilustrată a României. Pteridophyta et Spermatophyta*. Ediția a 3-a revizuită și adăugită. Ed. Ceres, București.

3. Sârbu I., Ștefan N., Oprea A. 2013. *Plante vasculare din România. Determinator ilustrat de teren*. Ed. Victor B Victor, București.

4. Săvulescu T. (ed). 1961. *Flora RPR*, vol. VIII. Ed. Academiei RPR, București.

5. https://dryades.units.it/saline/index.php?procedure=taxon_page&id=5097&num=9126

6. Bejenaru L.E., Mogoșanu G.D., Bejenaru C., Popescu H. 2015. *Farmacognozie-Fitoterapie*. Vol. I, ediția a III-a. Ed. Sitech, Craiova.

7. Adom M.B., Taher M., Mutalabisin M.F., Amri M.S., Abdul Kudos M.B., Wan Sulaiman M.W.A., Susanti D. 2017. Chemical constituents and medical benefits of *Plantago major*. *Biomedicine & Pharmacotherapy* 96:348–360.

8. Najafian Y., Hamedi S.S., Farshchi M.K., Feyzabadi Z. 2018. *Plantago major* in Traditional Persian Medicine and modern phytotherapy: A narrative review. *Electronic Physician* 10(2):6390–6399.

9. https://dryades.units.it/scuole/index.php?procedure=taxon_page&id=5115&num=1974

10. Fierascu R.C., Fierascu I., Ortan A., Paunescu A. 2021. *Plantago media* L.—Explored and potential applications of an underutilized plant. *Plants* 10(2):265.

11. Singh B., Kishor A., Singh S., Bhat M.N., Surmal O., Musarella C.M. 2020. Exploring plant-based ethnomedicine and quantitative ethnopharmacology: Medicinal plants utilized by the population of Jasrota Hill in Western Himalaya. *Sustainability* 12(18):7526.

12. Süntar I. 2020. Importance of ethnopharmacological studies in drug discovery: Role of medicinal plants. *Phytochemistry Reviews* 19:1199–1209.

13. Papp N., Sali N., Csepregi R., Tóth M., Gyergyák K., Dénes T., Bartha S.G. et al. 2019. Antioxidant potential of some plants used in folk medicine in Romania. *Farmacia* 67:323–330.

14. Mohsenzadeh S., Sheidai M., Ghahremaninejad F., Koohdar F. 2020. A palynological study of the genus *Plantago* (*Plantaginaceae*). *Grana* 59(6):454–465.

15. Ji X., Hou C., Guo X. 2019. Physicochemical properties, structures, bioactivities and future prospective for polysaccharides from *Plantago* L. (*Plantaginaceae*): A review. *International Journal of Biological Macromolecules* 135:637–646.

16. Lukova P., Karcheva-Bahchevanska D., Dimitrova-Dyulgerova I., Katsarov P., et al. 2018. A comparative pharmacognostic study and assesment of antioxidant capacity of three species from *Plantago* genus. *Farmacia* 66:609–614.

17. Volodymirivna K.T., Pavlyvna S.H., Kostyantynivna Y.O., Vladylenovych M.O., Oleksandrivna M.O. 2020. Mycostatic activity of extracts from leaves of *Plantago media* L. and *Plantago altissima* L. *Annals of Tropical Medicine and Public Health* 3:299–303.

18. Majkić T., Bekvalac K., Beara I. 2020. Plantain (*Plantago* L.) species as modulators of prostaglandin E2 and thromboxane A2 production in inflammation. *Journal of Ethnopharmacology* 262:113140.

19. Kreitschitz A., Haase E., Gorb S.N. 2020. The role of mucilage envelope in the endozoochory of selected plant taxa. *Science of Nature* 108:2.

20. Loizzo M.R., Tundis R. 2019. Plant antioxidant for application in food and nutraceutical industries. *Antioxidants* 8:453.

21. Petruk G., Del Giudice R., Rigano M.M., Monti D.M. 2018. Antioxidants from plants protect against skin photoaging. *Oxidative Medicine and Cellular Longevity* 2018:1454936.

22. Fierascu R.C., Fierascu I., Lungulescu E.M., Nicula N., Somoghi R., Dițu L.M., Ungureanu C., et al. 2020. Phytosynthesis and radiation-assisted methods for obtaining metal nanoparticles. *Journal of Materials Science* 55 :1915–1932.

23. https://dryades.units.it/lamone/index.php?procedure=taxon_page&id=5121&num=9129

24. EMA/HMPC/437859/2010. 2012. Committee on Herbal Medicinal Products (HMPC). Assessment report on *Plantago lanceolata* L., *folium.* 2012.

25. www.inaturalist.org/taxa/53178-Plantago-lanceolata

26. www.naturalherbs.ro/ro/project/patlagina-plantago-lanceolata/

27. Abate L., Bachheti R.K., Tadesse M.G., Bachheti A., 2022. Ethnobotanical uses, chemical constituents, and application of *Plantago lanceolata* L. *Journal of Chemistry* 2022(1532031), 17 pp.

28. www.genialvegetal.net/-Plantain-des-sables-

29. Hammami S., Debbabi H., Jlassi I., Joshi R.K., Mokni R.E. 2020. Chemical composition and antimicrobial activity of essential oil from the aerial parts of *Plantago afra* L. (*Plantaginaceae*) growing wild in Tunisia. *South African Journal of Botany* 132:410–414.

30. Hirst E.L., Percival E.G.V., Wylam C.B. 1954. Studies on seed mucilages. Part VI. The seed mucilage of *Plantago arenaria*. *Journal of the Chemical Society (Resumed)* 528:189–198.

31. Ramadan E.S., Elfedawy M.G., ElShamy M.M., Abdel-Mogib M. 2017. Phytochemical and biological evaluation of *Plantago arenaria*. *Research Journal of Pharmaceutical, Biological and Chemical Sciences* 8(4):849–857.

32. Kutluay V., Genç Y., Doğan Z., Inoue M., Saraçoğlu I. 2022. Nuclear receptor agonist activity studies on some *Plantago* species and *Scutellaria salviifolia* Benth.: A particular focus on liver x receptor alpha and retinoid x receptor alpha connected with the inflammation process. *Journal of Research in Pharmacy* 26(2):272–278.

33. Mogoşanu G.D., Grumezescu A.M. 2014. Natural and synthetic polymers for wounds and burns dressing. *International Journal of Pharmaceutics* 463(2):127–136.

34. Ozcan T. 2014. Fatty acid composition of seed oils in some sand dune vegetation species from Turkey. *Chemistry of Natural Compounds* 50:804–809.

25 *Polypodiaceae* Bercht. et J. Presl Family

Cornelia Bejenaru[1], Antonia Radu[1], George Dan Mogoşanu[2],
Ludovic Everard Bejenaru[2], Andrei Biţă[2], and Adina-Elena Segneanu[3]
[1] Department of Pharmaceutical Botany, Faculty of Pharmacy, University of Medicine
and Pharmacy of Craiova, 2 Petru Rareş Street, Craiova, Dolj County, Romania
[2] Department of Pharmacognosy & Phytotherapy, Faculty of Pharmacy, University of
Medicine and Pharmacy of Craiova, 2 Petru Rareş Street, Craiova, Dolj County, Romania
[3] Institute for Advanced Environmental Research–West University of
Timisoara (ICAM–WUT), 4 Oituz Street, Timisoara, Romania

BOTANICAL DESCRIPTION

This family comprises the most common and well-known ferns. In the flora of Romania, there are 44 species out of approximately 2800 found worldwide. These plants are perennial herbaceous with rhizomes and leaves of the trophosporophyll type (the leaves serve both in assimilation and bearing sporangia). The leaves are initially coiled and circinate at the tip. They are rarely entire, usually simple pinnate or compound pinnate, with both leaves and petioles covered with membranous, brown scales. The sori have a vertical, mechanical ring that opens through an oblique slit and are located on the lower surface of the leaves. They have various forms and group sporangia with spores. The shape and position of the sori, as well as the presence or absence of the indusium, have been the basis for dividing the family into subfamilies, sometimes leading to classifications where species are grouped into multiple botanical families. The sporangia contain isospores, from which monoecious protals develop, which are green, leafy, cordiform, and display antheridia and archegonia on the lower surface.[1-2]

The *Polypodiaceae* family includes the following genera: *Asplenium, Blechnum, Matteuccia, Cryptogramma, Notholaena, Pteridium, Polypodium, Cystopteris, Polystichum, Dryopteris, Athyrium, Gymnocarpium*, and *Lastrea* şi *Thelypteris*.[1,3-4]

Dryopteris filix-mas (L.) Schott (Male fern)

Scientific name: *Dryopteris filix-mas* (L.) Schott
Common name: Male fern
Romanian name: Ferigă
Taxonomic classification
Kingdom: *Plantae*
Subkingdom: *Cormobionta*
Phylum: *Pteridophyta*
Class: *Polypodiatae (Filicatae, Polypodiopsida)*
Order: *Polypodiales*
Family: *Polypodiaceae*
Genus: *Dryopteris*
Species: *filix-mas* (L.) Schott (Figure 25.1)[1,3-4]

In the genus *Dryopteris* se găsesc, there are 9 species found in the territory of Romania: *Dryopteris filix-mas* (L.) Schott, *Dryopteris cristata* (L.) A. Gray, *Dryopteris affinis* (Lowe) Fraser-Jenk., *Dryopteris submontana* (Fraser-Jenk. et Jermy) Fraser-Jenk., *Dryopteris pallida* (Bory) C. Chr. ex Maire et Petitmengin, *Dryopteris carthusiana* (Vill.) H.P. Fuchs (*D. spinulosa* Watt.), *Dryopteris expansa* (C. Presl) Fraser-Jenk. et Jermy (*D. assimilis* S. Walker), *Dryopteris dilatata* (Hoffm.) A.Gray (*D. austriaca* auct.), and *Dryopteris remota* (A. Braun ex Dőll) Druce.[3-4]

SPREAD

A common species throughout the country, found in forests.[1]

DESCRIPTION

The plant has a well-developed, thick rhizome, measuring up to 25 cm in length, covered by the leaf stalks from previous years. It has double-pinnate-pinnatifid leaves at the top, arranged in a bouquet manner, and in their young stage, they are circinate. The leaves can reach up to 1 m in length, with both the leaf stalk and rachis covered with reddish-brown scales. On the dorsal side,

Figure 25.1 *Dryopteris filix-mas* (L.) Schott (Male fern).

Source: 5.

along the median area of the secondary segments, there are round sori, partially covered by a reniform, membranous, flat, and non-recurved indusium.[1-2]

Main Phytoconstituents

The rhizome of the fern (*Filicis rhizoma*) contains: 5–15% oleoresin (*Filicis resina*); 7–8% catechic tannin (filicitannic acid or aspidotannic acid), which condenses into phlobaphenes called "fern red," "filicic red," or "filixic red"; 0.05% volatile oil; 10% starch; 5–6% fatty oil; waxes; simple sugars.

The oleoresin contains approximately 5% phloroglucinol derivatives, which structurally are phlorobutyrophenones or butanone-phloroglucinols. In the composition of the active principles are identified two phloroglucinolic structures: filicinic acid and methyl phloroglucinol.

The total of active principles from the rhizome of *Dryopteris filix-mas* are called *crude filicin* and represent phloroglucinol derivatives in which filicic acid or filixic acid predominates (14–18%). The brown color of the rhizome is determined by filixnegrine, which results from the degradation of phloroglucinol derivatives.[6]

The volatile oil, stored in glandular hairs located in secretory pouches in the rhizome, imparts a specific odor to the fresh vegetable product. It consists of esters of eucalyptol, *n*-hexyl alcohol, and *n*-octyl alcohol with butyric, valerianic, and pelargonic acids.[6]

The green-colored fatty oil is composed of glycerides of palmitic, oleic, and cerotic acids. For medicinal products, F.R. VIII specifies a minimum content of 1.6% crude filicin.[6]

Main phytoconstituents	Ref.
flavonoids (quercetin, quercitrin, kaempherol)	7-11
alkaloids	
tannins	
terpenoids (myrcene, linalool, selinadiene, nerolidol)	
sterols (stigmasterol)	

fatty acids (palmitic acid, stearic acid)
carotenoids (epoxy-α-ionone, 4-hydroxyepoxy-β-ionol, blumenol C)
alcohols (3-hexenol, I-tricosanol, docosanol)
aldehydes (benzaldehyde)
phenolic acids (caffeic acid)
tannins
saponins
misceleous (aspidin, aspidinol, albaspidin, flavaspidic acid, filicic acid, acetylfilicinic acid, propionylfilicinic acid)

Medicinal Use

The vegetable product *Filicis rhizoma* is administered in the form of capsules, powders, or emulsions, and it is recommended to ingest a saline purgative (sodium sulfate or magnesium sulfate) within a maximum of two hours to eliminate paralyzed intestinal worms. During the treatment, the administration of Castor oil (*Ricini oleum*) and/or alcohol is not recommended. The vegetable product should be taken in small doses to avoid side effects (nervous, digestive, or ocular disturbances) caused by overdosing due to increased solubility and absorption of phloroglucinols. To enhance the pharmacological activity of the vegetable product, it is advised to neutralize gastric acidity with sodium bicarbonate, taken 15 minutes before the anthelmintic.[1]

Biological Activity

The vegetable product *Filicis rhizoma* has a toxic helminthic effect (anthelmintic) and is mainly used in veterinary medicine against cestodes:

- *Taenia solium* (Linnaeus, 1758), the pork tapeworm;

- *Hymenolepis nana* (Siebold, 1852) syn. *Taenia nana* (Siebold, 1852), the dwarf tapeworm;

- *Diphyllobothrium latum* (Linnaeus, 1758) syn. *Bothriocephalus latus* (Bremser, 1819), the broad fish tapeworm;

- *Ascaris lumbricoides* (Linnaeus, 1758), the human roundworm;

- *Ancylostoma duodenale* (Dubini, 1843), the Old World hookworm;

- *Fasciola hepatica* (Linnaeus, 1758) syn. *Distoma hepaticum* (Linnaeus, 1758), the common liver fluke.

On cestodes, crude filicin or aspidin have a paralyzing effect. Aspidin is the most pharmacologically active component.[6]

Other Uses

A compost made from male fern leaves provides great benefits to tree seed beds, aiding in germination. The ashes of the plant are rich in potash and have been utilized in soap and glass production. It serves as an effective ground cover plant. Despite being mostly deciduous, its decaying fronds serve as a useful mulch in winter, effectively suppressing weed growth.[12]

Polypodium vulgare L. (Common polypody)

Scientific name: *Polypodium vulgare* L.
Common name: Common polypody
Romanian name: Feriguță
Taxonomic classification
Kingdom: *Plantae*
Subkingdom: *Cormobionta*
Phylum: *Pteridophyta*
Class: *Polypodiatae* (*Filicatae, Polypodiopsida*)
Order: *Polypodiales*
Family: *Polypodiaceae*
Genus: *Polypodium*
Species: *vulgare* L. (Figure 25.2)[1,3-4]

Figure 25.2 *Polypodium vulgare* L. (Common polypody).

Source: 13.

SPREAD

A common species found in forests and rocky areas, preferring shady regions, in the submontane and montane regions.[1-2]

DESCRIPTION

A perennial plant with a horizontal, fleshy rhizome covered in brown scales. The leaves are evergreen in winter, glabrous, and have long, naked petioles. The leaves measure 10–30 cm and have a linear-lanceolate blade, simple pinnate-partite, with 10–25 pairs of lanceolate segments, entire or slightly dentate, merging at the base. On the lower side of the leaves, there are large, round sori, about 2 mm in diameter, yellowish in color, nake,d and arranged in two parallel rows[1] on either side of the median segmental nerve.[1-2]

Main phytoconstituents   Ref.

flavonoids (hyperoside, rutin)   [14-17]

terpenoids (serratene, cycloartanol, cyclolaudenol, 31-norcycloartanol, 31-norcyclolaudenol, cycloeucalenol, 24-methylenecycloartanol)

saponins (osladin, polypodosaponin)

alkaloids

tannins (epicatechin, catechin)

sterols

glycosides (samambain)

fatty acids (lauric acid, palmitic acid, oleic acid, linoleic acid)

ecdysterone (polypodin A-B)

carbohydrates

phenolic acids (caffeic acid, gallic acid, shikimic acid)

esthers (butyric ester, isovaleric ester, α-methylbutyric ester)

organic acids (butyric acid, succinic acid, hexoic acid)

Medicinal Use

The rhizome (*Polypodii rhizoma*) has anthelmintic and laxative properties.[1]

The rhizome of *Polypodium vulgare* Linn. is utilized in various traditional medicinal practices. It is believed to be effective in treating jaundice, dropsy, scurvy, and, when combined with mallows, it can alleviate the hardness of the spleen. The distilled water from the roots and leaves is considered beneficial for ague (malarial fever), and the fresh or dried roots, when mixed with honey and applied to the nose, were used to treat polyps. The fresh root was prepared as a decoction or powder for treating melancholia and rheumatic swelling of the joints.[5]

Biological Activity

The rhizome extract exhibited antiepileptic effects. The rhizome contains ecdysones (0.07%–1% dry weight), which were observed to have topical effects on various arthropods, leading to abnormal molting and death. As a result, ecdysone analogues may have potential applications not only as insecticides but also as miticides. Moreover, the aqueous extract of *Polypodium vulgare* Linn. demonstrated analgesic properties, protective effects in various neurological and neurodegenerative disorders, stimulatory effects on adrenoceptors, and antioxidant properties.[5]

Other Uses

Polypodium vulgare rhizome was also used as a sweetener.[18]

Asplenium scolopendrium L. (*Phyllitis scolopendrium* (L.) Newman) (Hart's tongue fern)

Scientific name: *Asplenium scolopendrium* L. (*Phyllitis scolopendrium* (L.) Newman)
Common name: Hart's tongue fern
Romanian name: Năvalnic, Limba cerbului
Taxonomic classification
Kingdom: *Plantae*
Subkingdom: *Cormobionta*
Phylum: *Pteridophyta*
Class: *Polypodiatae* (*Filicatae, Polypodiopsida*)
Order: *Polypodiales*
Family: *Polypodiaceae*
Genus: *Asplenium*
Species: *scolopendrium* L. (Figure 25.3)[1,3-4]

SPREAD

It is found sporadically in the mountainous area, from the oak zone up to the beech zone, on shaded rocky limestone slopes.[1,3-4]

DESCRIPTION

It is a plant with an ascending or erect rhizome and lanceolate, persistent leaves, with an entire or undulate blade and a cordate base with rounded ears, reaching a length of up to 60 cm. On the lower side of the leaves, linear sori can be found, parallel to each other and obliquely arranged in relation to the leaf axis. Initially, they are covered by indusia that open longitudinally.[1-2]

Main phytoconstituents Ref.
flavonoids (quercetin, rutin, kaempferol, myrcitin) 18-22
terpenoids
saponins
aldehydes (2-heptenal, 2-decenal, benzaldehyde)
alkaloids
tannins
sterols
glycosides
fatty acids (steric acid, arachidic acid, pelargonic acid, capric acid, lauric acid, myristic acid, oleic acid)
ecdysterone
alcohols (octanol, benzyl alcohol)

Figure 25.3 *Asplenium scolopendrium* L. (*Phyllitis scolopendrium* (L.) Newman) (Hart's tongue fern).
Source: 19.

carbohydrates
phenolic acids
esthers
organic acids (benzoic acid)

Medicinal Use

The extracts from the leaves are used in the composition of syrups with a calming effect.[1]

Biological Activity

Antioxidant and cytogenotoxic activity. Studies have shown the ability of ethanolic extracts of mature leaves and rhizomes of *A. scolopendrium* L. to phytosynthesize Ag nanoparticles *in vitro*, revealing its antitumor potential.[23]

Other Uses

The fronds' decoction serves cosmetic purposes, being used as a hair wash to combat oily skin and as a facial pack for sensitive skin.[24]

Asplenium ceterach L. (*Ceterach officinarum* Willd.) (Rustyback)

Scientific name: *Asplenium ceterach* L. (*Ceterach officinarum* Willd.)
Common name: Rustyback
Romanian name: Unghia ciutei
Taxonomic classification
Kingdom: *Plantae*
Subkingdom: *Cormobionta*
Phylum: *Pteridophyta*

324

Figure 25.4 *Asplenium ceterach* L. (*Ceterach officinarum* Willd.) (Rustyback).

Source: 25.

Class: *Polypodiatae* (*Filicatae*, *Polypodiopsida*)
Order: *Polypodiales*
Family: *Polypodiaceae*
Genus: *Asplenium*
Species: *ceterach* L. (Figure 25.4)[1,3-4]

SPREAD

Species sporadically found on rocks, crevices, and rarely on the ground, from the steppe zone up to the beech zone.[1-3]

DESCRIPTION

Perennial herbaceous plant, with a short, ascending rhizome, small (10–20 cm), and numerous discolorous leaves (green on the upper surface, brown-silvery on the lower surface), forming tufts. The leaf blade, linear-lanceolate in shape, is pinnately divided with short lobes, oval-obtuse or rounded, entire or crenate, and on the lower side, it has linear sori without an indusium. The leaves are green and glabrous on the upper surface, while on the lower surface, they are covered with small, shiny, reddish-brown scales.[1-2]

Main phytoconstituents   Ref.
                                                                                 26-27
flavonoids (quercetin, rutin, kaempferol)
terpenoids
saponins
aldehydes (2-heptenal, 2-decenal, benzaldehyde)
alkaloids
tannins (epicatechin, catechin**)**
sterols
glycosides
fatty acids (palmitic acid, linoleic acid)
xanthones (mangiferin)
alcohols (octanol, benzyl alcohol)
carbohydrates
phenolic acids (*p*-coumaric acid, chlorogenic acid, caffeic acid, gentisic acid)
esthers
organic acids (benzoic acid)

Medicinal Use

Asplenium ceterach is used in traditional medicine against kidney stones, gallstones, to facilitate diuresis and to treat benign prostatic hyperplasia.[26]

Biological Activity

■ antioxidant, antimicrobial (on *Shigella dysenteriae* and *Staphylococcus aureus*) and DNA damage protection activities[27-28]

■ anticancerous activity against prostate cancer[29]

REFERENCES

1. Bejenaru C., Mogoșanu G.D., Bejenaru L.E., Popescu H. 2020. *Botanică farmaceutică. Cormobionta*, ediția a III-a. Ed. Sitech, Craiova.

2. Săvulescu T. (ed). 1952. *Flora RPR*, vol. I, Ed. Academiei RPR, București.

3. Ciocârlan V. 2009. *Flora ilustrată a României. Pteridophyta et Spermatophyta.* Ediția a 3-a revizuită și adăugită. Ed. Ceres, București.

4. Sârbu I., Ștefan N., Oprea A. 2013. *Plante vasculare din România, Determinator ilustrat de teren.* Ed. Victor B Victor, București.

5. https://dryades.units.it/rosandra_en/index.php?procedure=taxon_page&id=129&num=3707

6. Mogoșanu G.D., Bejenaru L.E., Bejenaru C., Popescu H. 2015. *Farmacognozie-Fitoterapie.* Vol. II, ediția a III-a. Ed. Sitech, Craiova.

7. Erhirhie E.O., Emeghebo C.N., Ilodigwe E.E., Ajaghaku D.L., Umeokoli B.O., Eze P.M., Ngwoke K.G., Gerald Chiedu Okoye F.B. 2019. *Dryopteris filix-mas* (L.) Schott ethanolic leaf extract and fractions exhibited profound anti-inflammatory activity. *Avicenna Journal of Phytomedicine* 9(4):396–409.

8. Nafees H., Nafees S., Khan I. 2019. Review on Sarakhs (*Dryopteris filix mas*) an essential unani medicine. *International Journal of Emerging Technologies and Innovative Research* 6(5):200–204.

9. Femi-Adepoju A., Oluyori A.P., Fatoba P.O., Adepoju A. 2021. Phytochemical and antimicrobial analysis of *Dryopteris filix-mas* (L.) Schott. *Rasayan Journal of Chemistry* 14(1):616–621.

10. Telfel Lawson T., Talrins A. 1962. Constituents of the oleoresin of the male fern (*Dryopteris filix mas* L.) Part I. alcohol and sterol fractions. *Canadian Journal of Chemistry* 40:1302–1309.

11. Fons F., Froissard D., Bessière J.M., Buatois B., Rapior S. 2010. Biodiversity of volatile organic compounds from five French ferns. *Natural Product Communications* 5(10):1655–1658.

12. https://pfaf.org/User/Plant.aspx?LatinName=Dryopteris+filix-mas

13. https://commons.wikimedia.org/wiki/File:Polypodium_vulgare_002.JPG

14. Khan A., Siddiqui A., Jafri M.A., Asif M. 2018. Ethnopharmacological studies of *Polypodium vulgare* Linn. A comprehensive review. *Journal of Drug Delivery and Therapeutics* 8(5):73–76.

15. Farràs A., Mitjans M., Maggi F., Caprioli G., Vinardell M.P., López V. 2021. *Polypodium vulgare* L. (*Polypodiaceae*) as a source of bioactive compounds: polyphenolic profile, cytotoxicity and cytoprotective properties in different cell lines. *Frontiers in Pharmacology* 12:727528.

16. Berti G., Bottari F., Marsili A., Morelli I., Mandelbaum A. 1967. The isolation of serratene from *Polypodium vulgare*. *Chemical Communications (London)* 1(1):50–51.

17. EMEA/HMPC/600669/2007. Assessment Report on *Polypodium vulgare* L., *rhizoma*.

18. Pervaiz D., Ghulamuddin S., Mustahsan J. 2012. *Polypodium vulgare* Linn. A versatile herbal medicine: A review. *International Journal of Pharmaceutical Sciences and Research* 9:1616–1620.

19. https://dryades.units.it/floritaly/index.php?procedure=taxon_page&tipo=all&id=103

20. Voronkov A., Ivanova T. 2022. Significance of lipid fatty acid composition for resistance to winter conditions in *Asplenium scolopendrium*. *Biology* 11(4):507.

21. Froissard D., Rapior S., Bessière J.M., Buatois B., Fruchier A., et al. 2015. *Asplenioideae* species as a reservoir of volatile organic compounds with potential therapeutic properties. *Natural Product Communications* 10(6):1079–1083.

22. Ismail A.M., Al-Khasreji T.O., Maulood B.K. 2019. Phytochemical and antioxidant activity of *Asplenium* species (spleenworts) extracts from Northern Districts of Iraq. *Engineering and Technology Journal* 37(1 Pt C):248–251.

23. Şuţan N.A., Fierăscu I., Fierăscu R.C., Manolescu D.S., Soare L.C. 2016. Comparative analytical characterization and in vitro cytogenotoxic activity evaluation of *Asplenium scolopendrium* L. leaves and rhizome extracts prior to and after Ag nanoparticles phytosynthesis. *Industrial Crops and Products* 83:379–386.

24. www.naturalmedicinalherbs.net/herbs/a/asplenium-scolopendrium=hart's-tongue-fern.php

25. www.fernsoftheworld.com/fow_species/asplenium-ceterach-l/

26. Tomou E.M., Skaltsa H. 2018. Phytochemical investigation of the fern *Asplenium ceterach* (*Aspleniaceae*). *Natural Product Communications* 13(7):849–950.

27. Živković S., Skorić M., Šiler B., Dmitrović S. et al. 2017. Phytochemical characterization and antioxidant potential of rustyback fern (*Asplenium ceterach* L.). *Lekovite Sirovine* 37:15–20.

28. Berk S., Tepe B., Arslan S. 2011. Screening of the antioxidant, antimicrobial and DNA damage protection potentials of the aqueous extract of *Asplenium ceterach* DC. *African Journal of Biotechnology* 10:8902–8908.

29. Murtaza I. 2020. Nutraceutical composition and anti-cancerous potential of an unexplored herb *Asplenium ceterach* from Kashmir Region. *Indian Journal of Pure & Applied Biosciences* 8:289–297.

26 *Ranunculaceae* Adans. Family

Cornelia Bejenaru[1], Antonia Radu[1], George Dan Mogoşanu[2],
Ludovic Everard Bejenaru[2], Andrei Biţă[2], and Adina-Elena Segneanu[3]

[1] Department of Pharmaceutical Botany, Faculty of Pharmacy, University of Medicine and Pharmacy of Craiova, 2 Petru Rareş Street, Craiova, Dolj County, Romania
[2] Department of Pharmacognosy & Phytotherapy, Faculty of Pharmacy, University of Medicine and Pharmacy of Craiova, 2 Petru Rareş Street, Craiova, Dolj County, Romania
[3] Institute for Advanced Environmental Research–West University of Timisoara (ICAM–WUT), 4 Oituz Street, Timisoara, Romania

BOTANICAL DESCRIPTION

The family includes herbaceous plants, which can be annual or perennial, rarely woody (*Clematis*). The plants have leaves with variously divided blades, arranged alternately (rarely opposite – *Clematis*) and typically lack stipules (rarely with stipules – *Thalictrum*). The flowers are hermaphroditic, actinomorphic (occasionally zygomorphic), arranged either singly or grouped in racemose or cymose inflorescences. The flowers may have a simple envelope (sepaloid or petaloid perigone) or a double perianth, with nectar glands located at its base or in a spur. The androecium consists of numerous stamens with longitudinally dehiscent anthers, while the gynoecium is superior, composed of numerous free carpels, sometimes fused (as in *Nigella*), or just a single carpel (*Consolida, Actaea*).

The arrangement of floral elements on the elongated, convex receptacle varies, with examples of spirocyclic (all floral elements arranged along a spiral line), hemicyclic (the perianth arranged in a verticil, while the androecium and gynoecium are spirocyclic), or all elements arranged in verticils. The fruit can be a polyachene or polyfollicle, occasionally a berry. The seeds are large, with a small, straight embryo and oleaginous endosperm.[1-4]

The family comprises approximately 1500 plant species, the majority of which grow in temperate regions. The family is divided into two subfamilies and 22 genera in the spontaneous flora of Romania.[1-4]

Aconitum L. (Aconite, Monkshood)

This genus includes perennial herbaceous plant species with tuberized or non-tuberized rhizomes that grow in mountainous areas and are highly toxic. The leaves are palmately divided, and the flowers are hermaphroditic, zygomorphic, non-spurred, grouped in terminal inflorescences. The tepals are yellow or blue in color. Nectaries are present on the superior tepal, resembling a hood or helmet. The fruit is polyfollicular.[1]

Aconitum callibotrion Rchb. (Aconite)

Scientific name: *Aconitum callibotrion* Rchb.
Common name: Aconite
Romanian name: Omag, Capişonul călugărului[5]
Taxonomic classification
Kingdom: *Plantae*
Subkingdom: *Cormobionta*
Phylum: *Spermatophyta*
Subphylum: *Magnoliophytina (Angiospermae)*
Class: *Magnoliatae (Magnoliopsida, Dicotyledonatae)*
Subclass: *Magnoliidae*
Order: *Ranunculales*
Family: *Ranunculaceae*
Subfamily: *Helleboroideae*
Genus: *Aconitum*
Species: *callibotrion* Rchb.[1-3]
Subspecies: – subsp. *callibotrion*, with glabrous inflorescence and perigone (Figure 26.1);

- subsp. *bucovinense* (Zapal.) G Grinţ., with hairy inflorescence and perigone.[2]

DOI: 10.1201/9781003270515-27

Figure 26.1 *Aconitum callibotrion* Rchb. (Aconite).

Source: 6.

SPREAD

The species is sporadically distributed from the spruce zone to the subalpine zone, through meadows, grassy areas, and rocky places.[2,4]

DESCRIPTION

Perennial herbaceous plant with a thickened, tuberous rhizome, reaching a height of 50–100 cm. The stem and floral pedicels are pubescent in youth, later becoming glabrescent or glabrous. The leaves are palmatisect, and the blue flowers have an arched superior petal (with a length of no more than 2.5 times the width) and are grouped in a raceme branched at the base, with entire bracteoles at the base (1–2 mm long). The seeds are longitudinally winged on three edges, without blades on their surfaces. The flowering period is from July to August.[1,2,4]

Aconitum firmum Rchb. (*A. skerisorae* Gáyer) (Strong aconite)

Scientific name: *Aconitum firmum* Rchb.
Common name: Strong aconite
Romanian name: Omag
Taxonomic classification
Kingdom: *Plantae*
Subkingdom: *Cormobionta*
Phylum: *Spermatophyta*
Subphylum: *Magnoliophytina (Angiospermae)*
Class: *Magnoliatae (Magnoliopsida, Dicotyledonatae)*
Subclass: *Magnoliidae*
Order: *Ranunculales*
Family: *Ranunculaceae*
Subfamily: *Helleboroideae*
Genus: *Aconitum*
Species: *firmum* Rchb. (Figure 26.2)[1-3]

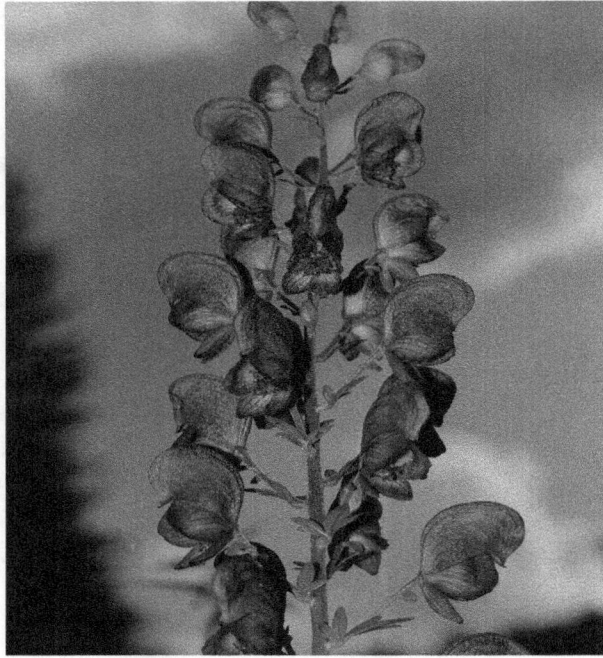

Figure 26.2 *Aconitum firmum* Rchb. (*A. skerisorae* Gáyer) (Strong aconite).

Source: 7.

SPREAD

The plant grows spontaneously in the Eastern Carpathians and the Western Carpathians.[1]

DESCRIPTION

Perennial herbaceous plant, up to 1.5 m tall, with middle palmate-partite leaves. The flowers are grouped in a glabrous raceme, and the bracteoles (2.5–15 mm long) are 2–3 times incised (they can be deeply divided). The gynoecium consists of 2 carpels. The flowering period is from July to August.[1,2]

Aconitum tauricum Wulf. (Monkshood)

Scientific name: *Aconitum tauricum* Wulf.
Common name: Aconite, Monkshood, Wolfsbane, Leopard's bane, Devil's helmet, Blue rocket
Romanian name: Omag
Taxonomic classification
Kingdom: *Plantae*
Subkingdom: *Cormobionta*
Phylum: *Spermatophyta*
Subphylum: *Magnoliophytina (Angiospermae)*
Class: *Magnoliatae (Magnoliopsida, Dicotyledonatae)*
Subclass: *Magnoliidae*
Order: *Ranunculales*
Family: *Ranunculaceae*
Subfamily: *Helleboroideae*
Genus: *Aconitum*
Species: *tauricum* Wulf.[1-3]
Subspecies: – *tauricum*, which has flowers of blue-violet color, with the superior petal having a rostrum (Figure 26.3);

– *huniadense* (Degen) Ciocârlan, whose flowers are white with a blue hue, and the superior petal does not have a rostrum. This subspecies is endemic to Romania.[2]

Figure 26.3 *Aconitum tauricum* Wulf. (Monkshood).

Source: 8.

SPREAD

Species commonly found from the spruce zone to the subalpine zone, on rocky soils, scree, and meadows.[1-3]

DESCRIPTION

Perennial herbaceous plant, of maximum 60 cm in height. Erect and thick stem, densely foliate up to under the inflorescence. Leaves are short-petioled, 5-7-divided, pedate, with the long middle segment attenuated at the base. Inflorescence is a dense, glabrous raceme, cylindrical, 10-20 cm long, with pedicels shorter than the flowers. Bracteoles, 3–8 mm, linear, are arranged in the superior part of the pedicels. The flowers have a glabrous perigone on the exterior, dark violet in color, and the superior petal is 12–20 mm, gradually narrowing into a rostrum. The flowering period is from August to September.[1-4]

Main phytoconstituents   Ref.

alkaloids (aconine, navirine, chellespontine, lycoctonine, aconitine, mesaconitine, hypaconitine, napelline, neoline, ephedrine, sparteine, aconosine, ipaconitine, eoline, benzaconitine, aneopelline, atidine, anopterine, coryphine, isoatisine, heteratisine, heterophylline, hetidine, hetsinone, benzoylheteratisine, jesaconitine)   9-11

flavonoids (rhamnetin, quercetin, pinocembrin, kaempferol, myricetin, apigenin, rutin, chrysin, naringin, liquiritigenin, liquiritin)

terpenoids

sterols

organic acid (succinic acid, malonic acid)

fatty acids (linoleic acid, palmitic acid, oleic acid)

phenolic acids (rosmarinic acid, gallic acid; chlorogenic acid, caffeic acid, ferulic acid, vanillic acid, syringic acid, sinapic acid, coumaric acid)

saponins
sterols (β-sitosterol, daucosterol, lupeol)
carbohydrates: (glucose, rhamnose, galactose, arabinose)

The main alkaloid is aconitine. *Aconitum* species also contain a similar alkaloid called hypaconitine, with a structure very close to aconitine and with useful pharmacological action. The alkaloid content (on average 1%) is influenced by chemical race, harvesting period, pedoclimatic factors, drying and conservation conditions. According to the Romanian Pharmacopoeia (F.R. X), the vegetable product must contain a minimum of 0.5% ether-soluble alkaloids, expressed as aconitine. According to O. Contz, the plant material harvested from the flora of Romania is richer in hypaconitine than in aconitine, but the author did not publish the results due to the economic implications for the Romanian *Aconiti tuber* export.[9]

Medicinal Use

The vegetable product *Aconiti tuber* is obtained by harvesting the tuber during the flowering period from species of the genus *Aconitum*. It is used in trigeminal neuralgia, dry and irritating cough, and laryngeal inflammations.[1]

The vegetal product is antitussive, antirheumatic, stomachic, diuretic, antihemorrhoidal, analgesic, diaphoretic, antipyretic, sedative, expectorant, emmenagogue, narcotic.[12-13]

Biological Activity

Aconitine acts on the bulbar nerve centers and peripheral nerve endings in two stages: in the first phase, it acts as a stimulant, causing excitement and hyperthermia; in the second phase, it manifests by inhibition, leading to anesthesia, a decrease in respiratory rate, and atrioventricular dissociation.[9]

antioxidant, anti-inflammatory, antiparasitic, analgesic, antibacterial, antiviral, antitumoral, antirheumatic, anticancer.[12-13]

Toxicity

Species of *Aconitum* are highly toxic. The lethal dose for an adult, when administered *per os*, is 1–5 mg of aconitine or 10 g of *Tinctura Aconiti*.

Aconitine intoxication is characterized by tingling of the lips, tongue, throat, and limbs, anxiety, dizziness, myasthenia, numbness, diarrhea, hypothermia, arterial hypotension, respiratory depression, altered heart rate, visual disturbances (seeing colored images in yellow or green), severe limb pain, asphyxiation, collapse. There is no antidote, and the prognosis is reserved. Currently, aconitine intoxications are very rare as its use has been abandoned.[9]

Other Uses

- ornamental

- insecticide[12-13]

Aconitum lycoctonum L. em. Koelle[3] (Wolf's bane)

Scientific name: *Aconitum lycoctonum* L. em. Koelle[3]
Common name: Wolf's bane
Romanian name: Omag
Taxonomic classification
Kingdom: *Plantae*
Subkingdom: *Cormobionta*
Phylum: *Spermatophyta*
Subphylum: *Magnoliophytina (Angiospermae)*
Class: *Magnoliatae (Magnoliopsida, Dicotyledonatae)*
Subclass: *Magnoliidae*
Order: *Ranunculales*
Family: *Ranunculaceae*
Subfamily: *Helleboroideae*
Genus: *Aconitum*
Species: *lycoctonum* L. em. Koelle (Figure 26.4)[2-3]

Figure 26.4 *Aconitum lycoctonum* L. (Wolf's bane).

Source: 14.

Subspecies: – *moldavicum* (Hacq.) Jalas (incl. *A. hosteanum* Schur.) is an endemic species in Romania, with blue flowers, found from the hilly region up to the subalpine zone, in meadows, forest edges, and shrublands.[3]

– *lasiostomum* (Rchb. ex Besser) K. Warncke has yellow flowers and is sporadically distributed from the hilly region up to the mountain zone in forest edges and shrublands.[3]

– *vulparia* (Rchb.) Nyman (*A. vulparia* Rchb.) — commonly known as "wolf's aconite," is sporadically distributed from the hilly region up to the spruce zone in clearings, shrublands, and forest edges.[3]

SPREAD
This species is distributed in the Bihor-Vlădeasa Mountains.[2]

DESCRIPTION
Perennial herbaceous plant, with a height of 60–150 cm. The leaves are palmately partite with wide laciniae, and the inflorescence is branched and covered with glandular and non-glandular hairs. The flowers have a superior tepal that is clearly wider at the base, measuring 18–25 mm in length and 3–5 mm in width. It blooms from June to July.[2-3]

Main phytoconstituents   Ref.
alkaloids (lycoctonine, aconitine, veatchine, delnudine, atidine, napelline, mesaconitine, anopterine, coryphine, isoatisine, hypaconitine, heteratisine, heterophylline, hetidine, hetsinone, benzoylheteratisine, jesaconitine)   [12-13,15-19]
flavonoids (rhamnetin, quercetin, pinocembrin, kaempferol, myricetin, apigenin, rutin, chrysin, naringin, liquiritigenin, liquiritin)

terpenoids
sterols (β-sitosterol, daucosterol, lupeol)
fatty acids (linoleic acid, palmitic acid, oleic acid)
phenolic acids (rosmarinic acid, gallic acid; chlorogenic acid, caffeic acid, ferulic acid, vanillic acid, syringic acid, sinapic acid, coumaric acid)
saponins
carbohydrates

Medicinal Use

■ anesthetic, antiarthritic, rheumatoid arthritis, diuretic, antitussive, diaphoretic, gout, sedative, stimulant[20]

Biological Activity

■ antiarrhythmic, analgesic, antimicrobial, anti-inflammatory, antitussive, antiepileptic, antidiabetic, muscle relaxant, antiproliferative[12-13,15-19]

Other Uses

■ ornamental

■ insecticide [12-13,15-19]

Helleborus purpurascens Waldst. et Kit. (Stinking hellebore)

Scientific name: *Helleborus purpurascens* Waldst. et Kit.
Common name: Stinking hellebore, Setterwort, Bear's foot, Stinkwort
Romanian name: Spânz
Taxonomic classification
Kingdom: *Plantae*
Subkingdom: *Cormobionta*
Phylum: *Spermatophyta*
Subphylum: *Magnoliophytina (Angiospermae)*
Class: *Magnoliatae (Magnoliopsida, Dicotyledonatae)*
Subclass: *Magnoliidae*
Order: *Ranunculales*
Family: *Ranunculaceae*
Subfamily: *Helleboroideae*
Genus: *Helleborus*
Species: *purpurascens* Waldst. et Kit. (Figure 26.5)[1-3]

SPREAD
Frequent in forest margins and shrubs, in hilly and mountainous areas.[1]

DESCRIPTION
The scientific name of the genus derives from the Greek words "helein," meaning to destroy or to invade, and "bora," meaning fodder or pasture, signifying that these are poisonous plants found in pastures and fodder.[4]

It is a perennial herbaceous plant with erect stems, reaching a height of 15–35 cm. It has bifurcated, leafy branches with short-petioled or sessile leaves, arranged alternately, but without hibernating leaves. The radical leaves, usually two in number, are large and palmatisect, with 5–7 lanceolate segments. These segments are broad, dentate along the margins, and pubescent on the veins or the entire inferior surface. The flowers are non-fragrant, large, weakly nutant, solitary, arranged on long peduncles. The perigone has a green interior and purple-violet exterior, appearing glaucous, and persistent during fruiting. The nectaries are yellow-green, with a slightly bilabiate cornet-like appearance, shorter than the numerous stamens, which number over 50. The follicles, numbering 4–6 and fused at the base, have a long violet-colored rostrum and contain numerous seeds. The flowering period is from February to April.[1-4]

Figure 26.5 *Helleborus purpurascens* Waldst. et Kit. (Stinking hellebore).

Source: 21.

Main phytoconstituents	Ref.
amino acids (asparagine, aspartic and glutamic acid, serine, lysine, arginine, glycine, histidine, threonine, alanine, proline, methionine, valine, leucine, isoleucine)	22-29

bufadienolides (hellebrine, hellebrigenin, deglucohellebrin)
terpenoids (phytol, squalene)
nucleoside (uridine)
phenolic acids (phenyllactic acid)
flavonoids (quercetin, kaempferol
sterols (β-sitosterol, stigmasterol)
thionins (heletionin)
furastanol saponins (furrostan, helleboroside)
lactones (protoanemonin)
carbohydrates (glucose, rhamnose)
fatty acids (palmitic acid, linoleic acids)
ecdysteroids

Medicinal Use

■ amenorrhea neuralgia anticancer, narcotic, antirheumatic, dermatological conditions (eczema, mycoses), anti-inflammatory, antidiabetic, antibacterial, antioxidant[23-29]

■ It has been used in traditional medicine in the treatment of osteoarthritis, polyarticular pain, joint rheumatism.[1]

Biological Activity

■ antibacterial, cardiotonic, immunostimulatory effect, antitumor[23-29]

Since ancient times, Hippocrates of Kos (460–370 BC) mentioned that *Hellebori rhizoma* is toxic and gastric irritant, causing strong vomiting reactions. It was a valuable emetic in those days.

The cardioactive effects of the medicinal product were highlighted around the year 1865, approximately at the same time as the introduction of *Strophanthus* seeds into therapy.

Hellebroside is a very potent cardioactive substance, with its aglycone (hellebrigenol) having a more pronounced cardiotonic/cardiotoxic action. Their action is rapid, similar to strophanthins, but they accumulate quickly in cardiac muscle, similar to digitoxin. Therefore, administration should be done with caution.

Hellebori rhizoma also has a strong purgative effect, emmenagogue properties, and immunostimulatory properties. It is rarely used in its raw form, in cardiac insufficiency, and is more frequently used as a source for pharmaceutical preparations containing hellebrosides. It is used in the treatment of polyarticular algia, osteoarthritis, and joint rheumatism, often in the form of ointments and local injections containing extracts without cardiotonic heterosides.[22]

Other Uses

- Veterinary medication. In folk medicine, the hellebore rhizome is harvested by peasants in hilly or mountainous areas and is used for veterinary purposes, to induce "fixation abscess" in animals. For example, to cure swine plague in pigs, the organism's immunity is increased through a technique called "helleboring," where the animal's ear is pierced, and a fragment of hellebore rhizome is inserted into the incision. This technique is only known in Muntenia, not in Transylvania. In the past, humans used to undergo "fixation abscess" through subcutaneous abdominal injections with turpentine oil.[22]

- ornamental [23-29]

Adonis vernalis L. (Pheasant's eye)

Scientific name: *Adonis vernalis* L.
Common name: Pheasant's eye, Spring pheasant's eye
Romanian name: Ruscuţă de primăvară
Taxonomic classification
Kingdom: *Plantae*
Subkingdom: *Cormobionta*
Phylum: *Spermatophyta*
Subphylum: *Magnoliophytina (Angiospermae)*
Class: *Magnoliatae (Magnoliopsida, Dicotyledonatae)*
Subclass: *Magnoliidae*
Order: *Ranunculales*
Family: *Ranunculaceae*
Subfamily: *Ranunculoideae*
Genus: *Adonis*
Species: *vernalis* L. (Figure 26.6)[1-3]

SPREAD

It commonly grows in dry and sunny meadows from the steppe zone to the sessile oak forest zone.[1,2]

DESCRIPTION

Perennial herbaceous species, 10–40 cm tall, with a thick, tough rhizome from which fibrous, brown-black roots detach, and an erect, glabrous, moderately branched stem. The plant also has sterile shoots with membranous, wide, brown scales at the base. The leaves are 2–4 times pinnatisect, with linear segments (the final ones are entire and up to 1 mm wide). The flowers are solitary, large (8–10 cm in diameter), hermaphroditic, actinomorphic, non-spurred, with the floral envelope consisting of calyx and corolla. The corolla consists of 10–24 glossy yellow petals without nectariferous pits. The androecium consists of numerous stamens with yellow anthers, and the gynoecium is apocarpous. The fruits are free. The flowering period is from April to May, sometimes it also blooms a second time.[1-4]

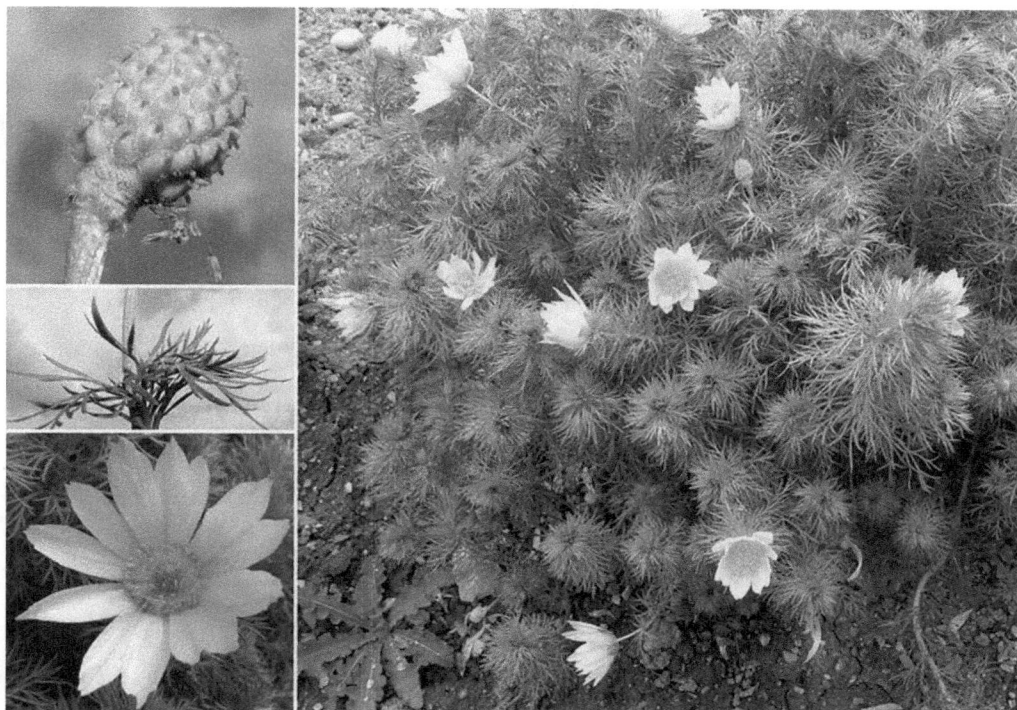

Figure 26.6 *Adonis vernalis* L. (Pheasant's eye).

Source: 30.

Main Phytoconstituents

Adonidis herba contains approximately 0.25% cardiotonic heterosides, numbering 15 in total; 1% flavonoids, including C-heterosides of luteolin, represented by adonivernite (80% of the total) and adonivernitin (10%); pregnane derivatives; volatile oil; lipids; organic acids; simple sugars; vitamins; and mineral salts.

The main component, approximately 15–20% of the total cardenolides, is adonitoxoside, with the aglycone adonitoxigenol (adonitoxigenin). Adonitoxigenol is a positional isomer of K-strofantigenol (convalatoxigenol): the -OH group at C5 in K-strofantigenol migrates to C16 in the case of adonitoxigenol.

In small proportions, the vegetable product also contains K-strofantosides (α, β), as well as other heterosides with K-strofantigenol (cimaroside), strofadogenol, and adonitoxologenol as their aglycones.[22]

Main phytoconstituents	Ref.
amino acids (asparagine)	31-32

alkaloids (berberine, cymarin, digitoxigenin)

cardiac glycosides (cymarin, digitoxigenin, adonitoxin, vernadigin, strophanthidin, adonitoxoside, acetyl-adonitoxoside, adonitoxol, adoniverdinase, adonivernozide, cymaroside, strophanthoside)

coumarins

glycosides (fukujusone, nicotinoylisoramanone, digipurprogenin II, lineolon)

flavonoids (quercitrin, luteolin, orientin, vitexin, adonivernite, homoadronivernite, isoorientin)

glycosides (strofantidine)

sterols (stimasterol)

fatty acids (linoleic acid, oleic acid, palmitic acid)

phenolic acids (chlorogenic acid, caffeic acid)

carbohydrates (rhamnose, pentosans)
organic acids (citric acid)
carotenoids

Medicinal Use

The aerial part of the plant (*Adonidis herba*), harvested during the flowering period, has a rapid and short-duration cardiotonic action, without accumulation, a strong diuretic effect, and coronary vasodilation properties.[1]

Biological Activity

The vegetable product is used in cases of cardiac insufficiency (interdigital pauses), angina pectoris, and hyperthyroidism.

From a QSAR (Quantitative Structure-Activity Relationship) analysis perspective, starting from the aglycon structures of adonitoxigenol and strofadogenol, the -CHO group at C10 compensates for the 1/4 reduction in the intensity of the cardiotonic effect induced by the -OH group at C16.[22]

Other Uses

■ ornamental

Anemone narcissiflora L. (*A. narcissifolia* L.) (Narcissus anemone)

Scientific name: *Anemone narcissiflora* L.
Common name: Narcissus anemone, Narcissus-flowered anemone
Romanian name: Oițe
Taxonomic classification
Kingdom: *Plantae*
Subkingdom: *Cormobionta*
Phylum: *Spermatophyta*
Subphylum: *Magnoliophytina (Angiospermae)*
Class: *Magnoliatae (Magnoliopsida, Dicotyledonatae)*
Subclass: *Magnoliidae*
Order: *Ranunculales*
Family: *Ranunculaceae*
Subfamily: *Ranunculoideae*
Genus: *Anemone*
Species: *narcissiflora* L.[2-3]
Subspecies: – *narcissiflora (narcissifolia)* the leaf lamina of the basal leaves is orbicular, palmately 3–5 divided (Figure 26.7);[2,3]

- *biarmiensis* (Juz.) Jalas has reniform basal leaves with 3 petiolated segments. Rarely found in the Piatra Craiului Mountains.[2,3]

SPREAD

The plant is frequently found from the subalpine region to the alpine zone, in meadows and on rocks.[2-4]

DESCRIPTION

Perennial herbaceous plant, with a height of 24–40 cm, featuring a short, brown, fibrous rhizome at the superior part due to the remains of dried petioles and numerous roots. The stem is erect, long, and patent hairy. The numerous basal leaves have long, divaricate, patent hairy petioles, are 3–5-palmately divided, with the margins and the inferior surface of the leaves covered in long, woolly hairs. The stem leaves are sessile and unevenly deeply divided. The flowers, with long and patent hairy peduncles, are grouped in clusters of 3–8 in umbels. The flowers are large, glabrous, with white tepals or with a reddish hue on the inferior surface. The fruits are winged, glabrous achenes. The flowering period is from May to July.[2-4]

Figure 26.7 *Anemone narcissiflora* L. (*A. narcissifolia* L.) (Narcissus anemone).

Source: 33.

Main phytoconstituents   Ref.
alkaloids   34-36
lactones (protoanemonin, anemonin, ranunculin)
terpenoid (ursane, lupane, farnesene, (hederagenin, hederagenin, lupane, ursane, oleanolic
    acid, *narcissiflorine*, narcissiflorinine, narcissifloridine, *raddeanoside, anemoside, p*ulsatiloside,
    ursolic acid, lupeol, caryophyllene)
flavonoids (quercetin, pelargonidine delphinidine, cyanidine
coumarins (coumarin, esculetin, fraxidin)
fatty acids (heptanoic acid, linoleic acid, palmic acid)
phenolic acids (ferulic acid, cinnamic acid, caffeic acid, coumaric acid, chlorogenic acid,
    syringic acid)
saponins
sterols (beta-sitosterol, daucosterol, ergosterol, stigmasterol)
nucleoside (adenine adenosine, urimine)

Medicinal Use

- insomnia, toothache, stomach pains, menstrual cramps, premenstrual syndrome, hyperactivity,
  neuralgia, rheumatism, cough, asthma[34-36]

Biological Activity

- antitumoral, immunomodulatory, anti-inflammatory, antioxidant, antimicrobial[37]

Other Uses

- ornamental[37]

Figure 26.8 *Pulsatilla alba* Rchb. (Pasque flower).

Source: 38.

Pulsatilla alba Rchb. (Pasque flower)[3]

Scientific name: *Pulsatilla alba* Rchb.
Common name: Pasque flower, Wind flower, Prairie crocus, Easter flower, Meadow anemone
Romanian name: Sisinei de munte[2-3]
Taxonomic classification
Kingdom: *Plantae*
Subkingdom: *Cormobionta*
Phylum: *Spermatophyta*
Subphylum: *Magnoliophytina (Angiospermae)*
Class: *Magnoliatae (Magnoliopsida, Dicotyledonatae)*
Subclass: *Magnoliidae*
Order: *Ranunculales*
Family: *Ranunculaceae*
Subfamily: *Ranunculoideae*
Genus: *Pulsatilla*
Species: *alba* Rchb. (Figure 26.8)[3]

SPREAD

The plant is commonly found from the subalpine to alpine zone, in meadows, rocky areas, and heath shrublands.[2-4]

DESCRIPTION

Perennial herbaceous plant, ranging in height from 6–30 cm, with a robust dark brown to blackish rhizome and an erect stem. The stem is initially softly long-hairy, later becoming glabrescent, and bears 1–3 long-petioled basal leaves. The stem leaves are short and broadly petiolate, with ovate-triangular blades that are 2–3 pinnately divided, with lanceolate laciniae, initially covered in long hairs, later becoming glabrous. The involucral leaves, numbering 3, have wide, slightly fused vaginae. The solitary flowers are nearly erect, provided with long peduncles that elongate during fruiting. The perigone consists of 6 petaloid tepals, white in color with a blue-violet exterior, and abundantly hairy. The flowers lack nectaries. The flowering period is from May to July.[2-4]

Main phytoconstituents   Ref.
[37,39]
sterols (beta-sitosterol, daucosterol, ergosterol, stigmasterol)
fatty acids (palmic acid, heptanoic acid, linoleic acid)
terpenoids (hederagenin, hederagenin, lupane, ursane, oleanolic acid, narcissiflorine, narcissiflorinine, narcissifloridine, raddeanoside, anemoside, pulsatiloside, ursolic acid, lupeol, caryophyllene)
phenolic acids (ferulic acid, cinnamic acid, caffeic acid, coumaric acid, chlorogenic acid, syringic acid)
flavonoids (quercetin, pelargonidine delphinidine, cyanidine)
coumarins (coumarin, esculetin, fraxidin)
nucleoside (adenine, adenosine, urimine)

Medicinal Use

■ sedative, rheumatism, neuralgia, migraines, paralysis, antispasmodic, various dermatological conditions (eczema, fungal infections, red spots, psoriasis), freckles[37,39]

Biological Activity

■ anti-inflammatory, antifungal, antioxidant, sedative, antispasmodic, analgesic, diaphoretic[37,39]

Other Uses

■ decorative[39]

REFERENCES

1. Bejenaru C., Mogoşanu G.D., Bejenaru L.E., Popescu H. 2020. *Botanică farmaceutică. Cormobionta*, ediția a III-a. Ed. Sitech, Craiova.

2. Ciocârlan V. 2009. *Flora ilustrată a României. Pteridophyta et Spermatophyta*. Ediția a 3-a revizuită și adăugită. Ed. Ceres, București.

3. Sârbu I., Ștefan N., Oprea A. 2013. *Plante vasculare din România, Determinator ilustrat de teren*. Ed. Victor B Victor, București.

4. Săvulescu T. (ed). 1953. *Flora RPR*, vol. II, Ed. Academiei RPR, București.

5. Drăgulescu C. 2018. *Dicționar de Fitonime Românești*. Ediția a 5-a completată, Editura Universității "Lucian Blaga" Sibiu.

6. www.hlasek.com/aconitum_callibotryon_4623.html

7. https://en.wikipedia.org/wiki/Aconitum_firmum

8. https://dryades.units.it/dolomitibellunesi/index.php?procedure=taxon_page&id=1019&num=4727

9. Bejenaru L.E., Mogoşanu G.D., Bejenaru C., Popescu H. 2015. *Farmacognozie-Fitoterapie*. Vol. I, ediția a III-a. Ed. Sitech, Craiova.

10. Yin T., Zhou H., Cai L., Ding Z. 2019. Non-alkaloidal constituents from the genus *Aconitum*: A review. *RSC Advances* 9:10184–10194.

11. Shyaula S. 2012. Phytochemicals, traditional uses and processing of *Aconitum* species in Nepal. *Nepal Journal of Science and Technology* 12:171–178.

12. Srivastava N., Sharma V., Kamal B., Dobriyal A.K., Jadon V.S. 2010. Advancement in research on *Aconitum* sp. (*Ranunculaceae*) under different area: A review. *Biotechnology* 9:411–427.

13. Nyirimigabo E., Xu Y., Li Y., Wang Y., Agyemang K., Zhang Y. 2015. A review on phytochemistry, pharmacology and toxicology studies of *Aconitum*. *Journal of Pharmacy and Pharmacology* 67(1):1–19.

14. https://dryades.units.it/dolomitibellunesi/index.php?procedure=taxon_page&id= 1012&num=3062

15. Karalija E., Paric A., Dahija S., Besta Gajevic R., Cavar Zeljkovic S. 2021. *Aconitum lycoctonum* L.: Phenolic compounds and their bioactivities. *Poisonous Plant Research (PPR)* 4:1–9.

16. Hao D.C., Gu X.J., Xiao P.G. 2015. Chemical and biological studies of *Aconitum* pharmaceutical resources. In: *Medicinal plants, chemistry, biology and omics*. Woodhead Publishing, Cambridge, UK, pp. 253–292. ISBN 978-0-08-100085-4

17. Mukesh Kr. Singh et al. 2012. Aconite: A pharmacological update. *International Journal of Research in Pharmaceutical Sciences* 3(2):242–246.

18. Povšnar M., Koželj G., Kreft S., Lumpert M. 2017. Rare tradition of the folk medicinal use of *Aconitum* spp. is kept alive in Solčavsko, Slovenia. *Journal of Ethnobiology and Ethnomedicine* 13:45.

19. Şuţan N.A. 2018. Phytochemical constituents and biological properties of extracts from *Aconitum* sp. – A short review. *Current Trends in Natural Sciences* 7(14):28–39.

20. https://pfaf.org/USER/Plant.aspx?LatinName=Aconitum+lycoctonum

21. https://dryades.units.it/floritaly/index.php?procedure=taxon_page&tipo=all&id=11823

22. Mogoșanu G.D., Bejenaru L.E., Bejenaru C., Popescu H. 2015. *Farmacognozie-Fitoterapie*, Vol. II, ediția a III-a. Ed. Sitech, Craiova.

23. Maior M.C., Dobrotă C. 2013. Natural compounds with important medical potential found in *Helleborus* sp. *Central European Journal of Biology* 8:272–285.

24. Stănescu U., Hăncianu M., Miron A., Aprotosoaie C. 2004. *Plante medicinale de la A la Z, monografii ale produselor de interes therapeutic*. Ed. Grigore T. Popa University of Medicine and Pharmacy, Iași, Romania, 1:280–282.

25. Balázs V.L., Filep R., Ambrus T. et al. 2020. Ethnobotanical, historical and histological evaluation of *Helleborus* L. genetic resources used in veterinary and human ethnomedicine. *Genetic Resources and Crop Evolution* 67:781–797.

26. Rosselli S., Maggio A., Bruno M., Spadaro V., Formisano C., Irace C., Maffettone C., Mascolo N. 2009. Furostanol saponins and ecdysones with cytotoxic activity from *Helleborus bocconei* ssp. *intermedius*. *Phytotherapy Research* 23:1243–1249.

27. Sasidharan S., Chen Y., Saravanan D., Yoga Latha L. 2011. Extraction, isolation and characterization of bioactive compounds from plants' extracts. *African Journal of Traditional, Complementary and Alternative Medicines* 8:1–10.

28. Segneanu A.E., Grozescu I., Cziple F., Berki D., Damian D., Niculite C.M., Florea A., Leabu M. 2015. *Helleborus purpurascens*—Amino acid and peptide analysis linked to the chemical and antiproliferative properties of the extracted compounds. *Molecules* 20(12):22170–22187.

29. Vochita G., Mihai C.T., Gherghel D., Iurea D., Roman G., Radu G.L., Rotinberg P. 2011. New potential antitumoral agents of polyphenolic nature obtained from *Helleborus purpurascens* by membranary micro- and ultrafiltration techniques. *Scientific Annals of the Al. I. Cuza University Iași* XII:41–51.

30. https://dryades.units.it/floritaly/index.php?procedure=taxon_page&tipo=all&id=1074

31. Paciana I., Butnariu M. 2021. Highlighting the compounds with pharmacological activity from some medicinal plants from the area of Romania. *Medicinal and Aromatic Plants (Los Angeles)* 10:370.

32. Shang X., Miao X., Yang F., Wang C., Li B., Wang W., Pan H., et.al. 2019. The genus *Adonis* as an important cardiac folk medicine: A review of the ethnobotany, phytochemistry and pharmacology. *Frontiers in Pharmacology* 10:407153.

33. https://dryades.units.it/floritaly/index.php?procedure=taxon_page&tipo=all&id=1047

34. Hao D.C. 2019. Genomics and evolution of medicinal plants, in *Ranunculales*. In: *Medicinal plants, biodiversity, chemodiversity and pharmacotherapy*. Academic Press–Elsevier, London, UK, pp. 1–33, ISBN: 978-0-12-814232

35. Kelemen C.D., Houdkova M., Urbanova K., Badarau S., Gurean D., Pamfil D., Kokoska L. 2019. Chemical composition of the essential oils of aerial parts of *Aconitum*, *Anemone* and *Ranunculus* (*Ranunculaceae*) species from Romania. *Journal of Essential Oil Bearing Plants* 22(3):728–745.

36. Kolayli S., Sivrikaya A., Marap A., Aliyazicioglu R., Karaoglu S., Coskuncelebi K., Kucuk M. 2008. Chemical composition and biological activities of some extracts from *Anemone narcissiflora* subsp. narcissiflora. *Journal of Pharmacy and Chemistry* 2(1):45–50.

37. Hao D.C., Gu X., Xiao P. 2017. *Anemone* medicinal plants: ethnopharmacology, phytochemistry and biology. *Acta Pharmaceutica Sinica B* 7(2):146–158.

38. http://plants.nature4stock.com/?page_id=2905

39. Hropot O.S., Konechny Y.T., Kolb Y.I., Konechna R.T., Gubytska I.I., Holota S.M., et al. 2019. Study of acute toxicity and anti-inflammatory activity of alcoholic extracts of sleep white grass (*Pulsatila alba*). *Pharmaceutical Journal* (2):60–66.

27 *Rhamnaceae* Juss. Family

Cornelia Bejenaru[1], Antonia Radu[1], George Dan Mogoşanu[2],
Ludovic Everard Bejenaru[2], Andrei Biţă[2], and Adina-Elena Segneanu[3]

[1] Department of Pharmaceutical Botany, Faculty of Pharmacy, University of Medicine and Pharmacy of Craiova, 2 Petru Rareş Street, Craiova, Dolj County, Romania
[2] Department of Pharmacognosy & Phytotherapy, Faculty of Pharmacy, University of Medicine and Pharmacy of Craiova, 2 Petru Rareş Street, Craiova, Dolj County, Romania
[3] Institute for Advanced Environmental Research–West University of Timisoara (ICAM–WUT), 4 Oituz Street, Timisoara, Romania

BOTANICAL DESCRIPTION

The *Rhamnaceae* family comprises 50 genera, with approximately 500 species, distributed in all regions of the world.[1]

It includes trees or shrubs often with thorns (rarely herbaceous plants), with deciduous or evergreen leaves, simple, with small stipules, which may be caduceous or transformed into spines, arranged alternately or opposite. The flowers, of type 4 or 5, are actinomorphic, hermaphroditic, or unisexual-monoecious, forming cymose inflorescences, axillary. The sepals are valvate, and the petals are small, concave, or hooded. The floral envelope elements are free and attached to a receptacle, hollow like a cup, which houses the inferior ovary free from it. The androecium is dialystemon, formed by epipetalous stamens. The fruit can be a berry, drupe, capsule, or achene. The seeds are exalbuminous or with little endosperm, with large, flat cotyledons, and the embryo is orthotropous.[1-2]

In the spontaneous flora of Romania, there are 4 genera: *Frangula, Rhamnus, Paliurus,* and *Zizyphus.* The genus *Rhamnus* comprises 2 species: *Rhamnus cathartica* L. and *Rhamnus saxatilis* Jacq. subsp. *tinctorius* (Waldst. et Kit.) Nyman.[3-4]

Frangula alnus Mill. (*Rhamnus frangula* L.) (Alder buckthorn)

Scientific name: *Frangula alnus* Mill. (*Rhamnus frangula* L.)
Common name: Alder buckthorn, Glossy buckthorn, Breaking buckthorn
Romanian name: Cruşân, Paţachină[1]
Taxonomic classification
Kingdom: *Plantae*
Subkingdom: *Cormobionta*
Phylum: *Spermatophyta*
Subphylum: *Magnoliophytina (Angiospermae)*
Class: *Magnoliatae (Magnoliopsida, Dicotyledonatae)*
Subclass: *Rosidae*
Order: *Rhamnales*
Family: *Rhamnaceae*
Genus: *Frangula*
Species: *alnus* Mill. (Figure 27.1)[2-4]

SPREAD

It is present from the plains up to the beech forest level, occurring in meadows, riverbanks, wet places, forests, and shrubs.[4]

DESCRIPTION

The plant is an arborescent species, reaching a height of 1–5 m, and it is non-thorny. The buds are naked, composed of brownish, hairy foliage, and the annual stems are initially pubescent, distinctly divergent, brown or grayish-brown and provided with long whitish, gray, or yellowish lenticels. The leaves are elliptical, entire, and slightly sinuate, with a narrowed base and an acute or acuminate tip. They are thin in youth, becoming rigid at maturity, and arranged alternately. It blooms from late spring to mid-summer, with hermaphroditic flowers of type 5, infundibuliform, pale greenish-white, grouped in axillary cymes. The sepals are elongated-triangular and acute, while the petals are concave, unguiculate, and emarginate. The stamens have short filaments and

DOI: 10.1201/9781003270515-28

Figure 27.1 *Frangula alnus* Mill. (*Rhamnus frangula* L.) (Alder buckthorn).

Source: 5.

large anthers, and the gynoecium has a single style. The fruit is a spherical drupe, reddish-purple or violet when mature, containing 2–3 compressed seeds. Germination is hypogeal. It flowers from May to July.[1-3]

Main Phytoconstituents

Frangulae cortex contains: 3–8% anthraquinone derivatives. The anthraquinone heterosides originate from the same aglycone, rhein-emodin (emodin, frangula-emodin), which is also found in a free state (approximately 0.1%). Depending on the glucidic radical, the heterosides can be: glucofrangulosides A and B, with the glucidic part composed of glucose and rhamnose, respectively glucose and apiose; and franguloside B, with the glucidic part consisting solely of apiose.[6]

The Romanian Pharmacopeea (F.R. X) specifies, for the medicinal product, a minimum content of 2.5% anthracenosides expressed as 1,8-dihydroxyanthraquinone.[6]

Main phytoconstituents — Ref. [6-10]

amino acids
alkaloids (frangulanine, frangulin)
flavonoids (quercetin, hyperoside, trifolin, isorhamnetin, rutin, kaempherol)
terpenoids (p-cymene, limonene, spathulenol, curcumene, cadinol, α-terpineol, geraniol, nerolidol, epicubebol, germacrene D, caryophyllene, linalool, phytol)
sterols (β-sitosterol)

fatty acids (linoleic acid, oleic acid)
phenolic acids (quinic acid, ferulic acid, caffeic acid, rosmarinic acid, p-coumaric acid)
tannins (catechin)
anthraquinones (emodin, glucofrangulin A-B, chrysophanol)
naphthalene derivatives
hydrocarbons (tricosane, tetracosane)

Medicinal Use

The bark of alder buckthorn is harvested from the trunk and branches of 3–4-year-old plants, then it is naturally dried or dried at 100°C for one hour, and it can be preserved for at least one year.[2]

The vegetable product is an officinal medicine in F.R. X, which recommends it for its purgative action. Depending on the dosage, it exhibits choleretic-cholagogue effects (at low doses), laxative effects (at moderate doses), or purgative effects (at high doses).

The laxative action is of the cathartic type, with a latency period of 8–10 hours. Preparations containing *Frangulae cortex* are usually administered in the evening, before bedtime, and the effect is observed in the morning. It is often combined with antispasmodics and a mechanical laxative. Prolonged use is not recommended.[6]

Externally, extracts of alder buckthorn have a cicatrizing and antieczematous effect.[6]

Biological Activity

■ antioxidant, anticancer, cyto/genotoxic, and antimicrobial activity.[9,11-13]

Other Uses

The leaves and bark yield a yellow dye, which is used in Russia and turns black when mixed with iron salts. Unripe fruits provide a green dye, while ripe berries give a blue or grey dye. Alder buckthorn charcoal is highly esteemed for gunpowder production and is considered the best wood for this purpose. It is particularly prized for time fuses due to its consistent burn rate. The wood's hardness, durability, and ability to sharpen well have led to its use in making wooden nails, shoes, shoe lasts, arrows, and skewers.[14]

Paliurus spina-christi Mill. (Christ's thorn)

Scientific name: *Paliurus spina-christi* Mill.
Common name: Christ's thorn, Jerusalem thorn, Garland thorn, Crown of thorns
Romanian name: Spinul lui Christos, Păliur
Taxonomic classification
Kingdom: *Plantae*
Subkingdom: *Cormobionta*
Phylum: *Spermatophyta*
Subphylum: *Magnoliophytina (Angiospermae)*
Class: *Magnoliatae (Magnoliopsida, Dicotyledonatae)*
Subclass: *Rosidae*
Order: *Rhamnales*
Family: *Rhamnaceae*
Genus: *Paliurus*
Species: *spina-christi* Mill. (Figure 27.2)[2-4]

SPREAD

The species grows on stony soils, in only a few places in the country (Dobrogea, Mehedinți, Arad, Giurgiu). In Dobrogea, it grows in isolated patches or in clusters, forming hedges.[1-2]

DESCRIPTION

This shrub, of Mediterranean origin, grows up to 3 meters tall. It is heavily branched, with annual geniculate branches that are reddish-brown or reddish-grey, finely tomentose, pubescent, or becoming glabrous. The plant has stipules transformed into thorns, two of them (one straight, the other short and curved), at the base of the broadly ovate-elliptical leaves, finely

Figure 27.2 *Paliurus spina-christi* Mill. (Christ's thorn).

Source: 15.

serrated, short-petiolate, with a rounded base, truncate, or abruptly narrowed. The flowers are hermaphroditic, small, greenish-yellow, grouped in racemes with axillary dichasiums. The fruit is dry, subglobose-compressed, yellowish-brown, with a monosamaroid appearance, having a circular and undulating wing. The seeds are obovate-compressed, one per locule. It blooms from May to August.[1-2]

Main phytoconstituents   Ref.
amino acids   16-20
coumarins
alkaloids
flavonoids (quercetin, isoquercetin, rutoside, isorhamnetin, rutin, hesperidine, naringerin, hyperoside, kaempherol, chrysin, apigenin, cosmosiin, myricetin, fisetin, taxifolin, astilbin, scoparin)
terpenoids (euphorbioside B)
Phospholipids (phosphatidylcholine, phosphatidylethanolamine, N-acylphosphatidyl-ethanolamine, N-acylisophosphatidylethanolamine, phosphatidylglyceride, lysophosphatidylcholine, lysophosphatidylinositol, phosphatidic acid, and lysophosphatidic acid)
sterols (campesterol, stigmasterol, β-sitosterol)
fatty acids (palmitic acid, linoleic acid, oleic acid, myristic acid, arachidic acid)
phenolic acids (gallic acid, vanillic acid, quinic acid, caffeic acid, p-coumaric acid, salicylic acid, ferulic acid, chlorogenic acid, rosmarinic acid, cinnamic acid, syringic acid)
tannins (catechin, epicatechin, gallocatechin)
organic acids (fumaric acid, malic acid)
anthraquinones
glycosides
hydrocarbons

Medicinal Use

Christ's thorn has been employed in alternative and traditional medicine for an extended period to address various health issues, including fever, diabetes, pain, urinary disorders, dandruff, wounds, ulcers, skin infections, inflammatory conditions, digestive issues, asthma, and the treatment of ophthalmological diseases.[21-22]

Biological Activity

Studies have revealed that *Paliurus spina-christi* possesses antibacterial, antifungal, anti-inflammatory, antioxidant, antihyperglycemic, antitumor, and antinociceptive properties.[21-23]

Other Uses

In garden settings, it is commonly utilized as a hedge or border plant because of its thorny branches, which serve as a deterrent for animals and intruders. Additionally, it is a favored option for xeriscaping, particularly in dry regions, as it has excellent drought tolerance and can flourish in poor soil conditions.[24]

Ziziphus jujuba Mill. (Common jujube)

Scientific name: *Ziziphus jujuba* Mill.
Common name: Common jujube, Indian jujube, Chinese date
Romanian name: Finap, Măslin dobrogean
Taxonomic classification
Kingdom: *Plantae*
Subkingdom: *Cormobionta*
Phylum: *Spermatophyta*
Subphylum: *Magnoliophytina (Angiospermae)*
Class: *Magnoliatae (Magnoliopsida, Dicotyledonatae)*
Subclass: *Rosidae*
Order: *Rhamnales*
Family: *Rhamnaceae*
Genus: *Ziziphus*
Species: *jujuba* Mill. (Figure 27.3)[2-4]

SPREAD

The Common jujube is native to the temperate region of Asia and is scarce on stony soils in Dobrogea.[2]

Figure 27.3 *Ziziphus jujuba* Mill. (Common jujube).

Source: 25.

DESCRIPTION

It is a shrub or tree that reaches a height of 1–8 m, with spiny, geniculate, glabrous branches of brown color, marked with numerous small lenticels. The leaves, arranged alternately, are ovate-elliptical, 3-nerved, glabrous, with a abruptly narrowed base, obtuse and mucronate tip, and crenate-serrated margin. At the base of the short-petiolate leaves, there are two spines, one straight and the other curved, each measuring 3 cm in length. The small, yellow flowers of type 5 are grouped in clusters of 2–5 in dense axillary cymes. The fruit is a spherical or elongated-ovoid drupe, edible, with a dark red, almost black color, and rich in vitamin C. It flowers in April-May.[1-2]

Main phytoconstituents   Ref.
amino acids   [26-30]
alkaloids (magnoflorine)
flavonoids (epiafzelechin, afzelechin, genistin, hesperidin, pinocembrin, quercetin, vitexin, saponarin, spinosyn, swertish, rutin, kaempherol, apigenin)
terpenoids (betulin, ursolic acid, maslinic acid, ceanothenic acid, betulinic acid, alphitolicacid, alphitolic acid methyl ester, zizyberenalic acid, pomonic acid, oleanolic acid. epiceanothic acid, ceanothic acid, platanic acid, myrcene, zizyberanal, beranalic acid, terpinene, elemene, cymene, β-phellandrene, β-ocimene, linalool, nerol, sabinene)
sterols (stigmasterol, daucosterol)
fatty acids
phenolic acids (ferulic acid, caffeic acid, cinnamic acid, gallic acid, vanillic acid, chlorogenic acid, p-coumaric acid)
tannins (catechin, epicatechin, ellagic acid)
organic acids (ascorbic acid)
glycosides (jujuphenoside, jujuboside A-B)
carbohydrates (rhamnose, arabinose, galactose, mannose, ribose, glucose, xylose, galacturonic acid)
esters (isobornyl acetate, hexyl butyrate, hexyl isovalerate, ethyl benzoate, methyl salicilate)
hydrocarbons (tetradecane)
aldehydes&ketones (decanal, undecanal, 7-epi-silphiperfol-5-ene, presilphiperfol-7-ene)
misceleous (eugenol, α-tocopherol, β-carotene, vitamin C)

Medicinal Use

Due to their emollient properties, extracts from the fruits (*Ziziphi fructus*) are used in the formulation of some cosmetic products.[2]
Ziziphus jujuba is commonly used in traditional medicine and herbal remedies as tonic and aphrodisiac. Morevover, it is believed to have various health benefits, including promoting relaxation, aiding digestion, and supporting the immune system.[31]

Biological Activity

Multiple studies have demonstrated the health benefits of *Ziziphus jujuba*, including anticancer, anti-inflammatory, antioxidant, immunostimulant, antiobesity, antifungal, antibacterial, hypnotic-sedative, anxiolytic, antispastic, and hepato- and gastrointestinal protective activities, which are due to its bioactive compounds.[32-35]

Other Uses

In addition to its medicinal uses, *Ziziphi fructus* is also consumed as a nutritious and flavorful food source in many cultures. It is very rich in vitamins, such as C, B1, and B2. Compared with other edible fruits, one fruit of common jujube per day would meet the diet requirements for vitamin C and vitamin B complex for an adult man, as recommended by the World Health Organisation.[36]
The genus *Rhamnus* comprises 2 species in the spontaneous flora of our country: *Rhamnus cathartica* L. and *Rhamnus saxatilis* Jacq.

Rhamnus cathartica L. (Common buckthorn)

Scientific name: *Rhamnus cathartica* L.
Common name: Common buckthorn
Romanian name: Verigariu, Spinul cerbului
Taxonomic classification

Figure 27.4 *Rhamnus cathartica* L. (Common buckthorn).

Source: 37.

Kingdom: *Plantae*
Subkingdom: *Cormobionta*
Phylum: *Spermatophyta*
Subphylum: *Magnoliophytina (Angiospermae)*
Class: *Magnoliatae (Magnoliopsida, Dicotyledonatae)*
Subclass: *Rosidae*
Order: *Rhamnales*
Family: *Rhamnaceae*
Genus: *Rhamnus*
Species: *Cathartica* L. (Figure 27.4)[2-4]

SPREAD
The species is commonly distributed from the steppe zone up to the oak forest level, through shrubs, forest edges, meadows, and dry locations, on heavy or rocky soils.[1-3]

DESCRIPTION
Shrub or short tree, reaching a height of 4–6 m, with divaricate branching, many short lateral branches ending in a thorn. Annual shoots have a gray or partially reddish-brown color, glabrous, shiny, with rare lenticels, terminating in a thorn. Biennial shoots are gray or reddish-brown, and the brachyblasts terminate in clusters of leaves. Buds are ovoid-conical, pressed against the stem, pointed, with a curved tip, covered by brown scales, and bordered with a wide, gray, ciliate stripe. Leaves are arranged oppositely, subrounded, ovate to elliptic, acute, shortly acuminate or obtuse, with a truncated base, rounded to slightly cordate, with crenate-serrulate margins and 3-4 (5) pairs of curved lateral veins. Leaves on short shoots are 2–2.5 times longer than the petioles, and petioles are 2–3 times longer than the thin, deciduous stipules. Small flowers are unisexual, in a 4-pattern, grouped in axillary fasciculate cymes. The calyx has triangular-lanceolate to lanceolate sepals, acute and reflexed at maturity. Petals are green, linear-lanceolate, twice as long as the sepals. The fruit starts green and then turns black, rarely yellow. Seeds are ovoid, three-edged, with a narrow longitudinal fissure on the dorsal side, widened at the ends, with cartilaginous-thickened margins. It blooms from May to July.[1-3]

Main phytoconstituents	Ref.
amino acids	[38-39]
flavonoids (quercetin, rutin, kaempherol)	
terpenoids	
sterols	
fatty acids	
phenolic acids	

tannins
organic acids
anthraquinones (emodin, glucofrangulin A, madagascin)
glycosides
lactones (geshoidin, β-sorigenin)
saponins
hydrocarbons
alkaloids
misceleous (dendrochrysanene, pruniflorone H)

Medicinal Use

The specific name of *Rhamnus cathartica* refers to its use as a cathartic, and it is also known as Purging buckthorn for the same reason. Additionally, historical records indicate that the syrup derived from this plant was once well-regarded as a hydragogue, which is a type of purgative inducing watery diarrhea. This syrup was employed in the treatment of conditions such as gout, rheumatism, and dropsy.[40]

Biological Activity

■ antimicrobial activity[38,41]

■ cytotoxicity and antimalarial activity[42]

Other Uses

While references to Common buckthorn as a medicinal plant are prevalent in historical texts, these same sources also note its utilization for creating dyes or tints. Fully matured berries yield a green dye/tint, while unripe berries produce a "fair yellow colour." Painters would combine juice from ripe berries with alum to craft a pigment recognized as sap green.[38]

Rhamnus saxatilis Jacq. subsp. *tinctorius* (Waldst. et Kit.) Nyman. (*R. tinctoria* Waldst. et Kit.)

Scientific name: *Rhamnus saxatilis* Jacq. subsp. *tinctorius* (Waldst. et Kit.) Nyman. (*R. tinctoria* Waldst. et Kit.)
Common name: Rock buckthorn, Avignon buckthorn
Romanian name: Verigariu
Taxonomic classification
Kingdom: *Plantae*
Subkingdom: *Cormobionta*
Phylum: *Spermatophyta*
Subphylum: *Magnoliophytina (Angiospermae)*
Class: *Magnoliatae (Magnoliopsida, Dicotyledonatae)*
Subclass: *Rosidae*
Order: *Rhamnales*
Family: *Rhamnaceae*
Genus: *Rhamnus*
Species: *saxatilis* Jacq. (Figure 27.5)
Subspecies: *tinctorius* (Waldst. et Kit.) Nyman.[2-4]

SPREAD

The species is sporadic from the oak forest level to the beech forest level, in dry areas, clearings, forest edges, rocky areas, stony places, shrubs, on limestone rock substrates.[1-3]

DESCRIPTION

Shrub with a maximum height of 2 m, with ascending or erect stems, rarely prostrate, often divaricately branched, having numerous short lateral branches terminating in a thorn and numerous brachyblasts terminating in clusters of leaves. Annual shoots are slender, long, pubescent, brown, reddish-brown, or yellow-brown in color, terminating in a thorn. 2–3-year-old shoots are brownish-yellow to gray-brown, exfoliating in scales. Buds are small, ovoid, brown,

Figure 27.5 *Rhamnus saxatilis* Jacq. subsp. *tinctorius* (Waldst. et Kit.) Nyman. (*R. tinctoria* Waldst. et Kit.) (Rock buckthorn).

Source: 43.

pressed against the stem. Leaves are arranged oppositely to alternate, fascicled on brachyblasts, with variable shapes: ovate, elliptic, ovate-elliptic, obovate-elliptic, elongated-elliptic, lanceolate, rarely subrounded. Leaves have obtuse, rounded, or slightly acute tips, while the base is truncate, narrowed, rounded, or weakly cordate. They have 2–4 pairs of curved lateral veins, are glabrous on the superior surface, sometimes pubescent only near the base on the inferior surface or along the midrib or in the axils of the veins. The leaf blade margin is crenate-serrulate, and the petioles are villose-pubescent to becoming glabrous. On short shoots, leaves have a blade 3–6 times longer than the petioles, and the petioles are equal to or slightly longer than the stipules. Flowers are small, in a 4-pattern, polygamous-dioecious, greenish-yellow, grouped in axillary pauciflorous cymes. Male flowers have linear petals of the same length as the stamens, while female flowers are apetalous. The spherical fruit is attached to a concave receptacle, initially green and then turning black. Seeds are ovoid, light brown, with a wide fissure on the dorsal side bordered by cartilaginous edges. The flowering period is from April to June.[1-3]

Main phytoconstituents   Ref.
amino acids   [44-47]
coumarins (umbelliferone)
flavonoids (quercetin, isorhamnetin, rhamnazin, rhamnetin, kaempherol, naringin, apigenin, hispidulin, rhamnocitrin)
alkaloid
terpenoids (carnosol, tymol, carvacrol, rosmanol)
sterols
fatty acids (palmitic acid, stearic acid, myristic acid)
phenolic acids (ferulic acid, caffeic acid, vanillic acid, p-coumaric acid)
tannins
organic acids

carbohydrates
anthraquinones (rhein, emodin, chrysophanol, physcion, madagascin)
esters
hydrocarbons
aldehydes & ketones
misceleous

Medicinal Use

Traditionally, this species was employed for the treatment of ailments such as wounds, jaundice, hepatitis, gonorrhea, as a laxative, for malaria, stomachaches, and snake bites.[46]

Biological Activity

The assessment of pharmacological activity for the extracts and individual compounds demonstrated anti-inflammatory, antioxidant, antimalarial, antibacterial, and hepatoprotective effects.[46]

Other Uses

The thorns of *Rhamnus saxatilis* have been employed as needles for extracting venom from animals that have been bitten by snakes or spiders. This is achieved by puncturing the affected area with the spine and subsequently applying pressure to the bitten site.[48]

REFERENCES

1. Săvulescu T. (ed). 1958. *Flora RPR*, vol. VI. Ed. Academiei RPR, București.

2. Bejenaru C., Mogoșanu G.D., Bejenaru L. E., Popescu H. 2020. *Botanică farmaceutică. Cormobionta*, ediția a III-a. Ed. Sitech, Craiova.

3. Ciocârlan V. 2009. *Flora ilustrată a României. Pteridophyta et Spermatophyta*. Ediția a 3-a revizuită și adăugită. Ed. Ceres, București.

4. Sârbu I., Ștefan N., Oprea A. 2013. *Plante vasculare din România, Determinator ilustrat de teren.* Ed. Victor B Victor, București.

5. www.infoflora.ch/en/flora/frangula-alnus.html

6. Mogoșanu G.D., Bejenaru L.E., Bejenaru C., Popescu H. 2015. *Farmacognozie-Fitoterapie*, Vol. II, ediția a III-a. Ed. Sitech, Craiova.

7. Arsenijević J., Drobac M., Slavkovska V., Kovačević N., Lakušić B., 2018. Anatomical analysis and phytochemical screening of *Frangula rupestris* (Scop.) Schur (*Rhamnaceae*). *Botanica Serbica* 42(2):231–239.

8. Roudbaraki S.J., Nori-Shargh D. 2016. Analysis of the volatile constituents of *Frangula alnus* Mill. from Iran. *Russian Chemical Bulletin* 65:2770–2772.

9. Elansary H.O., Szopa A., Kubica P., Ekiert H., Al-Mana F.A., El-Shafei A.A. 2020. Polyphenols of *Frangula alnus* and *Peganum harmala* leaves and associated biological activities. *Plants* 9(9):1086.

10. EMEA/HMPC/76306/2006. Assessment Report on *Rhamnus frangula* L., *cortex.*

11. Kremer D., Kosalec I., Locatelli M., Epifano F., Genovese S., Carlucci G., Zovko Končić M. 2012. Anthraquinone profiles, antioxidant and antimicrobial properties of *Frangula rupestris* (Scop.) Schur and *Frangula alnus* Mill. bark. *Food Chemistry* 131(4):1174–1180.

12. Brkanac S.R., Gerić M., Gajski G., Vujčić V., Garaj-Vrhovac V., Kremer D., Domijan A.M. 2015. Toxicity and antioxidant capacity of *Frangula alnus* Mill. bark and its active component emodin. *Regulatory Toxicology and Pharmacology* 73(3):923–929.

13. Đukanović S., Cvetkovic S., Lončarević B., Lješević M., Nikolić B., Simin N. et.al. 2020. Antistaphylococcal and biofilm inhibitory activities of *Frangula alnus* bark ethyl-acetate extract. *Industrial Crops and Products* 158:113013.

14. www.woodlandtrust.org.uk/trees-woods-and-wildlife/british-trees/a-z-of-british-trees/alder-buckthorn/

15. https://dryades.units.it/rosandra_en/index.php?procedure=taxon_page&id=3061&num=3447

16. Zor M., Aydin S., Güner N.D., Başaran N., Başaran A.A. 2017. Antigenotoxic properties of *Paliurus spina-christi* Mill. fruits and their active compounds. *BMC Complementary and Alternative Medicine* 17(1):229.

17. Kemertelidze É.P., Dalakishvili T.M., Gusakova S.D., Shalashvili K.G., Khatiashvili N.S. et al. 1999. Chemical composition and pharmacological activity of the fruits of *Paliurus spina-christi* Mill. *Pharmaceutical Chemistry Journal* 33(11):591–594.

18. Brantner A.H., Males Z. 1999. Quality assessment of *Paliurus spina-christi* extracts. *Journal of Ethnopharmacology* 66(2):175–179.

19. Takım K., Işık M. 2020. Phytochemical analysis of *Paliurus spina-christi* fruit and its effects on oxidative stress and antioxidant enzymes in streptozotocin-induced diabetic rats. *Applied Biochemistry and Biotechnology* 191(4):1353–1368.

20. Zengin G., Fernández-Ochoa Á., Cádiz-Gurrea L., Leyva-Jiménez F.J., et al. 2023. Phytochemical profile and biological activities of different extracts of three parts of *Paliurus spina-christi*: A linkage between structure and ability. *Antioxidants* 12(2):255.

21. Asgarpanah J., Haghighat E. 2012. Phytochemistry and pharmacologic properties of *Ziziphus spina christi* (L.) Willd. *African Journal of Pharmacy and Pharmacology* 6(31):2332–2339.

22. Abdulrahman M.D., Zakariya A.M., Hama H.A., Hamad S.W., Al-Rawi S.S., Bradosty S.W., Ibrahim A.H. 2022. Ethnopharmacology, biological evaluation, and chemical composition of *Ziziphus spina-christi* (L.) Desf.: A review. *Advances in Pharmacological and Pharmaceutical Sciences* 2022:4495688.

23. Hussein A.S. 2019. *Ziziphus spina-christi*: analysis of bioactivities and chemical composition. In: Mariod A.A. (ed). *Wild Fruits: Composition, Nutritional Value and Products*. Springer, Cham, Switzerland, pp. 175–197.

24. www.picturethisai.com/wiki/Paliurus_spina-christi.html

25. https://stock.adobe.com/se/search?k=ziziphus

26. Gao Q.H., Wu C.S., Wang M. 2013. The jujube (*Ziziphus jujuba* Mill.) fruit: A review of current knowledge of fruit composition and health benefits. *Journal of Agricultural and Food Chemistry* 61(14):3351–3363.

27. Wu Y., Chen M., Du M.B., Yue C.H., Li Y.Y. et al. 2014. Chemical constituents from the fruit of *Zizyphus jujuba* Mill. var. *spinosa*. *Biochemical Systematics and Ecology* 57:6–10.

28. Bai L., Zhang H., Liu Q., Zhao Y., Cui X., Guo S. et al. 2016. Chemical characterization of the main bioactive constituents from fruits of *Ziziphus jujuba*. *Food & Function* 7(6):2870–2877.

29. Bandeira Reidel R.V., Melai B., Cioni P. L., Pistelli L. 2017. Chemical composition of volatiles emitted by *Ziziphus jujuba* during different growth stages. *Plant Biosystems – An International Journal Dealing with All Aspects of Plant Biology* 152(4):825–830.

30. Li L.M., Liao X., Peng, S.L., Ding, L.S. 2005. Chemical constituents from the seeds of *Ziziphus jujuba* var. *spinosa* (Bunge) Hu. *Journal of Integrative Plant Biology* 47(4):494–498.

31. Mahajan R., Chopda M. 2009. Phyto-pharmacology of *Ziziphus jujuba* Mill. – A plant review. *Pharmacognosy Reviews* 3:320–329.

32. Tahergorabi Z., Abedini M.R., Mitra M., Fard M.H., Beydokhti H. 2015. *Ziziphus jujuba*: A red fruit with promising anticancer activities. *Pharmacognosy Reviews* 9(18):99–106.

33. Ji X., Guo J., Ding D. et al. 2022. Structural characterization and antioxidant activity of a novel high-molecular-weight polysaccharide from *Ziziphus jujuba* cv. Muzao. *Food Measure* 16:2191–2200.

34. Hossain M.A. 2019. A phytopharmacological review on the Omani medicinal plant: *Ziziphus jujuba*. *Journal of King Saud University – Science* 31(4):1352–1357.

35. Peng W.H., Hsieh M.T., Lee Y.S., Lin Y.C., Liao J. 2000. Anxiolytic effect of seed of *Ziziphus jujuba* in mouse models of anxiety. *Journal of Ethnopharmacology* 72(3):435–441.

36. Preeti, Tripathi S. 2014. *Ziziphus jujuba*: A phytopharmacological review. *International Journal of Research and Development in Pharmacy & Life Sciences* 3(3):959–966.

37. www.michigannatureguy.com/blog/tag/rhamnus-cathartica/

38. Hamed M.M., Refahy L.A., Abdel-Aziz M.S. 2015. Evaluation of antimicrobial activity of some compounds isolated from *Rhamnus cathartica* L. *Oriental Journal of Chemistry* 31(2):1133–1140.

39. Epifano F., Genovese S., Kremer D., Randic M., Carlucci G., Locatelli M. 2012. Re-investigation of the anthraquinone pool of *Rhamnus* spp.: madagascin from the fruits of *Rhamnus cathartica* and *R. intermedia*. *Natural Product Communications* 7(8):1029–1032.

40. Kurylo J., Endress A.G. 2012. *Rhamnus cathartica*: notes on Its early history in North America. *Northeastern Naturalist* 19(4):601–610.

41. Nayyeri S., Azadbakht M., Chabra A., Asgarirad H., Akbari J., Davoodi A., et al. 2020. Antibacterial activities of gel containing 5% hydroalcoholic extract of *Rhamnus cathartica* L. bark. *Journal of Mazandaran University of Medical Sciences* 29(182):106–110.

42. Bayat F., Motevalli Haghi A., Nateghpour M., Rahimi-Esboei B., et al. 2022. Cytotoxicity and anti-*Plasmodium berghei* activity of emodin loaded nanoemulsion. *Iranian Journal of Parasitology* 17(3):339–348.

43. https://dryades.units.it/floritaly/index.php?procedure=taxon_page&tipo=all&id=3065

44. Cuoco G., Mathe C., Vieillescazes C. 2014. Liquid chromatographic analysis of flavonol compounds in green fruits of three *Rhamnus* species used in stil de grain. *Microchemical Journal* 115:130–137.

45. Locatelli M., Tammaro F., Menghini L., Carlucci G., Epifano F., Genovese S. 2009. Anthraquinone profile and chemical fingerprint of *Rhamnus saxatilis* L. from Italy. *Phytochemistry Letters* 2(4):223–226.

46. Nigussie G., Melak H., Endale M. 2021. Traditional medicinal uses, phytochemicals, and pharmacological activities of genus *Rhamnus*: A review. *Journal of the Turkish Chemical Society Section A: Chemistry* 8(3):899–932.

47. Rocchetti G., Miras-Moreno M.B., Zengin G., Senkardes I., Sadeer N.B., Mahomoodally M.F., Lucini L. 2019. UHPLC-QTOF-MS phytochemical profiling and in vitro biological properties of *Rhamnus petiolaris* (*Rhamnaceae*). *Industrial Crops and Products* 142:111856.

48. Benítez G., González-Tejero M.R., Molero-Mesa J. 2012. Knowledge of ethnoveterinary medicine in the Province of Granada, Andalusia, Spain. *Journal of Ethnopharmacology* 139(2):429–439.

28 *Rosaceae* Juss. Family

Cornelia Bejenaru[1], Antonia Radu[1], George Dan Mogoşanu[2],
Ludovic Everard Bejenaru[2], Andrei Biţă[2], and Adina-Elena Segneanu[3]
[1] Department of Pharmaceutical Botany, Faculty of Pharmacy, University of Medicine and Pharmacy of Craiova, 2 Petru Rareş Street, Craiova, Dolj County, Romania
[2] Department of Pharmacognosy & Phytotherapy, Faculty of Pharmacy, University of Medicine and Pharmacy of Craiova, 2 Petru Rareş Street, Craiova, Dolj County, Romania
[3] Institute for Advanced Environmental Research–West University of Timisoara (ICAM–WUT), 4 Oituz Street, Timisoara, Romania

BOTANICAL DESCRIPTION

Herbaceous (perennial or annual) or woody (trees, shrubs, subshrubs) plants with simple or branched stems, equipped with thorns or spines and stipulate, simple or compound leaves, arranged alternately, deciduous, very rarely persistent. The flowers, solitary or arranged in inflorescences (racemes or cymes), are usually hermaphroditic, actinomorphic, rarely unisexual, of the pentamerous type (rarely unisexual or of the tetramerous type). The floral envelope consists of free elements (dialisepal and dialipetal). An epicalyx may rarely appear on the exterior, and the corolla may also be absent, albeit rarely. Numerous stamens are fixed, along with sepals and petals, to the receptacle, which can take the form of a disc, cone, or pitcher. The gynoecium can be monocarpellary or multicarpellary, with carpels either free or fused, uni- or polilocular, with one or many ovules, and it may be inferior or superior. The ovules are anatropous. Nectaries are present at the base of the receptacle, and pollination is primarily entomophilous. The fruit can vary greatly, being either simple or compound. When it is compound, it is made up of drupes, achenes, or follicles, partially united or free on the fleshy receptacle. In other cases, the partial (free) fruits may be enclosed in an enlarged and swollen receptacle resembling a pitcher, or attached to a large, fleshy, bulging receptacle, or concrescent with the receptacle, which becomes enlarged and fleshy. The seeds are exalbuminate. The extensive variability of the gynoecium and fruits has led to the classification of this family into three subfamilies: *Rosoideae*, *Maloideae* (*Pomoideae*), and *Prunoideae*.

Rosaceae is one of the most well-represented families in the Romanian Carpathians, and includes **36** genera of plants and the following grow spontaneously in the Romanian flora: *Spiraea* L., *Aruncus* L., *Rubus* L., *Fragaria* L., *Comarum* L., *Potentilla* L., *Geum* L., Dryas L., *Waldsteinia* Willd., *Aremonia* Neck. ex nestl., *Filipendula* Mill., *Aphanes* L., *Alchemilla* L., *Agrimonia* Willd., *Rosa* L., *Cotoneaster* Medik., *Crataegus* L., *Sorbus* L. em. Crantz, *Pyrus* L., *Malus* Mill., *Amelanchier* Medik., *Amygdalus* L., *Prunus* L., *Cerasus* Juss., and *Padus* Mill.

Only some of the most famous Romanian medicinal plants in this family were selected and presented in detail.[1-4]

Alchemilla L. (Lady's mantle)

Perennial herbaceous species, with a height of 20–50 cm, palmate leaves, and green flowers, of the tetramerous type (with a double calyx, without a corolla), clustered in terminal cymes.[2]

Alchemilla vulgaris L. (Lady's mantle)

Scientific name: *Alchemilla vulgaris* L. emend. Fröhner (*A. acutiloba* Opiz; *A. vulgaris* L. subsp. *acutangula* (Buser) Palitz)
Common name: Lady's mantle
Romanian name: Creţişoară
Taxonomic classification
Kingdom: *Plantae*
Subkingdom: *Cormobionta*
Phylum: *Spermatophyta*
Subphylum: *Magnoliophytina (Angiospermae)*
Class: *Magnoliatae (Magnoliopsida, Dicotyledonatae)*
Subclass: *Rosidae*
Order: *Rosales*
Family: *Rosaceae*

Figure 28.1 *Alchemilla vulgaris* L. (Lady's mantle).

Source: 5.

Subfamily: *Rosoideae*
Genus: *Alchemilla*
Species: *vulgaris* L. (Figure 28.1)[2-4]

SPREAD

This species is commonly found from the beech forest zone to the subalpine zone in meadows, shrubs, on steep slopes, rocky areas, and occasionally in forest edges.[1,4]

DESCRIPTION

Perennial plant, with brown rhizome and numerous adventitious roots, as well as stipular remnants and petiole rudiments. The stems are erect, ascending, or creeping, bluish-green, glabrous, or sometimes with spreading hairs at the base, and they reach a height of 20–50 (70) cm. The leaves have lobed blades (9–11 elongated triangular lobes) with few hairs on the superior surface. The basal leaves are long-petiolate and reniform, forming a rosette. The leaves have elongated stipules, and those at the base of the stem leaves have dentate margins. The inflorescence is glabrous, and the flowers have long pedicels and are yellowish-green. It blooms from June to August.[1-4]

Main phytoconstituents Ref.
[6-9]

coumarins (umbelliferone, scopoletin, esculetin)
flavonoids (quercetin, genistein, morin, apigenin, kaempherol, luteolin, hyperoside, rutin, avicularin, cyanidin, tiliroside, arbutin)
terpenoids (cymene, limonene, farnesene, linalool, terpineol, thymol, carvone, caryophyllene)
sterols
fatty acids
phenolic acids (chlorogenic acid, ellagic acid, caffeic acid, gallic acid, p-coumaric acid)

saponins
tannins (gallotannin, agrimoniin)
hydrocarbons (decane, dodecane, tetradecane)
aldehydes & ketones (pentanal, benzaldehyde, hexanal, heptanal, ionone)

Medicinal Use

skin problems (wounds, ulcers, eczema), gynecological disorders (fibroids, cysts, endometriosis, infertility, dysmenorrhea, menorrhagia), diuretic, antianemic and antidiabetic, antihemorrhagic, antispasmodic, regulation of thyroid and reproductive hormones[6-9]

Biological Activity

antibacterial, antioxidant, anticancer, neuroprotective, antifungal, antiviral, gastroprotective, anti-inflammatory, wound healing, antiproliferative, enzyme inhibitory activity, gynecological disorders[6-9]

Other Uses

- ornamental

- culinary use (salads, pesto)[6-9]

Alchemilla monticola Opiz (Hairy lady's mantle)

Scientific name: *Alchemilla monticola* Opiz (*A. pastoralis* Buser; *A. vulgaris* L. subsp. *pastoralis* (Buser) Palitz)
Common name: Hairy lady's mantle
Romanian name: Crețișoară
Taxonomic classification
Kingdom: *Plantae*
Subkingdom: *Cormobionta*
Phylum: *Spermatophyta*
Subphylum: *Magnoliophytina (Angiospermae)*
Class: *Magnoliatae (Magnoliopsida, Dicotyledonatae)*
Subclass: *Rosidae*
Order: *Rosales*
Family: *Rosaceae*
Subfamily: *Rosoideae*
Genus: *Alchemilla*
Species: *monticola* Opiz (Figure 28.2)[2-4]

SPREAD

It is commonly found from hilly areas to the subalpine zone in meadows, forest edges, clearings, sheepfolds, and weedy places.[3-4]

DESCRIPTION

Perennial plant, reaching a height of 20–50 cm, covered with spreading hairs on the stems, petioles, and inflorescence branches. It has a subround lobed leaf blade with equal teeth and dense hairs on both sides. It blooms from June to August.[2-4]

Main phytoconstituents   Ref.
amino acids (alanine, threonine, valine)   [11-12]
flavonoids (astragalin, kaempferol, quercetin, isoquercitrin, rutin, vitexin)
terpenoids (achillicin, matricarin)
sterols
organic acid (fumaric acid, succinic acid)
fatty acids (myristic acid, palmitic acid, stearic acid, linoleic acid, arahnidic acid, oleic acid)
phenolic acids (vanillic acid)
carbohydrates (glucose, sucrose, fructose)
misceleous (Inositol)

Figure 28.2 *Alchemilla monticola* Opiz (Hairy lady's mantle).

Source: 10.

Medicinal Uses

Hepatites, gastric ulcer, obesity[11-12]

Biological Activity

anti-inflammatory, hepatoprotective, skin photoprotective, antiadipogenic[11-12]

Alchemilla crinita Buser

Scientific name: *Alchemilla crinita* Buser (*A. vulgaris* L. subsp. *crinita* (Buser) Palitz)
Common name: -
Romanian name: Crețișoară
Taxonomic classification
Kingdom: *Plantae*
Subkingdom: *Cormobionta*
Phylum: *Spermatophyta*
Subphylum: *Magnoliophytina (Angiospermae)*
Class: *Magnoliatae (Magnoliopsida, Dicotyledonatae)*
Subclass: *Rosidae*
Order: *Rosales*
Family: *Rosaceae*
Subfamily: *Rosoideae*
Genus: *Alchemilla*
Species: *crinita* Buser (Figure 28.3)[2-4]

SPREAD

It is commonly found from hilly areas to the subalpine zone in meadows, forest edges, clearings, sheepfolds, and weedy places.[2-4]

Figure 28.3 *Alchemilla crinita* Buser.

Source: 13.

Description

Perennial plant, with a height ranging from 10 to 50 cm. It has reniform leaves that are lobed with uneven teeth, and the receptacle is glabrous. It typically blooms from June to August.[2-4]

Main phytoconstituents
amino acids
flavonoids
terpenoids (achillicin, matricarin)
sterols
organic acid
fatty acids
phenolic acids
carbohydrates
misceleous

Ref.
[14]

Medicinal Uses

asthma, bronchitis, cough[15]

Biological Activity

anti-inflammatory[15]

Other Uses

■ ornamental[15]

Alchemilla mollis (Buser) Rothm. (Garden lady's mantle)

Scientific name: *Alchemilla mollis* (Buser) Rothm.
Common name: Garden lady's mantle
Romanian name: Crețișoară
Taxonomic classification
Kingdom: *Plantae*
Subkingdom: *Cormobionta*

Figure 28.4 *Alchemilla mollis* (Buser) Rothm. (Garden lady's mantle).

Source: 16.

Phylum: *Spermatophyta*
Subphylum: *Magnoliophytina (Angiospermae)*
Class: *Magnoliatae (Magnoliopsida, Dicotyledonatae)*
Subclass: *Rosidae*
Order: *Rosales*
Family: *Rosaceae*
Subfamily: *Rosoideae*
Genus: *Alchemilla*
Species: *mollis* (Buser) Rothm (Figure 28.4)[2-4]

SPREAD

This plant is sporadically distributed from the beech forest zone to the subalpine zone in meadows and weedy places.[1-3]

DESCRIPTION

Robust plant, reaching a height of 20–50 cm, covered with dense, spreading hairs. It has a glabrous inflorescence and achenes that partially protrude from the receptacle. It typically blooms from May to July.[1-3]

Main phytoconstituents	Ref.
coumarins	12,17-19

flavonoids (apigenin, luteolin, rutin, miquelianin, hyperoside, isoquercetin, tiliroside, sinocrassoside D2, rhodiolgin, gossypetin)
terpenoids (terpineol, linalool)
sterols (sitosterol)

fatty acids (lauric acid, myristic acid, palmitic acid, stearic acid, linoleic acid, arahnidic acid, lignoceric acid)

phenolic acids (gallic acid, caffeic acid, gentisic acids)

tannins (epicatechin, pedunculagin,agrimoniin, sanguiin)

saponins

organic acids

misceleous (vitamins)

Medicinal Use

The vegetable product *Alchemillae herba* has been used in treating skin problems (wounds, ulcers, eczema), gynecological disorders (fibroids, cysts, endometriosis, infertility, dysmenorrhea, menorrhagia), diuretic, antianemic and antidiabetic, antihemorrhagic, antispasmodic, regulation of thyroid and reproductive hormones.[6-9,19-20] It is used in the prevention of spontaneous abortion by strengthening the endometrium, as a uterine antihemorrhagic (reduces coagulation time and increases the number of platelets), in female infertility, and as a favorable remedy in the treatment of hemophilia (Prof. Teodor Goina, Pharm. PhD László Tuka).[20]

Biological Activity

antibacterial, antioxidant, anticancer, neuroprotective, antifungal, antiviral, gastroprotective, anti-inflammatory, wound healing, antiproliferative, enzyme inhibitory activity, gynecological disorders[6-9,19-20]

Other Uses

- Ornamental
- culinary use (salads, pesto)[6-9,20]

POTENTILLA L. (CINQUEFOIL)

Perennial herbaceous plants (occasionally annual or shrubs), with trailing stems or stolons. The leaves are pinnate or palmate, and the flowers, with both sepals and petals, have a generally pentamerous yellow, white, or red corolla. The fruit is polyachene. There are 31 species of this genus in Romania.[1-2]

Potentilla anserina L. (Silverweed)

Scientific name: *Potentilla anserina* L.
Common name: Silverweed, Silverweed cinquefoil, Goosegrass
Romanian name: Coada racului
Taxonomic classification
Kingdom: *Plantae*
Subkingdom: *Cormobionta*
Phylum: *Spermatophyta*
Subphylum: *Magnoliophytina (Angiospermae)*
Class: *Magnoliatae (Magnoliopsida, Dicotyledonatae)*
Subclass: *Rosidae*
Order: *Rosales*
Family: *Rosaceae*
Subfamily: *Rosoideae*
Genus: *Potentilla*
Species: *anserina* L. (Figure 28.5)[2-4]

SPREAD

This species commonly grows from the lowland to the mountain regions in wet meadows or along watercourses, on alluvial soils, gravelly areas, and ruderal sites.[2-4]

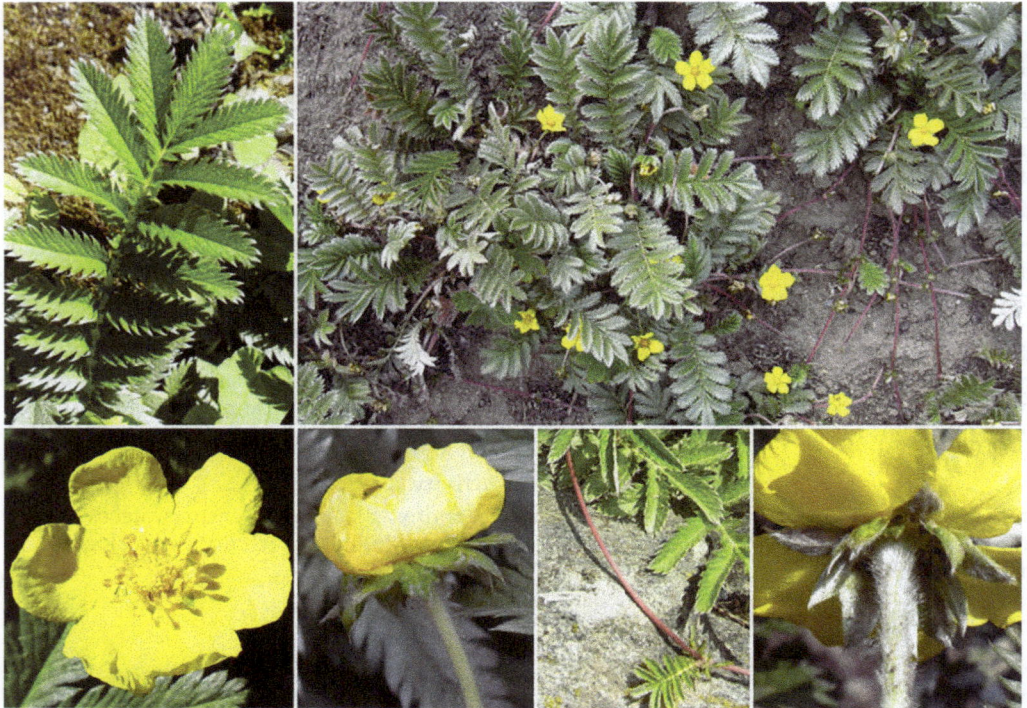

Figure 28.5 *Potentilla anserina* L. (Silverweed).

Source: 21.

DESCRIPTION

Perennial herbaceous plant, reaching a height of 15–50 cm, with a thick and multicauline basal axis. The stems are trailing, elongated, slender, and rooted at the nodes. The leaves are pinnately partite to pinnately sect, with uniformly serrated dentate leaflets. The flowers have long pedicels, are of the pentamerous type, hermaphroditic, actinomorphic, with yellow petals longer than the sepals. The androecium consists of 20 stamens, and the gynoecium has a lateral, virgulate style and a dilated stigma. The achenes are glabrous. It typically blooms from May to August.[1-4]

Main Phytoconstituents

Anserinae herba contains approximately 5–10% gallic tannin and ellagic tannin, flavonoids, sterols, choline, organic acids, simple sugars, and mineral salts.[20]

Main phytoconstituents   Ref.
amino acids (alanine, valine, proline, leucine, serine, aspartic acid, phenylalanine)   [22-25]
coumarins
flavonoids (myricetin, quercetin, acacetin, genistein, kaempherol, astragalin, rutin)
carbohydrates (xylose, arabinose, glucose, manitol, fructose, sucrose)
terpenoids (ursolic acid, euscaphic acid, madengaisu A, B, and C, tormentic acid)
sterols
fatty acids
phenolic acids (chlorogenic acid, ferrulic acid, caffeic acid,)
organic acids (glycolic acid, malic acid, citric acid)
tannins (gallocatechin, agrimoniin, ellagic acid)
hydrocarbons
aldehydes & ketones

Medicinal Use

The aerial part is harvested during flowering, forming the medicinal product *Anserinae herba*, which contains gallic and ellagic tannins. *Anserinae herba* has anti-inflammatory, antibacterial, astringent, antidiarrheal, and antihemorrhagic properties.[2]

Biological Activity

Immunomodulatory, antioxidant, α-glucosidase and tyrosinase inhibitory, antiradiation, antitussive and expectorant, hepatoprotective, nephroprotective, and antimicrobial activities, along with the potential to ameliorate pulmonary edema.[25-33]

Other Uses

The roots can be consumed either in their raw or cooked form, while the stems and leaves can be enjoyed as a salad.[34]

Potentilla erecta (L.) Raeusch. (Tormentil)

Scientific name: *Potentilla erecta* (L.) Raeusch.
Common name: Tormentil, Septfoil, Erect cinquefoil
Romanian name: Sclipeți
Taxonomic classification
Kingdom: *Plantae*
Subkingdom: *Cormobionta*
Phylum: *Spermatophyta*
Subphylum: *Magnoliophytina (Angiospermae)*
Class: *Magnoliatae (Magnoliopsida, Dicotyledonatae)*
Subclass: *Rosidae*
Order: *Rosales*
Family: *Rosaceae*
Subfamily: *Rosoideae*
Genus: *Potentilla*
Species: *erecta* (L.) Raeusch. (Figure 28.6)[2-4]

SPREAD

Commonly found from the oak forest zone to the subalpine region in wet meadows and peat bogs.[2-4]

DESCRIPTION

Perennial herbaceous plant with a thick basal axis, measuring 1–3 cm in diameter, red on the inside and woody brown on the exterior. Numerous stems, 10–30 cm tall, arise from the basal axis; they are erect, ascending, branched, and sparsely hairy. The leaves are trifoliate, with stipules resembling the leaflets. The flowers are long-pedicellate, solitary or grouped in dichasia, of the tetramerous type, with elongated linear sepals on the external cycle, narrower than those on the internal cycle, and a corolla composed of yellow petals. It blooms from May to August.[1-4]

Main Phytoconstituents

Tormentillae rhizoma contains: 15–22% mixed tannins and catechic tannins; flavonoids (camferol-glycosides); proanthocyanidols – dimers and trimers; pentacyclic triterpenoids glycosidated (tormentozide, chinovozide); organic acids; simple sugars; mineral salts.[36]

Main phytoconstituents	Ref.
amino acids (alanine, valine, proline, leucine, serine, aspartic acid, phenylalanine)	25
flavonoids (kaempherol, eugenin)	
carbohydrates (xylose, arabinose, glucose, manitol, fructose, sucrose)	
terpenoids (ursolic acid, tormentic acid, pomolic acid, arjunetin, chinovic acid)	
fatty acids	
phenolic acids (gallic acid, p-coumaric acid, gentisic acid, syringic acid, caffeic acid)	
tannins (agrimoniin (tormentillin, ellagic acid, epicatechin, catechin, gallocatechin)	
anthicyanidins (cyanidin)	

Figure 28.6 *Potentilla erecta* (L.) Raeusch. (Tormentil).

Source: 35.

Medicinal Use

The vegetable product *Tormentillae rhizoma* is astringent, antidiarrheal, antihemorrhagic, wound healing, dysentery, intestinal infections, oral antiseptic.[2,25]

Biological Activity

- antioxidant, antimicrobial, wound healing, anticariogenic, and antidiabetic properties[37-39]

Other Uses

Tormentil is an excellent choice for border edges, bed plantings, and also for rock, cottage, and herb gardens. Its blossoms are a magnet for butterflies and various pollinators, making it a valuable addition to gardens designed to attract wildlife, pollinators, or butterflies.[40]

Potentilla reptans L. (Creeping cinquefoil)

Scientific name: *Potentilla reptans* L.
Common name: Creeping cinquefoil, European cinquefoil, Creeping tormentil
Romanian name: Cinci degete
Taxonomic classification
Kingdom: *Plantae*
Subkingdom: *Cormobionta*
Phylum: *Spermatophyta*
Subphylum: *Magnoliophytina (Angiospermae)*
Class: *Magnoliatae (Magnoliopsida, Dicotyledonatae)*
Subclass: *Rosidae*
Order: *Rosales*
Family: *Rosaceae*

Figure 28.7 *Potentilla reptans* L. (Creeping cinquefoil).

Source: 41.

Subfamily: *Rosoideae*
Genus: *Potentilla*
Species: *reptans* L. (Figure 28.7)[2-4]

SPREAD

Pioneer plant, frequently found in flood-prone areas and ruderalized environments, as well as in wet meadows, ranging from lowland to mountain regions.[2]

DESCRIPTION

Perennial herbaceous plant with a thick, multicauline basal axis and trailing stems, measuring 30–100 cm in length, which root at the nodes. The leaves are petiolate, pentafoliate, with obovate leaflets. The superior surface of the leaves is glabrous, while the inferior surface is densely covered with appressed hairs. The leaf margins are sharply dentate and ciliate. The flowers are solitary, rarely in pairs, large and yellow, of the pentamerous type, and they are axillary in arrangement. It typically blooms from June to August.[1-4]

Main phytoconstituents   Ref.
amino acids   [42-46]
flavonoids (rutin, naringin, quercetin)
carbohydrates
terpenoids
fatty acids
phenolic acids (gallic acid, chlorogenic acid, vanilic acid, p-coumaric acid, sinapic acid, syringic acid, cinnamic acid)
tannins (catechin)
organic acid (benzoic acid)

Medicinal Use

In traditional medicine, the plant is used for its diuretic and antidiarrheal properties.[2]

Biological Activity

- antioxidant, anti-inflammatory, antimicrobial, cardioprotective, antinociceptive, and cytotoxic effects.[43-46]

Other Uses

- culinary uses — the raw young leaves are a nutritional addition to salads.[47]

Geum urbanum L. (Wood avens)

Scientific name: *Geum urbanum* L.
Common name: Wood avens, Herb bennet, Colewort, St. Benedict's herb
Romanian name: Cerențel
Taxonomic classification
Kingdom: *Plantae*
Subkingdom: *Cormobionta*
Phylum: *Spermatophyta*
Subphylum: *Magnoliophytina (Angiospermae)*
Class: *Magnoliatae (Magnoliopsida, Dicotyledonatae)*
Subclass: *Rosidae*
Order: *Rosales*
Family: *Rosaceae*
Subfamily: *Rosoideae*
Genus: *Geum*
Species: *urbanum* L. (Figure 28.8)[2-4]

Figure 28.8 *Geum urbanum* L. (Wood avens).
Source: 48.

SPREAD

A plant commonly found in weedlands, forest edges, clearings, and ruderal areas, ranging from lowlands to mountains.[2]

DESCRIPTION

Perennial herbaceous plant with a strong, thick, oblique, cylindrical rhizome from which numerous adventitious roots emerge. The stem is slender, simple or branched, erect, curved at the base, roughly hairy, and 20–60 cm in height. The basal leaves are short-petiolate and form a basal rosette, while the stem leaves are alternate. The leaves are irregularly odd-pinnate with irregularly doubly serrated margins on both sides, and they are covered with dispersed and joined glandular hairs, especially along the veins. The stipules resemble the leaflets. The flowers are yellow, erect, hermaphroditic, actinomorphic, of the pentamerous type, with long, simple, and glandular-hairy pedicels, an external linear-lanceolate calyx, and 4–7 mm yellow, round-obovate petals that are caducous. The elongated receptacle is brown, hairy, and the achenes (nuts) are shortly setose. These plants can bloom from May to September.[1-2]

Main Phytoconstituents

Gei rhizoma contains: 10–20% catechic tannins (catechol and derivatives) and gallic tannins (galoil-glucose), along with free gallic acid and ellagic acid; approx. 1% geoside, which is the glucoside of eugenol with vicianose (glucosyl-arabinosyl); polyphenolcarboxylic acids; volatile oil; polyuronides; resins.

According to Romanian Pharmacopeae (F.R. VII), it specifies a minimum tannin concentration of 10%.

Geoside plays an analytical role because in the presence of geoside, it breaks down into eugenol and imparts a characteristic odor to the vegetable product.[20]

Main phytoconstituents

Ref.
49-51

amino acids
flavonoids (rubuside, isorhamnetin, naringin, quercetin, diosmetin)
carbohydrates
terpenoids (oleanoic acid, ursolic acid, pomonic acid, myrtanol, geniposide, asiatic acid)
fatty acids
phenolic acids (gallic acid, vanilic acid. caffeic acid, chlorogenic acid)
sterols (β-sitosterol)
tannins (ellargic acid, tellimagrandin, stachyurin, casuarynin, and gemin A)
misceleous (eugenol)

Medicinal Use

The medicinal product *Gei rhizoma* (*Caryophyllatae rhizoma*) is used for diarrhea, stomatitis, and gingivitis.[20]

- sedative, hemostatic, astringent and anti-inflammatory, skin diseases, inflammations of the bladder and urinary tract[50]

Biological Activity

Geum urbanum has astringent, antidiarrheal, antibacterial, anti-inflammatory, and antihemorrhagic properties.[2]

- antiinflammatory, antimicrobial, antioxidant, neuroprotective, Parkinsonian, hypotesive and many other effects.[50]

Other Uses

Fresh leaves and roots can be prepared by cooking. They serve as a spice to enhance the flavor of soups, stews, and even ale, imparting a taste reminiscent of cloves with a subtle hint of cinnamon. For culinary purposes, it is most suitable during the spring season. Additionally, the roots can be boiled to create a beverage.[52]

Agrimonia eupatoria L. (Common agrimony)

Scientific name: *Agrimonia eupatoria* L.
Common name: Common agrimony, Church steeples, Sticklewort
Romanian name: Turiță mare
Taxonomic classification
Kingdom: *Plantae*
Subkingdom: *Cormobionta*
Phylum: *Spermatophyta*
Subphylum: *Magnoliophytina (Angiospermae)*
Class: *Magnoliatae (Magnoliopsida, Dicotyledonatae)*
Subclass: *Rosidae*
Order: *Rosales*
Family: *Rosaceae*
Subfamily: *Rosoideae*
Genus: *Agrimonia*
Species: *eupatoria* L.[2-4]
Subspecies:

- *eupatoria* non-pubescent, with broadly obovate leaflets and a basal leaf rosette (Figure 28.9).
- *grandis* (Andrz.) Bornm. (var. *major* Boiss.) pubescent, with elliptical leaflets and typically without a basal rosette.[2-4]

SPREAD
Common in shrubs and meadows, in ruderal areas with loose soils, ranging from lowlands to mountains.[2]

Figure 28.9 *Agrimonia eupatoria* L. (Common agrimony).
Source: 53.

DESCRIPTION

Perennial species with a short, trailing rhizome, either simple or branched, and a tall stem (which can reach up to 150 cm in height). The stem is covered in long and short tector hairs. The leaves have elliptical to ovate leaflets with rounded bases, and the inferior surfaces are entirely covered in tector hairs. The stipules are ovate and measure 1–2(3) cm. The flowers have short pedicels, a subrotund receptacle, densely hairy, with ovate-triangular sepals and golden-yellow petals. The flowering period is from June to August.[1-2]

Main Phytoconstituents

Agrimoniae herba contains approximately 20% catechic tannins, flavonoids (rutoside, hyperoside), isocoumarins (agrimonolide), volatile oil, polyuronides, pentacyclic triterpenoids, organic acids, simple sugars, and mineral salts.[20]

Main phytoconstituents	Ref.
amino acids	54-55

flavonoids (luteolin, apigenin, quercetin, kaempferol, hyperoside, rutin, quercetin)
terpenoids (ursolic acid, euscapic acid, tormentic acid, cedrol, α-pinene, linalool, α-terpineol, bornol, eucalyptol)
sterols (cholesterol, campesterol, stigmasterol, sitosterol, avenasterol)
fatty acids (palmitic acid, stearic acid)
phenolic acids (chlorogenic acid, ferulic acid, caffeic acid, vanillic acid, gentisic acid, p-coumaric acid)
tannins (ellagic acid)
organic acids (acid ascorbic, citric acid, malic acid, nicotinic acid)
coumarins
glycosides
esters
aldehydes & ketones
miscellaneous (vitamins)

Medicinal Use

The aerial part, harvested during flowering, forms the vegetable product *Agrimoniae herba*. It possesses astringent, antidiarrheal, antiviral, anti-inflammatory, cicatrizing, cholagogic, diuretic, hypotensive, antidiabetic, hepatoprotective properties.[2,20,50-51]

Biological Activity

Compounds derived from this species have demonstrated various beneficial properties, including antioxidant, antimicrobial, antidiabetic, neuroprotective, antinociceptive, antitumoral, and anti-inflammatory effects. Notably, studies have indicated the hepatoprotective properties and the protective role against cardiovascular diseases, metabolic disorders, and diabetes.[56-58]

Other Uses

A revitalizing herbal tea can be prepared from either fresh or dried leaves, flowers, and stems, suitable for consumption both hot and cold. In the past, it enjoyed widespread popularity either as a standalone beverage or when added to Chinese tea, contributing a distinct delicacy and fragrance. Additionally, the seeds can be dried and ground into a meal, serving as a last resort in times of famine.[59]

ROSA L. (ROSE)

This genus comprises a large number of species (over 100), of which 23 are native to Romania. Plants of this genus are erect or climbing shrubs that grow on the edges of forests, in shrubby areas, meadows, and on hills. The leaves are odd-pinnately compound, and the flowers are grouped in inflorescences, although solitary flowers are less common. The floral envelope is double, with the corolla sometimes involute, and it can be red (in various shades), yellow, or white.[2]

Rosa canina L. s.l. (Dog rose)

Scientific name: *Rosa canina* L. s.l.
Common name: Dog rose
Romanian name: Măceş
Taxonomic classification
Kingdom: *Plantae*
Subkingdom: *Cormobionta*
Phylum: *Spermatophyta*
Subphylum: *Magnoliophytina (Angiospermae)*
Class: *Magnoliatae (Magnoliopsida, Dicotyledonatae)*
Subclass: *Rosidae*
Order: *Rosales*
Family: *Rosaceae*
Subfamily: *Rosoideae*
Genus: *Rosa*
Species: *canina* L. s.l. (Figure 28.10)[2-4]

SPREAD

A pioneering species, frequently found on the edges of roads, in forests, shrubby areas, or meadows, ranging from lowlands to mountain regions.[2]

DESCRIPTION

A spiny shrub that forms bushes reaching a height of 3 meters (up to 5 meters). The stems are elongated, branched, and armed with strong, curved thorns. The leaves are green, with non-leathery leaflets that are oval or elliptical, with a pointed tip and a narrow base. They are either

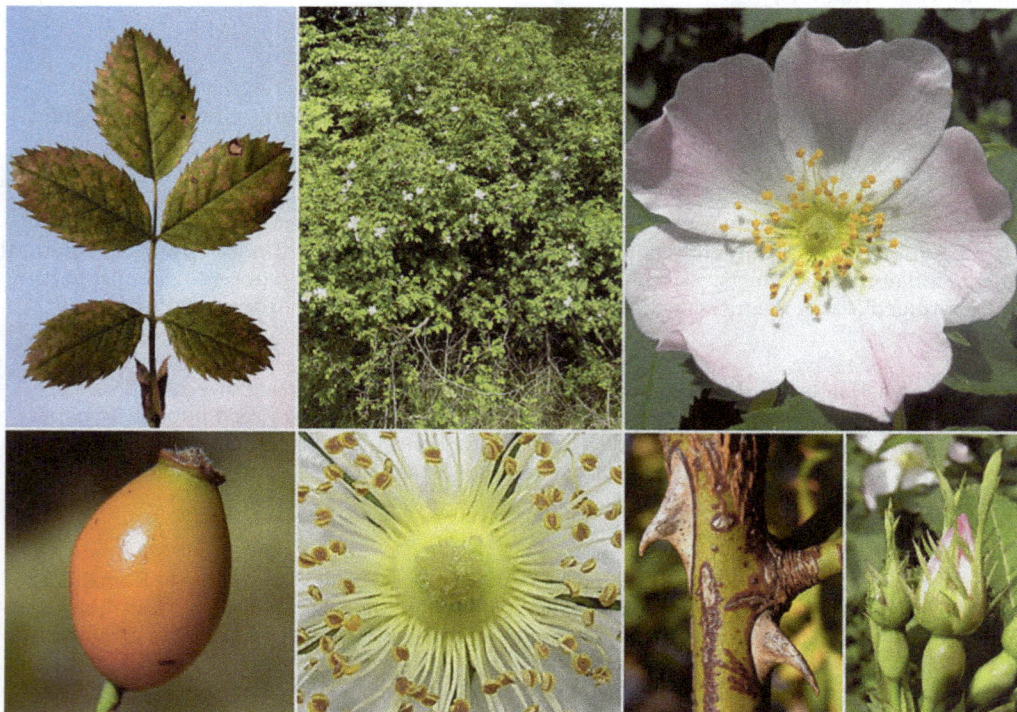

Figure 28.10 *Rosa canina* L. s.l. (Dog rose).

Source: 60.

singly serrate or imperfectly doubly serrate. The stipules are small, broader on the flowering branches, and they can be either glabrous or stipitate-glandular on the margins. The receptacle is globose, ovoid, or ellipsoidal, non-glandular or glandular-stipitate at the base. The flowers are small, with reflexed sepals that are decidous before the fruit matures, and large petals that can be either white or red. These plants bloom in June and July and are cultivated for both food and medicinal purposes.[1-2]

Main Phytoconstituents

Cynosbati fructus contains: 0.5–1% ascorbic acid, 0.5–10% carotenoids, vitamins B2, E, K, and PP; flavonoids (hyperoside, rutin), approx. 15–25% carbohydrates, 25% pectins, D-sorbitol, tannins, glycosides of oleanolic acid, and *β*-sitosterol; a volatile oil rich in linalool, geraniol, citronelol, nerol; organic acids such as malic acid and citric acid; and 1.5–3.5% mineral salts.[20]

Main phytoconstituents   Ref.

amino acids   [61-67]

flavonoids (quercetin, tilroside hyperoside, rutin, myrcetin)

terpenoids (humulene, pinene, elemene, limonene, citronellol, geraniol, linalool, α-terpineol, phytol)

sterols (sitosterol)

fatty acids (linoleic, oleic, linolenic, palmitic, stearic acid, linolenic acid arachidonic acid)

phenolic acids (gallic acid, quinic acid, chlorogenic acid, coumaric acid, rosmarinic acid)

tannins (catechin, astragalin)

organic acids (ascorbic acid, malic acid, citric acid, fumaric acid)

carbohydrates (fructose, glucose,trehalose, sucrose)

glycosides

esters

carotenoids (β -carotene, licopene, andisomeres of rubixanthin)

hydrocarbons

aldehydes & ketones

miscellaneous (chalcone, carotenoids)

Medicinal Use

The vegetable product *Rosae pseudofructus*, also known as *Cynosbati fructus*, has antiscorbutic, antioxidant, astringent, antidiarrheal, astringent, colagogue, choleretic, vasoprotective, and diuretic properties.[2,20,68]

Biological Activity

Utilized for addressing avitaminosis or hypovitaminosis, digestive ailments (such as anorexia, indigestion, and cholecystitis), vascular conditions like capillary fragility, convalescence, asthenia, and oxidative stress. In phytotherapy, as advocated by Dr. Jean Valnet (1920–1995), this product is regarded as a source of assimilable silicon, similar to *Equiseti herba*.[20]

■ anti-inflammatory, antioxidant, and antimutagen[68]

Other Uses

The fruit can be consumed either raw or cooked, and it is often used to prepare delicious jams and syrups. The syrup, in particular, serves as a nutritional supplement, particularly suitable for babies. Additionally, the fruit can be dried and utilized in the preparation of tea. The seeds are a rich source of vitamin E and can be ground to be mixed with flour or added to other foods as a dietary supplement. It is important to remove the seed hairs before consumption. Dried leaves can serve as a substitute for tea, and according to another report, they can even be used as a coffee substitute. The petals of the plant, whether raw or cooked, are edible, although it's advisable to remove the bitter base of the petal. In China, they are consumed as a vegetable and are also used to create uniquely scented jams.[69]

Figure 28.11 *Dryas octopetala* L. (Siverware).

Source: 71.

Dryas octopetala L. (Siverware)

Scientific name: *Dryas octopetala* L.
Common name: Siverware, Mountain avens[70]
Romanian name: Argințică
Taxonomic classification
Kingdom: *Plantae*
Subkingdom: *Cormobionta*
Phylum: *Spermatophyta*
Subphylum: *Magnoliophytina (Angiospermae)*
Class: *Magnoliatae (Magnoliopsida, Dicotyledonatae)*
Subclass: *Rosidae*
Order: *Rosales*
Family: *Rosaceae*
Subfamily: *Rosoideae*
Genus: *Dryas*
Species: *octopetala* L. (Figure 28.11)[2-4]

SPREAD

This species is commonly encountered from the beech forest zone to the alpine zone, particularly on rocky and limestone soils.[2-4]

DESCRIPTION

The name derives from the Greek word "dryas," meaning woodland nymphs. It is a dwarf, creeping shrub, reaching a height of 5–20 cm, with a taproot that is fusiform and long. The stems are highly branched, brownish-red, and terminated with sterile leaf buds. The leaves are simple, coriaceous, short-petioled, elliptical, sempervirent, deeply crenate, with slightly curled margins. The superior surface is glabrous or pubescent, wrinkled and dark, while the inferior surface is densely white-tomentose. The stipules are linear-lanceolate, acute, hairy, and fused for 2/3 of the petiole's length. The flowers are solitary, hermaphroditic, or polygamous, with tomentose, glandular-hairy pedicels and no external calyx. The corolla consists of 8 petals, occasionally more, which are white and deciduous. The flowering period is from June to August.[2-4]

Main phytoconstituents
Ref.
[72-74]

flavonoids (kaempferol, quercetin, avicularin)
terpenoids (oleanolic acid, ursolic acid, hederagenin)
sterols
organic acid
fatty acids
phenolic acids
saponide
sterols
carbohydrates

Medicinal Use

stomach tonic, mouth and throat inflammation (gingivitis), hemostatic[75]

Biological Activity

antibacterial, astringent, tonic digestive, antidiarrheic, anti-inflammatory[72-74]

Other Uses

■ tea substitute[72-74]

Spiraea crenata L. (Scalloped spirea)

Scientific name: *Spiraea crenata* L.
Common name: Scalloped spirea
Romanian name: Tavalgă
Taxonomic classification
Kingdom: *Plantae*
Subkingdom: *Cormobionta*
Phylum: *Spermatophyta*
Subphylum: *Magnoliophytina (Angiospermae)*
Class: *Magnoliatae (Magnoliopsida, Dicotyledonatae)*
Subclass: *Rosidae*
Order: *Rosales*
Family: *Rosaceae*
Subfamily: *Rosoideae*
Genus: *Spiraea*
Species: *crenata* L. (Figure 28.12)[3-4]

SPREAD

Rarely distributed from the lowland areas up to the beech forest zone, typically found on sunny rocks and slopes and in forest clearings. It is known to grow in several counties, including Dolj, Brașov, Cluj, Iași, and Tulcea.[3-4]

DESCRIPTION

A shrub that can reach a height of up to 1 meter. It has an erect main stem and oblique or pendulous secondary branches, with the younger ones being shortly hairy. The leaves are greenish-grey, shortly hairy when young, elongated-obovate to obovate-lanceolate, narrowing at the base, and either entire or slightly crenate. They have three nearly parallel median veins in the superior half. The flowers are long-peduncled and grouped in umbels, with rounded white petals. The fruits are follicles with an acuminate tip. It blooms from May to June.[2,4]

Figure 28.12 *Spiraea crenata* L. (Scalloped spirea).

Source: 76.

Main phytoconstituents   Ref.

terpenoid (β-amyrin, phytol, linalool, erpineol, nerolidol, β-ionone, lupeone, betulinic acid,
   taraxerol, germanicol, oleanolic acid, ursolic acid)   77-79

flavonoids (quercetin, apigenin, isorhamnetin, hyperoside, avicularin, vitexin, vicenin,
   sparin, luteolin, kaempferol, astragalin)

organic acid (ascorbic acid, succinic acid, fumaric acid)

fatty acids (linoleic acid, palmitic acid, arachidic acid, behenic acid, tricosanoic acid,
   lignoceric acid)

phenolic acids (ferulic acid, cinnamic acid, caffeic acid, coumaric acid, gentisic acid, vanilic
   acid, syringic acid)

sterols (beta-sitosterol, stigmasterol)

saponins

Medicinal Use

insomnia, toothache, stomach pains, menstrual cramps, premenstrual syndrome, hyperactivity,
neuralgia, rheumatism, cough, asthma, malaria[77-79]

Biological Activity

anti-inflammatory, antioxidant, antimicrobial, antibacterial, antitumoral, antifungal, analgesic,
neuroprotective[77-79]

Other Uses

■ ornamental[77-79]

Crataegus laevigata (Poir.) DC. (***C. oxyacantha*** auct. non L.) (English hawthorn)

Scientific name: *Crataegus laevigata* (Poir.) DC. (*C. oxyacantha* auct. non L.)
Common name: English hawthorn
Romanian name: Păducel, Mărăcine[1]
Taxonomic classification
Kingdom: *Plantae*
Subkingdom: *Cormobionta*
Phylum: *Spermatophyta*
Subphylum: *Magnoliophytina (Angiospermae)*
Class: *Magnoliatae (Magnoliopsida, Dicotyledonatae)*
Subclass: *Rosidae*
Order: *Rosales*
Family: *Rosaceae*
Subfamily: *Maloideae*
Genus: *Crataegus*
Species: *laevigata* (Poir.) DC. (Figure 28.13)[2-4]

SPREAD

It sporadically grows from the lowland areas up to the beech forest zone, particularly in forest clearings and forest edges. [2-4]

DESCRIPTION

A spiny tree or shrub that reaches a height of 2–5 meters. Its branches are initially hairy in youth, later becoming glabrous and brown, equipped with thorns measuring 0.6–1.5 cm in length, and also with small and sparse lenticels. The leaves are simple, round, obovate, rhomboidal, and broadly

Figure 28.13 *Crataegus laevigata* (Poir.) DC. (*C. oxyacantha* auct. non L.) (English hawthorn).
Source: 80.

cuneate at the base, with 3–5 short lobes. The stipules are ovate to linear-lanceolate, glabrous, and glandular-serrated. The small flowers have 2(3) styles and are grouped in multiflower, erect, and glabrous corymbs. The fruit is a red drupe, occasionally yellow or white, with 2 pits that have 2 oblique grooves, lacking a crusty covering around the pit, and elongated, flattened, and yellow-brown seeds. It blooms in the months of May to June.[1-4]

Main phytoconstituents   **Ref.**
amino acids   [81-82]
terpenoids (oleanic acid, ursolic acid, crataegolic acid)
phenolic acids (chlorogenic acid, caffeic acid)
flavonoids (rutin, vitexin, isovitexin, hyperoside, quercetin, spiraeoside, isoquercitrin)
tanins (catechin, epicatechin)
sterols
carbohydrates (glucose, sucrose, fructose, xylose)
fatty acids
miscellaneous

Crataegus pentagyna Waldst. et Kit. (Small-flowered black hawthorn)

Scientific name: *Crataegus pentagyna* Waldst. et Kit.
Common name: Small-flowered black hawthorn
Romanian name: Păducel, Gherghinar
Taxonomic classification
Kingdom: *Plantae*
Subkingdom: *Cormobionta*
Phylum: *Spermatophyta*
Subphylum: *Magnoliophytina (Angiospermae)*
Class: *Magnoliatae (Magnoliopsida, Dicotyledonatae)*
Subclass: *Rosidae*
Order: *Rosales*
Family: *Rosaceae*
Subfamily: *Maloideae*
Genus: *Crataegus*
Species: *pentagyna* Waldst. et Kit. (Figure 28.14)[2-4]

Figure 28.14 *Crataegus pentagyna* Waldst. et Kit. (Small-flowered black hawthorn).
Source: 83.

SPREAD

Sporadically found in the lowland and hilly regions, particularly in the south and east of the country, along forest edges and shrubby areas.[2-4]

DESCRIPTION

Tree or shrub with spines, 1–5 m tall, with pubescent young branches. The leaves, 3–7 lobed, are initially densely hairy on the inferior side, becoming glabrous on both sides when mature. The flowers, with a 2–3 carpel gynoecium and 2–3 styles, are grouped in a compound corymb. The pedicels are densely tomentose, the sepals very short, tomentose, folded at the tip, the petals white, and the anthers red. The fruit is a purplish-black, matte drupe with 5 seeds. Flowering occurs in May and June.[2-4]

Main phytoconstituents	Ref.
amino acids	81-82

amino acids
terpenoids (oleanic acid, ursolic acid, crataegolic acid)
phenolic acids (chlorogenic acid, caffeic acid)
flavonoids (rutin, vitexin, isovitexin, hyperoside, quercetin, isoquercitrin)
tanins (catechin, epicatechin)
sterols
carbohydrates (glucose, sucrose, fructose, xylose)
fatty acids
misceleous

Crataegus monogyna Jacq. (Hawthorn)

Scientific name: *Crataegus monogyna* Jacq.
Common name: Hawthorn, Aubépine
Romanian name: Păducel, Gherghinar[1-3]
Taxonomic classification
Kingdom: *Plantae*
Subkingdom: *Cormobionta*
Phylum: *Spermatophyta*
Subphylum: *Magnoliophytina (Angiospermae)*
Class: *Magnoliatae (Magnoliopsida, Dicotyledonatae)*
Subclass: *Rosidae*
Order: *Rosales*
Family: *Rosaceae*
Subfamily: *Maloideae*
Genus: *Crataegus*
Species: *monogyna* Jacq.
Subspecies: *subsp. azarella* (Griseb.) Franco

  subsp. brevispina (G. Kunze) Franco

  subsp. monogyna (Figure 28.15)[1-3]

SPREAD

Frequent in forest margins and shrubby areas, from the lowlands to the mountains.[2]

DESCRIPTION

Thorny shrub or small tree, 2–5(10) m tall, with young branches that are green-brown to brownish-red, glabrous or initially hairy, with rare and small lenticels. The leaves are pinnately lobed to pinnately fid, 3–9 lobed, petiolate, with subcordate to lanceolate stipules, entire, glabrous. The flowers are white, 8–15 mm in size, with a monocarpellar gynoecium, a single style, grouped in hairy corymbs. The fruit is red-purple, 6–9 mm, with reflexed sepals, with a single ovate seed, light brown in color. Blooming occurs in May to June.[1-4]

Figure 28.15 *Crataegus monogyna* L. (Hawthorn).

Source: 84.

Main phytoconstituents   Ref.

terpenoids (camphene, pinene, myrcene, terpinene, cymene, limonene, eucalytol,
   terpinolene, caryophyllene, carene, farnesene, humulene, crataegolic acid, ursolic acid,
   oleanic acid)   20,81-82, 85-89

aldehydes&ketones (butyraldehyde, valeraldehyde, hexenal, benzaldehyde,
   enanthaldehyde, hept-5-en-2-one, heptandienal, phenylacetaldehyde, pelargonaldehyde)

phenolic acids (chlorogenic acid, caffeic acid, gentisic acid, isovanillic acid)

flavonoids (quercetin, orientin, kaempferol, catechin, naringerin, myricetin, rutin, vitexine,
   hyperoside, apigenin, arbutin, hyperoside, epicatechin, rhamnoside, schaftoside,
   neoschaftoside, isoschaftoside, vicenin)

sterols (β-sitoserol, stigmasterol, ergosterol)

carbohydrates (glucose, sucrose, fructose, xylose)

fatty acids (palmitic acid, palmitoleic acid, stearic acid, oleic acid, linoleic acid, eicosenoic
   acid, behenic acid, tricosanoic acid)

amines (phenethylamine, acetylcoline, ethylamine)

others (lutidine, ethylvinyl carbinol, cyanidin-3-O-galactoside, tocopherol, ascorbic acid, carotene,
purine derivatives)

Medicinal Use

The vegetable products are represented by the inflorescences with their basal leaves, as well as
the mature fruits harvested from *Crataegus monogyna* Jacq. and *C. laevigata* (Poiret) DC., synonym
C. oxyacantha auct. non L.[20]

The vegetable product has the following actions:

- Vasodilatory, especially coronary, with favorable effects at the cardiac level;

- Direct cardiac tonic, with vigorous positive inotropic effect, by strengthening and restoring normal cardiac contractions. The vegetable product is considered a minor cardiac tonic and should not be administered in emergencies or severe, chronic cases;

- Sedative and antispasmodic, when combined with *Passiflorae herba* and *Valerianae rhizoma cum radicibus*;

- Antitumoral and diuretic (pentacyclic triterpenoid acids).

The medicinal product is practically non-toxic, and the total extracts are more active than the sum of the actions of the components (synergistic potentiation).

Romanian Pharmacopeae (F.R. X) mentions that *Crataegi folium cum flore* has sedative and bradycardic effects. It is recommended for early-stage heart failure, rhythm disturbances (palpitations), vascular spasms, coronary artery disease, myocardial disorders following infectious diseases, atherosclerosis, arterial hypertension, diarrhea, neurosis, menopause, obesity, and spasmodic conditions.[20]

Biological Activity

antioxidant, anti-inflammatory, gastroprotective, neuroprotective, diuretic, antiarrythmic, cytotoxic, anti-HIV, antimicrobial, hepatoprotective, cardio-protective, nephroprotective, antispasmodic, cardio-tonic, astringent, antihypertensive, hypoglycaemic antihyperlipidemic, immunostimulating activity, antiatherosclerotic[85-89]

Other Uses

- ornamental

- culinary (jam, syrup, jelly)

- wine (hawthorn wine) or for flavoring cognac

- dried fruits, ground for flavoring in bread and cakes.

- substitute for coffee (powder resulting from dried and ground fruits)

- substitute for green tea (leaves)

- young shoots have a walnut-like aroma, and are used raw in salads

- flowers are used in the preparation of teas, syrups, jams[85-89]

Sorbus aucuparia L. (European mountain ash)

Scientific name: *Sorbus aucuparia* L.
Common name: European mountain ash
Romanian name: scoruș de munte
Taxonomic classification
Kingdom: *Plantae*
Subkingdom: *Cormobionta*
Phylum: *Spermatophyta*
Subphylum: *Magnoliophytina (Angiospermae)*
Class: *Magnoliatae (Magnoliopsida, Dicotyledonatae)*
Subclass: *Rosidae*
Order: *Rosales*
Family: *Rosaceae*
Subfamily: *Maloideae*
Genus: *Sorbus*
Species: *aucuparia* L.
Subspecies: *subsp. aucuparia* – the buds, the inferior surface of the leaves, and the inflorescences are hairy, while the fruits are subglobose (Figure 28.16).

subsp. glabrata (Wimm. et Graebn.) – the buds, leaves, and inflorescences are glabrous, while the fruits are longer than wide[2-4]

381

Figure 28.16 *Sorbus aucuparia* L. (European mountain ash).

Source: 90.

SPREAD

A pioneer plant, frequently found from the oak forest level up to the subalpine level, in clearings and forest clearings, on rocky soils.[2-4]

DESCRIPTION

A non-spiny shrub, 5–10 meters tall, with a round crown and gray or reddish-brown branches that are tomentose in youth, becoming glabrescent, glossy, and smooth later. The rhytidome is blackish with long furrows. The leaves are odd-pinnate compound, arranged alternately, with elongated elliptical or elongated lanceolate leaflets, sessile, acute or obtuse, with a sharply serrated margin. The white flowers, with yellow stamens and 3(4) styles, are grouped in compound corymbs. The fruit is globose, red, with a diameter of 6–9 mm. It blooms from May to June.[1-4]

Main phytoconstituents   Ref.

flavonoids (rhamnetin, quercetin, astragalin, apigenin, rutin, isoquercetin, hyperoside, isorhamnetin, pelargonidin, catechin, cyanidin, sexangularetin, scutellarin, baicalin)   [91-96]

terpenoids (squalene, geranylacetone, amyrin, betulinic acid, oleanolic acid)

organic acid (malonic acid, azelaic acid, succinic acid, benzoic acid, oxalic acid. citric acid, pimelic acid)

fatty acids (myristic acid, tricosanoic acid, behenic acid, palmitic acid, lignoceric acid, palmitoleic acid, lauric acid, margaric acid)

phenolic acids (gallic acid; chlorogenic acid, caffeic acid, ferulic acid, vanillic acid, syringic acid, gentisic acid, coumaric acid)

sterols (β-sitosterol)

carbohydrates: (glucose, sorbitol, fructose)

hydrocarbons (heptacosane, nonacosane)

aldehydes & ketones (hexenal, nonanal, dodecanal, dodecan-2-one)

other (parasorbic acid, eriobofuran, noreriobofuran)

Medicinal Use

disorders of the respiratory tract (fever, infections, colds, flu), rheumatism, gout, cancer, diabetes; diuretic, cholagogue, hemostatic[91-96]

The fruits have therapeutic uses in scurvy and diarrhea, and as a local hemostatic for superficial wounds.[2]

Biological Activity

antioxidant, hepatoprotective, anti-inflammatory, antiparasitic, analgesic, antibacterial, antiviral, antitumoral, antirheumatic, anticancer, cardioprotective[91-96]

Other Uses

- culinary (jam, jellies, conserves, marmalades, vinegar, wines, spirits, confectionery, ketchup, pies, soups, turkish delight)

- coffee substitute

- woodcraft (baskets, cartwheels, kitchen utensils, crates, furniture, hoops for barrels)

- cosmetics (face-mask)

- black dye

- decorative[91-86]

REFERENCES

1. Săvulescu T. (ed). 1956. *Flora RPR*, vol. IV. Ed. Academiei RPR, București.

2. Bejenaru C., Mogoşanu G.D., Bejenaru L.E., Popescu H. 2020. *Botanică farmaceutică. Cormobionta*, ediția a III-a. Ed. Sitech, Craiova.

3. Ciocârlan V. 2009. *Flora ilustrată a României. Pteridophyta et Spermatophyta*. Ediția a 3-a revizuită și adăugită. Ed. Ceres, București.

4. Sârbu I., Ștefan N., Oprea A. 2013. *Plante vasculare din România, Determinator ilustrat de teren*. Ed. Victor B Victor, București.

5. https://dryades.units.it/dolomitifriulane/index.php?procedure=taxon_page&id=20073&num=5081

6. Boroja T., Mihailović V., Katanić J., Pan S.P., Nikles S., Imbimbo P., Monti D.M., Stanković N., Stanković M.S., Bauer R. 2018. The biological activities of roots and aerial parts of *Alchemilla vulgaris* L. *South African Journal of Botany* 116:175–184.

7. Duckstein S.M., Lotter E.M, Meyer U., Lindequist U., Stintzing F.C. 2012. Phenolic constituents from *Alchemilla vulgaris* L. and *Alchemilla mollis* (Buser) Rothm. at different dates of harvest. *Zeitschrift für Naturforschung C Journal of Biosciences* 67(11–12):529–540.

8. Renda G., Tevek F., Korkmaz B., Yayli N. 2017. Comparison of the *Alchemilla* L. samples from Turkish herbal market with the European Pharmacopoeia 8.0. *FABAD Journal of Pharmaceutical Sciences* 42(3):167–177.

9. Vlaisavljević S., Jelača S., Zengin G., Mimica-Dukić N., Berežni S., Miljić M., Stevanović Z.D. 2019. *Alchemilla vulgaris* agg. (Lady's mantle) from central Balkan: Antioxidant, anticancer and enzyme inhibition properties. *RSC Advances* 9:37474–37483.

10. https://dryades.units.it/floritaly/index.php?procedure=taxon_page&tipo=all&id=2102

11. Mladenova S.G., Vasileva L.V., Savova M.S., Marchev A.S., Tews D., Wabitsch M., Ferrante C., Orlando G., Georgiev M.I. 2021. Anti-adipogenic effect of *Alchemilla monticola* is mediated via PI3K/AKT signaling inhibition in human adipocytes. *Frontiers in Pharmacology* 12:707507.

12. Ayaz F.A., Hayirlioglu-Ayaz S., Beyazoglu O. 1999. Fatty acid composition of leaf lipids of some *Alchemilla* L (*Rosaceae*) species from Northeast Anatolia (Turkey). *Grasas y Aceites* 50(5):341–344.

13. https://dryades.units.it/floritaly/index.php?procedure=taxon_page&tipo=all&id=2103

14. Mohammadhosseini M., Sarker S.D., Akbarzadeh A. 2017. Chemical composition of the essential oils and extracts of *Achillea* species and their biological activities: A review. *Journal of Ethnopharmacology* 199:257–315.

15. Polat R., Cakilcioglu U., Kaltalioğlu K., Ulusan M.D., Türkmen Z. 2015. An ethnobotanical study on medicinal plants in Espiye and its surrounding (Giresun-Turkey). *Journal of Ethnopharmacology* 163:1–11.

16. https://dryades.units.it/floritaly/index.php?procedure=taxon_page&tipo=all&id=8653

17. Şeker Karatoprak G., Ilgün S., Koşar M. 2017. Phenolic composition, anti-Inflammatory, antioxidant, and antimicrobial activities of *Alchemilla mollis* (Buser) Rothm. *Chemistry & Biodiversity* 14(9):e1700150.

18. Kurtul E., Eryilmaz M., Sarialtin S.Y., Tekin M., Bahadir Acikara Ö., Çoban T. 2022. Bioactivities of *Alchemilla mollis*, *Alchemilla persica* and their active constituents. *Brazilian Journal of Pharmaceutical Sciences* 58:e18373.

19. Tadic V., Krgovic N., Žugic A. 2020. Lady's mantle (*Alchemilla vulgaris* L., *Rosaceae*): A review of traditional uses, phytochemical profile, and biological properties. *Natural Medicinal Materials* 40:66–74.

20. Bejenaru L.E., Mogoşanu G.D., Bejenaru C., Popescu H. 2015. *Farmacognozie-Fitoterapie.* Vol. I, ediţia a III-a. Ed. Sitech, Craiova.

21. https://dryades.units.it/dolomitibellunesi/index.php?procedure=taxon_page&id=2067&num=4175

22. Mari A., Lyon D., Fragner L., Montoro P., Piacente S., Wienkoop S., et. al. 2012. Phytochemical composition of *Potentilla anserina* L. analyzed by an integrative GC-MS and LC-MS metabolomics platform. *Metabolomics* 9(3):599–607.

23. Olennikov D.N., Kashchenko N.I., Chirikova N.K., Kuz'mina S.S. 2015. Phenolic profile of *Potentilla anserina* L. (*Rosaceae*) herb of Siberian origin and development of a rapid method for simultaneous determination of major phenolics in *P. anserina* pharmaceutical products by microcolumn RP-HPLC-UV. *Molecules* 20(1):224–248.

24. Wu J., Zhang Q., Zhou D., Yao Y., Chen L., Chu L., Yu H., Yang P., Li B., Wang W. 2022. New terpenoids from *Potentilla freyniana* Bornm. and their cytotoxic activities. *Molecules* 27(12):3665.

25. Tomczyk M., Latté K.P. 2009. *Potentilla*—A review of its phytochemical and pharmacological profile. *Journal of Ethnopharmacology* 122(2):184–204.

26. Chen J.R., Yang Z.Q., Hu T.J., Yan Z.T., Niu T.X., Wang L., Cui D.A., Wang M. 2010. Immunomodulatory activity in vitro and in vivo of polysaccharide from *Potentilla anserina*. *Fitoterapia* 81(8):1117–1124.

27. Zhao B., Zhang J., Yao J., Song S., Yin Z., Gao Q. 2013. Selenylation modification can enhance antioxidant activity of *Potentilla anserina* L. polysaccharide. *International Journal of Biological Macromolecules* 58:320–328.

28. Yang D., Wang L., Zhai J., Han N., Liu Z., Li S., Yin J. 2021. Characterization of antioxidant, α-glucosidase and tyrosinase inhibitors from the rhizomes of *Potentilla anserina* L. and their structure–activity relationship. *Food Chemistry* 336:127714.

29. Shi J., Guo J., Chen L., Ding L., Zhou H., Ding X., Zhang J. 2023. Characteristics and anti-radiation activity of different molecular weight polysaccharides from *Potentilla anserina* L. *Journal of Functional Foods* 101:105425.

30. Guo T., Wei Jun Q., Ma J.P. 2016. Antitussive and expectorant activities of *Potentilla anserina*. *Pharmaceutical Biology* 54(5):807–811.

31. Morikawa T., Imura K., Akagi Y. et al. 2018. Ellagic acid glycosides with hepatoprotective activity from traditional Tibetan medicine *Potentilla anserina*. *Journal of Natural Medicines* 72:317–325.

32. Rong S., Disheng L., Chenchen H., Di L., Lixia Z., Cheng J., Degui W., Decheng B. 2017. Protective effect of *Potentilla anserina* polysaccharide on cadmium induced nephrotoxicity in vitro and in vivo. *Food & Function* 8(10):3636–3646.

33. Shi J., Liu Z., Li M., Guo J., Chen L., Ding L., Ding X., Zhou T., Zhang J. 2021. Polysaccharide from *Potentilla anserina* L. ameliorate pulmonary edema induced by hypobaric hypoxia in rats. *Biomedicine & Pharmacotherapy* 139:111669.

34. www.webmd.com/vitamins/ai/ingredientmono-52/potentilla

35. https://dryades.units.it/ampezzosauris/index.php?procedure=taxon_page&id=2053&num=688

36. Mogoșanu G.D., Bejenaru L.E., Bejenaru C., Popescu H. 2015. *Farmacognozie-Fitoterapie*, Vol. II, ediția a III-a. Ed. Sitech, Craiova.

37. Kaltalioglu K., Balabanli B., Coskun-Cevher S. 2020. Phenolic, antioxidant, antimicrobial, and in-vivo wound healing properties of *Potentilla erecta* L. root extract in diabetic rats. *Iranian Journal of Pharmaceutical Research* 19(4):264–274.

38. Wiater R.A., Pleszczyńska M., Próchniak K.E., Szczodrak J. 2008. Anticariogenic activity of the crude ethanolic extract of *Potentilla erecta* (L.) Raeusch. *Herba Polonica* 54(2):41–45.

39. Petrović A, Madić V, Stojanović G., Zlatanović I., Zlatković B., Vasiljević P., Đorđević L. 2023. Antidiabetic effects of polyherbal mixture made of *Centaurium erythraea, Cichorium intybus* and *Potentilla erecta*. *Journal of Ethnopharmacology* 319(Pt 1):117032.

40. www.picturethisai.com/wiki/Potentilla_erecta.html

41. https://dryades.units.it/ampezzosauris/index.php?procedure=taxon_page&id=2055&num=689

42. Uysal S., Zengin G., Locatelli M., Bahadori M.B., Mocan A., Bellagamba G., et.al. 2017. Cytotoxic and enzyme inhibitory potential of two *Potentilla* species (*P. speciosa* L. and *P. reptans* Willd.) and their chemical composition. *Frontiers in Pharmacology* 8:290.

43. Tumbarski Y., Lincheva V., Petkova N., Nikolova R., Vrancheva R., Ivanov I. 2017. Antimicrobial activity of extract from aerial parts of potentilla (*Potentilla reptans* L.). *Industrial Technologies* IV(1):37–43.

44. Enayati A., Khori V., Saeedi Y., Yassa N. 2018. Antioxidant activity and cardioprotective effect of *Potentilla reptans* L. via ischemic preconditioning (IPC). Research Journal of Pharmacognosy 6(1):19–27.

45. Ahangar N., Bakhshi Jouybari H., Davoodi A., Shahani S. 2021. Phytochemical screening and antinociceptive activity of the hydroalcoholic extract of *Potentilla reptans* L. *Pharmaceutical and Biomedical Research* 7(4):271–278.

46. Radovanovic A.M., Cupara S.M., Popovic S.L., Tomovic M.T., Slavkovska V.N., Jankovic S.M. 2013. Cytotoxic effect of *Potentilla reptans* L. rhizome and aerial part extracts. *Acta Poloniae Pharmaceutica* 70(5):851–854.

47. https://pfaf.org/user/Plant.aspx?LatinName=Potentilla+reptans

48. https://dryades.units.it/torlonia/index.php?procedure=taxon_page&id=2015&num=695

49. Quang T.T., Truong Van N.T., Huu P.D., Ngoc L.H., Loan K.T.V., Minh H.D.N., Nhan T.N., Thien S.N., Hansen P.E. 2018. Chemical constituents of *Geum urbanum* L. roots. *Natural Product Research* 32(21):2529–2534.

50. Al-Snafi A.E. 2019. Constituents and pharmacology of *Geum urbanum* – A review. *IOSR Journal of Pharmacy* 9(5):28–33.

51. Bunse M., Lorenz P., Stintzing F.C., Kammerer D.R. Insight into the secondary metabolites of *Geum urbanum* L. and *Geum rivale* L. seeds (Rosaceae). *Plants* 10(6):1219.

52. https://pfaf.org/User/Plant.aspx?LatinName=Geum+urbanum

53. https://dryades.units.it/lamone/index.php?procedure=taxon_page&id=1995&num=2676

54. Karlińska E., Romanowska B., Kosmala M. The aerial parts of *Agrimonia procera* Wallr. and *Agrimonia eupatoria* L. as a source of polyphenols, and especially agrimoniin and flavonoids. *Molecules* 26(24):7706.

55. EMA/HMPC/680595/2013. Assessment Report on *Agrimonia eupatoria* L., *herba*.

56. Malheiros J., Simões D.M., Figueirinha A., Cotrim M.D., Fonseca D.A. 2022. *Agrimonia eupatoria* L.: An integrative perspective on ethnomedicinal use, phenolic composition and pharmacological activity. *Journal of Ethnopharmacology* 296:115498.

57. Muruzović M.Ž., Mladenović K.G., Stefanović O.D., Vasić S.M., Čomić L.R. 2016. Extracts of *Agrimonia eupatoria* L. as sources of biologically active compounds and evaluation of their antioxidant, antimicrobial, and antibiofilm activities. *Journal of Food and Drug Analysis* 24(3):539–547.

58. Malheiros J., Simões D.M., Figueirinha A., Cotrim M.D., Fonseca D.A. 2022. *Agrimonia eupatoria* L.: An integrative perspective on ethnomedicinal use, phenolic composition and pharmacological activity. *Journal of Ethnopharmacology* 296:115498.

59. https://dryades.units.it/lamone/index.php?procedure=taxon_page&id=1995&num=2676

60. https://dryades.units.it/floritaly/index.php?procedure=taxon_page&tipo=all&id=1972

61. Roman I., Stănilă A., Stănilă S. 2013. Bioactive compounds and antioxidant activity of *Rosa canina* L. biotypes from spontaneous flora of Transylvania. *Chemistry Central Journal* 7:73.

62. Kayahan S., Ozdemir Y., Gulbag F. 2023. Functional compounds and antioxidant activity of *Rosa species* grown in Turkey. *Erwerbs-Obstbau* 65:1079–1086.

63. Demir F., Özcan M. 2001. Chemical and technological properties of rose (*Rosa canina* L.) fruits grown wild in Turkey. *Journal of Food Engineering* 47(4):333–336.

64. Peña F., Valencia S., Tereucán G., Nahuelcura J., Cornejo P., Ruiz A. 2023. Bioactive compounds and antioxidant activity in the fruit of rosehip (*Rosa canina* L. and *Rosa rubiginosa* L.). *Molecules* 28(8):3544.

65. Guimarães R., Barros L., Carvalho A.M., Ferreira I.C.F.R. 2010. Studies on chemical constituents and bioactivity of *Rosa micrantha*: An alternative antioxidants source for food, pharmaceutical, or cosmetic applications. *Journal of Agricultural and Food Chemistry* 58(10):6277–6284.

66. Ahmad N., Anwar F., Gilani A.H. 2016. Rose hip (*Rosa canina* L.) oils. In: Preedy V.R. (ed). *Essential oils in food preservation, flavor and safety*. Academic Press–Elsevier, London, UK, pp. 667–675.

67. Öz M., Deniz I., Okan O.T., Baltaci C., Karatas S.M. 2021. Determination of the chemical composition, antioxidant and antimicrobial activities of different parts of *Rosa canina* L. and *Rosa pimpinellifolia* L. essential oils. *Journal of Essential Oil Bearing Plants* 24(3):519–537.

68. Roman I., Stănilă A., Stănilă S. 2013. Bioactive compounds and antioxidant activity of *Rosa canina* L. biotypes from spontaneous flora of Transylvania. *Chemistry Central Journal* 7:73.

69. https://pfaf.org/user/plant.aspx?LatinName=Rosa+canina

70. Govaerts R. 2020. *Dryas octopetala* L. Plants of the World Online. Board of Trustees of the Royal Botanic Gardens, Kew. Retrieved 14 December 2020.

71. https://dryades.units.it/ampezzosauris/index.php?procedure=taxon_page&id= 2009&num=4076

72. Pangon J.F., Jay M., Voirin B. 1974. Les flavonoides du *Dryas octopetala*. *Phytochemistry* 13(9):1883–1885.

73. Petelka J., Plagg B., Säumel I., Zerbe S. 2020. Traditional medicinal plants in South Tyrol (northern Italy, southern Alps): biodiversity and use. *Journal of Ethnobiology and Ethnomedicine* 16(1):74.

74. Servettaz O., Colombo M.L., De Bernardi M., Uberti E., Vidari G., Vita-Finzi P. 1984. Flavonol glycosides from *Dryas octopetala*. *Journal of Natural Products* 47(5):809–814.

75. https://pfaf.org/user/plant.aspx?latinname=Dryas+octopetala

76. https://species.wikimedia.org/wiki/Spiraea_crenata

77. Kiss T. 2017. *Phytochemical, pharmacological and toxicological studies of alkaloid- and sesquiterpene lactone-containing medicinal plants*. PhD Thesis, Department of Pharmacognosy, Faculty of Pharmacy, University of Szeged, Hungary.

78. Kiss T., Bence Cank K., Orbán-Gyapai O., Liktor-Busa E., Zomborszki Z.P., Rutkovska S., Pučka I., Németh A., Csupor D. 2017. Phytochemical and pharmacological investigation of *Spiraea chamaedryfolia*: a contribution to the chemotaxonomy of *Spiraea* genus. *BMC Research Notes* 10:762.

79. Kostikova V.A, Petrova N.V. 2021. Phytoconstituents and bioactivity of plants of the genus *Spiraea* L. (*Rosaceae*): A review. *International Journal of Molecular Sciences* 22(20):11163.

80. https://dryades.units.it/rosandra_it/index.php?procedure=taxon_page&id= 2207&num=3598

81. Edwards J.E., Brown P.N., Talent N., Dickinson T.A., Shipley P.R. 2012. A review of the chemistry of the genus *Crataegus*. *Phytochemistry* 79:5–26.

82. EMA/HMPC/159076/2014. Assessment Report on *Crataegus* spp., *folium cum flore*.

83. https://powo.science.kew.org/taxon/urn:lsid:ipni.org:names:723990-1

84. https://dryades.units.it/valerio/index.php?procedure=taxon_page&id=2208&num=2996

85. Nabavi S.F., Habtemariam S., Ahmed T., et al. 2015. Polyphenolic composition of *Crataegus monogyna* Jacq.: from chemistry to medical applications. *Nutrients* 7(9):7708–7728.

86. Nazhand A., Lucarini M., Durazzo A., Zaccardelli M., Cristarella S., Souto S.B., Silva A.M., Severino P., Souto E.B., Santini A. 2020. Hawthorn (*Crataegus* spp.): An updated overview on its beneficial properties. *Forests* 11:564.

87. Dinesh K., et al. 2012. The genus *Crataegus*: chemical and pharmacological perspectives. *Revista Brasileira de Farmacognosia* 22(5):1187–1200.

88. Abuashwashi M.A., Palomino O.M., Gómez-Serranillos M.P. 2016. Geographic origin influences the phenolic composition and antioxidant potential of wild *Crataegus monogyna* from Spain. *Pharmaceutical Biology* 54(11):2708–2713.

89. Deveci E., Tel Cayan G., Karakurt S., Duru M.E. 2020. Antioxidant, cytotoxic, and enzyme inhibitory activities of *Agropyron repens* and *Crataegus monogyna* species. *European Journal of Biology* 79(2):98–105.

90. https://dryades.units.it/ampezzosauris/index.php?procedure=taxon_page&id=2178&num=3070

91. Bobinaitė R., Grootaert C., Van Camp J., Šarkinas A., Liaudanskas M., Žvikas V., Viškelis P., Rimantas Venskutonis P. 2020. Chemical composition, antioxidant, antimicrobial and antiproliferative activities of the extracts isolated from the pomace of rowanberry (*Sorbus aucuparia* L.). *Food Research International* 136:109310.

92. Šavikin K.P., Zdunić G.M., Krstić-Milošević D.B., Šircelj H.J., Steševć D.D., Pljevljakušić D.S. 2017. *Sorbus aucuparia* and *Sorbus aria* as a source of antioxidant phenolics, tocopherols, and pigments. *Chemistry & Biodiversity* 14(12):e1700329.

93. Tovchiga O.V., Markin O.M., Shtrygol' S.Yu., Kryvoruchko O.V. 2020. The study of the chemical composition of extracts from *Sorbus aucuparia* L. leaves and their influence on glucose metabolism and the excretory renal function in rats. *Clinical Pharmacy* 24(1):26–34.

94. Olszewska M.A., Kolodziejczyk-Czepas J., Rutkowska M., Magiera A., Michel P., Rejman M.W., Nowak P., Owczarek A. 2019. The effect of standardised flower extracts of *Sorbus aucuparia* L. on proinflammatory enzymes, multiple oxidants, and oxidative/nitrative damage of human plasma components in vitro. *Oxidative Medicine and Cellular Longevity* 2019, Article ID 9746358.

95. Sołtys A., Galanty A., Podolak, I. 2020. Ethnopharmacologically important but underestimated genus *Sorbus*: a comprehensive review. *Phytochemistry Reviews* 19: 491–526.

96. Krivoruchko E., Markin A., Samoilova V., Ilina T., Koshovyi O. 2018. Research in the chemical composition of the bark of *Sorbus aucuparia*. *Ceska a Slovenska Farmacie* 67(3):113–115.

29 *Rubiaceae* Juss. Family

Cornelia Bejenaru[1], Antonia Radu[1], George Dan Mogoşanu[2],
Ludovic Everard Bejenaru[2], Andrei Biţă[2], and Adina-Elena Segneanu[3]

[1] Department of Pharmaceutical Botany, Faculty of Pharmacy, University of Medicine and Pharmacy of Craiova, 2 Petru Rareş Street, Craiova, Dolj County, Romania

[2] Department of Pharmacognosy & Phytotherapy, Faculty of Pharmacy, University of Medicine and Pharmacy of Craiova, 2 Petru Rareş Street, Craiova, Dolj County, Romania

[3] Institute for Advanced Environmental Research–West University of Timisoara (ICAM–WUT), 4 Oituz Street, Timisoara, Romania

BOTANICAL DESCRIPTION

Although predominantly woody plants (trees or shrubs) are classified in this botanical family, in Romania, the family is represented solely by herbaceous species. The leaves are simple, entire, feature interpetiolar stipules, and are arranged in opposite-decussate pairs. When the stipules resemble leaves due to their positioning between leaf petioles, they create the impression of verticils of 4–9 leaves at nodes. The flowers, grouped in cymose or racemose inflorescences, have a 4- or 5-parted perianth and can be hermaphroditic or unisexual, as well as actinomorphic or zygomorphic. The floral envelope consists of a dialysepalous calyx (which may occasionally be absent or rudimentary), and the corolla is gamopetalous, tubular, or infundibuliform. The androecium comprises stamens inserted alternately on the petals and having the same number as them. The flowers possess a nectariferous disc with 2 lobes or tubular form. The gynoecium is inferior (rarely superior or semi-inferior), bicarpellate, bilocular, containing 1–2 or more ovules. The fruit is a dicaryopsis, rarely a drupe, capsule, or berry. The seeds have a secondary endosperm.[1-4]

The *Rubiaceae* family includes 343 genera with approximately 4,500 species, widely distributed in the tropical regions.[4]

In the spontaneous flora of Romania, the following genera can be found: *Sherardia* L., *Crucianella* L., *Asperula* L., *Galium* L., and *Cruciata* Mill.[2-3]

Plants with therapeutic properties demonstrated by recent research on metabolites profile and biological activities were selected.

Galium verum L. (Yellow bedstraw)

Scientific name: *Galium verum* L.
Common name: Yellow bedstraw
Romanian name: Sânziene, Drăgaică
Taxonomic classification
Kingdom: *Plantae*
Subkingdom: *Cormobionta*
Phylum: *Spermatophyta*
Subphylum: *Magnoliophytina (Angiospermae)*
Class: *Magnoliatae (Magnoliopsida, Dicotyledonatae)*
Subclass: *Asteridae*
Order: *Rubiales*
Family: *Rubiaceae*
Genus: *Galium*
Species: *verum* L. (Figure 29.1)[1-3]

SPREAD

A frequent species in the sylvosteppe zone up to the beech forest zone, growing in meadows, pastures, shrublands, forest edges, and ruderal areas.[1-4]

DESCRIPTION

A perennial herbaceous species, ranging from 30 to 80 cm in height. The stems are erect or ascending, tetragonal, smooth, glabrous, sometimes retrorsely hairy in the inferior part. The strongly revolute leaves are ovate or elliptical, trinerved, with short cilia on the lamina margin and veins, otherwise glabrous, arranged in groups of 4 in verticils, accompanied by similar stipules, forming verticils of 8–12 elements. The flowers have glabrous, unbracteate pedicels that reflex after flowering. They are

DOI: 10.1201/9781003270515-30

Figure 29.1 *Galium verum* L. (Yellow bedstraw).

Source: 5.

yellow in color, have a honey-like scent, and are grouped in axillary cymose inflorescences. The fruit is smooth and glabrous achene. The flowering period is from June to August.[1-4]

Main phytoconstituents   Ref.
amino acids   [6-11]
flavonoids (quercetin, fisetin, chrysin astragalin, hyperoside, diosmetin, isoquercitrin, hesperetin, hispidulin, rutoside, isorhamnetin, rutin, kaempherol, luteolin)
terpenoids (ursolic acid, rubifolic acid, betulalbuside A, cyclocitral, camphor, borneol, α-copaene, α-terpineol, geraniol, β-bourbonene, bisabolene, δ-cadinene, nerolidol, squalene, germacrene D, caryophyllene linalool, eugenol, phytol)
sterols (cholesterol, campesterol, stigmasterol, sitosterol, avenasterol)
fatty acids (palmitic acid, stearic acid, myristic acid)
phenolic acids (ferulic acid, caffeic acid, rosmarinic acid, p-coumaric acid)
tannins (catechin, epicatechin)
organic acids
anthraquinones (physcion)
glycosides
esters
irinoids (asperuloside, monotropein)
hydrocarbons
aldehydes&ketones
miscellaneous (chalcone, carotenoids)

Medicinal Use

Galii veri herba is recommended as: sudorific, diuretic, antispasmodic, diaphoretic, depurative, astringent, hemostatic, sedative, tissue regeneration, skin diseases (rash, wounds, boils, panaritium, keratosis, erysipelas), biliary diseases, kidney diseases (kidney stones, prostatitis, urinary retention).[8,12-13]

Biological Activity

■ antioxidant, antitumor, antibacterial, hepato-protector, antimicrobial, antifungal, immunomodulatory, antihaemolytic[12]

Other Uses

■ cosmetics

■ ornamental

■ cheese preparation.[12]

Asperula cynanchica L. (Squinancywort)

Scientific name: *Asperula cynanchica* L.
Common name: Squinancywort
Romanian name:
Taxonomic classification
Kingdom: *Plantae*
Subkingdom: *Cormobionta*
Phylum: *Spermatophyta*
Subphylum: *Magnoliophytina (Angiospermae)*
Class: *Magnoliatae (Magnoliopsida, Dicotyledonatae)*
Subclass: *Asteridae*
Order: *Rubiales*
Family: *Rubiaceae*
Genus: *Asperula*
Species: *cynanchica* L. (Figure 29.2)[2-3]

SPREAD

This species is commonly distributed from the sylvosteppe zone up to the oak forest zone, growing in dry meadows, rocky areas, and rocky slopes.[2,3]

DESCRIPTION

A perennial herbaceous plant with a slender, multicapitate rhizome and numerous procumbent or ascending, glabrous, branched stems, reaching a height of 20–40 cm. The plant produces sterile shoots at the time of flowering. The leaves are linear, narrow, flat, distinctly revolute, glabrous, with an aristate-acuminate apex and a width of 0.8–1.5 mm. The superior leaves are shorter, grouped in pairs on the stem, while the others are arranged in vericils of four. The flowers are grouped in terminal paucifloral cymes, with a glabrous, sometimes densely hairy corolla that is white on the internal side and pink on the external side, measuring 2–2.5 mm in length. The fruits are small and dispersely granulose-scabrous. The flowering period is from June to July.[2-4]

Main phytoconstituents	Ref.
amino acids	15-18,19-27

flavonoids
terpenoids (cyclocitral, α-copaene, α- terpineol, geraniol, β-bourbonene, bisabolene, δ-cadinene, nerolidol, squalene, germacrene D, linalool, eugenol, phytol, caryophyllene
sterols
fatty acids
phenolic acids
irinoids (asperuloside)
tannins
aldehydes & ketones

Figure 29.2 *Asperula cynanchica* L. (Squinancywort).
Source: 14.

Medicinal Uses

sore throat, cough, fever[21]

Biological Activity

antioxidant, anti-inflammatory, analgesic, antibacterial, mutagenic, antiviral[16-18]

Other Uses

- ornamental

- dye (roots)

- drinks[17,22]

Cruciata laevipes Opiz (*Galium cruciata* (L.) Scop.) (Crosswort)

Scientific name: *Cruciata laevipes* Opiz
Common name: Crosswort, Smooth bedstraw
Romanian name: Smântânică
Taxonomic classification
Kingdom: *Plantae*
Subkingdom: *Cormobionta*
Phylum: *Spermatophyta*
Subphylum: *Magnoliophytina (Angiospermae)*
Class: *Magnoliatae (Magnoliopsida, Dicotyledonatae)*
Subclass: *Asteridae*
Order: *Rubiales*
Family: *Rubiaceae*

Figure 29.3 *Cruciata laevipes* Opiz (*Galium cruciata* (L.) Scop.) (Crosswort).

Source: 28.

Genus: *Cruciata*
Species: *laevipes* Opiz (Figure 29.3)[2-3]

SPREAD

The plant is commonly distributed from the plains up to the beech forest zone, growing in meadows, shrublands, grassy areas, and forest edges.[2-4]

DESCRIPTION

A perennial herbaceous plant, reaching a height of 20–60 cm, with an ascending or erect stem, tetrahedral, patent-hairy, with 1–2 mm long hairs. The leaves are ovate-elongated, acute, trinerved, pubescent, yellowish-green in color, arranged in groups of 4 in verticils. The flowers have bracteate, hairy pedicels, are yellow in color, and are grouped in axillary cymes. The fruits are rough and glabrous. It blooms from April to May.[2-4]

Main phytoconstituents	Ref
amino acids	7,29-30

glycosides
coumarins (daphnin, daphnetin)
flavonoids
terpenoids (borneol,verbenone, α-copaene, α-terpineol, geraniol, δ-cadinene, eugenol, farnesyl, squalene, p-cymene, nerol, thymol, germacrene D, ionone, linalool, humulene, phytol, caryophyllene)
sterols
fatty acids
phenolic acids

iridoids (deacetylasperulosidic acid, scandoside, asperuloside, asperulosidic acid, daphylloside, geniposidic acid, 10-hydroxyloganin, deacetylasperuloside, iridoid V3)
hydrocarbons
esters
aldehydes

Medicinal Uses

■ edema, rheumatism, headaches, wound healing[7]

Biological Activity

■ anti-inflamatory, sedative, antimicrobial, antioxidant[7,29-30]

Other Uses

■ culinary (salads, dishes similar to spinach)[31-32]

REFERENCES

1. Bejenaru C., Mogoşanu G.D., Bejenaru L.E., Popescu H. 2020. Botanică farmaceutică. Cormobionta, ediția a III-a. Ed. Sitech, Craiova.

2. Ciocârlan V. 2009. *Flora ilustrată a României. Pteridophyta et Spermatophyta.* Ediția a 3-a revizuită și adăugită. Ed. Ceres, București.

3. Sârbu I., Ştefan N., Oprea A. 2013. *Plante vasculare din România, Determinator ilustrat de teren.* Ed. Victor B Victor, București.

4. Săvulescu T. (ed). 1961. *Flora RPR*, vol. VIII. Ed. Academiei RPR, București.

5. https://dryades.units.it/floritaly/index.php?procedure=taxon_page&tipo=all&id=4125

6. Zaichikova S.G., Bokov D.O., Kiselevskii M.V., Antsyshkina A.M., Bondar A.A., Prostodusheva T.V., Shchepochkina O.Yu., Gegechkori V.I. 2020. Determination of the chemical composition of Lady's Bedstraw (*Galium verum* L.) herb extract by GC-MS. *Pharmacognosy Journal* 12(4):857–863.

7. Tava A., Biazzi E., Ronga D., Avato P. 2020. Identification of the volatile components of *Galium verum* L. and *Cruciata leavipes* Opiz from the Western Italian Alps. *Molecules* 25(10):2333.

8. Bradic J., Petkovic A., Tomovic M. 2018. Phytochemical and pharmacological properties of some species of the genus *Galium* L. (*Galium verum* and *Mollugo*). *Serbian Journal of Experimental and Clinical Research* 22(3):1–7.

9. Farcas A.D., Mot A.C., Zagrean-Tuza C., Toma V., Cimpoiu C., Hosu A., Parvu M., Roman I., Silaghi-Dumitrescu R. 2018. Chemo-mapping and biochemical-modulatory and antioxidant/prooxidant effect of *Galium verum* extract during acute restraint and dark stress in female rats. *PLoS One* 13(7):e0200022.

10. Ilina T., Kashpur N., Granica S., Bazylko A., Shinkovenko I., Kovalyova A., Goryacha O., Koshovyi O. 2019. Phytochemical profiles and *in vitro* immunomodulatory activity of ethanolic extracts from *Galium aparine* L. *Plants (Basel)* 8(12):541.

11. Al-Snafi A.E. 2018. *Galium verum* – A review. *Indo American Journal of Pharmaceutical Sciences* 05(04):2142–2149.

12. Turcov D., Barna A.S., Trifan A., Blaga A.C., Tanasa A.M., Suteu D. 2022. Antioxidants from *Galium verum* as ingredients for the design of new dermatocosmetic products. *Plants (Basel)* 11(19):2454.

13. https://www.sfatulmedicului.ro/plante-medicinale/sanzienele-galbene-galium-verum_14574

14. https://dryades.units.it/floritaly/index.php?procedure=taxon_page&tipo=all&id=4105

15. Oroian S., Sămărghiţan M. 2018. Flora from Fărăgău Area (Mureş County) as potential source of medicinal plants. *Acta Biologica Marisiensis* 1(1):60–70.

16. Minareci E., Ergönül B., Kayalar H., Kalyoncu F. 2011. Chemical compositions and antioxidant activities of five endemic *Asperula* taxa. *Archives of Biological Sciences* 63(3):537–543.

17. Martins D., Nunez C. 2015. Secondary metabolites from *Rubiaceae* species. *Molecules* 20(7):13422–13495.

18. Kooiman P. 1969. The occurrence of asperulosidic glycosides in the *Rubia*. *Acta Botanica Neerlandica* 18(1):124–137.

19. www.muntesiflori.ro/lista-rubiaceae/

20. https://eunis.eea.europa.eu/species/181904

21. www.plantlife.org.uk/uk/discover-wild-plants-nature/plant-fungi-species/squinancywort

22. www.picturethisai.com/wiki/Asperula_cynanchica.html

23. https://pfaf.org/user/Plant.aspx?LatinName=Asperula+cynanchica

24. www.maltawildplants.com/RUBI/Asperula_cynanchica.php

25. https://luontoportti.com/en/t/1834/dyers-woodruff

26. www.missouribotanicalgarden.org/PlantFinder/PlantFinderDetails.aspx?taxonid=366920&isprofile=0&%22

27. www.uksouthwest.net/wildflowers/rubiaceae/asperula-cynanchica.html

28. https://dryades.units.it/floritaly/index.php?procedure=taxon_page&tipo=all&id=4180

29. Il'ina T.V., Kovaleva A.M., Goryachaya, O.V., Vinogradov B.A. 2013. Terpenoids and aromatic compounds from essential oils of *Cruciata laevipes* and *C. glabra*. *Chemistry of Natural Compounds* 48(6):1106–1108.

30. Mitova Iv.M., Anchev M.E., Panev S.G., Handjieva N.V., Popov S.S. 1996. Coumarins and iridoids from *Crucianella graeca*, *Cruciata glabra*, *Cruciata laevipes* and *Cruciata pedemontana* (*Rubiaceae*). *Zeitschrift für Naturforschung C* 51(9–10):631–634.

31. www.first-nature.com/flowers/cruciata-laevipes.php

32. https://uk.inaturalist.org/taxa/132627-Cruciata-laevipes

30 *Saxifragaceae* Juss. Family

Cornelia Bejenaru[1], Antonia Radu[1], George Dan Mogoşanu[2],
Ludovic Everard Bejenaru[2], Andrei Biţă[2], and Adina-Elena Segneanu
[1] Department of Pharmaceutical Botany, Faculty of Pharmacy, University of Medicine and Pharmacy of Craiova, 2 Petru Rareş Street, Craiova, Dolj County, Romania
[2] Department of Pharmacognosy & Phytotherapy, Faculty of Pharmacy, University of Medicine and Pharmacy of Craiova, 2 Petru Rareş Street, Craiova, Dolj County, Romania
[3] Institute for Advanced Environmental Research–West University of Timisoara (ICAM–WUT), 4 Oituz Street, Timisoara, Romania

BOTANICAL DESCRIPTION

Herbaceous or woody plants with leaves of various forms, non-stipulate, arranged alternately or in a basal rosette. The flowers are hermaphroditic, actinomorphic, generally of type 5, with a double floral envelope and a bicarpellary syncarpous gynoecium. The fruit is a capsule, with numerous small albuminous seeds, having a small straight embryo and a seed coat that is finely prickly, granulated, or reticulated.[1-3]

In Romania, the *Saxifragaceae* family includes 5 genera with species found in the spontaneous flora: *Saxifraga, Chrysosplenium, Astilbe, Bergenia*, and *Heuchera*.[2,4]

Saxifraga L.

This genus comprises over 400 species of perennial herbaceous plants, with or without stolons. The basal leaves are arranged in a rosette. The flowers are arranged in racemose inflorescences (occasionally solitary), and they can be red, yellow, white, or green. In Romania, there are 23 spontaneous species spread from the hilly region up to the alpine zone[1]: *Saxifraga aizoides* L., *Saxifraga oppositifolia* L., *Saxifraga retusa* Gouan (*S. baumgarteni* Schott), *Saxifraga carpathica* Sternb. (*S. carpathica* Rchb.), *Saxifraga bulbifera* L. (*S. russii* J. Presl et C. Presl), *Saxifraga cernua* L., *Saxifraga cymbalaria* L., *Saxifraga rotundifolia* L., *Saxifraga marginata* Sternb. (*S. rocheliana* Sternb.), *Saxifraga paniculata* Mill. (*S. aizoon* Jacq.), *Saxifraga corymbosa* Boiss. (*S. luteo-viridis* Schott et Kotschy), *Saxifraga mutata* L., *Saxifraga demissa* Schott et Kotschy, *Saxifraga tridactylites* L., *Saxifraga adscendens* L., *Saxifraga hieracifolia* Waldst. et Kit. ex Willd., *Saxifraga bryoides* L., *Saxifraga hirculus* L., *Saxifraga pedemontana* All. subsp. *cymosa* Engler, *Saxifraga moschata* Wulfen, *Saxifraga androsacea* L., *Saxifraga cuneifolia* L., and *Saxifraga stellaris* L.[2,4]

Saxifraga aizoides L. (Yellow mountain saxifrage)

Scientific name: *Saxifraga aizoides* L.
Common name: Yellow mountain saxifrage
Romanian name: Ochii şoricelului[5]
Taxonomic classification
Kingdom: *Plantae*
Subkingdom: *Cormobionta*
Phylum: *Spermatophyta*
Subphylum: *Magnoliophytina (Angiospermae)*
Class: *Magnoliatae (Magnoliopsida, Dicotyledonatae)*
Subclass: *Rosidae*
Order: *Saxifragales*
Family: *Saxifragaceae*
Genus: *Saxifraga*
Species: *aizoides* L. (Figure 30.1)[2-3]

SPREAD

The species is commonly found from the beech zone up to the alpine zone, in rocky and moist herbaceous places, near springs and streams, and on gravelly soils.[2-4]

DESCRIPTION

It is a perennial plant with thin, horizontal rhizomes and branched, leafy, ascending, or prostrate shoots. It has a flowering stem ranging from 3 to 20 (30) cm in length, erect or ascending, glabrous

DOI: 10.1201/9781003270515-31

Figure 30.1 *Saxifraga aizoides* L. (Yellow mountain saxifrage).

Source: 6.

or slightly rough and short-hairy, and branched in the upper part. The upper leaves are linear or linear-elongated, simple, fleshy, glabrous, with an entire margin, rigidly ciliate or distantly dentate, mucronate, flat on the upper surface, and convex on the lower surface. The flowers, numbering 2 to 12 (15), are grouped in a raceme, with densely and finely hairy pedicels. The calyx has glabrous sepals, and the corolla is composed of yellow petals, sometimes with orange dots, while the androecium has the same number of stamens as petals. The fruits are ovate-rounded capsules containing brown, elongated-fusiform, weakly papillose seeds. This species is frequently found throughout the entire Carpathian Mountains range.[3]

Main phytoconstituents   Ref
                                                              7-9
coumarin (bergenin, umbelliferone**)**
flavonoids (quercetin, naringenin, tricin, luteolin, myricetin, eriodictyol, taxifolin, rutin)
sterols (sitosterol)
diarylheptanoids (hirsutenone)
fatty acids (palmitic acid)
tanins (gallocatechin, catechin, epigallocatechin)
phenolic acids (gallic acid)
sterols (sitosterol)
misceleous (tyrosol)

Medicinal Use

Its name derives from the Latin word "saxifraga" that means "stone-breaker," which is traditionally used to indicate a medicinal use for treatment of urinary calculi.[10]

Biological Activity

The anti-proliferative role of *Saxifraga aizoides* extract in human ovarian, colon, prostate and breast cancer, evaluating its toxicity and anti-inflammatory activity, has been demostrated.[10]

Saxifraga carpathica Sternb.

Scientific name: *Saxifraga carpathica* Sternb.[2] (*S. carpathica* Rchb.)[4]
Common name: -
Romanian name: Iarba surzilor[11]
Taxonomic classification
Kingdom: *Plantae*

397

Figure 30.2 *Saxifraga carpathica* Sternb.

Source: 12.

Subkingdom: *Cormobionta*
Phylum: *Spermatophyta*
Subphylum: *Magnoliophytina (Angiospermae)*
Class: *Magnoliatae (Magnoliopsida, Dicotyledonatae)*
Subclass: *Rosidae*
Order: *Saxifragales*
Family: *Saxifragaceae*
Genus: *Saxifraga*
Species: *carpathica* Sternb. (Figure 30.2)[2,4]

SPREAD

Rare species in the alpine zone through meadows, on rocky places, on wet soil. In our country, it can be found in the Făgăraș Mountains, Piatra Craiului, Țibleș, Rodna, Călimani, Bucegi, and Retezat.[2-4]

DESCRIPTION

Perennial plant with erect or ascending, slender, loosely foliated stems, sparsely hairy or glabrous at the inferior part, reaching a height of 5–15 (20) cm. The basal and inferior stem leaves are glabrescent, rounded cordate or reniform, palmately lobed, with 5–9 ovate-triangular or subrounded, obtuse, or acute lobes. The leaves have long petioles that are expanding towards the base. Middle leaves have shorter petioles, palmately lobed, with 3-5 more acute lobes, while superior leaves are entire, lanceolate, or ovate. Basal leaves have axillary bulblets, while stem leaves do not. The flowers, with short pedicels that are densely glandular-hairy, are grouped in terminal clusters of 1–3 (5) at the tips of the stems. The calyx has elongated, obtuse lobes, with the base and edges sparsely hairy. The corolla consists of petals measuring 5–7 mm, elongated-obovate, with rounded tips, trinervate, shortly unguiculate, twice as long as the calyx, and ranging in color from white to pinkish. The gynoecium has a superiorly positioned ovoid ovary. The fruits are ovate capsules with divaricate styles and contain elongated, blackish seeds. It blooms from July to August.[2-4]

Main phytoconstituents Ref.
alkaloids [7-9]
coumarin (bergenin, umbelliferone)
flavonoids (quercetin, naringenin, tricin, luteolin, myricetin, eriodictyol, taxifolin, rutin)
sterols (sitosterol)
fatty acids (palmitic acid)
tanins (gallocatechin, catechin, epigallocatechin)
phenolic acids (gallic acid)
sterols (sitosterol)
miscellaneous (tyrosol)

Medicinal Use

urinary disorders[13]

Biological Activity

anti-inflamatory, antibacterial, antioxidant[13]

Saxifraga bulbifera L.

Scientific name: *Saxifraga bulbifera* L. (*S. russii* J. Presl et C. Presl)
Common name: -
Romanian name: Floare buboasă[5]
Taxonomic classification
Kingdom: *Plantae*
Subkingdom: *Cormobionta*
Phylum: *Spermatophyta*
Subphylum: *Magnoliophytina (Angiospermae)*
Class: *Magnoliatae (Magnoliopsida, Dicotyledonatae)*
Subclass: *Rosidae*
Order: *Saxifragales*
Family: *Saxifragaceae*
Genus: *Saxifraga*
Species: *bulbifera* L. (Figure 30.3)[2,4]

SPREAD

The species is sporadically distributed from the oak forest zone up to the beech zone, through meadows and forest clearings.[2-4]

DESCRIPTION

Perennial species with simple stems, that are strictly erect, foliate all the way to the tip, with branched, densely and shortly glandular-hairy cymose corymbs at the top, reaching a height of 15–50 cm. Basal leaves are long-petioled, glandular-pubescent, subround-reniform, lobate-crenate, with larger, obtuse terminal lobes and a cordate base. Inferior stem leaves are short-petioled, cuneate-ovate, lobate, while superior stem leaves are sessile, entire, linear-lanceolate. The leaves have axillary fleshy, reddish, ovate-acute bulblets. The flowers are grouped in terminal inflorescences, densely and shortly glandular-hairy, pauciflorus, with 3–7 flowers, pedicellate, arranged in an umbelliferous fashion. The flowers have a dense and shortly glandular calyx with acute, ovate lobes, and a corolla with elongated-obovate petals, entire tips, white in color, trinervate, non-unguiculate, narrowed at the base, 2–3 times longer than the calyx, and glandular-hairy on the superior surface. The gynoecium has an ovoid semi-inferior ovary. The fruits are subround capsules with divaricate styles and erect sepals. The seeds are blackish, long, and finely papillose. The flowering period is from June to July.[2-4]

Figure 30.3 *Saxifraga bulbifera* L.

Source: 14.

Main phytoconstituents Ref.
alkaloids 7-9
coumarin (bergenin, umbelliferone)
flavonoids (quercetin, naringenin, tricin, luteolin, myricetin, eriodictyol, taxifolin, rutin)
sterols (sitosterol)
fatty acids (palmitic acid)
tanins (gallocatechin, catechin, epigallocatechin)
phenolic acids (gallic acid)
sterols (sitosterol)
miscellaneous (tyrosol)

Medicinal Use

urinary disorders[13]

Biological Activity

anti-inflamatory, antibacterial, antioxidant[13]

Chrysosplenium alternifolium L. (Golden Saxifrage)

Scientific name: *Chrysosplenium alternifolium* L.
Common name: Golden saxifrage
Romanian name: Splină
Taxonomic classification
Kingdom: *Plantae*
Subkingdom: *Cormobionta*
Phylum: *Spermatophyta*

Figure 30.4 *Chrysosplenium alternifolium* L. (Golden saxifrage).

Source: 15.

Subphylum: *Magnoliophytina (Angiospermae)*
Class: *Magnoliatae (Magnoliopsida, Dicotyledonatae)*
Subclass: *Rosidae*
Order: *Saxifragales*
Family: *Saxifragaceae*
Genus: *Chrysosplenium*
Species: *alternifolium* L. (Figure 30.4)[1-2,4]

SPREAD
Frequent in mountainous regions, in shady and moist places, floodplains, forests, and along stream banks.[1,3]

DESCRIPTION
It is a delicate plant (5–20 cm), succulent, with thin, creeping, brown-colored rhizomes, and stolons covered in scales. The stems are three-edged, erect, fleshy, and glabrous, with white or reddish hairs on the lower part, and the leaves are arranged alternately. The basal rosette leaves are reniform-cordate with deep crenations. The flowers are of type 4, apetalous, and arranged in flat cymose inflorescences. The fruits are capsules that open into two valves, releasing smooth, glossy, oval-elongated glabrous seeds. The plant blooms from April to June.[1,3]

Main phytoconstituents
alkaloids
coumarin
flavonoids (quercetin, kaempferol, chrysosplenol B)
sterols (sitosterol, sigmaterol, campesterol)
fatty acids (lignoceric acid)
tanins
phenolic acids
ecdysteroids

Ref.
16-19

Medicinal Use

In traditional medicine, the species has been utilized for treating different hepatobiliary disorders, including cholecystitis, cholelithiasis, acute icteric hepatitis, and acute liver necrosis.[18]

Biological Activity

Numerous studies have demonstrated that extracts and isolated components from *Chrysosplenium* possess diverse pharmacological activities, including antitumor, antibacterial, antiviral, hepatoprotective, and insecticidal properties.[18]

Other Uses

■ raw, young spring leaves are used as food ingredients and incorporated into salads[19]

REFERENCES

1. Bejenaru C., Mogoșanu D.G., Bejenaru L.E., Popescu H. 2020. *Botanică farmaceutică. Cormobionta*, ediția a III-a. Ed. Sitech, Craiova.

2. Ciocârlan V. 2009. *Flora ilustrată a României. Pteridophyta et Spermatophyta*. Ediția a 3-a revizuită și adăugită. Ed. Ceres, București.

3. Săvulescu T. (ed). 1956. *Flora RPR*, vol. IV. Ed. Academiei RPR, București.

4. Sârbu I., Ștefan N., Oprea A. 2013. *Plante vasculare din România, Determinator ilustrat de teren*. Ed. Victor B Victor, București.

5. Drăgulescu C. 2018. *Dicționar de fitonime românești*. Ediția a 5-a. Ed. Universității "Lucian Blaga", Sibiu.

6. www.monaconatureencyclopedia.com/saxifraga-aizoides/?lang=en

7. Chen Z., Liu Y.-M., Yang, S., Song, B.-A., Xu, G.-F., Bhadury, P. S., et al. 2008. Studies on the chemical constituents and anticancer activity of *Saxifraga stolonifera* (L) Meeb. *Bioorganic & Medicinal Chemistry* 16(3):1337–1344.

8. Yang A., Qi G., Zheng Z., Wu R., Zhang F., Li C., Han N., Shang Q. 2018. Chemical constituents of *Saxifraga umbellulata*. *Chemistry of Natural Compounds* 54(4):757–759.

9. Badral D., Odonbayar B., Murata T., Munkhjargal, T., Tuvshintulga B., Igarashi I., et al. 2017. Flavonoid and galloyl glycosides isolated from *Saxifraga spinulosa* and their antioxidative and inhibitory activities against species that cause piroplasmosis. *Journal of Natural Products* 80(9):2416–2423.

10. Pezzani R. 2016. *Saxifraga aizoides* extract: novel potential effects on tumor cell models. *Research Ideas and Outcomes* 2:e9632.

11. https://ziardebusteni.ro/2014/10/08/flori-in-bucegi-sit-natura-2000-partea-a-ii-a/

12. https://ceb.wikipedia.org/wiki/Saxifraga_carpatica

13. www.dreamstime.com/white-saxifraga-flowers-white-saxifraga-flowers-against-void-backdrop-flowers-have-medicinal-used-to-dissolve-kidney-image190574695

14. https://dryades.units.it/floritaly/index.php?procedure=taxon_page&tipo=all&id=1802

15. www.flickr.com/photos/khianti/6975469984

16. Gudej J., Czapski P. 2009. Components of the petroleum ether and chloroform extracts of *Chrysosplenium alternifolium*. *Chemistry of Natural Compounds* 45(5):717–719.

17. Lafont R., Harmatha J., Marion-Poll F., Dinan L., Wilson I.D. 2007. Compilation of the literature reports for the screening of vascular plants, algae, fungi and nonarthropod invertebrates for the presence of ecdysteroids In: *The Ecdysone Handbook,* 3rd edition. Cybersales, Prague, Czech Republic.

18. Zhao J., Qiu X., Zhao Y., Wu R., Wei P., Tao C., Wan L. 2022. A review of the genus *Chrysosplenium* as a traditional Tibetan medicine and its preparations. *Journal of Ethnopharmacology* 290:115042.

19. Olszewska M.A., Gudej J. 2009. Quality evaluation of golden saxifrage (*Chrysosplenium alternifolium* L.) through simultaneous determination of four bioactive flavonoids by high-performance liquid chromatography with PDA detection. *Journal of Pharmaceutical and Biomedical Analysis* 50(5):771–7.

31 *Scrophulariaceae* Juss. Family

Cornelia Bejenaru[1], Antonia Radu[1], George Dan Mogoşanu[2],
Ludovic Everard Bejenaru[2], Andrei Biţă[2], and Adina-Elena Segneanu[3]

[1] Department of Pharmaceutical Botany, Faculty of Pharmacy, University of Medicine and Pharmacy of Craiova, 2 Petru Rareş Street, Craiova, Dolj County, Romania
[2] Department of Pharmacognosy & Phytotherapy, Faculty of Pharmacy, University of Medicine and Pharmacy of Craiova, 2 Petru Rareş Street, Craiova, Dolj County, Romania
[3] Institute for Advanced Environmental Research–West University of Timisoara (ICAM–WUT), 4 Oituz Street, Timisoara, Romania

BOTANICAL DESCRIPTION

This family encompasses annual, biennial, and perennial herbaceous plants (less commonly, woody plants), with a simple or branched stem. Some are parasitic or semi-parasitic on superior plants. The leaves are simple, without stipules, either entire or divided, arranged alternately or oppositely, with hairy venation and indumentum. The indumentum, which can consist of simple or multicellular hairs, branched or unbranched, glandular or non-glandular, may appear star-like. From the perspective of gynoecium and fruit composition, the family is homogeneous, but the arrangement of the corolla and androecium is nonhomogeneous. The flowers are hermaphroditic, of the 5-type, forming racemose inflorescences, occasionally solitary. The floral envelope is both gamosepalous and gamopetalous, with a persistent calyx, 5- or 4-laciniate (*Veronica, Euphrasia*). The corolla is zygomorphic, deciduous, 5-lobed or 5-parted, rarely 4-parted due to the fusion of the two posterior lobes (*Veronica*), typically forming two labia (superior labium with 2 petals, inferior labium with 3), and it can be tubular, campanulate, or urceolate. The androecium consists of 4 stamens (didynamous), occasionally 5 or 2, with bilocular anthers that dehisce longitudinally or transversely. The gynoecium is superior, with a bicarpellary and bilocular ovary, bearing numerous anatropous ovules. It features a hypogynous nectariferous disc, either annular, fringed, lobed, or entire. The fruit is a capsule or, more rarely, a berry. The seeds are numerous, small, with fleshy endosperm. The family comprises 200 genera with approximately 3000 species distributed worldwide.[1-2]

The family is divided into 3 subfamilies:

- The subfamily *Verbascoideae* includes the genus *Verbascum*.

- The subfamily *Scrophularioideae* (*Antirrhinoideae*) encompasses the following genera: *Linaria* Mill., *Antirrhium* L., *Misopates* Rafin., *Cymbalaria* Hill, *Kickxia* Dumort., *Scrophularia* L., *Mimulus* L., *Gratiola* L., *Limosella* L., *Lindernia* All., *Veronica* L., and *Digitalis* L.

- The subfamily *Rhinanthoideae* comprises the following genera: *Bartsia* L., *Tozzia* L., *Euphrasia* L., *Pedicularis* L., *Rhinanthus* L., *Melampyrum* L., *Lathraea* L., *Pestemon* Schmidel, and *Bellardia* All.[3-4]

Verbascum L. (Mullein)

This genus comprises robust biennial or perennial herbaceous plants, with a taproot, simple or branched stems, erect and covered with multi-layered tector hairs. The plants have large basal leaves arranged in a rosette, and smaller stem leaves that are ovate or elliptical, arranged alternately. The flowers have a deeply 5-parted calyx with approximately equal laciniae, a yellow or violet corolla, slightly zygomorphic, rotated and with a short tube, and 5 stamens with lanate-hairy filaments (sometimes the two inferior ones are glabrous). The flowers form terminal racemose inflorescences. The fruit is an ovoid septicidal capsule.[1-2]

The genus *Verbascum* is represented in the spontaneous flora of Romania by 17 species: *Verbascum blattaria* L., *Verbascum purpureum* (Janka) Huber-Morath (*V. glanduligerum* Velen.), *Verbascum ovalifolium* Donn, *Verbascum phoeniceum* L., *Verbascum phlomoides* L., *Verbascum thapsus* L., *Verbascum crassifolium* Lam. (*V. thapsus* L. subsp. *crassifolium* (Lam.)), *Verbascum vandasii* (Rohlena) Rohlena (*V. heuffelii* Neilr.), *Verbascum densiflorum* Bertol. (*V. thapsiforme* Schrad.), *Verbascum banaticum* Rochel, *Verbascum alpinum* Turra (*V. lanatum* Schrad.), *Verbascum glabratum* Friv., *Verbascum nigrum* L., *Verbascum chaixii* Vill., *Verbascum lychnitis* L., *Verbascum pulverulentum* Vill., and *Verbascum speciosum* Schrad.[3-4]

DOI: 10.1201/9781003270515-32

Figure 31.1 *Verbascum phlomoides* L. (Orange mullein).

Source: 5.

Verbascum phlomoides L. (Orange mullein)

Scientific name: *Verbascum phlomoides* L.
Common name: Orange mullein
Romanian name: Lumânărică, Coada vacii
Taxonomic classification
Kingdom: *Plantae*
Subkingdom: *Cormobionta*
Phylum: *Spermatophyta*
Subphylum: *Magnoliophytina (Angiospermae)*
Class: *Magnoliatae (Magnoliopsida, Dicotyledonatae)*
Subclass: *Asteridae*
Order: *Scrophulariales (Solanales)*
Family: *Scrophulariaceae*
Subfamily: *Verbascoideae*
Genus: *Verbascum*
Species: *phlomoides* L. (Figure 31.1)[1,3-4]

SPREAD

The species is commonly found from the steppe zone up to the beech forest level, in sunny and dry places, shrubs, meadows, cultivated fields, uncultivated areas, ruderal places, gravelly areas, rocky slopes, forest clearings, and streamside woods.[2-4]

DESCRIPTION

A tall herbaceous plant, reaching 1.5 meters in height, this biennial species has an erect, cylindrical, simple stem that is leafy and densely hairy, with rough indumentum, covered in long, branching, gray or grayish-yellow hairs. All its leaves are gray-green or greenish-yellow, with a velvety tomentum on the inferior side and less tomentose on the superior side. The basal leaves are elongated-elliptical, with a cuneate base and petiolate. The inferior stem leaves are shortly petiolate or sessile, elongated obovate, sharp, crenate, or crenate-dentate. The middle leaves are sessile, broadly or narrowly ovate, acute, crenate-dentate, not or slightly decurrent, and the superior leaves are broadly ovate, long-cuspidate, denticulate, fused with the stem. The yellow flowers are grouped in pairs of 2 to 7 in the axils of the bracts. The superior stamens are hairy and white or yellow, the inferior ones are glabrous, and the anthers are irregular. It blooms from June to August.[1-4]

Verbascum thapsus L. (Common mullein)[6]

Scientific name: *Verbascum thapsus* L.
Common name: Common mullein
Romanian name: Lumânarea râului,[6] Corovatic[2]
Taxonomic classification
Kingdom: *Plantae*
Subkingdom: *Cormobionta*
Phylum: *Spermatophyta*
Subphylum: *Magnoliophytina (Angiospermae)*
Class: *Magnoliatae (Magnoliopsida, Dicotyledonatae)*
Subclass: *Asteridae*
Order: *Scrophulariales (Solanales)*
Family: *Scrophulariaceae*
Subfamily: *Verbascoideae*
Genus: *Verbascum*
Species: *thapsus* L. (Figure 31.2)[1,3-4]

Figure 31.2 *Verbascum thapsus* L. (Common mullein).

Source: 7.

SPREAD

A common species along forest edges, rocky and sandy sunny areas, ranging from oak forests to the spruce forest zone.[1,3-4]

DESCRIPTION

A herbaceous species, biennial, reaching a height of 2 m, equipped with persistent tomentum, gray or white-gray, rarely light yellow, non-glandular. The stem is erect, simple, stout, foliate, narrowly winged. The basal rosette leaves are petiolate, elongated, obovate or broadly elliptical, crenate, attenuated into a petiole. The inferior stem leaves are similar to the basal ones, narrowed towards the base into long decurrent wings, and the superior ones are elongated lanceolate or elongated obovate, crenate, with a decurrent base that extends to the next leaf. The flowers are grouped in pairs of 2–7 in the axil of a bract with an ovate base and an acuminate tip, having a yellow corolla with clear, transparent dots, densely tomentose on the exterior. The stamens have reniform anthers, not decurrent, with the inferior ones having glabrous filaments, and the gynoecium has a capitate stigma. The plants flower from June to August.[1-4]

Verbascum densiflorum Bertol. (*V. thapsiforme* Schrad.) (Denseflower mullein)

Scientific name: *Verbascum densiflorum* Bertol.
Common name: Denseflower mullein
Romanian name: Lumânărică[2,6]
Taxonomic classification
Kingdom: *Plantae*
Subkingdom: *Cormobionta*
Phylum: *Spermatophyta*
Subphylum: *Magnoliophytina (Angiospermae)*
Class: *Magnoliatae (Magnoliopsida, Dicotyledonatae)*
Subclass: *Asteridae*
Order: *Scrophulariales (Solanales)*
Family: *Scrophulariaceae*
Subfamily: *Verbascoideae*
Genus: *Verbascum*
Species: *densiflorum* Bertol. (Figure 31.3)[1,3-4]

SPREAD

The plants grow sporadically in shrubs, vineyards, riverbanks, ruderal places, forest clearings, on rocky places or meadows, from lowland areas up to the beech forest zone.[1,3-4]

DESCRIPTION

Plants reaching a height of 1.2 m, biennial, covered with a soft, gray or grayish-yellow felt. The stem is unbranched, erect, thick, foliate, winged due to decurrent leaves, gray or yellow-green, less tomentose on the superior side and densely tomentose on the inferior side. The basal leaves have a broad-winged petiole, they are ovate, elongated-ovate, more rarely elongated-elliptical, acute, with an attenuated base, and the stem leaves are long-decurrent. The flowers are grouped in pairs of 2–7 in the axil of a non-membranous bract, the corolla is yellow, rotated, stelar tomentose on the exterior, and the stigma is spatulate, decurrent along the style. Flowering occurs in the months of July–August.[1-4]

Verbascum speciosum Schrad. (Hungarian mullein)

Scientific name: *Verbascum speciosum* Schrad.
Common name: Hungarian mullein
Romanian name: Coada vacii[4]
Taxonomic classification
Kingdom: *Plantae*
Subkingdom: *Cormobionta*
Phylum: *Spermatophyta*
Subphylum: *Magnoliophytina (Angiospermae)*

Figure 31.3 *Verbascum densiflorum* Bertol. (*V. thapsiforme* Schrad.) (Denseflower mullein).

Source: 8.

Class: *Magnoliatae (Magnoliopsida, Dicotyledonatae)*
Subclass: *Asteridae*
Order: *Scrophulariales (Solanales)*
Family: *Scrophulariaceae*
Subfamily: *Verbascoideae*
Genus: *Verbascum*
Species: *speciosum* Schrad. (Figure 31.4)[1,3-4]

SPREAD

It grows sporadically in seedbeds, gravelly places, clearings, and forest openings, dry pastures, sunny spots, shrubs, or meadows, ranging from the sylvosteppe zone up to the oak forest zone.[1-4]

DESCRIPTION

A robust biennial herbaceous species, with a height of 1–2 m, branching, covered in dense gray tomentum, non-glandular. The vigorous stem is erect, branching in a panicle towards the top, with simple, slightly curved, almost straight branches. The large basal leaves are elongated-obovate lanceolate, with entire margins, acute tips, slightly attenuated into the short, non-winged or narrowly winged petiole. The inferior stem leaves resemble the basal ones but are smaller, sessile, amplexicaul, with an auriculate base, while the superior leaves are cordate-ovate, acuminate, and have rounded auricles at the base. The flowers are grouped in pairs of 2–7 in the axil of an ovate-lanceolate bract, shorter than the flowers. The calyx has gray and persistent indumentum, with acute, narrow lanceolate laciniae, and the corolla is yellow, rotated, without transparent dots, stelar tomentose on the exterior. The stamens have orange filaments covered up to the anthers with white or yellowish hairs, the anthers are not decurrent, reniform, mediofix. The style is long, and the stigma is obovate. The fruit is a cylindrical capsule, stelar hairy, longer than the calyx. It flowers from June to August.[1-4]

Figure 31.4 *Verbascum speciosum* Schrad. (Hungarian mullein).

Source: 9.

Main Phytoconstituents

Verbasci flos contains: 3% mucilage, composed of sugars (D-glucose, D-galactose, L-arabinose, D-xylose) and uronic acids (D-glucuronic acid, D-galacturonic acid); flavonoids, approximately 3–4% hesperidoside or hesperidin (hesperetol-7-O-rutinoside); carotenoids, hydrocarbons (*β*-carotene) and oxygenated derivatives (crocoside or crocin – the ester of crocetin with gentiobiose); iridoids (aucuboside, ajugol, catalpol, and derivatives); saponins; sterols; glycosylated lignans; polyphenolcarboxylic acids (caffeic acid, chlorogenic acid); volatile oil; lipids; 10% simple sugars; mineral salts.[10]

Main phytoconstituents   Ref.
amino acids   [11-14]
flavonoids (apigenin, chrysoeriol, luteolin, kaempherol, tamarixetin, quercetin, hesperidin, rutin, acacetin)
irinoids (verbacoside, aucubin, catalpol)
terpenoids (buddlindeterpene A-C)
sterols (β-sitosterol)
fatty acids (oleanolic acid, behemic acid)
phenolic acids (caffeic acid, rosmarinic acid, ferulic acid, p-coumaric acid, chlorogenic acid)
tannins (catechin, epicatechin)
organic acids
carbohydrates (glucose, galactose, arabinose, xylose)
saponins
glycosides (acteoside)
carotenoids (*β*-carotene, zeaxanthin, cryptoxanthin)
miscellaneous (δ-tocopherol)

Medicinal Use

Verbasci flos presents the following actions and uses: demulcent (due to mucilage content), in gastrointestinal irritations and various types of coughs; expectorant (mucilage, saponins); anti-inflammatory, in joint conditions; anticatarrhal, in stomatitis, pharyngitis, laryngitis; analgesic (iridoids); diaphoretic.

Empirically, due to its flavonoid content, the vegetable product is used as a diuretic.[10]

Biological Activity

Verbascum speciosum exhibits various biological activities, demonstrated by numerous studies, such as antibacterial, antioxidant, antifungal, anticancer, and wound healing properties.[16-21]

Other Uses

■ insecticidal effects against insect pests[22]

Digitalis L. (Foxgloves)

Species of biennial herbaceous plants (rarely perennial), which form a rosette of basal leaves, from the center of which, in the second year, the stem with leaves and flowers develops. The leaves are entire or with small incisions, elliptical or lanceolate in shape, with the stem leaves arranged alternately. The flowers are gamopetalous, zygomorphic, with a thimble-like appearance, hanging in a pendant raceme. The persistent calyx at fruiting has 5 dentiform divisions, almost equal. The corolla is tubular or urceolate-campanulate, without a spur, bilabiate with reflected teeth, and can be purple, yellowish, brownish-yellow, or reddish-brown in color. The androecium is didynamous, composed of four stamens, with filaments fused to the corolla tube and anthers with joined lobes at the base. The style is long, and the stigma is bilobed. The fruit is a bilocular valvicidal capsule. The seeds are small, numerous, ovoid, albuminous, with a finely reticulate, alveolate seminal tegument, and a cylindrical embryo.[1-2]

The genus comprises 36 species widespread in the northern hemisphere, and in the spontaneous flora of Romania, 3 species are found: *Digitalis lanata* Ehrh., *Digitalis grandiflora* Mill., and *Digitalis ferruginea* L.[2,4]

Digitalis lanata Ehrh. (Grecian foxglove)

Scientific name: *Digitalis lanata* Ehrh.
Common name: Grecian foxglove
Romanian name: Cucență
Taxonomic classification
Kingdom: *Plantae*
Subkingdom: *Cormobionta*
Phylum: *Spermatophyta*
Subphylum: *Magnoliophytina (Angiospermae)*
Class: *Magnoliatae (Magnoliopsida, Dicotyledonatae)*
Subclass: *Asteridae*
Order: *Scrophulariales (Solanales)*
Family: *Scrophulariaceae*
Subfamily: *Scrophularioideae*
Genus: *Digitalis*
Species: *lanata* Ehrh. (Figure 31.5)[1,3-4]

SPREAD

Found spontaneously in Romania, in meadows, pastures, rocky areas, thin soils, clearings, and forest edges, ranging from the steppe zone up to the oak forest zone.[1-4]

DESCRIPTION

A biennial or perennial herbaceous species, with a height of 30–70 cm. It has a taproot and an erect stem, simple, sometimes branching in the superior part. The stem is glabrescent in the inferior part and hairy in the superior part. The leaves are glabrous (at least on the inferior side), glossy on the superior side, oblong-lanceolate in shape. The basal leaves are grouped in a rosette, while the stem leaves are sessile, semi-amplexicaul, and acuminate. The flowers are pale yellowish (with brown or violet reticulations), measuring 1.5–3 cm, forming dense terminal racemose inflorescences. The axis of the inflorescence is glandular hairy. The calyx has lanceolate, glandular hairy laciniae, and the corolla has a much longer median lobe (of the lower labium) than the others. It flowers from June to August.[1-4]

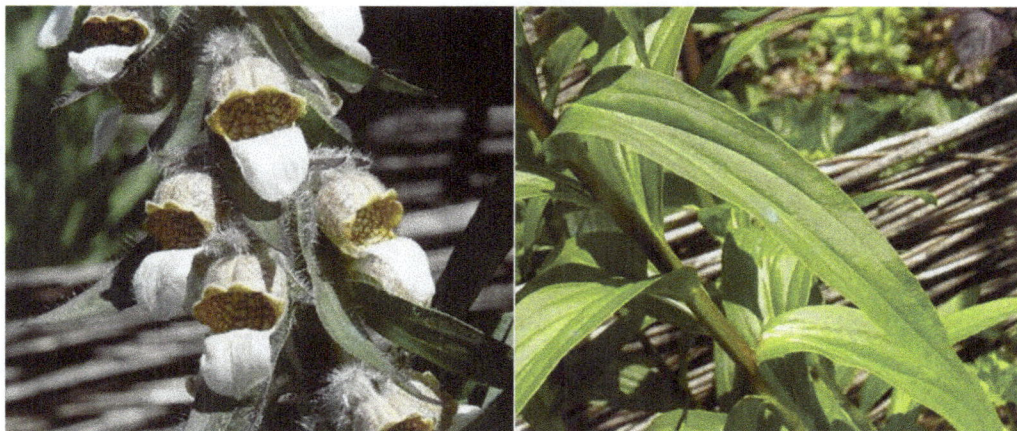

Figure 31.5 *Digitalis lanata* Ehrh. (Grecian foxglove).

Source: 23.

Main phytoconstituents   Ref.

cardenolides (lanatoside A, digitoxin, digitoxigenin, lanatoside B-C, diginatin, diginatigenin,
   digifolein, glucodigifolein, diginin, digipronin, lanafolein, gitonine)   24-25

steroid saponins (digitoxigenin, lanatoside A, purpureaglycoside A, acetyldigitoxin,
   digitoxin, odoroside H, gitoxigenin, lanatoside B, gitoxin, lanatoside C, deslanoside,
   digoxin, diginatin, lanatoside E, gitaloxin, verodoxin, tigonin, tigogenin, digalogenin,
   digitogenin, gitogenin)

flavonoids (luteolin and apigenin)

sterols (β-sitosterol, stigmasterol, cholesterol)

phenolic acids (caffeic acid, chlorogenic acid)

organic acids (ascorbic acid)

anthraquinones

glycosides

miscellaneous (choline, acetylcholine)

The lanatosides from *Digitalis lanatae folium* encompass five series, based on the structure of the
aglycones:

Series A, with the aglycone digitoxigenol (digitoxigenin):

■ Primary heteroside: lanatoside A

■ Secondary heterosides:

   ■ Desacetyllanatoside A (purpureaglycoside A)

   ■ Acetyldigitoxoside (acetyldigitoxin)

   ■ Digitoxoside (digitoxin, *digitalinum* Nativelle — 1881)

   ■ Odoroside H

Series B, with the aglycone gitoxigenol (gitoxigenin):

■ Primary heteroside: lanatoside B

■ Secondary heterosides:

   ■ Desacetyllanatoside B (purpureaglycoside B

   ■ Acetylgitoxoside (acetylgitoxin)

   ■ Gitoxoside (gitoxin)

- Strospeside
- *Digitalinum verum* (Schmiedeberg, 1885)

Series C, with the aglycone digoxigenol (digoxigenin):

- Primary heteroside: lanatoside C
- Secondary heterosides:

  - Desacetyllanatoside C (deslanoside)
  - Acetyldigoxoside (acetyldigoxin)
  - Digoxoside (digoxin)

Series D, with the aglycone diginatigenol (diginatigenin):

- Primary heteroside: lanatoside D
- Secondary heterosides:

  - Desacetyllanatoside D
  - Acetyldiginatoside (acetyldiginatin)
  - Diginatoside (diginatin)

Series E, with the aglycone gitaloxigenol (gitaloxigenin):

- Primary heteroside: lanatoside E
- Secondary heterosides:

  - Desacetyllanatoside E (purpureaglycoside E)
  - Acetylgitaloxoside (acetylgitaloxin)
  - Gitaloxoside (gitaloxin)
  - Verodoxoside (verodoxin)
  - Glucoverodoxoside (glucoverodoxin)

The content of lanatosides is variable, generally with a predominance of heterosides from series A (40–55%) and C (25–35%).

Cardiotonic heterosides from series C are more active, more stable, crystallize, and purify more easily. The glucidic part of primary heterosides is composed of two molecules of β-D-digitoxose (*dtox*), one molecule of acetyl-β-D-digitoxose (*Acdtox*), and one molecule of β-D-glucose (*gluc*) positioned terminally. Under the action of digilanidase, primary heterosides lose the glucose molecule and transform into secondary heterosides. By removing the acetyl group from the structure of acetyl-β-D-digitoxose, primary heterosides (lanatosides A–E) transform into secondary heterosides (desacetyllanatosides A–E). For series A, B, and E, these secondary heterosides are indeed the respective purpureaglycosides.

The glucidic part of the structure of *digitalinum verum* and glucoverodoxoside consists of one molecule each of β-D-digitalose (*dtal*) and β-D-glucose (*gluc*).

Odoroside H, strospeside, and verodoxoside have one molecule of β-D-digitalose (*dtal*) as their glucidic component.

The digitanol heterosides and spirosteranic saponins are similar to those found in *Digitalis purpureae folium*.[25]

Medicinal Use

The vegetable product *Digitalis lanatae folium* presents both cardiotonic and diuretic actions. The cardiotonic glycosides are administered in the form of pharmaceutical products, including industrial formulations such as Lanatoside C tablets or dragees of 0.25 mg, Deslanoside (Lanimerck) ampoules of 2 ml, containing 0.4 mg of the active substance, Digoxin tablets of 0.25 mg, and injectable solution ampoules of 2 ml, with 0.5 mg of the active substance.[1]

Biological Activity

Digitalis lanata exhibits a range of effects including cardiovascular, cytotoxic, antidiabetic, antioxidant, antiviral, insecticidal, immunological, hepato-, neuro-, and cardioprotective activities.[26-28]

Toxicity

Digitalis lanata is a highly toxic species, approximately 3-4 times more potent than *D. purpurea*.[16]

Other Uses

- ornamental[28]

The genus *Scrophularia* is represented in the spontaneous flora of Romania by 5 species: *Scrophularia nodosa* L., *Scrophularia vernalis* L., *Scrophularia heterophylla* Willd., *Scrophularia umbrosa* Dumort (*S. alata* auct.), and *Scrophularia scopolii* Hoppe.[3-4]

Scrophularia nodosa L. (Figwort)

Scientific name: *Scrophularia nodosa* L.
Common name: Figwort, Woodland figwort, Common figwort
Romanian name: Bubernic, Iarbă neagră
Taxonomic classification
Kingdom: *Plantae*
Subkingdom: *Cormobionta*
Phylum: *Spermatophyta*
Subphylum: *Magnoliophytina (Angiospermae)*
Class: *Magnoliatae (Magnoliopsida, Dicotyledonatae)*
Subclass: *Asteridae*
Order: *Scrophulariales (Solanales)*
Family: *Scrophulariaceae*
Subfamily: *Scrophularioideae*
Genus: *Scrophularia*
Species: *nodosa* L. (Figure 31.6)[1-3]

SPREAD

Frequently found in moist areas, meadows, alongside fences, forest edges, and streamside coppices, ranging from lowland to the beech forest zone.[1-4]

DESCRIPTION

Perennial herbaceous species, reaching heights of 30–80 cm, with a tuberized, horizontal, knobby rhizome. The aerial stems are four-edged, glabrous, and simple. The leaves are petiolate, ovate or elongated-ovate, with a base narrowed and rounded, truncate, or slightly cordate, arranged opposite, glabrous, undivided, and with a doubly serrated margin. The flowers are small, with membranous-edged sepals, and a tubular, inflated corolla with two short, rounded lobes, of greenish color (the superior labium is brownish-red). The androecium is didynamous, and there are also scale-like staminodes present (sessile stamens). The fruit is a septicidal capsule, and the seeds are dark brown, ellipsoidal. It blooms from June to August.[1-2]

Main phytoconstituents	Ref.
amino acids	30-32

amino acids
coumarine (spinoside)
flavonoids (centaurein, serpyllin, kaempherol, sakuranetin)
phenolic acids (cinnamic acid, ferulic acid, caffeic acid, p-coumaric acid)
glycosides
irinoids (scrophuloside, aucubin, angoroside, harpagoside)

413

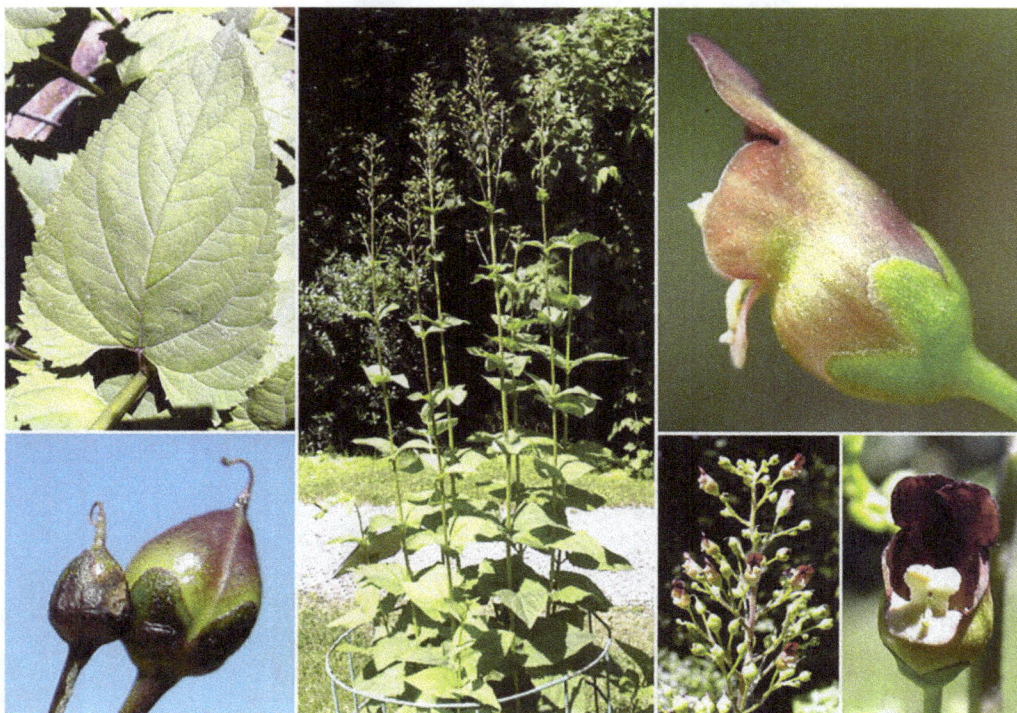

Figure 31.6 *Scrophularia nodosa* L. (Figwort).

Source: 29.

Medicinal Use

The aerial part of the species (*Scrophulariae herba*) primarily exhibits antitumoral activity.[1]

Biological Activity

Scrophularia nodosa possesses wound healing, spasmolytic, diuretic, antibacterial cardioprotective, anti-inflammatory action, and antioxidant activities.[31,33-34]

Other Uses

- edible uses: The root can be prepared by cooking. Despite its disagreeable aroma and flavor, it has been consumed during periods of famine.

- agroforestry uses: The flowers serve as a valuable nectar source.[35]

Veronica officinalis L. (Common speedwell)

Scientific name: *Veronica officinalis* L.
Common name: Common speedwell, Heath speedwell, Common gypsyweed, Paul's betony
Romanian name: Ventrilică, Strătorică
Taxonomic classification
Kingdom: *Plantae*
Subkingdom: *Cormobionta*
Phylum: *Spermatophyta*
Subphylum: *Magnoliophytina (Angiospermae)*
Class: *Magnoliatae (Magnoliopsida, Dicotyledonatae)*
Subclass: *Asteridae*
Order: *Scrophulariales (Solanales)*
Family: *Scrophulariaceae*

Figure 31.7 *Veronica officinalis* L. (Common speedwell).

Source: 36.

Subfamily: *Scrophularioideae*
Genus: *Veronica*
Species: *officinalis* L. (Figure 31.7)[1,3-4]

SPREAD
Common in our country, ranging from oak forest zones to the spruce forest zone, found in meadows and shrubs, as well as along forest edges or clearings, and ruderal places.[1-4]

DESCRIPTION
A perennial species with a cylindrical, creeping rhizome and long trailing, radicant stems, featuring secondary branches and erect or ascending flowering branches covered in hirsute hair. The leaves are shortly petiolate, elliptic to obovate-elliptic, with an obtuse tip, narrow base, serrated margins, and scattered simple hairs on both leaf surfaces. The flowers are of type 4, forming dense axillary racemes that are erect. The blue corolla is rotate with a short tube, and the androecium comprises two fertile stamens without staminodes. The fruit is a long, ovoid capsule, nearly twice as long as the calyx, glandular-pubescent. The seeds are flat-convex, lenticular. This species blooms from June to August.[1-4]

Main phytoconstituents   Ref. [37-39]
flavonoids (quercetin, apigenin, quercitrin, luteolin, hispidulin, eupatorin)
terpenoids (ursolic acid, α-thrujone, pinene, phellandrene, cineole, terpinene, linalool, ocimene, α-terpineol, camphor, borneol, α-copaene, caryophyllene, germacrene D, δ-cadinene, selinene, spathulenol, eudismol, α-muurolol, α-cadinol, bisabolene)
sterols (campesterol, stigmasterol, sitosterol, ergosterol)
fatty acids (palmitic acid)
phenolic acids (gentisic acid, chlorogenic acid, ferulic acid, caffeic acid, p-coumaric acid)

irinoids (aucubin, catalpol, catalposide, and veronicoside)
glycosides
esters
hydrocarbons (ecosane, docosane, tricosane)
aldehydes&ketones (nonanal, benzaldehyde)
miscellaneous (ionone, damascone)

Medicinal Use

The flowering aerial part of the plant is recommended in folk medicine for the treatment of renal and biliary lithiasis, hepato-biliary diseases, respiratory catarrh, rheumatism, and skin lesions.[1]

Biological Activity

Veronica officinalis exhibits antioxidant, antimicrobial, hypocholesterolaemic, anticancerogenic and antiproliferative properties. Applied topically, it demonstrated collagen synthesis and antiwrinkle activities.[37-38,40-41]

Other Uses

■ natural-based antiphytoviral agent[42]

Gratiola officinalis L. (Gratiole)

Scientific name: *Gratiola officinalis* L.
Common name: Gratiole, Common hedgehyssop, Grace of God, Hedge hyssop, Herb of grace
Romanian name: Veninariță
Taxonomic classification
Kingdom: *Plantae*
Subkingdom: *Cormobionta*
Phylum: *Spermatophyta*
Subphylum: *Magnoliophytina (Angiospermae)*
Class: *Magnoliatae (Magnoliopsida, Dicotyledonatae)*
Subclass: *Asteridae*
Order: *Scrophulariales (Solanales)*
Family: *Scrophulariaceae*
Subfamily: *Scrophularioideae*
Genus: *Gratiola*
Species: *officinalis* L. (Figure 31.8)[1,3-4]

SPREAD

It sporadically grows from the steppe zone up to the oak forest zone, in wet places or along water edges, ditches, and marshy areas.[1-4]

DESCRIPTION

A perennial plant with a creeping, articulate rhizome from which adventitious roots begin. The erect stem (20–60 cm tall) is fistulous, glabrous, simple or slightly branched, cylindrical at the base and tetragonous towards the top. The leaves are sessile, lanceolate, serrated, glabrous, acute, semi-amplexicaul, trinervate, and glandular-punctate on both surfaces. The flowers are tubular, bilabiate, white, solitary and axillary, with two fertile stamens and 2–3 staminodes. The fruit is an ovoid-spherical capsule that opens into 4 valves, containing small reticulate seeds. The flowering period is from June to August.[1-4]

Main phytoconstituents Ref.
amino acids 44-47
flavonoids (apigenin, vitexin, saponaretin, cosmosin, luteolin)
terpenoids (gratiogenin, cucurbitacin-E, soyaspogenol B)
phenolic acids (caffeic acid, gallic acid, ferrulic acid, chlorogenic acid)
tannins (catechin, ellagic acid)
carbohydrates (mannitol)

Figure 31.8 *Gratiola officinalis* L. (Gratiole).

Source: 43.

glycosides
esters
irinoids (gratioside, elaterinide, verbascoside, arenarioside, sophoroside, forsythoside B, samioside, linariifolioside)
aldehydes&ketones
miscellaneous

Medicinal Use

The plant is used in the treatment of gout and rheumatism.[1]

It finds application in herbal remedies to address a range of conditions such as scrofula, cystitis, colic, specific stomach and menstrual disorders, skin and liver ailments, along with the enlargement of the spleen, dropsy, jaundice, and intestinal worms. The dried upper part of *G. officinalis* acts as a diuretic and an emetic. Additionally, *G. officinalis* is utilized as a biostimulating tablet for hematopoietic, liver, and respiratory issues. The root and flowering herb function as a cardiac tonic, diuretic, strongly purgative, and vermifuge.

Despite its recognized toxicity, it is still used in homeopathy as an anthelmintic. The taste is notably bitter, attributed to glycosides of cucurbitacin, and excessive use may result in side effects such as nausea, acrid poisoning, excessive leanness, abortion, kidney damage, and bowel hemorrhage.[44]

Biological Activity

Possessing antibacterial, anti-inflammatory, antiallergic, anticarcinogenic, and adjuvant qualities in osmoregulation, *Gratiola officinalis* extract has been observed to exert specific cytotoxic and cytostatic effects on tumor cells. The plant also displays antioxidant attributes, along with anti-inflammatory, antipyretic, and antimicrobial capabilities.[44-47]

Figure 31.9 *Euphrasia officinalis* L. (Eyebright).

Source: 48.

Other Uses

■ ornamental[47]

The genus *Euphrasia* comprises 90 species distributed across all continents. In the spontaneous flora of Romania, there are 8 species: *Euphrasia officinalis* L., *Euphrasia stricta* D. Wolff ex J.F. Lehm., *Euphrasia illyrica* Wettst., *Euphrasia salisburgensis* Funck ex Hoppe, *Euphrasia arctica* Lange ex Rostrup subsp. *slovaca* Yeo, and *Euphrasia hirtella* Jord. ex Reut.[2-3]

Euphrasia officinalis L. (Eyebright)

Scientific name: *Euphrasia officinalis* L.
Common name: Eyebright, Eyewort
Romanian name: Silur
Taxonomic classification
Kingdom: *Plantae*
Subkingdom: *Cormobionta*
Phylum: *Spermatophyta*
Subphylum: *Magnoliophytina (Angiospermae)*
Class: *Magnoliatae (Magnoliopsida, Dicotyledonatae)*
Subclass: *Asteridae*
Order: *Scrophulariales (Solanales)*
Family: *Scrophulariaceae*
Subfamily: *Rhinanthoideae*
Genus: *Euphrasia*
Species: *officinalis* L. (Figure 31.9)[1,3-4]

SPREAD
Frequently encountered in wet meadows and shrubs, between the oak and subalpine zones.[1]

DESCRIPTION
An annual plant with stems up to 35 cm tall. The leaves are opposite, uniformly and sharply dentate, densely hairy. The bilabiate flowers are white or pale lilac in color, arranged in a multilateral manner at the base of bracts. The calyx and bracts are densely hairy, and the androecium is didynamous.[1]

The species has 2 subspecies:

Main phytoconstituents	Ref.
	49-51

amino acids

flavonoids (apigenin, chrysoeriol, luteolin, rutin, kaempherol, quercetin)

terpenoids (α-terpineol, thujone, borneol, linalool, thymol, carvacrol, sabinene, germacrene D, caryophyllene, β-bisabolene, eudesmol, safrole, phytol)

sterols (stigmasterol, sitosterol)

fatty acids (linolenic acid, lauric acid, linolenic acid, capric acid)

phenolic acids (ferulic acid, caffeic acid, vanillic acid, gentisic acid, chlorogenic acid, gallic acid, myristic acid)

tannins

organic acids

glycosides

esters

irinoids (aucubin, catapol, euphroside, eurostoside, geniposide, ixoside, adoxosid)

hydrocarbons

aldehydes & ketones

miscellaneous

- Subsp. *rostkoviana* (Hayne) F. Towns. (*E. rostkoviana* Hayne, *E. montana* Jord., *E. officinalis* subsp. *monticola* Silverside, *E. pratensis* Fr., *E. campestris* Jord.) has a branched stem from the base, with internodes up to three times longer than the leaves, covered with glandular hairs. The inferior flowers begin at nodes 6–10, flowering from June to October.

- Subsp. *officinalis*.[3]

Medicinal Use
The aerial part (*Euphrasiae herba*) is recommended for gastrointestinal disorders, diabetes, conjunctivitis, and blepharitis.[1]

Biological Activity

- beneficial effects in ophtalmology; antioxidant and anti-inflammatory activities.[51-52]

- *Euphrasia officinalis* extract reduces UV-induced skin photoaging by inhibiting oxidative stress, proinflammatory activity, and cell apoptosis.[53]

Other Uses

- The raw leaves are consumed in salads for their slightly bitter flavour.[54]

Euphrasia stricta D. Wolff ex J.F. Lehm. (Strict eyebright)

Scientific name: *Euphrasia stricta* D. Wolff ex J.F. Lehm.
Common name: Strict eyebright
Romanian name: Bucuriţă[6]
Taxonomic classification
Kingdom: *Plantae*
Subkingdom: *Cormobionta*
Phylum: *Spermatophyta*

419

Figure 31.10 *Euphrasia stricta* D. Wolff ex J.F. Lehm.

Source: 55.

Subphylum: *Magnoliophytina (Angiospermae)*
Class: *Magnoliatae (Magnoliopsida, Dicotyledonatae)*
Subclass: *Asteridae*
Order: *Scrophulariales (Solanales)*
Family: *Scrophulariaceae*
Subfamily: *Rhinanthoideae*
Genus: *Euphrasia*
Species: *stricta* D. Wolff ex J.F. Lehm. (Figure 31.10)[3-4]

SPREAD

The species is widespread from the hilly area to the subalpine zone, found in meadows, forest clearings, rocky slopes, grassy cliffs, pastures, shrubs, and orchards.[2-3]

DESCRIPTION

Annual species, with a height of 5–40 cm, having erect, flexuous, or straight stems, simple or slightly branched, with short branches, covered with non-glandular, erect hairs, devoid of leaves in the inferior part during flowering. The inferior leaves are opposite, cuneate, obtuse, with 1–2 obtuse teeth on each side. The middle and superior leaves are ovate or suborbicular, acute, with a cuneate base, 3–5 sharp and aristate teeth, green, smooth on the inferior surface or with prominent veins, glabrous, sometimes with a few short, rough hairs on the margins. The nearly sessile flowers have a glabrous or slightly rough-pubescent calyx, unchanged or slightly enlarged at fruiting, a corolla 7–10 mm long with a tube that does not exceed or slightly exceeds the calyx after flowering, pale violet, rarer blue or whitish, externally hairy. The superior labium has dentate lobes, the inferior one lighter in color with a yellow spot, rarely whitish and with blue or purple stripes, and emarginate lobes. The stigma is recurved. The fruit is an ovoid capsule, narrow, shorter than the calyx. It blooms from June to September.[1-4]

This species has 3 subspecies:

- subsp. *brevipila* (Burnat et Gremli) Ciocârlan has leaves, bracts, and calyx covered with short glandular hairs. Rarely encountered from the beech forest zone to the subalpine zone through dry bushes and meadows.[2]

- subsp. *stricta* has stems with 2–6 pairs of branches and leaves, bracts, and calyx without glandular hairs, almost glabrous, only slightly hairy on the margins. Frequent from the oak forest zone to the subalpine zone through bushes and meadows.[2,3]

- subsp. *pectinata* (Ten.) P. Fourn. (*E. tatarica* Fisch. ex Spreng.) has stems with 3–6 pairs of branches, leaves, bracts, and calyx are densely and shortly rough-haired. Rarely distributed from the oak forest zone to the subalpine zone through meadows and bushes.[2,4]

Main phytoconstituents   Ref.
amino acids   [56-57]
flavonoids (luteolin, quercetin, rutin, kaempherol, apigenin, rhamnetin)
terpenoids (α-terpineol, thujone, geranial, borneol, menthone, camphor, lavandulol, linalool, thymol, carvacrol, eugenol, α-copaene, viridiflorene, silenene, curcumene, humulene, limonene, spathulenol, α-muurolene caryophyllene, sabinene, β-bisabolene, germacrene D, phytol)
sterols (cholesterol, β sitosterol)
fatty acids (linolenic acid, myristic acid, palmitic acid, margaric acid, stearic acid, pentadecanoic acid)
phenolic acids (caffeic acid, p-coumaric, protocatechuic acid, salicylic acid)
hydrocarbons (pentacosane, tricosane, tetracosane)
organic acids (oxalic acid, aconitic acid, citric acid, malic acid, quinic acid, acetic acid, fumaric acid)
esters
aldehydes&ketones
miscellaneous (anethole, estragole, damascenone)

Medicinal Use

Euphrasia stricta has a long history in traditional folk medicine, where it has been employed to address a range of eye issues including cataracts, conjunctivitis, as well as red, inflamed, irritated, and painful eyes.[57]

Biological Activity

Euphrasia stricta exhibits antioxidant, antibacterial, hypotensive, anti-inflammatory, antimicrobial, and hepatoprotective activities.[56,58-60]

The genus *Melampyrum* comprises 9 species spread across the spontaneous flora of Romania: *Melampyrum bihariense* A. Kern., *Melampyrum cristatum* L., *Melampyrum arvense* L., *Melampyrum barbatum* Waldst. et Kit., *Melampyrum nemorosum* L., *Melampyrum sylvaticum* L., *Melampyrum pratense* L., *Melampyrum saxosum* Baumg., and *Melampyrum herbichii* Woloszczak.[3-4]

Melampyrum bihariense A. Kern. (Cow wheat)

Scientific name: *Melampyrum bihariense* A. Kern.
Common name: Cow wheat
Romanian name: Grâul pădurii[6]
Taxonomic classification
Kingdom: *Plantae*
Subkingdom: *Cormobionta*
Phylum: *Spermatophyta*
Subphylum: *Magnoliophytina (Angiospermae)*
Class: *Magnoliatae (Magnoliopsida, Dicotyledonatae)*
Subclass: *Asteridae*
Order: *Scrophulariales (Solanales)*
Family: *Scrophulariaceae*
Subfamily: *Rhinanthoideae*

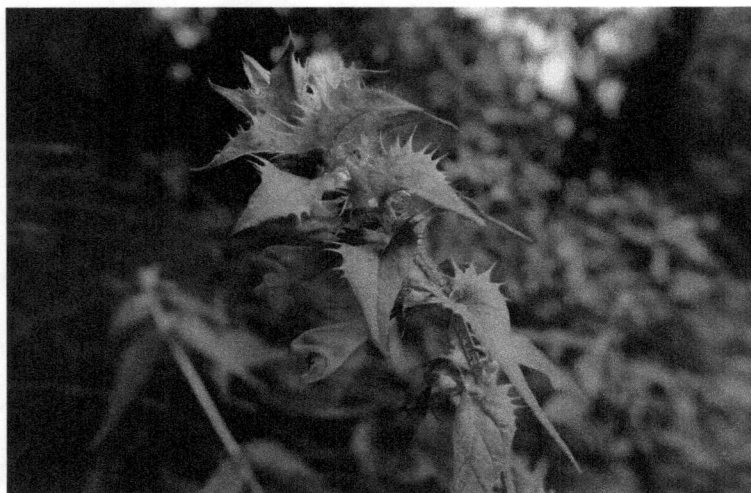

Figure 31.11 *Melampyrum bihariense* A. Kern. (Cow wheat).

Source: 61.

Genus: *Melampyrum*
Species: *bihariense* A. Kern. (Figure 31.11)[3-4]

SPREAD

The species is commonly distributed from the oak forest zone to the spruce forest zone, through forest edges, meadows, thickets, and forest clearings.[2-4]

DESCRIPTION

Annual plant with an erect stem, reaching a height of 20–50 cm, simple or branched, surrounding or biseriate pubescent. The leaves are lanceolate or ovate-lanceolate, with a sharply attenuated base, long acuminate, glabrous or slightly hairy along the veins. The inflorescence is inserted from nodes 4–10. The inferior bracts resemble the leaves, while the superior ones are azure, cordate-ovate, deeply pinnatisect-dentate. The calyx is tubular, with teeth of 8 mm, linear or linear-lanceolate, separated by obtuse sinuses, finely pubescent or glabrous, bearing slightly longer, whitish hairs only on the veins. The corolla is yellow-golden, twice the length of the calyx, with a broad inferior labium, trilobate, and a superior labium rounded like a hood. It blooms from June to September.[2-4]

Main phytoconstituents	Ref.
amino acids	62-64

flavonoids (apigenin, chrysoeriol, cinaroside, kaempherol, luteolin, quercetin, taxifolin myricetin)
terpenoids (ursolic acid, oleanolic acid)
sterols
fatty acids
phenolic acids (chlorogenic acid, ferulic acid, caffeic acid)
glycosides
esters
irinoids (aucubin, loganic acid, mussaenoside)
hydrocarbons
aldehydes & ketones
miscellaneous

Medicinal Use

Melampyrum bihariense, a species incorporated into traditional Romanian medicine, has been historically utilized for managing rheumatic disorders and skin infections. It was employed for therapeutic baths targeting rheumatism and, in mixture with other botanicals, for preparing tea remedies to address jaundice.[62-63]

Biological Activity

Melampyrum bihariense exhibits antioxidant, diuretic, anti-inflammatory, antibacterial, and antifungal actions.[62-63,65]

Other Uses

- Larvae of certain *Lepidoptera* species, such as the mouse moth (*Amphipyra tragopoginis*), feed on *Melampyrum* species as part of their diet.[66]

Genus *Pedicularis* presents the following species in the spontaneous flora of Romania: *Pedicularis verticillata* L., *Pedicularis sceptrum-carolinum* L., *Pedicularis oederi* Vahl, *Pedicularis exaltata* Besser, *Pedicularis hacquetii* Graf, *Pedicularis comosa* L., *Pedicularis baumgarteni* Simonk, *Pedicularis limnogena* A. Kern., *Pedicularis rostratocapitata* Crantz, *Pedicularis palustris* L., and *Pedicularis sylvatica* L.[3-4]

Pedicularis verticillata L. (Whorled lousewort)

Scientific name: *Pedicularis verticillata* L.
Common name: Whorled lousewort
Romanian name: Vârtejul pământului
Taxonomic classification
Kingdom: *Plantae*
Subkingdom: *Cormobionta*
Phylum: *Spermatophyta*
Subphylum: *Magnoliophytina (Angiospermae)*
Class: *Magnoliatae (Magnoliopsida, Dicotyledonatae)*
Subclass: *Asteridae*
Order: *Scrophulariales (Solanales)*
Family: *Scrophulariaceae*
Subfamily: *Rhinanthoideae*
Genus: *Pedicularis*
Species: *verticillata* L. (Figure 31.12)[3-4]

SPREAD

The species is commonly distributed from the subalpine to alpine zone, through limestone valleys, rocky slopes, cliffs, wet meadows, and shrublands.[3-4]

DESCRIPTION

Perennial plant, with a height of 5–15 cm, glabrous or occasionally woolly hairy stems, with dense hairs at the nodes, solitary or branched. The basal leaves are long-petiolate, lanceolate, pinnatifid, dentate, and the rachis is winged. Stem leaves are arranged in whorls of three. Flowers are arranged in a whorled pattern. The inferior bracts are pinnatifid, while the superior ones are crenate, smaller than the flowers. Calyx is swollen, long-hairy along the veins, and the corolla is red, glabrous, with an upright and rounded superior labium and a flat inferior labium, oriented horizontally. The fruit is an ovate-lanceolate capsule, twice as long as the calyx. It blooms from June to August.[2-4]

Main phytoconstituents   Ref.

amino acids (alanine, aspartic acid, arginine, cysteine, serine, valine glycine, glutamic acid,
   isoleucine, prolineleucine, threonine, tyrosine, methionine, phenylalanine,)   [68-70]
coumarins (scopoletin)
alkaloids (lupanine, pediculine, plantagonine, platagonin, indicainine, indicine, senecionine)
flavonoids (quercetin, apigenin, kaempherol, luteolin)

Figure 31.12 *Pedicularis verticillata* L. (Whorled lousewort).

Source: 67.

sterols (sitosterol, daucosterol)
lignans (verticillatoside A, verticillatoside B)
glycosides (verbascoside, wiedemannioside C)
irinoids (loganic acid, densispicnin B, caryoptoside, boschnaloside, geniposidic acid, kansuenin B, ligustroside, pediverticilatasin A-C, plantarenaloside, scyphiphin A1, scyphiphin A2, aucubin, euphroside, monomelittoside, mussaenosidic acid)

Medicinal Use

Pedicularis verticillata is used in traditional medicine, addressing conditions including leucorrhoea, fevers, sterility, rheumatism, general debility, collapse, and urinary issues. It is also utilized to enhance blood circulation, support digestion, and sustain overall vitality.[70]

Biological Activity

Research indicates that *Pedicularis verticillata* exhibits a wide array of properties, including antioxidant, immunomodulatory, anti-inflammatory, antidiabetic, antibacterial, antifungal, analgesic, antitumor, hepatoprotective, neuroprotective, muscle-relaxant, antifatigue, diuretic, antipyretic, antithrombotic, antihemolytic, and DNA-repairing effects.[70-72]

Other Uses

- Industrial use: Dried flower stalks are utilized to produce an olive-green dye.

- Culinary use: The roots or young flowering stems of *Pedicularis verticillata* can be consumed either in their raw form or as a potherb in cooking.[73]

REFERENCES

1. Bejenaru C., Mogoșanu G.D., Bejenaru L.E., Popescu H. 2020. *Botanică farmaceutică. Cormobionta*, ediția a III-a. Ed. Sitech, Craiova.

2. Săvulescu T. (ed). 1960. *Flora RPR*, vol. VII. Ed. Academiei RPR, București.

3. Ciocârlan V. 2009. *Flora ilustrată a României. Pteridophyta et Spermatophyta*. Ediția a 3-a revizuită și adăugită. Ed. Ceres, București.

4. Sârbu I., Ștefan N., Oprea A. 2013. *Plante vasculare din România. Determinator ilustrat de teren*. Ed. Victor B Victor, București.

5. https://dryades.units.it/valerio/index.php?procedure=taxon_page&id=4749&num=1600

6. Drăgulescu C. 2018. *Dicționar de Fitonime Românești*. Ediția a 5-a completată, Editura Universității "Lucian Blaga" Sibiu.

7. https://dryades.units.it/floritaly/index.php?procedure=taxon_page&tipo=all&id=11404

8. https://dryades.units.it/floritaly/index.php?procedure=taxon_page&tipo=all&id=4750

9. https://botany.cz/cs/verbascum-speciosum/

10. Bejenaru L.E., Mogoșanu G.D., Bejenaru C., Popescu H. 2015. *Farmacognozie-Fitoterapie*. Vol. I, ediția a III-a. Ed. Sitech, Craiova.

11. Bhuvad M.A., Samant L.R. 2019. *In-silico* docking analysis of phytochemicals from *Verbascum phlomoides* L. as an antiviral agent against herpes simplex virus type I and type II. *International Journal of Pharmaceutical Sciences and Research* 10(3):1241–1245.

12. Grigore A., Colceru-Mihul S., Litescu S., Panteli M., Rasit I. 2013. Correlation between polyphenol content and anti-inflammatory activity of *Verbascum phlomoides* (Mullein). *Pharmaceutical Biology* 51(7):925–929.

13. Gvazava L.N., Kikoladze V.S. 2009. Iridoids from *Verbascum phlomoides* and *V. densiflorum*. *Chemistry of Natural Compounds* 45(5):751–752.

14. Riaz M., Zia-Ul-Haq M., Jaafar H.Z. 2013. Common mullein, pharmacological and chemical aspects. *Revista Brasileira de Farmacognosia* 23(6):948–959.

15. Nofouzi K., Mahmudi R., Tahapour K., Amini E., Yousefi K. 2016. *Verbascum speciosum* methanolic extract: phytochemical components and antibacterial properties. *Journal of Essential Oil Bearing Plants* 19(2):499–505.

16. Nofouzi K. 2015. Study on the antioxidant activity and *in vitro* antifungal activity of *Verbascum speciosum* methanolic extract. *Journal of Mycology Research* 2(2):97–103.

17. Mousavi-Kouhi S.M., Beyk-Khormizi A., Mohammadzadeh V. et al. 2022. Biological synthesis and characterization of gold nanoparticles using *Verbascum speciosum* Schrad. and cytotoxicity properties toward HepG2 cancer cell line. *Research on Chemical Intermediates* 48(1):167–178.

18. Kayir S., Demirci Y., Demirci S., Ertürk E., Ayaz E., Doğan A., Sahin F. 2018. The *in vivo* effects of *Verbascum speciosum* on wound healing. *South African Journal of Botany* 119:226–229.

19. Amirnia R., Ghiyasi M. 2011. Antimicrobial activity of *Verbascum speciosum* against three bacteria strains. *Fresenius Environmental Bulletin* 20(3a):690–693.

20. Selseleh M., Samad N.E., Atousa A., Ali S., Mirjalili M. 2020. Metabolic profiling, antioxidant, and antibacterial activity of some Iranian *Verbascum* L. species. *Industrial Crops and Products* 153:112609.

21. Demirci S., Alp, C. Akşit, H. Ulutaş, Y. Altay, A. Yeniçeri E., Köksal E., Yaylı N. 2023. Isolation, characterization and anticancer activity of secondary metabolites from *Verbascum speciosum*. *Chemical Biology & Drug Design* 101:1273–1282.

22. Khoshnoud H., Ghiyasi M., Amirnia R., Sadig Fard S., Tajbakhsh M., Salehzadeh H., Alahyary P. 2008. The potential of using insecticidal properties of medicinal plants against insect pests. *Pakistan Journal of Biological Sciences* 11:1380–1384.

23. https://luirig.altervista.org/schedenam/fnam.php?taxon=Digitalis+lanata

24. Al-Snafi A.E. 2017. Phytochemical constituents and medicinal properties of *Digitalis lanata* and *Digitalis purpurea* – A review. *Indo American Journal of Pharmaceutical Sciences* 4(02):225–234.

25. Mogoşanu G.D., Bejenaru L.E., Bejenaru C., Popescu H. 2015. *Farmacognozie-Fitoterapie*. Vol. II, ediţia a III-a. Ed. Sitech, Craiova.

26. Bertol J.W., Rigotto C., de Pádua R.M., Kreis W., Barardi C.R., Braga F.C., Simões C.M. 2011. Antiherpes activity of glucoevatromonoside, a cardenolide isolated from a Brazilian cultivar of *Digitalis lanata*. *Antiviral Research* 92(1):73–80.

27. Yang H.Y., Chen Y.X., Luo S., He Y.L., Feng W.J., Sun Y., Chen J.J., Gao K. 2022. Cardiac glycosides from *Digitalis lanata* and their cytotoxic activities. *RSC Advances* 12(36):23240–23251.

28. Jograna M.B., Patil D.S., Kotwal S.V. 2020. *Digitalis* species a potent herbal drug: A review on their pharmacognosy and pharmacological activities. *Journal of Current Pharma Research* 10(4):3821–3831.

29. https://dryades.units.it/floritaly/index.php?procedure=taxon_page&tipo=all&id=4780

30. Miyase T., Mimatsu A. 1999. Acylated iridoid and phenylethanoid glycosides from the aerial parts of *Scrophularia nodosa*. *Journal of Natural Products* 62(8):1079–1084.

31. Ahmad M., Muhammad N. et al. 2012. Spasmolytic effects of *Scrophularia nodosa* extract on isolated rabbit intestine. *Pakistan Journal of Pharmaceutical Sciences* 25(1):267–275.

32. de Santos Galíndez J., Díaz Lanza A.M., Fernández Matellano L. 2002. Biologically active substances from the genus *Scrophularia*. *Pharmaceutical Biology* 40(1):45–59.

33. Stevenson P.C., Simmonds M.S.J., Sampson J., Houghton P.J., Grice, P. 2002. Wound healing activity of acylated iridoid glycosides from *Scrophularia nodosa*. *Phytotherapy Research* 16(1):33–35.

34. Soualeh N., Stiévenard A., Baudelaire E., Bouayed J., Soulimani R. 2018. Powders with small microparticle size from *Hedera helix* and *Scrophularia nodosa* exhibited high preventive antioxidant activity against H_2O_2-induced oxidative stress in mouse primary spleen cells. *International Journal for Vitamin and Nutrition Research* 88(3–4):208–218.

35. https://temperate.theferns.info/plant/Scrophularia+nodosa

36. https://dryades.units.it/scuole/index.php?procedure=taxon_page&id=4887&num=1575

37. Mocan A., Vodnar D.C., Vlase L., Crişan O., Gheldiu M., Crişan G. 2015. Phytochemical characterization of *Veronica officinalis* L., *V. teucrium* L. and *V. orchidea* Crantz from Romania and their antioxidant and antimicrobial properties. *International Journal of Molecular Sciences* 16(9):21109–21127.

38. Nazlić M, Fredotović Ž, Vuko E, Fabijanić L, Kremer D, Stabentheiner E, Ruščić M, Dunkić V. 2021. Wild species *Veronica officinalis* L. and *Veronica saturejoides* Vis. ssp. *saturejoides*— Biological potential of free volatiles. *Horticulturae* 7(9):295.

39. Raclariu A.C., Mocan A., Popa M.O., Vlase L., Ichim M.C., Crisan G., Brysting A.K., De Boer H. 2017. *Veronica officinalis* product authentication using DNA metabarcoding and HPLC-MS reveals widespread adulteration with *Veronica chamaedrys*. *Frontiers in Pharmacology* 8:275574.

40. Crişan G, Tămaş M, Miclăuş V, Krausz T, Sandor V. 2007. A comparative study of some *Veronica species*. *Revista Medico-chirurgicala a Societatii de Medici si Naturalisti din Iasi* 111(1):280–284.

41. Lee H., Ghimeray A., Yim J., Chang M. 2015. Antioxidant, collagen synthesis activity in vitro and clinical test on anti-wrinkle activity of formulated cream containing *Veronica officinalis* extract. *Journal of Cosmetics, Dermatological Sciences and Applications* 5:45–51.

42. Suhinina T.V., Petrichenko V.M. 2009. Chemical composition and biological activity of some species of the genus *Euphrasia* (*Scrophulariaceae*) of the flora of Russia. *Rastitel'nye Resursy* 45(4):129–136.

43. https://dryades.units.it/floritaly/index.php?procedure=taxon_page&tipo=all&id=4737

44. Zia-Ul-Haq M., Kausar A., Shahid S. A., Qayum M., Ahmad S., Khan I. 2012. Phytopharmacological profile of *Gratiola officinalis* Linn.: A review. *Journal of Medicinal Plants Research* 6(16):3087–3092.

45. Šliumpaitė I., Venskutonis P.R., Murkovic M., Pukalskas A. 2013. Antioxidant properties and polyphenolics composition of common hedge hyssop (*Gratiola officinalis* L.). *Journal of Functional Foods* 5(4):1927–1937.

46. Navolokin N.A., Polukonova N.V., Maslyakova G.N., Bucharskaya A.B., Durnova N.A. 2012. Effect of extracts of *Gratiola officinalis* and *Zea mays* on the tumor and the morphology of the internal organs of rats with trasplanted liver cancer. *Russian Open Medical Journal* 1(2):0203.

47. Polukonova N.V., Navolokin N.A., Răkova S.V., et al. 2015. Anti-inflammatory, antipyretic and antimicrobial activity of flavonoid-containing extract of *Gratiola officinalis* L. *Eksperimental'naia i Klinicheskaia Farmakologiia* 78(1):34–38.

48. https://dryades.units.it/floritaly/index.php?procedure=taxon_page&tipo=all&id=4937

49. Committe for Veterinary Medicinal Products. 1999. *Euphrasia officinalis* Summary Reports. EMEA/MRL/667/99-Final.

50. Novy P., Davidova H., Serrano-Rojero C.S., Rondevaldova J., Pulkrabek J., Kokoska L. 2015. Composition and antimicrobial activity of *Euphrasia rostkoviana* Hayne essential oil. *Evidence-Based Complementary and Alternative Medicine* 2015:734101.

51. Paduch R., Woźniak A., Niedziela, P., Rejdak R. 2014. Assessment of eyebright (*Euphrasia officinalis* L.) extract activity in relation to human corneal cells using *in vitro* tests. *Balkan Medical Journal* 31(1):29–36.

52. Bigagli E, Cinci L, D'Ambrosio M., Luceri C. 2017. Pharmacological activities of an eye drop containing *Matricaria chamomilla* and *Euphrasia officinalis* extracts in UVB-induced oxidative stress and inflammation of human corneal cells. *Journal of Photochemistry and Photobiology B* 173:618–625.

53. Liu Y., Hwang E., Ngo H.T.T., Perumalsamy H., Kim Y.J., Li L., Yi T.-H. 2018. Protective effects of *Euphrasia officinalis* extract against ultraviolet B-induced photoaging in normal human dermal fibroblasts. *International Journal of Molecular Sciences* 19:3327.

54. https://pfaf.org/User/Plant.aspx?LatinName=Euphrasia+officinalis

55. https://dryades.units.it/FVG/index.php?procedure=taxon_page&id=4943&num=5064

56. Teixeira R., Silva L.R. 2013. Bioactive compounds and *in vitro* biological activity of *Euphrasia rostkoviana* Hayne extracts. *Industrial Crops and Products* 50:680–689.

57. Miladinovic D.L., Ilic B.S., Nikolic D.M, Markovic M.S., Nikolic N.D., Miladinovic L.C., Miladinovic M.D. 2014. Volatile constituents of *Euphrasia stricta*. *Chemistry of Natural Compounds* 49(6):1146–1147.

58. Khalid J.S.A.A., Khan Z.M., Zakryya M., Ullah J.N. 2022. Evaluation of phytochemical and antioxidant potential of various extracts from traditionally used medicinal plants of Pakistan. *Open Chemistry* 20(1):1337–1356.

59. Suhinina T.V., Petrichenko V.M. 2009. Chemical composition and biological activity of some species of the genus *Euphrasia* (*Scrophulariaceae*) of the flora of Russia. *Rastitel'nye Resursy* 45(4):129–136.

60. Weryszko-Chmielewska E., Matysik-Woźniak A., Sadowska D. 2012. The structure and distribution of glandular trichomes on the stems and leaves of drug eyebright (*Euphrasia stricta* D. Wolff ex J. F. Lehm.). *Acta Agrobotanica* 63:13–23.

61. www.pinterest.com/pin/374502525248633247/

62. Munteanu M.F., Vlase L. 2011. The determination of the iridoids from the *Melampyrum* species by modern chromatographic methods. *Notulae Botanicae Horti Agrobotanici Cluj-Napoca* 39(1):79–83.

63. Háznagy-Radnai E., Wéber E., Czigle S., Berkecz R., Csedő K., Hohmann J. 2014. Identification of iridoids, flavonoids and triterpenes from the methanolic extract of *Melampyrum bihariense* A. Kern. and the antioxidant activity of the extract. *Chromatographia* 77(17–18):1153–1159.

64. Galishevskaya E.E., Petrichenko V.M. 2010. Phenolic compounds from two *Melampyrum* species. *Pharmaceutical Chemistry Journal* 44(9):497–500.

65. Mogosan C., Munteanu M.-F. 2008. A comparative study on antiinflammatory effect of the tinctures from *Melampyrum bihariense* Kern and *Melapyrum cristatum* L. (*Scrophulariaceae*). *Farmacia* 56(4):389–392.

66. https://en.wikipedia.org/wiki/Melampyrum

67. https://dryades.units.it/dolomitifriulane/index.php?procedure=taxon_page&id=4980&num=3929

68. Frezza C., Venditti A., Toniolo C., Vita D.D., Serafini I., Ciccòla A., Franceschin M., Ventrone A., Tomassini L., Foddai S., Guiso M., Nicoletti M., Bianco A., Serafini M. 2019. *Pedicularis*

L. genus: systematics, botany, phytochemistry, chemotaxonomy, ethnopharmacology, and other. *Plants (Basel)* 8(9):306.

69. Venditti A., Frezza C., Sciubba F., Foddai S., Serafini M., Nicoletti M., Bianco A. 2016. Secoiridoids and other chemotaxonomically relevant compounds in *Pedicularis*: phytochemical analysis and comparison of *Pedicularis rostratocapitata* Crantz and *Pedicularis verticillata* L. from Dolomites. *Natural Product Research* 30(15):1698–1705.

70. Yatoo M.I., Dimri U., Gopalakrishnan A., Karthik K., Gopi M., Khandia R., et al. 2017. Beneficial health applications and medicinal values of *Pedicularis* plants: A review. *Biomedicine & Pharmacotherapy* 95:1301–1313.

71. Shao M.H., Dai W., Yuan S.W., Lu Y., Chen D.F., Wang, Q. 2018. Iridoids from *Pedicularis verticillata* and their anti-complementary activity. *Chemistry & Biodiversity* 15:e1800033.

72. Li M.X., He X.R., Tao R., Cao X. 2014. Phytochemistry and pharmacology of the genus *Pedicularis* used in traditional Chinese medicine. *American Journal of Chinese Medicine* 42(05):1071–1098.

73. www.flora.dempstercountry.org/0.Site.Folder/Species.Program/Species2.php?species_id= Pedi.verti

32 *Solanaceae* Adans. Family

Cornelia Bejenaru[1], Antonia Radu[1], George Dan Mogoşanu[2],
Ludovic Everard Bejenaru[2], Andrei Biţă[2], and Adina-Elena Segneanu[3]
[1] Department of Pharmaceutical Botany, Faculty of Pharmacy, University of Medicine and Pharmacy of Craiova, 2 Petru Rareş Street, Craiova, Dolj County, Romania
[2] Department of Pharmacognosy & Phytotherapy, Faculty of Pharmacy, University of Medicine and Pharmacy of Craiova, 2 Petru Rareş Street, Craiova, Dolj County, Romania
[3] Institute for Advanced Environmental Research–West University of Timisoara (ICAM–WUT), 4 Oituz Street, Timisoara, Romania

BOTANICAL DESCRIPTION

This botanical family mainly encompasses herbaceous plants, annual, biennial, or perennial, with rarely occurring subshrubs. Their leaves are unstipulate, simple or compound, entire or divided, glabrous or hairy, arranged alternately, often opposite on flowering branches. The flowers are hermaphroditic, actinomorphic, pentamerous in arrangement, forming cymose inflorescences, axillary or terminal; solitary flowers are only rarely present. The floral envelope is gamosepalous, pentadentate, parted or laciniate, persistent or accrescent, and gamopetalous, having a campanulate, rotate, infundibuliform, or hypocrateriform corolla, pentadentate or lobed, becoming deciduous after fertilization. The androecium comprises 5 stamens fused with the corolla, each with introse anthers, bilocular and poricidal dehiscence, transversal or longitudinal. The gynoecium is composed of 2 carpels (rarely more), with a superior ovary containing numerous campylotropous or anatropous ovules attached to the thickened separating placental wall. The fruit is a berry or capsule, containing numerous albuminous seeds, reniform, compressed, with a curved or bent embryo, rarely straight.[1-2]

The species of the *Solanaceae* family amount to 1700, classified into 70 genera, and are distributed in temperate climate regions as well as in warm zones.[1-2]

Atropa belladonna L. (Belladonna)

Scientific name: *Atropa belladonna* L.
Common name: Belladonna, Deadly nightshade
Romanian name: Mătrăgună,[1] Cireaşa lupului[2]
Taxonomic classification
Kingdom: *Plantae*
Subkingdom: *Cormobionta*
Phylum: *Spermatophyta*
Subphylum: *Magnoliophytina (Angiospermae)*
Class: *Magnoliatae (Magnoliopsida, Dicotyledonatae)*
Subclass: *Asteridae*
Order: *Scrophulariales (Solanales)*
Family: *Solanaceae*
Genus: *Atropa*
Species: *belladonna* L. (Figure 32.1)[1,3-4]

SPREAD

Belladonna is a wild species found in damp and shady areas, being common in mountainous regions through beech forests.[1]

DESCRIPTION

Perennial herbaceous plant, vigorous, tall (50–150 cm), with a thick, cylindrical, multicapitate, branched rhizome, and an erect stem, green or dirty reddish-violet in color, branched and glandular hairy in the superior part. Leaves are entire, petiolate, with an ovate-elliptic blade, acuminate, occasionally acute, slightly narrowed and decurrent into the petiole, with an entire margin. Leaves are green, glabrous or glandular hairy, equipped with very fine, sparse hairs, the inferior ones arranged alternately, and the superior ones opposite, unequal, one small and ovate, the other large and elliptic. Flowers, grouped 1–2 in the leaf axils, nodding, with glandular pubescent pedicels,

DOI: 10.1201/9781003270515-33

Figure 32.1 *Atropa belladonna* L. (Belladonna).

Source: 5.

have a persistent calyx at the base of the fruit, without enclosing it, deeply pentapartite, with ovate or elongated-ovate, foliaceous, acuminate, glandular pubescent lobes. Corolla is tubular-campanulate, pentalobate, with broadly triangular ovate lobes, acute or obtuse, slightly reflexed, with a length of 2–3 cm and a color ranging from brown-violet to reddish-brown. Stamens with slightly dilated filaments, hairy at the base, and large yellow dorsifixed anthers, with longitudinal dehiscence. The fruit is a spherical, slightly flattened berry, initially green, then shiny black, with a violet, toxic juice. Seeds are numerous, dark brown-blackish, with a consistent seminal tegument, reticulate, and a curved embryo. It blooms from June to August.[1-4]

Main Phytoconstituents

Belladonnae radix contains: 0.3–1% tropane alkaloids, primarily (90%) hyoscyamine, alongside hyoscyamine N-oxide, scopolamine, belladonine, atropine, tropanol, scopanol; volatile nitrogenous bases, coumarins (scopoletol, umbelliferone); organic acids; enzymes; mineral salts. [6]

Swiss researchers Steinegger and Phokas (1954) demonstrated that the anticonvulsant agent from the root is a dimeric N-methyl-pyrrolidine alkaloid, which they named belaradine. This alkaloid was already known since 1893–1895, when Liebermann isolated it from the leaves of *Erythroxylon coca* and named it cuscohigrine (cuskhigrine).[6]

Atropine or (±)-hyoscyamine is found in very small amounts, appearing during the extraction and processing of vegetable products with solvents due to the racemization of hyoscyamine. Also, from hyoscyamine, in an alkaline medium during extraction, apoatropine results (5-7% of the mixture), an ester of tropanol with atropic acid, pharmacodynamically inactive and highly toxic. Atropine is much more soluble in benzene at 15°C than hyoscyamine, and this property forms the basis for the separation of the two alkaloids.

Scopolamine represents approximately 5% of the total alkaloids.

Belladonine originates from the esterification of tropanol with isatropic acid.

According to Romanian Pharmacopeea (F.R. X.), belladonna leaves must contain at least 0.3% total alkaloids expressed as hyoscyamine. If the powder of *Belladonnae folium* contains more than 0.3% total alkaloids, it will be diluted with lactose to achieve a concentration of 0.3%. The blue-green fluorescence under UV light of coumarins (scopoletol, scopolin) has applications in the differential diagnosis of *Belladonnae folium* products because leaves from other medicinal species within the *Solanaceae* family (*Hyoscyamus niger*, *Datura stramonium*) contain only traces of coumarins. Through evaporation, upon heating in an etuve at 105°C, volatile nitrogenous bases (N-methyl-pyrrolidine, N-methyl-pyrrolidone) are removed from the acidic extract obtained from the roots and leaves of belladonna, because they hold no pharmacodynamic significance but rather analytical, as they interfere with the quantitative determination of hyoscyamine and other tropane alkaloids. Through induced polyploidization with colchicine or acenaphthene, the quantity of active principles (tropane alkaloids) is increased to 0.3–0.5%, with significant variations in content. This method is used to obtain the "of Vrancea" variety.[6]

Main phytoconstituents   Ref.

alkaloids (atropamine, atropine, scopolin, hyoscine, tropic acid, noratropine, belladonine,
   tropanol, scopanol)   6-9
coumarines (scopolin, esculin, umbelliferone)
amino acids
flavonoids (quercetin, i**cariside, rutin,** kaempherol)
terpenoids (phytol)
nitrogenous bases (pyridine, nicotine, N-methyl-pyrrolidine, N-methyl-pyrrolidone, belaradine)
sterols (daucosterol)
fatty acids (oleic acid, palmitic acid, stearic acid, cis-vaccenic acid, capric acid, linoleic acid)
phenolic acids (chlorogenic acid, caffeic acid, neochlorogenic acid)
glycosides
misceleous (pyridine, ascorbic acid)

Medicinal Use

The Romanian pharmacopoeia (F.R. X) recommends *Belladonnae folium* as an antispasmodic due to its anticholinergic (parasympatholytic) action, attributed to hyoscyamine: it relaxes smooth muscles, inhibits exocrine secretions, and induces mydriasis.[6]

As an antispasmodic, hyoscyamine alleviates gastric pains, intestinal, biliary, or renal colic, and acts as an antiemetic in motion sickness (car, plane, ship sickness). It is also used in cases of bronchial asthma and nocturnal enuresis. Through the inhibition of gastric secretions, hyoscyamine can be administered for conditions like gastric hyperacidity or night sweats, symptoms associated with pulmonary tuberculosis or other diseases.[6]

Atropine presents only half of the pharmacological effects of hyoscyamine but twice its toxicity, hence its use is limited, mainly as an antiemetic and in ophthalmology as a mydriatic, for funduscopic examination (*Oculoguttae atropini sulfatis* 1%, F.R. X). Inside the heart, following temporary bradycardia, atropine increases heart rate by suppressing the inhibitory action of the vagus nerve. In therapeutic doses, it has a depressive effect (sedative) on the central nervous system.[6]

The medicinal products *Belladonnae radix* and *Belladonnae folium* exhibit the pharmacological effects of hyoscyamine, but additionally, from the root, an extract is obtained with anticonvulsant properties due to belaradine (cuscohigrine),with applications in epilepsy, Parkinson's disease, and sequelae of epidemic encephalitis.[6]

Toxicity

In cases of overdose, tropane alkaloids lead to the appearance of characteristic symptoms: facial flushing, dryness of the mouth and mucous membranes, intense thirst, tachycardia, mydriasis, central nervous system stimulation, difficulty in speech, hyperthermia, restlessness, hallucinations, delirium, motor incoordination, insomnia, extreme fatigue, sleep, and coma. Hallucinations and delirium are distinctive: the individual speaks to themselves and may start crawling on all fours or running in circles around the room. Death occurs due to respiratory function impairment.[6]

The lethal dose for atropine is 100 mg. Consumption of belladonna berries, 10–15 for adults and 2–5 for children, often mistaken for black cherries, is lethal. The diagnosis of intoxication considers the specific fluorescence under UV light of urine or feces, caused by coumarins.

The treatment of intoxication with tropane alkaloids includes gastric lavage; administration of activated charcoal; cold compresses to reduce body temperature; oxygen therapy; slow intravenous administration of physostigmine as an antidote; intravenous diazepam or chlorpromazine for severe excitation cases.[6]

Biological Activity

- antimicrobial and antioxidant activities[10-11]

- analgesic, anti-inflammatory and neuropharmacological effects[12]

- antispasmodic, anticholinergic, sedative, analgesic, and mydriatic properties[13]

- neurotoxicity[14]

Other Uses

- In the past, it was used as a recreational drug, poison, and female attractiveness enhancer.[15]

In the spontaneous flora of Romania, the genus *Hyoscyamus* comprises two species: *Hyoscyamus niger* L. and *Hyoscyamus albus* L.[3-4]

Hyoscyamus niger L. (Henbane)

Scientific name: *Hyoscyamus niger* L.
Common name: Henbane, Black henbane, Stinking nightshade
Romanian name: Măselariţă
Taxonomic classification
Kingdom: *Plantae*
Subkingdom: *Cormobionta*
Phylum: *Spermatophyta*
Subphylum: *Magnoliophytina (Angiospermae)*
Class: *Magnoliatae (Magnoliopsida, Dicotyledonatae)*
Subclass: *Asteridae*
Order: *Scrophulariales (Solanales)*
Family: *Solanaceae*
Genus: *Hyoscyamus*
Species: *niger* L. (Figure 32.2)[1,3-4]

SPREAD

Henbane is a common plant found from plains to mountains, in ruderal areas, ditches, rubble, as a weed in fields, meadows, and uncultivated places.[1-2]

DESCRIPTION

This biennial herbaceous plant, glandular hairy, emits an unpleasant, heavy odor which fades upon drying. The tap root is thick, napiform in the superior part. The stem is erect (30–80 cm tall), obtusely angled, and branched. Stem leaves are sessile, semi-amplexicaul, elongated-ovate or triangular-lanceolate, sinuate-lobate or divided, with an acute or acuminate tip, while basal leaves are long-petiolate, elliptic or elongated-ovate, with a sinuate-lobed or divided blade, pinnate venation, dark green on the superior surface, and grayish-green on the inferior surface, densely and long-hairy, especially along the veins. The flowers are sessile, arranged at the apex of the stem or branches, with elongated or narrow lanceolate sessile bracts. The calyx is accrescent, campanulate, pentadentate, with triangular, acuminate teeth and the corolla is dirty yellow with reddish-violet reticulate venation, infundibuliform, glabrous inside and hairy outside. The androecium consists of 5 stamens, 2 shorter and 3 longer, epipetalous, with finely hairy filaments at the base and violet

Figure 32.2 *Hyoscyamus niger* L. (Henbane).

Source: 16.

anthers. The fruit is a capsule, dilated at the base, dehiscent through a superior cap (pyxis), covered by the accrescent calyx, longer than the pyxis. The pyxis contains numerous small, reniform brownish seeds. The flowering period is from June to August.[1-2]

Main phytoconstituents   Ref.
alkaloids (hyoscyamine, scopolamine, atropine, hyoscine, apoatropine, tropine),
    hyoscypikrin, cuscohygrine, daturamine)   6, 17-18
glycosides (hyoscyamoside A-F, atroposide, petunioside, pyranoside)
coumarines (hyosgerin, venkatasin, cleomiscosin A-B)
amino acids
volatile nitrogenous bases (tetramethylputrescine)
flavonoids (**rutin,** hyosgerin, venkatasin, cleomiscosin)
sterols (β-sitosterol, daucosterol)
fatty acids (oleic acid, linoleic acid)
phenolic acids (vannilic acid)
lignans (hyosmin, hyoscyamal, balanophonin)
tannins (catechin**)**
misceleous (cannabisin A, cannabisin G, glycerol)

Medicinal Use

The medicinal vegetable product *Hyoscyami folium* has a parasympatholytic and CNS sedative action similar to belladonna leaves but of lower intensity due to its lower alkaloid content.[6]

Toxicity

Due to scopolamine, Henbane leaves possess mild hypnotic, CNS depressant, and psychotropic effects, which at high doses can induce dizziness, amnesia, and hallucinations.[6]

Biological Activity

Hyoscyami folium is a source of tropane alkaloids (hyoscyamine) and exhibits relaxing action on smooth muscles (antispasmodic), inhibitory effects on exocrine gland secretion, mydriatic action, and CNS sedation.[1]

■ antihistaminic, antimicrobial, antispasmodic, analgesic, antiinflammatory, antiallergic andsedative[18]

Other Uses

Plasters infused with henbane extract, applied behind the ear, are said to alleviate discomfort in individuals prone to motion sickness. Henbane oil is employed for therapeutic massages. The dried and chopped leaves and aerial parts of henbane, excluding the roots, find use in medicinal applications, as well as in incense and smoking mixtures. They are also utilized in brewing beer and tea, and as a seasoning for wine.[19]

Datura stramonium L. (Thorn apple)

Scientific name: *Datura stramonium* L.
Common name: Thorn apple, Jimsonweed, Devil's snare, Devil's trumpet
Romanian name: Ciumăfaie
Taxonomic classification
Kingdom: *Plantae*
Subkingdom: *Cormobionta*
Phylum: *Spermatophyta*
Subphylum: *Magnoliophytina (Angiospermae)*
Class: *Magnoliatae (Magnoliopsida, Dicotyledonatae)*
Subclass: *Asteridae*
Order: *Scrophulariales (Solanales)*
Family: *Solanaceae*
Genus: *Datura*
Species: *stramonium* L. (Figure 32.3)[1,3-4]

SPREAD

A cosmopolitan species, native to North America, found spontaneously in cultivated and ruderal areas, from plains to the oak forest zone, where it is frequently encountered.[1,3]

DESCRIPTION

An annual weed, tall (120 cm), with a white, fusiform, and branched taproot, and erect, glabrous, fistulous, and dichasially branched stems. The leaves are petiolate, narrowed at the petiole, with an ovate and irregularly sinuate blade, with acute lobes. The pedicellate flowers are solitary and erect, with a tubular calyx (falls after flowering) and a white infundibuliform corolla, 5–10 cm long. The fruit is a dehiscent capsule with four valves. The capsule is ovoid and spiny, containing numerous small, black, reniform seeds, with a reticulate dotted surface. Flowering occurs between June and September.[1-2]

Datura stramonium L. var. *tatula* (L.) Torrey has stems and flowers of violet color.[4]

Main phytoconstituents	Ref.
alkaloids (hyoscyamine, scopolamine, atropine, solanine, cuscohygrine, apoatropine, norscopolamine, noratropine, scopanol, genhyoscyamine, genscopolamine)	6,21-23
volatile nitrogenous bases (pyridine, N-methyl-pyrrolidine, N-methyl-pyrrolidone, belaradine)	
coumarines (esculetin, scopolin)	
amino acids	
flavonoids (rutin, hyperoside)	
terpenoids (phytol, ionone, geraniol, citral, linalool, α-bisabolol, farnesol, caryophyllene, levomenthol)	
sterols (stigmasterol)	
fatty acids (daturic acid)	

Figure 32.3 *Datura stramonium* L. (Thorn apple).

Source: 20.

phenolic acids (chlorogenic acid, caffeic acid, ferrulic acid)
tnnins (catechin)

Medicinal Use

The medicinal product is a parasympatholytic, nervous sedative, and hypnotic. It is used as an industrial source for extracting atropine and scopolamine. In the form of fumigations, it can be used as a bronchodilator in the treatment of certain respiratory conditions.[6]

Biological Activity

Datura stramonium exhibits properties such as antimicrobial, analgesic, anticholinergic, anticancer, anti-inflammatory, antifungal, vibriocidal, and antiasthmatic activities.[24-25]

Other Uses

Datura stramonium also possesses acaricidal, repellent, and oviposition deterrent attributes. It demonstrates larvicidal and mosquito repellent effects. It shows potential as a biopesticide with antifungal properties and serves as a protective agent in cases of severe organophosphate toxicity.[25]

Scopolia carniolica Jacq. (Henbane bell)

Scientific name: *Scopolia carniolica* Jacq.
Common name: Henbane bell
Romanian name: Mutulică
Taxonomic classification
Kingdom: *Plantae*
Subkingdom: *Cormobionta*
Phylum: *Spermatophyta*
Subphylum: *Magnoliophytina (Angiospermae)*

Figure 32.4 *Scopolia carniolica* Jacq. (Henbane bell).

Source: 26.

Class: *Magnoliatae (Magnoliopsida, Dicotyledonatae)*
Subclass: *Asteridae*
Order: *Scrophulariales (Solanales)*
Family: *Solanaceae*
Genus: *Scopolia*
Species: *carniolica* Jacq. (Figure 32.4)[1,3-4]

SPREAD

The plant is sporadic from the oak forest zone to the spruce forest zone, occurring in damp and shaded areas, on humus-rich soils, thickets, forests, and forest clearings.[2-4]

DESCRIPTION

Perennial species with a thick, fleshy rhizome, horizontally positioned. Erect stem, reaching a height of 60 cm, glabrous or sparsely hairy, with petiolate, elliptic, entire leaves narrowing at the petiole. The flowers are solitary and nodding, with long, filiform pedicels, a bluish-campanulate calyx, slightly lobed, and a cylindrical corolla, violet on the exterior and brownish-yellow on the interior. The fruit is a pyxis. The seeds are reniform, alveolate-reticulate, and yellowish-brown. These plants bloom from April to June.[1-4]

Main Phytoconstituents

Scopoliae rhizoma contains approximately 0.1–1% tropane alkaloids, predominantly hyoscyamine, along with scopolamine and belaradine.

Tropane alkaloids of medicinal importance, namely anisodine (the ester of scopanol with 2'-hydroxy-tropic acid) and anisodamine (the ester of 6β-hydroxy-tropanol with tropic acid), have also been isolated from the roots of the species *Scopolia tangutica* Maxim. syn. *Anisodus tanguticus* (Maxim.) Pasch (*Solanaceae*). The root powder (*zang qie*) is included in the formulation of traditional Chinese preparations with sedative and anesthetic effects.[6]

Main phytoconstituents   Ref.
alkaloids (scopolamine, atropine, hyoscyamine)   6,27-29
glycosides
coumarines (fabiatrin)
amino acids
flavonoids (quercetin, kaempferol, rutin)
terpenoids
sterols
fatty acids
phenolic acids (chlorogenic acid)
lignans
miscellaneous

Medicinal Use

Scopoliae rhizoma is industrially processed to obtain scopolamine. Scopolamine has parasympatholytic action (antispasmodic, mydriatic, exocrine antisecretory), depressant effects on the central nervous system (hypnotic, psychotropic, hallucinogenic), and analgesic properties.

Anisodine acts as a depressant on the central nervous system and is used in the treatment of migraine headaches. Anisodamine exhibits CNS stimulant, parasympatholytic, and vasodilatory effects. It is used in the treatment of acute enteritis and septic shock caused by bacillary dysentery.[6]

Biological Activity

Acetylcholinesterase inhibitory activity – the presence of high levels of atropine and scopolamine in *Scopolia carniolica* leads to its anticholinergic effects on muscarinic receptors, acting as competitive antagonists of acetylcholine.[30-32]

Other Uses

In Romania, the species is found in limited areas within the native flora, and from some of these regions, the plant is collected for industrial extraction of scopolamine and atropine, alkaloids that are contained in the rhizomes (*Scopoliae rhizoma*).[33]

Physalis alkekengi L. (Chinese lantern)

Scientific name: *Physalis alkekengi* L.
Common name: Chinese lantern
Romanian name: Păpălău
Taxonomic classification
Kingdom: *Plantae*
Subkingdom: *Cormobionta*
Phylum: *Spermatophyta*
Subphylum: *Magnoliophytina (Angiospermae)*
Class: *Magnoliatae (Magnoliopsida, Dicotyledonatae)*
Subclass: *Asteridae*
Order: *Scrophulariales (Solanales)*
Family: *Solanaceae*
Genus: *Physalis*
Species: *alkekengi* L. (Figure 32.5)[1,3-4]

SPREAD

Common from the sylvosteppe zone to the beech forest zone, occurring in thickets, forest clearings, vineyards, and floodplains, as well as in acacia plantations.[2,4]

DESCRIPTION

A perennial plant with a slender, cylindrical, creeping rhizome and fibrous adventitious roots. Erect stems (60 cm tall), obtusely edged, short-pubescent, simple or branched. Leaves are long-petiolate, simple, with an ovate blade and narrowed base, finely hairy on both surfaces. Solitary flowers, pedicellate, axillary, have a vesiculate calyx, persistent, swollen ovoid, becoming reticulate when fruiting, reddish-orange in color, and a rotate corolla with ciliated edges, dirty white in color. The

Figure 32.5 *Physalis alkekengi* L. (Chinese lantern).

Source: 34.

fruit is a spherical berry (with a diameter of at least 1 cm), glossy, reddish-orange, completely surrounded by the accrescent calyx (3–4 cm long). Seeds are elongated, reniform, finely verrucose, and yellow in color. Flowering occurs from June to August.[1-4]

Main phytoconstituents   Ref.
amino acids (aspartic acid, glutamic acid, phenylalanine, serine, histidine, arginine, threonine, alanine, valine, methionine, lysine, isoleucine)   [35-41]
withanolides (physalin A, physalin G, physalin O, physalin L, isophysalin,neophysalin)
alkaloids (tropan. pyrrolidine alkaloids, phenylalkylamine alkaloids)
flavonoids (catechin, quercetin, rutin, luteolin, naringenin, kaempferol, apigenin)
terpenoids (pinene, camphene, sabinene, myrcene, cymene, limonene, germacrene D, selinene, phytol, squalene)
sucrose esters (physakengoses B, E, F, G, H, K, L, M, N, O)
fatty acids (capric, lauric, myristic acid, margaric acid, palmitic acid, oleic acid, linoleic acid)
phenolic acids (chlorogenic acid, caffeic acid, ferulic acid, gallic acid)
carbohydrates (mannose, sucrose, glucose, galactose, fructose, xylose)
organic acids (succinic acid, malonic acid, citric acid, fumaric acid, oxalic acid)
miscellaneous (ascorbic acid, β-carotene, physakengose G)[16-1]

Medicinal Use

The fruits (*Physalidis fructus*) are used as uricosuric diuretics and laxatives.[1]

Biological Activity

Physalidis fructus has demonstrated positive impacts on hyperglycemia. This is achieved through the inhibition of oxidative stress, inflammation, and apoptosis, making it a valuable supplementary approach for managing diabetes.[38]

439

Numerous research studies have revealed a range of effects associated with physalins, encompassing antitumor, anti-inflammatory, antiparasitic, antimicrobial, immunomodulatory, antinociceptive, and antiviral activities.[39-41]

Other Uses

■ Cultivated for ornamental purposes.[2]

■ Edible uses: The fruit can be consumed either raw or cooked. It is abundant in vitamins, containing twice the amount of vitamin C found in lemons. However, it lacks strong flavor. The fruit is juicy but possesses a bitter and acrid taste. It can be used to enhance the taste of salads. It is important to note that the calyx of the fruit is toxic and should not be ingested. Young leaves can be cooked for consumption. Caution is necessary, as the leaves are poisonous, especially when consumed raw.[42]

The *Solanum* genus is represented in the spontaneous flora of Romania by the following species: *Solanum rostratum* Dunal, *Solanum triflorum* Nutt., *Solanum retroflexum* Dunal, *Solanum nigrum* L., *Solanum alatum* Moench (*S. luteum* Miller subsp. *alatum* (Moench) Dostál), *Solanum villosum* Mill. s. str. (*S. luteum* Mill. subsp. *luteum*), *Solanum dulcamara* L., and *Solanum carolinense* L.[3,4]

Solanum nigrum L. (Black nightshade)

Scientific name: *Solanum nigrum* L.
Common name: Black nightshade, Makoi
Romanian name: Zârnă
Taxonomic classification
Kingdom: *Plantae*
Subkingdom: *Cormobionta*
Phylum: *Spermatophyta*
Subphylum: *Magnoliophytina (Angiospermae)*
Class: *Magnoliatae (Magnoliopsida, Dicotyledonatae)*
Subclass: *Asteridae*
Order: *Scrophulariales (Solanales)*
Family: *Solanaceae*
Genus: *Solanum*
Species: *nigrum* L. (Figure 32.6)[1,3-4]

SPREAD

A common species found from the steppe zone to the beech forest zone through vineyards, pastures, floodplains, cultivated fields, vegetable gardens, seedbeds, roadside verges, and ruderal areas.[2-3]

DESCRIPTION

An annual segetal weed, reaching a height of 60 cm, with dark green color and scattered, curved and adherent hairs. It has erect, branched stems that are cylindrical in the inferior part and edged in the superior part. The leaves are petiolate, ovate-lanceolate, dentate to lobed, with an acute tip and a cuneate base decurrent into the petiole. The white flowers are grouped in clusters of 5–10, with a persistent glabrous calyx, a rotate corolla, and stamens with short filaments and yellow-golden anthers. The fruit is a spherical berry, usually black and shiny, although occasionally it can be green. The seeds are reniform and ovate-lenticular. It blooms from June to October.[1-3]

Solanum nigrum L. has two subspecies:

■ subsp. *nigrum* includes non-glandular plants, glabrous to sparsely pubescent, with short and adherent hairs;[3-4]

■ subsp. *schultesii* (Opiz) Wessely are glandular plants, hairy especially on the stem, with long and spreading hairs mixed with simple hairs.[3-4]

Figure 32.6 *Solanum nigrum* L. (Black nightshade).

Source: 43.

Main phytoconstituents   Ref.
44-45
amino acids (phenylalanine, glycine, proline, tyrosine)
steroidal saponins (diosgenin degalactotigonin, solanigroside)
alkaloids (solasonine, solamargine, solaoiacid, solasodine, tomatidenol, solanocapsine,
   solanaviol, leptinine)
coumarine (scopoletin)
nucleoside (adenosine, uracil, uridine)
glycoside (cinnacassoside A)
flavonoids (kaempferol, quercitrin, quercetin, rutin, narigenin)
terpenoids (ursolic acid, Linolenic acid)
sterols (stigmaterol, daucosterol, beta-sitosterol**)**
fatty acids (palmitic acid, oleic acid, linoleic acid)
phenolic acids (chlorogenic acid, caffeic acid, ferulic acid, gallic acid, vanillic acid, salicylic acid)
lignans (syringaresinol, acanthoside D, medioresinol, pinoresinol)
carbohydrates (mannose, glucose, galactose, arabinose, rhamnose, xylose)
organic acids (succinic acid, formic acid, fumaric acid)
miscellaneous (cannabisin F, choline, betaine, dihydrocapsaicin, carotenoids, xanthophyll)

Medicinal Use

In traditional practices, *Solanum nigrum* has been employed to address a range of conditions such as different types of cancers, acute nephritis, urethritis, leucorrhea, sore throat, toothache, dermatitis, eczema, carbuncles, and furuncles.[44]

Biological Activity

Studies exploring the pharmacological effects of *Solanum nigrum* have demonstrated its various therapeutic effects, such as antitumor, anti-inflammatory, antiproliferative, antiseizure, hepatoprotective, antioxidant, antibacterial, and neuroprotective actions.[44-45]

Other Uses

Culinary applications: The fruit is prepared by cooking and is commonly used in making preserves, jams, and pies. It offers a delightful musky flavor that somewhat resembles a tomato. The taste tends to improve slightly after exposure to frost. It is important to utilize fully ripe fruits only, as the unripe ones contain the toxin solanine. The fruit contains approximately 2.5% protein, 0.6% fat, 5.6% carbohydrate, and 1.2% ash. Additionally, young leaves and fresh shoots can be consumed either raw or cooked as a potherb or added to soups.[46]

REFERENCES

1. Bejenaru C., Mogoșanu G.D., Bejenaru L.E., Popescu H. 2020. *Botanică farmaceutică. Cormobionta*, ediția a III-a. Ed. Sitech, Craiova.

2. Săvulescu T. (ed). 1960. *Flora RPR*, vol. VII. Ed. Academiei RPR, București.

3. Ciocârlan V. 2009. *Flora ilustrată a României. Pteridophyta et Spermatophyta*. Ediția a 3-a revizuită și adăugită. Ed. Ceres, București.

4. Sârbu I., Ștefan N., Oprea A. 2013. *Plante vasculare din România. Determinator ilustrat de teren*. Ed. Victor B Victor, București.

5. www.infoflora.ch/en/flora/atropa-bella-donna.html

6. Bejenaru L.E., Mogoșanu G.D., Bejenaru C., Popescu H. 2015. *Farmacognozie-Fitoterapie.* Vol. I, ediția a III-a. Ed. Sitech, Craiova.

7. Jiang X., Chi J., Xu E.P., Wang Z.M., Dai L.P. 2023. Chemical constituents from *Atropa belladonna* roots. *Chemistry of Natural Compounds* 59:556–558.

8. Danaie E., Masoudi S., Masnabadi N. 2023. Chemical composition analysis of *Atropa belladonna* grown in Iran and evaluation of antibacterial properties of extract-loaded nanofibers. *Iranian Journal of Pharmaceutical Research* 22(1):e137839.

9. Rowson J.M. 1950. The pharmacognosy of *Atropa belladonna* Linn. *Journal of Pharmacy and Pharmacology* 2(1):201–216.

10. Öz M., Fidan M.S., Baltaci C., Ücüncü O., Karatas S.M. 2021. Determination of antimicrobial and antioxidant activities and chemical components of volatile oils of *Atropa belladonna* L. growing in Turkey. *Journal of Essential Oil Bearing Plants* 24(5):1072–1086.

11. Munir N., Iqbal A.S., Altaf I., Bashir R., Sharif N., Saleem F., Naz S. 2014. Evaluation of antioxidant and antimicrobial potential of two endangered plant species *Atropa belladonna* and *Matricaria chamomilla*. *African Journal of Traditional, Complementary and Alternative Medicines* 11(5):111–117.

12. Owais F., Anwar S., Saeed F., Muhammad S., Ishtiaque S., Mohiuddin O. 2014. Analgesic, anti-inflammatory and neuropharmacological effects of *Atropa belladonna*. *Pakistan Journal of Pharmaceutical Sciences* 27(6 Spec No.):2183–2187.

13. Sepehri E., Hosseini B., Hedayati A. 2022. The effect of iron oxide nano-particles on the production of tropane alkaloids, *h6h* gene expression and antioxidant enzyme activity in *Atropa belladonna* hairy roots. *Russian Journal of Plant Physiology* 69:122.

14. Kwakye G.F., Jiménez J, Jiménez J.A., Aschner M. 2018. *Atropa belladonna* neurotoxicity: implications to neurological disorders. *Food and Chemical Toxicology* 116(Pt B):346–353.

15. https://en.wikipedia.org/wiki/Atropa_belladonna

16. https://dryades.units.it/rosandra_de/index.php?procedure=taxon_page&id=4699&num=3592

17. Li J., Shi J., Yu X.W., Sun J.K., Men Q.M., Kang T.G. 2011. Chemical and pharmacological researches on *Hyoscyamus niger*. *Chinese Herbal Medicines* 3(2):117–126.

18. Aparna K., Joshi Abhishek J., Mahesh V. Phyto-chemical and pharmacological profiles of *Hyoscyamus niger* Linn (Parasika Yavani) — A review. *Pharma Science Monitor* 6(1):153–158.

19. https://en.wikipedia.org/wiki/Hyoscyamus_niger

20. https://dryades.units.it/floritaly/index.php?procedure=taxon_page&tipo=all&id=4729

21. Sharma M., Dhaliwal I., Rana K., Delta A.K., Kaushik P. 2021. Phytochemistry, pharmacology, and toxicology of *Datura* species—A review. *Antioxidants* 10(8):1291.

22. Gaire B.P., Subedi L. 2013. A review on the pharmacological and toxicological aspects of *Datura stramonium* L. *Journal of Integrative Medicine* 11(2):73–79.

23. Aboluwodi A.S., Avoseh O.N., Lawal O.A., Ogunwande I.A., Giwa A.A. 2017. Chemical constituents and anti-inflammatory activity of essential oils of *Datura stramonium* L. *Journal of Medicinal Plants Studies* 5(1):21–25.

24. Eftekhar F., Yousefzadi M., Tafakori V. 2005. Antimicrobial activity of *Datura innoxia* and *Datura stramonium*. *Fitoterapia* 76(1):118–120.

25. Soni P., Siddiqui A.A., Dwivedi J., Soni V. 2012. Pharmacological properties of *Datura stramonium* L. as a potential medicinal tree: an overview. *Asian Pacific Journal of Tropical Biomedicine* 2(12):1002–1008.

26. https://dryades.units.it/FVG/index.php?procedure=taxon_page&id=4698&num=4265

27. Fatur K., Ravnikar M., Kreft S. 2021. *Scopolia carniolica* var. *hladnikiana*: Alkaloidal analysis and potential taxonomical implications. *Plants (Basel)* 10(8):1643.

28. Nowak S., Wolbiś M. 2002. Flavonoids from some species of genus *Scopolia* Jacq. *Acta Poloniae Pharmaceutica* 59(4):275–280.

29. Wolbiś M., Nowak S., Kicel A. 2007. Polyphenolic compounds in *Scopolia caucasica* Kolesn. ex Kreyer (*Solanaceae*). *Acta Poloniae Pharmaceutica* 64(3):241–246.

30. Rollinger J.M., Hornick A., Langer T., Stuppner H., Prast H. 2004. Acetylcholinesterase inhibitory activity of scopolin and scopoletin discovered by virtual screening of natural products. *Journal of Medicinal Chemistry* 47(25):6248–6254.

31. Ştefănescu C., Vlase L., Tămaş M., Crişan G. 2013. The analysis of coumarins from *Scopolia carniolica* Jacq. (*Solanaceae*) of Romanian spontaneous flora. *Studia Universitatis Babes-Bolyai Chemia* 58:71–79.

32. Vitjan F., Samo K. 2021. Determining the content of tropane alkaloids in traditional preparations of carniolan belladonna (*Scopolia carniolica*). Master's Thesis, University of Ljubljana, Slovenia. Retrieved from: https://repozitorij.uni-lj.si/IzpisGradiva.php?lang=eng&id=133162.

33. Stefanescu C., Tamas M., Barbu-Tudoran L. 2006. Anatomical studies on *Scopolia carniolica* Jacq. vegetative organs. *Notulae Botanicae Horti Agrobotanici Cluj-Napoca* 34(1):12–17.

34. https://dryades.units.it/floritaly/index.php?procedure=taxon_page&tipo=all&id=10293

35. Popova V., Petkova Z., Mazova N., Ivanova T., Petkova N., Stoyanova M., Stoyanova A., Ercisli S., Okcu Z., Skrovankova S. et al. 2022. Chemical composition assessment of structural parts (seeds, peel, pulp) of *Physalis alkekengi* L. fruits. *Molecules* 27(18):5787.

36. Popova V., Ivanova Tanya, Stoyanova M., Mazova N., Dimitrova-Dyulgerova I., Stoyanova A. et al. 2022. Phytochemical analysis of leaves and stems of *Physalis alkekengi* L. (*Solanaceae*). *Open Chemistry* 20(1):1292–1303.

37. Popova V., Mazova N., Ivanova T., Petkova N. et al. 2022. Phytonutrient composition of two phenotypes of *Physalis alkekengi* L. fruit. *Horticulturae* 8:373.

38. Vicas L.G., Jurca T., Baldea I., Filip G.A. et al. 2020. *Physalis alkekengi* L. extract reduces the oxidative stress, inflammation and apoptosis in endothelial vascular cells exposed to hyperglycemia. *Molecules* 25:3747.

39. Fu Y., Zhu F., Ma Z. et al. 2021. *Physalis alkekengi* var. *franchetii* extracts exert antitumor effects on non-small cell lung cancer and multiple myeloma by inhibiting STAT3 signaling. *OncoTargets and Therapy* 14:301–314.

40. Yang J., Sun Y., Cao F., Yang B., Kuang H. 2022. Natural products from *Physalis alkekengi* L. var. *franchetii* (Mast.) Makino: A review on their structural analysis, quality control, pharmacology, and pharmacokinetics. *Molecules* 27(3):695.

41. Meira C.S., Soares J.W.C., Dos Reis B.P.Z.C., Pacheco L.V., Santos I.P. et al. 2022. Therapeutic applications of physalins: Powerful natural weapons. *Frontiers in Pharmacology* 13:864714.

42. https://pfaf.org/user/Plant.aspx?LatinName=Physalis+alkekengi

43. https://dryades.units.it/floritaly/index.php?procedure=taxon_page&tipo=all&id=4707

44. Chen X., Dai X., Liu Y., Yang Y., Yuan L., He X., Gong G. 2022. *Solanum nigrum* Linn.: An insight into current research on traditional uses, phytochemistry, and pharmacology. *Frontiers in Pharmacology* 13:918071.

45. Gabrani R., Jain R., Sharma A., Sarethy I.P., Dang S., Gupta S. 2012. Antiproliferative effect of *Solanum nigrum* on human leukemic cell lines. *Indian Journal of Pharmaceutical Sciences* 74(5):451–453.

46. https://pfaf.org/user/plant.aspx?LatinName=Solanum+nigrum

33 *Thymelaeaceae* Adans. Family

Cornelia Bejenaru[1], Antonia Radu[1], George Dan Mogoşanu[2],
Ludovic Everard Bejenaru[2], Andrei Biţă[2], and Adina-Elena Segneanu[3]

[1] Department of Pharmaceutical Botany, Faculty of Pharmacy, University of Medicine and Pharmacy of Craiova, 2 Petru Rareş Street, Craiova, Dolj County, Romania
[2] Department of Pharmacognosy & Phytotherapy, Faculty of Pharmacy, University of Medicine and Pharmacy of Craiova, 2 Petru Rareş Street, Craiova, Dolj County, Romania
[3] Institute for Advanced Environmental Research–West University of Timisoara (ICAM–WUT), 4 Oituz Street, Timisoara, Romania

BOTANICAL DESCRIPTION

Annual herbaceous or woody plants (shrubs, subshrubs) with simple, linear, sublanceolate or acicular, glabrous leaves, with entire margins, not stipulate, arranged alternately or opposite. The flowers are generally hermaphroditic and actinomorphic, with a 4-parted arrangement, solitary or in racemose inflorescences (raceme, spike, umbel). The floral envelope is inserted on the edge of a tubular-campanulate receptacle, often with reduced or absent petals (in the plants that grow in Romania, the floral envelope is simple, with an androecium of 8 stamens in two whorls, and an inferior gynoecium). The flowers have the axis transformed into a campanulate, cylindrical, or urceolate hypanthium. The fruit is a drupe, berry, or capsule. The seeds have a thin and hard seed coat, a straight embryo, and fleshy and thick cotyledons. Among the approximately 500 species, encompassing around 50 genera of this family, there are five species found in our country belonging to two genera (*Daphne* and *Thymelaea*).[1-3]

The genus *Daphne* in our country's spontaneous flora includes the following species: *Daphne blagayana* Freyer, *Daphne cneorum* L., *Daphne laureola* L., and *Daphne mezereum* L. The genus *Thymelaea* has only one species: *Thymelaea passerina* (L.) Coss. et Germ.[2,4]

Daphne blagayana Freyer (Balkan daphne)

Scientific name: *Daphne blagayana* Freyer
Common name: Balkan daphne
Romanian name: Iederă, Iederă albă[2], Floarea călugărilor[5]
Taxonomic classification
Kingdom: *Plantae*
Subkingdom: *Cormobionta*
Phylum: *Spermatophyta*
Subphylum: *Magnoliophytina (Angiospermae)*
Class: *Magnoliatae (Magnoliopsida, Dicotyledonatae)*
Subclass: *Rosidae*
Order: *Thymelaeales*
Family: *Thymelaeaceae*
Genus: *Daphne*
Species: *blagayana* Freyer (Figure 33.1)[1-2,4]

SPREAD

The species is rarely found from the beech zone to the spruce zone, on limestone screes and rocky areas, clearings of deciduous or coniferous forests. It is found in the Ciucaş Mountains, Piatra Mare, Postăvarul, Gârbova, Bucegi, Cozia, Vâlcan, Buila, Piatra Cloşanilor, and Bihor. It is a protected species by law.[2-4]

DESCRIPTION

Dwarf shrub, with a height of 15–35 cm, sparsely branched, with creeping and erect stems. The branches have leaves only towards the apex. The leaves are sessile, evergreen, obovate or elongated-elliptical, cuneate at the base, and with an obtuse tip. The flowers are short-pedicellate, yellowish-white in color, with a narrow, cylindrical hypanthium, on which 4–5 perigonial folioles, broadly ovate and obtuse, are fused, emitting a pleasant and intense fragrance. The flowers are grouped in large terminal capitula and accompanied by silky-hairy hypsophylls of a light color.

Figure 33.1 *Daphne blagayana* Freyer (Balkan daphne).

Source: 6.

The fruit is a glabrous, juicy, fleshy, golden-yellow berry. The flowering period is from April to May.[2,3]

Main phytoconstituents   Ref.
coumarins (daphnetin, daphnin, daphnesid, umbelliferone, esculetin, triumbellin)   [7]
flavonoids (daphnodorins)
terpenoids (daphauranol A-C, acutilobins, ursolic acid, taraxerol, taraxerone, β-viscol, amirin)
sterols (sitosterol)
fatty acids
phenolic
tannins
organic acids
lignans (sesamin, pinoresinol, lariciresinol)
glycosides
misceleous

Medicinal Use

Plants belonging to this species have been utilized for their medicinal properties due to their antioxidant effects, which make them valuable in treating skin diseases, toothaches, and malaria. Moreover, these plants can serve as natural laxatives or anticoagulants. The reduction of certain cytokines directly linked to the inflammatory response contributes to these beneficial effects.[7,8]

Biological Activity

antioxidant and antimicrobial activity[7,8]

Other Uses

- ornamental

Daphne mezereum L. (Paradise plant)

Scientific name: *Daphne mezereum* L.
Common name: Paradise plant
Romanian name: Tulichină
Taxonomic classification

Figure 33.2 *Daphne mezereum* L. (Paradise plant).

Source: 9.

Kingdom: *Plantae*
Subkingdom: *Cormobionta*
Phylum: *Spermatophyta*
Subphylum: *Magnoliophytina (Angiospermae)*
Class: *Magnoliatae (Magnoliopsida, Dicotyledonatae)*
Subclass: *Rosidae*
Order: *Thymelaeales*
Family: *Thymelaeaceae*
Genus: *Daphne*
Species: *mezereum* L. (Figure 33.2)[1,2,4]

SPREAD

The species is found isolated or in scattered clusters in the beech zone up to the subalpine zone, through forest clearings, shrubbery, and glades.[2-4]

DESCRIPTION

It is a short shrub (0.5–1.5 m) with erect or ascending branched stems. The young branches are yellowish or brownish-green, finely downy, while the older ones are thick, gray-brown, and bear leaf scars from previous years. Leaves are present only towards the top of the branches. They appear late after flowering and fall in autumn. The leaves are short-petioled, ovate, obovate, or obovate-lanceolate, glabrous, soft, finely hairy along the edges, intensely green on the upper surface, and brownish-green on the lower surface. The flowers appear before the leaves, are pleasantly fragrant, sessile, and grouped axillary towards the top of the branches, with a pink-purple color (rarely white). The hypanthium is silky-hairy and fuses with the sepals at the top,

447

later becoming deciduous. The fruit is a red, juicy, ovate, or ovate-lanceolate drupe. The seeds are brown, globular, or broadly ovate. It blooms in the months of February to April.[1-3]

Toxic plant (especially the fruits and bark) and also used as a dye (tinctorial).[1]

Main phytoconstituents	Ref.
coumarins (daphnetin, daphnin, daphnesid, umbelliferone, esculetin, triumbellin)	7-8, 10-11

flavonoids (daphnodorins, dismeotin)
terpenoids (daphnetoxin, mezerein, gniditrin, ursolic acid, amirin)
sterols (sitosterol)
fatty acids
phenolic
tannins
organic acids
lignans (sesamin, pinoresinol, lariciresinol)
glycosides
miscellaneous

Medicinal Use

From the bark, a tincture is prepared and used in folk medicine for treating chronic dermatitis and rheumatic pains.[1]

Biological Activity

An extract derived from *Daphne mezereum* L., commonly used in traditional medicine for cancer treatment, demonstrated antileukemic effects against lymphocytic leukemia in mice. Through systematic fractionation of the extract, researchers have successfully isolated and identified mezerein as the primary antileukemic compound.[12,13]

Other Uses

The aqueous leaf extract of *Daphne mezereum* enables the eco-friendly synthesis of iron oxide nanoparticles, which can serve as an innovative material for removing dyes.[14]

Thymelaea passerina (L.) Coss. et Germ.

Scientific name: *Thymelaea passerina* (L.) Coss. et Germ.
Common name: Spurge flax, Sparrow weed
Romanian name: Limba vrăbiei
Taxonomic classification
Kingdom: *Plantae*
Subkingdom: *Cormobionta*
Phylum: *Spermatophyta*
Subphylum: *Magnoliophytina (Angiospermae)*
Class: *Magnoliatae (Magnoliopsida, Dicotyledonatae)*
Subclass: *Rosidae*
Order: *Thymelaeales*
Family: *Thymelaeaceae*
Genus: *Thymelaea*
Species: *passerina* (L.) Coss. et Germ. (Figure 33.3)[2,4]

SPREAD

The species is commonly found from the steppe zone up to the oak forest level, in fields, vineyards, meadows, wet sands, gravel, saline places, limestone areas, ruderal places, fallow fields, dry and sunny locations.[2-4]

DESCRIPTION

Annual herbaceous plant with a fusiform, slender, light brown root. Erect, glabrous stem, green-yellow, thin, round, with lateral branches in the superior part not exceeding the height of the main stem, 15–40 cm tall. Leaves are arranged alternately, narrow lanceolate-linear, with a narrow base,

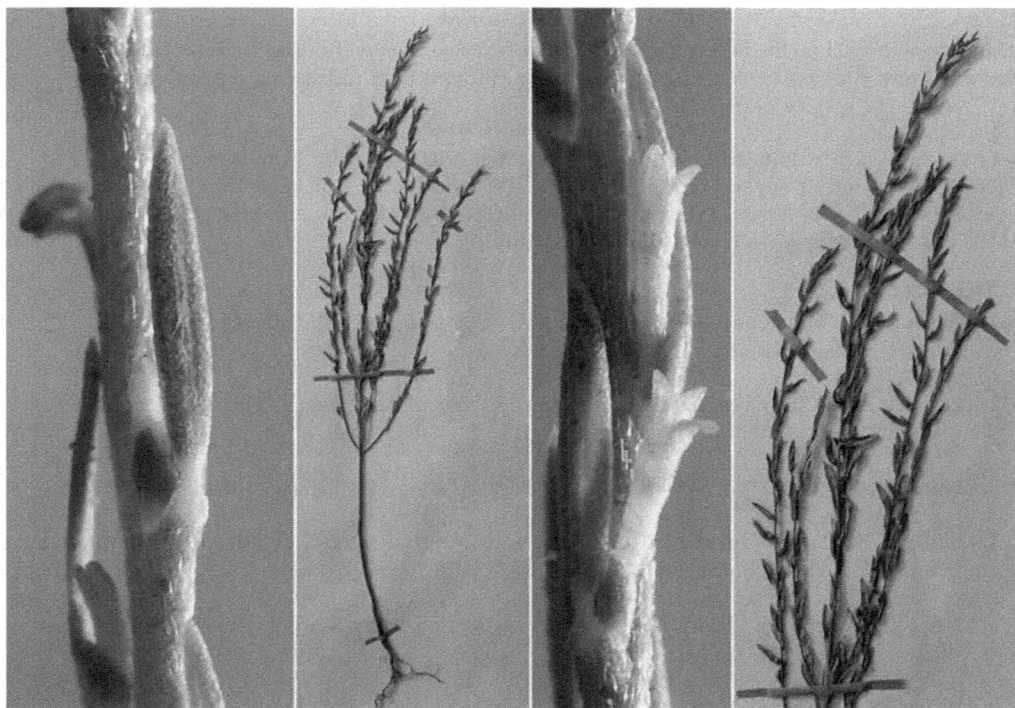

Figure 33.3 *Thymelaea passerina* (L.) Coss. et Germ.

Source: 15.

acute, erect, leathery, slightly glandular-punctate, smaller towards the tip of the branches. Flowers are sessile, small, hermaphroditic, solitary or in groups of 2–3 in the axils of the superior leaves, each with 2 lateral bracts resembling leaves at the base. The receptacle is concave, urceolate, silky hairy at the base, persistent. The perigonium consists of 4 yellowish, erect, ovate tepals, obtuse or rounded at the tip, shorter than the receptacle. Stamens are (2) 4–6 (8), arranged in two overlapping circles, with short filaments, rudimentary connectives, and elongated, small, yellow-reddish anthers. The gynoecium is elongated, unilocular, with a short style and globular stigma. The fruit is a pyriform capsule with irregular dehiscence. Seeds have a tough tegument, and the embryo is provided with fleshy, thick cotyledons, black-brown or black-blue in color. It blooms from June to July.[2-4]

Main phytoconstituents   Ref.
coumarins (daphenone, daphnelone, daphnoretin, umbelliferone)   [11,18]
flavonoids (kaempferol, baicalein, myricetin, vicenin, tiliroside, rutarensin, neochamaejasmin, ladanein)
terpenoids (geraniol, linalool, caryophyllene, copaene, murolene, germacrene D, humulene, selinene, cadinene)
sterols (sitosterol)
fatty acids
phenolic acid (caffeic acid, ferulic acid, vanillic acid, hydrocinnamic acid, m-hydroxybenzoic acid, p-coumaric acid)
tannins
organic acids
lignans (sesamin)
glycosides
misceleous

Medicinal Use

In traditional medicine, the aerial organs of *Thymelaea passerina* were used as decoctions or infusions, brewed as a tea, for treating respiratory ailments and addressing cold symptoms.[16]

Biological Activity

Thymelaea passerina exhibits antimicrobial activity, demonstrating effect on methicillin-resistant *Staphylococcus aureus*, several Gram (+), Gram (-) bacteria and *Candida albicans*. Along with other species from the *Thymelaeaceae* family, it exhibits antitumor, anti-inflammatory, neuroprotective, antidiabetic, antihypertensive, and antioxidant activities.[16-18]

REFERENCES

1. Bejenaru C., Mogoșanu G.D., Bejenaru L. E., Popescu H. 2020. *Botanică farmaceutică. Cormobionta*. Ediția a III-a. Ed. Sitech, Craiova.

2. Ciocârlan V. 2009. *Flora ilustrată a României. Pteridophyta et Spermatophyta*. Ediția a 3-a revizuită și adăugită. Ed. Ceres, București.

3. Săvulescu T. (ed). 1956. *Flora RPR*, vol. IV. Ed. Academiei RPR, București.

4. Sârbu I., Ștefan N., Oprea A. 2013. *Plante vasculare din România. Determinator ilustrat de teren*. Ed. Victor B Victor, București.

5. Drăgulescu C. 2018. *Dicționar de fitonime românești*. Ediția a 5-a, Ed. Universității "Lucian Blaga", Sibiu.

6. http://cristeaviorica.blogspot.com/2017/07/iedera-alba-daphne-blagayana.html

7. Sovrlić M.M. Manojlović N.T. 2017. Plants from the genus *Daphne*: A review of its traditional uses, phytochemistry, biological and pharmacological activity. *Serbian Journal of Experimental and Clinical Research* 18(1):69–79.

8. Manojlović N.T., Sovrlić M.M. Maskovic P., Vasiljevic P., Juskovic M. 2014. Phenolic and flavonoid content and antioxidant activity of *Daphne blagayana* growing in Serbia. *Serbian Journal of Experimental and Clinical Research* 15(1):21–27.

9. www.infoflora.ch/en/flora/daphne-mezereum.html

10. Nie W., Li Y., Luo L., Zhang Y., Fan W., Gu Y., Shi R., Zhai X., Zhu Y. 2021. Phytochemistry and pharmacological activities of the diterpenoids from the genus *Daphne*. *Molecules* 26(21):6598.

11. Mehdiyeva N., Alizade V., Batsatsashvili K. et al. 2017. *Daphne mezereum* L. *Thymelaeaceae*. In: Bussmann R.W. (ed). *Ethnobotany of the Caucasus*. Series European Ethnobotany, Springer International Publishing, Cham, Switzerland, pp. 1–4.

12. Kupchan S.M., Baxter R.L. 1975. Mezerein: antileukemic principle isolated from *Daphne mezereum* L. *Science* 187(4177):652–653.

13. Ulubelen A., Terem B., Tuzlaci E. 1986. Coumarins and flavonoids from *Daphne gnidioides*. *Journal of Natural Products* 49(4):692–694.

14. Beheshtkhoo N., Kouhbanani M.A.J., Savardashtaki A. et al. 2018. Green synthesis of iron oxide nanoparticles by aqueous leaf extract of *Daphne mezereum* as a novel dye removing material. *Applied Physics A* 124:363.

15. https://dryades.units.it/duino_en/index.php?procedure=taxon_page&id=3155&num=3234

16. Arı S., Temel M., Konuk M., 2017. An ethnobotanical approach to MRSA (methicillin-resistant *Staphylococcus aureus*) in Western Anatolia: A case of Afyonkarahisar. *Indian Journal of Traditional Knowledge* 16(1):35–43.

17. Gökbulut A., Özkal N., Yıldız S. 2007. Antimicrobial activity studies on *Thymelaea passerina* (L.) Cosson & Germ. *Journal of Faculty of Pharmacy of Ankara University* 35(3):171–176.

18. Marmouzi I., Bouchmaa N., Kharbach M., Ezzat S.M., Merghany R.M., Berkiks I., El Jemli M. 2021. *Thymelaea* genus: Ethnopharmacology, Chemodiversity, and Bioactivities. *South African Journal of Botany* 142:175–192.

34 *Violaceae* Batsch Family

Cornelia Bejenaru[1], Antonia Radu[1], George Dan Mogoşanu[2],
Ludovic Everard Bejenaru[2], Andrei Biţă[2], and Adina-Elena Segneanu[3]
[1] Department of Pharmaceutical Botany, Faculty of Pharmacy, University of Medicine and Pharmacy of Craiova, 2 Petru Rareş Street, Craiova, Dolj County, Romania
[2] Department of Pharmacognosy & Phytotherapy, Faculty of Pharmacy, University of Medicine and Pharmacy of Craiova, 2 Petru Rareş Street, Craiova, Dolj County, Romania
[3] Institute for Advanced Environmental Research–West University of Timisoara (ICAM–WUT), 4 Oituz Street, Timisoara, Romania

BOTANICAL DESCRIPTION

In this family, herbaceous species (rarely woody) are grouped, spread all over the Globe. The species of the *Violaceae* family that grow in Romania are herbaceous, perennial, belonging to a single genus (*Viola* L.). The plants have simple leaves, petiolate, and are arranged alternately. The leaf blade has a dentate margin. The leaves have stipules. The flowers are solitary, rarely grouped in inflorescences, of type 5, hermaphroditic, and generally zigomorphic. The floral envelope consists of a persistent dialisepal calyx, with appendices at the base, and a corolla with unequal, free petals, with the inferior petal having a straight, swollen, or thin spur, pointing upwards. The androecium has stamens with short filaments, adnate anthers, and connectives with apical appendices, while the two inferior stamens have nectariferous spurs inserted into the spur of the inferior petal. The gynoecium is tricarpellary, superior, and the ovary is unilocular, continuing with a curved style. The fruit is a loculicidal capsule, with numerous seeds with elaiosomes.[1-3]

The genus *Viola* includes the following spontaneous species: *Viola biflora* L., *Viola alpina* Jacq., *Viola arvensis* Murray, *Viola kitaibeliana* Schult., *Viola hymettia* Boiss. et Heldr., *Viola tricolor* L., *Viola declinata* Waldst. et Kit., *Viola dacica* Borbás, *Viola jordanii* Hanry, *Viola canina* L., *Viola pumila* Chaix, *Viola elatior* Fr., *Viola stagnina* Kit. ex Schult. (*V. persicifolia* Schreb.), *Viola mirabilis* L., *Viola riviniana* Rchb., *Viola sieheana* W. Becker, *Viola reichenbachiana* Jord. ex Boreau (*V. sylvestris* Lam.), *Viola rupestris* F. W. Schmidt, *Viola joói* Janka, *Viola epipsila* Ledeb., *Viola palustris* L., *Viola suavis* M. Bieb. (*V. cyanea* Celak.; *V. austriaca* A. Et J. Kern.; *V. pontica* W. Becker; *V. ignobilis* G. Grint.), *Viola odorata* L., *Viola alba* Besser, *Viola ambigua* Waldst. et Kit., *Viola hirta* L., and *Viola collina* Besser.[2,4]

Viola tricolor L. (Wild pansy)

Scientific name: *Viola tricolor* L.
Common name: Wild pansy, Heartsease
Romanian name: Trei fraţi pătaţi
Taxonomic classification
Kingdom: *Plantae*
Subkingdom: *Cormobionta*
Phylum: *Spermatophyta*
Subphylum: *Magnoliophytina (Angiospermae)*
Class: *Magnoliatae (Magnoliopsida, Dicotyledonatae)*
Subclass: *Dilleniidae*
Order: *Violales*
Family: *Violaceae*
Genus: *Viola*
Species: *tricolor* L. (Figure 34.1)[1-2,4]

SPREAD

It is frequently distributed from the plain area up to the subalpine zone, through shrubs, meadows, grassy cliffs, forest edges, wet meadows, slopes, and mountain pastures.[3]

DESCRIPTION

Herbaceous plant, without sterile shoots, annual, biennial, or perennial, with a height of 15–40 cm and uniformly pubescent with short and straight downward-pointing hairs. The aerial stems are leafy, arched from the base or ascending, simple or branched. The leaves at the inferior part of

DOI: 10.1201/9781003270515-35

Figure 34.1 *Viola tricolor* L. (Wild pansy).

Source: 5.

the plant are broadly ovate, elongated-ovate, or ovate-triangular, slightly concave, with a slightly rounded base, rarely cuneate, long-petiolate (the petiole exceeds the length of the blade); those in the middle part of the plant are ovate-lanceolate or elongated-ovate, with a cuneate base and short-petiolate, while the superior ones are elongated-lanceolate, with an obtuse-acuminate tip, gradually narrowed at the base, with an obtuse-serrate margin, and very short-petiolate. The stipules are 3–30 mm long, pennate-partite, rarely lirate, with a large foliaceous terminal lobe. The flower peduncles are 3–12 cm long, short-pubescent in the inferior part, glabrous in the superior part, and equipped with 2 membranous bracteoles, alternately arranged below the flowers. The sepals are elongated-lanceolate or linear, long-ciliate on the margins. The corolla, significantly larger than the calyx, is composed of petals whose colors are blue, yellow, and white. The superior petals are ovate, violet in color, rarely light violet, white, or yellow, the lateral petals are oval, yellow, white, violet-blue, or ochre-colored, and the inferior one is triangular, with rounded edges, yellow base, and 5–7 dark-colored stripes, pale violet margins, and a blue or blue-violet spur. The fruit is an elongated-ovate capsule, measuring 7–10 mm in length, and the seeds are small, provided with an elaiosome. The flowering period is from May to August.[1-4]

The species has two subspecies:

- subsp. *tricolor* L. includes annual or biennial herbaceous plants that grow from the plain area up to the mountain zone, in meadows, pastures, or cultivated places, being frequently encountered. These plants have a height of 10–25 cm, and the flowers have a diameter of up to 2 cm, with the superior petals violet (rarely yellow) and the inferior ones yellow;[1]

- subsp. *saxatilis* (F. W. Schmidt) Arcang. (subsp. *subalpina* (Latourr) Gaudin, *V. tricolor* L. var. *luteola* Schur.) comprises perennial plants, sporadically found in the mountain zone, on meadows and rocky areas, reaching a height of up to 40 cm, and their flowers have a diameter larger than 2 cm.[1-2,4]

Main phytoconstituents   Ref.

amino acids (alanine, glycine, serine, leucine, phenylalanine, methionine, lysine, valine, isoleucine, proline, tryptophan, tyrosine, threonine, asparagine, glutamine, arginine, histidine, aspartic acid, glutamic acid)   6-11

alkaloids (violaine)

coumarins (umbelliferone)

cyclotides (vitri A, varv peptides A—H)

flavonoids (quercitin, rutin, isorhamnetin, apigenin, vitexin, isovitexin, orientin, scoparin, saponarin, violanthin, violarvensin, chrysoeriol, kaempferol, quercetin, luteolin,)

terpenoids (cadinene, menthone, phytol, spathulenol, eucalyptol, β-bisabolene, farnesene, sabinene, β-amyrin, elemene, α-muurolene, copaene, caryophyllene, terpineol, ionone, carvone)

sterols (β-sitosterol)

fatty acids (margaric acid, lauric acid, palmitic acid, linolenic acid, myristic acid)

carbohydrates (galactose, glucose, arabinose, xylose, rhamnose, uronic acid)

phenolic acids (vanillic acid, gentisic acid, caffeic acid, p-cumaric)

carotenoids (zeaxanthin, violaxanthin, antheraxanthin, auroxanthin luteoxanthin, lutein, β-carotene)

antocyans (violanin, delphinidin, cyanidin)

tannins

saponins

miscellaneous (vitamin C)

Medicinal Use

The vegetable product *Violae herba cum flore* (Ph.Eur. 10), or *Violae tricoloris herba*, is obtained by harvesting the flowering aerial parts in the months of May to June.[1] It has diuretic and depurative actions (flavonoids and saponins), expectorant effects (saponins), analgesic and anti-inflammatory properties (methyl salicylate).[6]

External application for mild seborrheic conditions, including seborrhea of the scalp in nursing infants. In traditional medicine, it is used both externally and internally as a complementary treatment for various skin conditions like eczema, impetigo, acne, and pruritus. Internally, it is also used as a supportive agent to enhance metabolism.[12]

Biological Activity

It is recommended for various health conditions, including urinary (nephritis, edema), respiratory (dry cough, bronchitis), articular (rheumatism), dermatological (eczema, acne, pruritus).[6]

- antimicrobial activity[12]

- antiprotozoal and antioxidant activities[13-14]

- immunosuppressive activity[15]

- anti-inflammatory activity[7]

- antiangiogenic activity[16]

Other Uses

- ornamental

Viola odorata L. (Sweet violet)

Scientific name: *Viola odorata* L.
Common name: Sweet violet, English violet, Common violet
Romanian name: Toporaşi[1-2,4], Tămâioară, Viorea[3]
Taxonomic classification
Kingdom: *Plantae*
Subkingdom: *Cormobionta*
Phylum: *Spermatophyta*
Subphylum: *Magnoliophytina (Angiospermae)*
Class: *Magnoliatae (Magnoliopsida, Dicotyledonatae)*
Subclass: *Dilleniidae*
Order: *Violales*
Family: *Violaceae*
Genus: *Viola*
Species: *odorata* L. (Figure 34.2)[1-2,4]

Figure 34.2 *Viola odorata* L. (Sweet violet).

Source: 17.

SPREAD
A common species found from the plain area up to the beech forest zone, through shrubs, groves, wet meadows, the edge of deciduous forests, clearings, and along watercourses in hilly and lowland regions.[1-4]

DESCRIPTION
Perennial plant with a thick, articulated rhizome, and branched aerial stolons, which can be short or long, thick, creeping, and with adventitious roots at the nodes. It reaches a height of 5–15 cm. The stems are scapiform, and the leaves are subround, obtuse, reniform-cordate, broadly ovate, large, glossy, glabrous, with a deeply cordate base, and with lanceolate, glabrous, fringed internal stipules, with dense, short, brown glandular fringes. The floral pedicels are slightly appressed, pubescent, or glabrous, and the bracteoles are positioned halfway up the peduncle or higher. The large, fragrant flowers can be violet or white and have free sepals. They appear in early spring. The sepals are ovate with an obtuse tip, and the petals are ovate, emarginate, or entire, with the lateral ones being lanceolate, while the inferior one has an obtuse, straight, or upwardly bent spur, violet in color. The style is laterally flattened, and the stigma is bent with a hook-like appearance, and the rostrum is almost equal in diameter to the style. The fruit is a spherical, hexagonal, or trihedral capsule, with faintly pronounced edges, green or purplish in color. They bloom from March to April.[1-4]

Main phytoconstituents   Ref.
amino acids   18-20
alkaloids (violaine)
saponins
coumarins
cyclotides
flavonoids (quercitrin, rutin, vitexin, isovitexin, rutin, apigenin, violanthin, kaempferol, quercetin, luteolin)
terpenoids (ursolic acid, limonene, linalool, cadinene, α-cadinol, globulol, viridiflorol, phytol)
sterols (stigmasterol)
fatty acids
carbohydrates
phenolic acids (vanillic acid, gentisic acid, caffeic acid, p-cumaric acid, salicylic acid, chlorogenic, syringic acid, ferrulic acid, cinnamic acid)
carotenoids
antocyans
tannins
saponins
miscellaneous (vitamin C, eugenol, melatonin, resorcinol, hydrochinone)

Medicinal Use

The roots have emetic properties and the aerial parts of the plant have diaphoretic properties.[1]

Viola odorata finds wide applications in traditional medicine, particularly for managing cough, fever, common cold, headache, insomnia, epilepsy, constipation, palpitation, dyspnea, dysuria, and various skin diseases.[19]

Biological Activity

In modern phytotherapy, this plant has demonstrated anti-inflammatory, analgesic, antioxidant, diuretic, antihypertensive, and antibacterial properties.[19-22]

Other Uses

The flowers, due to their volatile oil content, are used in the perfume industry.[1]

Viola arvensis L. (Field pansy)

Scientific name: *Viola arvensis* L.
Common name: Field pansy
Romanian name: Viorele de ogoare
Taxonomic classification
Kingdom: *Plantae*
Subkingdom: *Cormobionta*
Phylum: *Spermatophyta*
Subphylum: *Magnoliophytina (Angiospermae)*
Class: *Magnoliatae (Magnoliopsida, Dicotyledonatae)*
Subclass: *Dilleniidae*
Order: *Violales*
Family: *Violaceae*
Genus: *Viola*
Species: *arvensis* L. (Figure 34.3)[1-2,4]

SPREAD

The species is common, ranging from the plain area up to the beech forest zone, and can be found in cultivated and ruderal areas, mountain meadows, forest edges, grasslands, riverbanks, rocky areas, and shrubs.[1-2,4]

Figure 34.3 *Viola arvensis* L. (Field pansy).

Source: 23.

Description

They are annual herbaceous plants, with a maximum height of 20 cm, and their stems are branched, ascending, rarely erect, glabrous, or slightly pubescent. The inferior leaves are round-elliptic or shortly ovate, with a rounded or slightly emarginate base, and the petiole is as long as the leaf blade. The middle leaves are ovate-elliptic, with a cuneate base, decurrent into a short petiole, and the superior leaves are elongated-lanceolate with a short petiole. The leaf margin is crenate or crenate-serrate, with short pubescence on the edge and on the inferior surface along the veins. The stipules are 2–40 mm long, pennate-partite, with a foliaceous terminal lacinia, elliptic to narrow-lanceolate, crenate, or entire. The floral peduncles are long and equipped with 2 lanceolate, glabrous bracteoles arranged at the top, curving around the peduncle. The flowers are small, about 1–1.5 cm in size, with oblong-lanceolate sepals irregularly dentate, and the petals are light yellow (sometimes with violet tinges), at most the length of the calyx. The fruits are capsules, 6–10 mm long, containing elongated-ovate seeds. These plants bloom throughout the year, from April to September.[1,3]

Main phytoconstituents	Ref.
amino acids	[7]

alkaloids
coumarins
cyclotides (varv peptide A)
flavonoids (violarvensin, isorhamnetin, kaempferol)
terpenoids (cyclocitral, geranyl, ionone)
sterols
fatty acids (palmitic acid, myristic acid, lauric acid, linoleic acid)
carbohydrates
phenolic acids

carotenoids
hydrocarbons (tricosane, pentacosane)
antocyans
saponins
miscellaneous (vitamin C)

Medicinal Use

The aerial parts of *Viola arvensis* are utilized in traditional medicine for their anti-inflammatory, expectorant, and diuretic properties. They are also employed for treating skin conditions, bronchitis, cystitis, and rheumatism.[24]

Biological Activity

■ haemolytic, antioxidant and cytotoxic activities.[8,25-26]

Viola canina L.

Scientific name: *Viola canina* L.
Common name: Dog violet
Romanian name: Viorele sălbatice
Taxonomic classification
Kingdom: *Plantae*
Subkingdom: *Cormobionta*
Phylum: *Spermatophyta*
Subphylum: *Magnoliophytina (Angiospermae)*
Class: *Magnoliatae (Magnoliopsida, Dicotyledonatae)*
Subclass: *Dilleniidae*
Order: *Violales*
Family: *Violaceae*
Genus: *Viola*
Species: *canina* L. (Figure 34.4)[1-3]

Figure 34.4 *Viola canina* L. (Dog violet).

Source: 27.

SPREAD

The species is commonly distributed from the oak forest zone up to the fir forest zone, through meadows, shrubs, open areas, forest edges, especially on sandy and dry soils.[2-4]

DESCRIPTION

The plant is perennial, with a short rhizome and solitary, ascending, rarely erect stems, with short inferior internodes and elongated middle internodes, reaching a height of 5–15 cm. The leaves are ovate or elongated-ovate, with a cordate or subcordate base, with a crenate margin, glabrous, or slightly pubescent towards the base of the leaf. The petioles are wingless, as long as the leaf blade or slightly longer. The stipules are linear or subulate, with a smooth margin or with a few teeth of different sizes, significantly shorter than half of the petiole. The floral peduncles are axillary and have 2 bracteoles at the top. The flowers are unscented, with a length of 1–2 cm, with falcate sepals, obovate, blue petals, only white at the base, with the inferior ones smaller than the lateral ones. The spur is cylindrical, obtuse, whitish-yellowish. The stigma has papillae at the tip. The fruits are elongated-ovate, glabrous capsules. They bloom from May to June.[2-4]

The species has three subspecies:

- subsp. *canina* has leaves with a cordate base, small stipules, and flowers with a yellow spur. The plants are procumbent or ascending and are frequently found in oak forests up to the fir forest zone, through meadows;[2,4]

- subsp. *ruppii* (All.) Schűbl. et G. Martens (*V. montana* auct.) has leaves with a slightly cordate base, larger stipules, and flowers with obovate petals and a straight, white spur. The plants are erect and are commonly found in oak forests up to the beech forest zone;[2,4]

- subsp. *schultzii* (Billot) Kirschl. has flowers with narrowly elliptic petals and a curved spur. It is present in oak forests up to the hornbeam forest zone. It is rare in Dâmbovița County (Teiș and Văcărești).[2,4]

Main phytoconstituents	Ref.
amino acids	11,28

alkaloids
coumarins
cyclotides
flavonoids (apigenin, vitexin, hyperoside, hesperidin, rutin)
terpenoids (cadinene, phytol, germacrene D, α-farnesene, caryophyllene, β -pinene)
sterols
fatty acids (palmitic acid, myristic acid)
carbohydrates
phenolic acids (ferulic acid, salicylic acid)
tannins
carotenoids
antocyans
saponins
miscellaneous (vitamin C, resorcinol)

Medicinal Use

Viola canina has been used in traditional medicine. Some of its medicinal applications include expectorant (promotes the expulsion of mucus from the respiratory tract, making it useful for conditions like coughs and bronchitis), diuretic, antirheumatic. The plant is employed to treat skin disorders such as eczema, acne, and pruritus.[11]

Biological Activity

Some of the reported biological activities of *Viola canina* include anti-inflammatory, antioxidant, antimicrobial, analgesic, and antiviral properties.[11]

Other Uses

The young leaves and flower buds of *Viola canina* can be consumed raw or cooked. When added to soups, they act as a natural thickening agent, similar to okra. Additionally, the leaves can be used to prepare tea. Moreover, it plays a significant role in supporting wildlife. It serves as a food source for rare butterflies, while other butterflies use the plant as a host for their eggs.[29]

REFERENCES

1. Bejenaru C., Mogoșanu G.D., Bejenaru L.E., Popescu H. 2020. *Botanică farmaceutică. Cormobionta*, ediția a III-a. Ed. Sitech, Craiova.

2. Ciocârlan V. 2009. *Flora ilustrată a României. Pteridophyta et Spermatophyta*. Ediția a 3-a revizuită și adăugită. Ed. Ceres, București.

3. Săvulescu T. (ed). 1955. *Flora RPR*, vol. III. Ed. Academiei RPR, București.

4. Sârbu I., Ștefan N., Oprea A. 2013. *Plante vasculare din România. Determinator ilustrat de teren.* Ed. Victor B Victor, București.

5. https://dryades.units.it/floritaly/index.php?procedure=taxon_page&tipo=all&id=3209

6. Bejenaru L.E., Mogoșanu G.D., Bejenaru C., Popescu H. 2015. *Farmacognozie-Fitoterapie*. Vol. I, ediția a III-a. Ed. Sitech, Craiova.

7. Toiu A., Philippe V., Oniga I., Tamas M. 2009. Composition of essential oils of *Viola tricolor* and *V. arvensis* from Romania. *Chemistry of Natural Compounds* 45(1):91–92.

8. EMA/HMPC/131735/2009. Committee on Herbal Medicinal Products (HMPC) Assessment Report on *Viola tricolor* L. and/or subspecies *Viola arvensis* Murray (Gaud) and *Viola vulgaris* Koch (Oborny), *herba cum flore*.

9. Dziągwa-Becker M., Weber R., Zajączkowska O., Oleszek W. 2018. Free amino acids in *Viola tricolor* in relation to different habitat conditions. *Open Chemistry* 16(1):833–841.

10. Batiha G.E.S., Beshbishy A.M., Alkazmi L. et al. 2020. Gas chromatography-mass spectrometry analysis, phytochemical screening and antiprotozoal effects of the methanolic *Viola tricolor* and acetonic *Laurus nobilis* extracts. *BMC Complementary Medicine and Therapies* 20:87.

11. Zhang Q., Wang Q., Chen S. 2023. A comprehensive review of phytochemistry, pharmacology and quality control of plants from the genus *Viola*. *Journal of Pharmacy and Pharmacology* 75(1):1–3.

12. Witkowska-Banaszczak E., Bylka W., Matławska I., Goślińska O., Muszyński, Z. 2005. Antimicrobial activity of *Viola tricolor* herb. *Fitoterapia* 76(5):458–461.

13. Batiha G.ES., Beshbishy A.M., Alkazmi L. et al. 2020. Gas chromatography-mass spectrometry analysis, phytochemical screening and antiprotozoal effects of the methanolic *Viola tricolor* and acetonic *Laurus nobilis* extracts. *BMC Complementary Medicine and Therapies* 20:87.

14. Gonçalves A.F.K., Friedrich R.B., Boligon A.A., Piana M., Beck R.C.R., Athayde M.L. 2012. Anti-oxidant capacity, total phenolic contents and HPLC determination of rutin in *Viola tricolor* (L) flowers. *Free Radicals and Antioxidants* 2(4):32–37.

15. Hellinger R., Koehbach J., Fedchuk H., Sauer B., Huber R., Gruber C.W., Gründemann C. 2014. Immunosuppressive activity of an aqueous *Viola tricolor* herbal extract. *Journal of Ethnopharmacology* 151(1):299–306.

16. Sadeghnia H.R., Ghorbani Hesari T., Mortazavian S.M., Mousavi S.H., Tayarani-Najaran Z., Ghorbani A. 2014. *Viola tricolor* induces apoptosis in cancer cells and exhibits antiangiogenic activity on chicken chorioallantoic membrane. *Biomed Research International* 2014:625792.

17. www.infoflora.ch/en/flora/viola-odorata.html

18. Singh A., Dhariwal S., Navneet. 2018. Traditional uses, antimicrobial potential, pharmacological properties and phytochemistry of *Viola odorata*: A mini review. *Journal of Phytopharmacology* 7(1):103–105.

19. Feyzabadi Z., Ghorbani F., Vazani Y., Zarshenas M.M. 2017. A critical review on phytochemistry, pharmacology of *Viola odorata* L. and related multipotential products in traditional Persian medicine. *Phytotherapy Research* 31(11):1669–1675.

20. Dhiman S., Singla S., Kumar I., Palia P., Kumar P., Goyal S. 2023. Protection of *Viola odorata* L. against neurodegenerative diseases: potential of the extract and major phytoconstituents. *Clinical Complementary Medicine and Pharmacology* 3(3):100105.

21. Akhbari M., Batooli H., Kashi F.J. 2012. Composition of essential oil and biological activity of extracts of *Viola odorata* L. from central Iran. *Natural Product Research* 26(9):802–809.

22. Rosenhech Murray E., Lobiuc A., Zamfirache M.M. 2016. Phenolic contents and antioxidant activity in *Viola odorata* L., *V. tricolor* L. and *V. arvensis* (L.). *Analele Stiintifice ale Universitatii "Al. I. Cuza" din Iasi: Biologie Vegetala, Serie Noua. Sectiunea II A* 62(1):132–133.

23. https://dryades.units.it/floritaly/index.php?procedure=taxon_page&tipo=all&id=3212

24. Chandra D., Kohli G., Prasad K., Bisht G., Punetha V.D., Khetwal K.S., Devrani M.K., Pandey H.K. 2015. Phytochemical and ethnomedicinal uses of family *Violaceae*. *Current Research in Chemistry* 7:44–52.

25. Kucekova Z., Mlcek J., Humpolicek P., Rop O. 2013. Edible flowers — antioxidant activity and impact on cell viability. *Open Life Sciences* 8(10):1023–1031.

26. Lindholm P., Göransson U., Johansson S., Claeson P., Gullbo J., Larsson R., Bohlin L., Backlund A. 2002. Cyclotides. A novel type of cytotoxic agents. *Molecular Cancer Therapeutics* 1(6):365–369.

27. www.orchid-nord.com/Flore-France/Violaceae/Viola%20canina/Viola-canina.html

28. Kirillov V., Stikhareva, T., Atazhanova, G., Makubayeva, A., Aleka, V., Rakhimzhanov A., Adekenov S. 2021. Composition of essential oil of the aerial parts of *Viola canina* L. growing wild in Northern Kazakhstan. Natural Product Research 35(13):2285–2288.

29. www.naturalmedicinalherbs.net/herbs/v/viola-canina=dog-violet.php

Regulatory Aspects of Traditional Medicines and Commercial Products

Diana Obistioiu[1], Cornelia Bejenaru[2], and Antonia Radu[2]

[1] Banat's University of Agriculture and Veterinary Medicine, Faculty of Veterinary Medicine Timisoara, 119 Calea Aradului Street, Timisoara, Romania

[2] Department of Pharmaceutical Botany, Faculty of Pharmacy, University of Medicine and Pharmacy of Craiova, 2 Petru Rareş Street, Craiova, Dolj County, Romania

Medicinal and aromatic products are derived from both spontaneous and cultivated plant species. Herbs have been cultivated since ancient times and have been used as medicines, in cosmetics or as ingredients for various culinary preparations. In Romania, as everywhere in the world, there is currently increasing interest in the use of medicinal and aromatic plants. The medicinal flora of Romania comprises 800 species, among which 283 possess specific therapeutic properties. Medicinal and aromatic plants are represented by annual, biennial and perennial species, whose products (vegetable raw materials) – flowers (*Flores*), leaves (*Folium*), grass or the entire aerial vegetative mass (*Herba*), fruits (*Fructus*), seeds (*Semen*), roots (*Radix*), etc., serve various purposes, becoming primary sources of raw materials for:

- the extraction of active ingredients and volatile oils;
- the medicinal and pharmaceutical industry;
- the cosmetics industry (perfumes, syrups, detergents); and
- the food industry (flavors, dyes).

A National Report detailing the regulations concerning the processing, marketing, and traditional utilization of aromatic and medicinal plants is accessible.[1]

In Romania, medicinal and aromatic plants are cultivated in relatively small parcels when compared to other agricultural crops. Romania's strategy aims to promote the establishment and expansion of commercial crops, which encompass medicinal, aromatic, and tinctorial species. This includes the cultivation of organic varieties, which are increasingly in demand for export purposes, and the development of alternative economic activities beyond rural agriculture. Moreover, expanding the range of forest-related products, aside from timber, such as ecologically supportive activities and agrotourism, can boost the earnings of rural households.

By aligning public policies with industrial and environmental initiatives, alongside research and development activities, a valuable source of natural raw materials can be efficiently exploited and the management of locally and nationally cultivated or sustainably harvested medicinal, aromatic, and tinctorial plant resources can be significantly improved.[1]

THE REGULATORY LEGISLATIVE FRAMEWORK FOR MEDICINAL AND AROMATIC PLANTS

European Union Regulatory Legislative Framework

Romania became a member of the European Union (EU) on January 1, 2007. In the EU, a complex regulatory system has been established to regulate (traditional) herbal treatment products. It is governed by the following principle: when a drug is placed on a controlled market, a sales permit approved by the competent authority is required. The necessary documents, approved procedures, instructions, and guidelines used in each country/region are all approved by law and implemented by the main authority of the governing body in each country/region.[2]

Within the European Medicines Agency (EMA), the Institution's Herbal Products Committee (HMPC) provides scientific advice on herbal-related issues, community herbal monographs on traditional and mature herbs, and lists substances, preparations and combinations for traditional use, that offer guidance on plant substances; In addition to assessing complaints and arbitrations about herbal medicinal products.[3]

In the primary legal framework within the EU which consists of Directive 2001/83/EC and Directive 2003/63/EC, the basic principles governing the quality assurance of EU medicines are mainly defined. The scientific guideline also specifies the quality requirements of herbal medicines.[4,5,6]

The scientific guidelines provide a practical and uniform basis for how the competent authorities of EU member states interpret and apply the detailed requirements for quality certification stipulated in regulations and directives. The detailed quality requirements for herbal products on

DOI: 10.1201/9781003270515-36

the European market are contained in the EU Drug Regulations. They include a manufacturing authorization system that ensures that only authorized manufacturers manufacture/import all herbal products on the European market, and the manufacturer's activities are subject to regular inspections by the competent authority.[7,8]

The uniqueness of the European drug regulatory system is that it is based on a network of approximately 50 regulatory agencies from 31 EEA countries (28 EU member states as well as Iceland, Liechtenstein and Norway), the European Commission and EMA. The network benefits from the expertise of thousands of experts from across Europe, tapping into the EU's most valuable regulatory medical knowledge and delivering the utmost quality in scientific guidance.[9]

According to the regulations of the European Medicines Agency (EMA), the European Union's regulations in the field of human medicines are included in the ten volumes of the "EU Regulations on Drug Administration." This collection introduces the content of legislation, notices for applicants and regulatory guidelines, scientific guidelines, GMP guidelines, maximum residue limits, pharmacovigilance rings and clinical trial guidelines.[10]

EMA and each Member state work closely together to evaluate new drugs and analyze the latest information about their safety. The EMA and the Member states also rely on each other to exchange information, such as reporting adverse drug reactions, monitoring clinical trials, inspecting drugs, or following good practices. These standards encompass Clinical Practice (GCP), Good Manufacturing Practice (GMP), Good Distribution Practice (GDP), and Good Pharmacovigilance Practice. EU regulations mandate that all Member states must grant authorization for and oversee pharmaceuticals in compliance with these consistent requirements and criteria.[11]

The European Parliament and the Council of the European Union presented (in the Official Journal of the European Union), the 2004/24/EC Directive of March 31, 2004. This report revises 2001/83/EC Directive regarding the traditional herbal medicinal products, within the framework of the Community code relating to medicinal products for human use.[12]

The Romanian Regulatory Legislative Framework
Standards for Cultivation, Harvesting, and Primary Processing of Medicinal and Aromatic Plants

The cultivation and marketing of medicinal plants, both cultivated and spontaneous flora, are emerging as significant economic endeavors, offering a lucrative income source for producers, traders, and processors.

This activity must be carried out in accordance with the principles of free competition and in compliance with the technical conditions for cultivation and harvesting. The European Herb Growers Association (EUROPAM) has developed good practice guidelines for harvesting and cultivating medicinal and aromatic plants.[13]

These standards establish the principles governing the cultivation, harvesting, and primary processing of medicinal and aromatic plants marketed in the European Union. These regulations are applicable to the production of all botanical raw materials used in the food industry, pharmaceuticals, flavorings, and fragrances, encompassing various methods, including organic production, in compliance with European provisions. By adhering to these standards, those involved in cultivating and processing medicinal and aromatic plants can guarantee the quality and quantity of plants harvested from crops and spontaneous flora, and can ensure raw materials and herbal products meet consumer hygiene and quality requirements. They also establish minimum microbiological load standards and help mitigate negative factors that affect plant processing during harvesting and storage.[1,14]

The European standards translated by The National Standardization Body – ASRO and adopted as Romanian standards have a voluntary applicability character. ASRO is a legal Romanian association, of public interest, non-profit, non-governmental, and apolitical, which operates based on the provisions of Law no. 163/2015 on national standardization and on Government Ordinance no. 26/2000 on associations and foundations.[15]

ASRO is the national platform for the development and adoption of European and international standards and at European level, the standardization activity is regulated by Regulation 1025_2012.[15,16]

- SR 13479: 2003 – Medicinal and aromatic plants. Good practice guide for cultivation;

- SR 13480:2003 – Medicinal and aromatic plants. Good harvesting practice;

- SR 1631-1: 2003-Medicinal and aromatic plants. Determination of organoleptic quality, which establishes the notions and methods for determining the organoleptic quality of medicinal and aromatic plants. The determination of the organoleptic quality of medicinal and aromatic plants consists in establishing their characters, which can be observed with the naked eye or with a magnifying glass (macroscopic characters, color and dimensions), as well as those that can be determined by perceiving smell or taste;

- SR 1632-1: 2003- Medicinal and aromatic plants. Botanical nomenclature, standard made in accordance with Flora Europaea and the latest publications of the International Seed Testing Association (ISTA). The standard helps to know and identify medicinal and aromatic plants marketed in the European Union, including both the name of the plant (in Latin and Romanian), and the name of the species and family.[1,15,16]

The Primary Romanian Regulatory Framework

As an acceding state and subsequently a member of the European Union, Romania has adopted a regulatory framework for the production and marketing of herbal and aromatic plant products, which provides a high level of consumer protection.

The principal legislation[17] includes:

- Law no.491/2003 on medicinal and aromatic plants (re-published);

- MAFRD (Ministry of Agriculture, Forestry and Rural Development) and MH (Ministry of Health) common Order no. 243/2005 on the approval of the technical rules of manufacturing, processing and marketing of medicinal and aromatic plants;

- Joint MAFRD and MH common Order no. 244/401/2005 on the processing, processing and marketing of medicinal and aromatic plants used as such, partially processed or processed as pre-measured food supplements;

- Joint Order MAFRD / MH / NSVFSA (The National Sanitary Veterinary and Food Safety Authority) no. 1228/244/63/2005 for the approval of the Technical Norms regarding the marketing of pre-dosed food supplements of animal and vegetable origin and / or their mixtures with vitamins, minerals and other nutrients;

- MAFRD and MH and NSVFSA nr. 7/2008 on drawing up a list of claims of national nutrition and health associated to medicinal and aromatic plant products;

- Law no. 239 of 2010 regarding medicinal and aromatic plant products.

Alongside the aforementioned laws and orders, there is the EU regulation.

Concerning the Romanian authority, the National Agency for Medicines and Medical Devices (NAMMD) is a public institution subordinated to the Ministry of Health, established by GEO no. 72 of June 30, 2010, because of the merger of the National Medicines Agency and the Technical Office of Medical Devices. NAMMD is the national competent authority in the field of medicine for human use, medical devices and evaluation of medical technologies and for more than 50 years has represented Romania's drug regulatory agency.[17-19]

The name of the agency was changed in 1960 to become the State Institute of Drug Control and Drugs (ICSMCF).

ICSMCF is the first organization in Romania that meets the definition of a modern drug regulatory agency. Its main responsibilities are drug authorization and registration, annual development of product indexes, complex control of drugs produced at home and abroad, drug inspections, the formulation of the Romanian Pharmacopoeia and its supplements, devise national standards and reference materials, etc.

Starting in 2000, through the Order of the Minister of Health No. 802/1999, the structure of the National Medicines Agency included the Centre for the State Control of Biological Products for Human Use, when the Agency further took upon itself the duty as Romanian competent authority in biological products for human use, resulting in additional duties and an extended range of stakeholders.[18]

Notification/authorization of products based on medicinal and aromatic plants for marketing through Law no. 491/2003 updated in 2011 (on medicinal and aromatic plants, as well as hive products) published in the Official Gazette 52/2011 of 20 January 2011, establishes the general framework on the production, processing, and organization of the market of medicinal plants, aromatics, and hive products, relations between producers, processors, and finished products based on medicinal plants, aromatic plants, and hive products that fall as drugs in the definition provided by art. 695 point 1 of Law no. 95/2006 on health care reform, with subsequent amendments and completions.[1,18,19]

Finished products based on medicinal, aromatic and hive products falling within the scope of foodstuffs, food supplements and products for internal or external use, excluding cosmetics shall be placed on the market in accordance with the rules laid down by the competent authority and notified by the operators concerned at the National Service for Aromatic Medicinal Plants and Hive Products (SNPMAPS) within the National Research-Development Institute for Food Bioresources – IBA Bucharest. This body shall ensure the notification, supervision and control of food supplements and products for external use obtained from medicinal plants, aromatics and hive products, with the exception of cosmetics.[1,13]

Traditional herbal medicinal products can be authorized by NAMMD for marketing in Romania under a simplified 120-days authorization procedure, if they meet certain conditions, such as:

■ they have indications exclusively appropriate to traditional herbal medicinal products which, by virtue of their composition and purpose;

■ are intended and designed for use without the supervision of a medical practitioner for diagnostic purposes or for prescription or monitoring of treatment;

■ are exclusively for administration in accordance with a specified strength and posology;

■ are an oral, external and/or inhalation preparation;

■ the period of traditional use as laid down has elapsed;

■ the data on the traditional use of the medicinal product are sufficient; in particular the product proves not to be harmful in the specified conditions of use and the pharmacological effects or efficacy of the medicinal product are plausible on the basis of long-standing use and experience.[19,20]

In certain cases, the 120-day period necessary for the issuance of the marketing authorization for traditional herbal medicinal products and for the homeopathic medicinal products may be extended by NAMMD, up to 210 days.[20,21]

If these product categories do not meet the legal requirement for the simplified authorization procedures, they can be authorized by NAMMD under the general authorization procedure for medicinal products.

The distribution and storage of medicinal products is governed by the Healthcare Reform Law (95/2006) and by Ministry of Health Order 131/2016 on authorisation for en-gross distribution units for human-use medicines.[20-22]

According to these, the holder of a manufacturing authorisation is entitled to distribute the permitted medicinal products. Aside from this implicit authorisation, any person applying to obtain a distribution license must be able to prove to the National Agency for Medicines and Medical Devices that it has appropriate storage space, installations, and equipment, and that it benefits from the services of qualified personnel. In addition, the applicant must be able to prove that both its suppliers and its clients are authorized to manufacture, distribute or place medicines on the market.[21,23]

The complete authorization file contains:

■ information regarding the applicant and the producer of the medicine;

■ the qualitative and quantitative characteristics of the medicine;

■ an evaluation of its potential impact on the environment;

■ a description of the manufacturing method;

■ information regarding how to safely store it;

■ a description of the manufacturing methods used;

■ the results of the pharmaceutical, preclinical and clinical testing;

■ a summary of the applicant's pharmacovigilance system;

■ a summary of the product's characteristics;

- a copy of the manufacture authorization;

- copies of any marketing authorization previously obtained.

The applicant must also submit proof of the qualifications of the person compiling the technical summaries listed above.

The mutual recognition procedure and the decentralized authorization procedure are available to applicants and are stated in the EU Directives 2001/83/EC and 2004/27/EC and in the national Romanian law, employing Law 95/2006.[5,12,21]

The primary criteria assessed when granting a marketing authorization include a positive risk-to-benefit ratio of the medication, a comprehensive description of the medicine's therapeutic efficacy, and alignment of its composition with the information provided in the application form.

A marketing authorization is valid for five years and can be renewed. The renewed authorization will be valid indefinitely, although in certain circumstances the National Agency for Medicines and Medical Devices may decide to renew it for another five-year term only.[8,10,23]

Marketing authorization holders are required to maintain records of all side effects, reported either by patients or by healthcare professionals, regardless of whether these records originate from within the European Union or from countries outside the EU. Holders are also required to convey such data to the Eudravigilance database within 90 days of receiving it.[21,23]

Holders must also comply with the requirement to submit periodic reports to the European Medicines Agency containing information related to their pharmacovigilance obligations. In certain circumstances they may be required by the National Agency for Medicines and Medical Devices to conduct post-authorization studies on the medicine's safety or efficacy.[8,18,21]

According to Order 368/2017, the holder of a newly issued marketing authorization must request approval from the Ministry of Health for the medicine's price.[21] Such a request must be accompanied by a number of documents, including:

- a copy of the marketing authorization;

- an excerpt from the National Agency for Medicines and Medical Devices' website with the medicine's details;

- catalogues of manufacturer prices applied in certain European countries (ie, Austria, Belgium, Bulgaria, the Czech Republic, Germany, Greece, Hungary, Italy, Lithuania, Poland, Slovakia and Spain);

- an affidavit regarding the accuracy of the information provided.

The marketing authorization holder must also propose a maximum price in RON. This must be equal to or lower than the lowest price at which the medicine is listed as being sold in the countries of reference. If the medicine is not priced in any of these countries, the price at which it is sold in its country of origin may be used as a guide.[21,23]

In conclusion, we highlight the economic significance of medicinal and aromatic plants in Romania and their regulation within the European Union. Romania's strategy aims to foster the cultivation of these plants, particularly organic varieties, as a means of economic development, including agrotourism and ecologically supportive activities. The European Union's complex regulatory framework ensures quality and safety in the production and marketing of herbal products. Romania, as an EU member, has its regulatory framework, and the National Agency for Medicines and Medical Devices plays a pivotal role in authorizing and overseeing these products. We emphasize the importance of adhering to stringent regulatory standards and pharmacovigilance to maintain the quality and safety of medicinal and aromatic plant products, ultimately contributing to their responsible and sustainable management.

REFERENCES

1. Report on the regulations for the processing and marketing of aromatic and medicinal plants and the traditional way of using them, POSDRU Program Nr. contract POSDRU / 110 / 5.2 / G / 88727 Project code ID 88727 Project title: "Entrepreneurship in the aromatic and medicinal plants sector – alternative for sustainable rural development," www.slideshare.net/apostoltudor/baza-legislativa-cercetare-utilizare

2. Kroes B.H. 2014. The legal framework governing the quality of (traditional) herbal medicinal products in the European Union. *Journal of Ethnopharmacology* 158(Pt B):449–453. doi: 10.1016/j.jep.2014.07.044. Epub 2014 Jul 31. PMID: 25086408.

3. www.ema.europa.eu/en/committees/committee-herbal-medicinal-products-hmpc

4. Shah R.R., Raymond A.S. 2009. Regulation of Human Medicinal Products in the European Union. In *The Textbook of Pharmaceutical Medicine*, J.P. Griffin (Ed.). https://doi.org/10.1002/9781444317800.ch18

5. Directive 2001/83/EC of the European Parliament and of the Council of 6 November 2001 on the Community code relating to medicinal products for human use, *Special edition in Romanian*: Chapter 13 Volume 033 P. 3–64

6. Commission Directive 2003/63/EC of 25 June 2003 amending Directive 2001/83/EC of the European Parliament and of the Council on the Community code relating to medicinal products for human use (Text with EEA relevance), *Special edition in Romanian:* Chapter 13 Volume 039 P. 235–283.

7. Regulation (EU) 2017/745 of the European Parliament and of the Council of 5 April 2017 on medical devices, amending Directive 2001/83/EC, Regulation (EC) No 178/2002 and Regulation (EC) No 1223/2009 and repealing Council Directives 90/385/EEC and 93/42/EEC (Text with EEA relevance)

8. DECISION No. 5/07.03.2012 on approval of the Guideline on the Good Manufacturing Practice for Medicinal Products for human use, The Scientific Council of the National Agency for Medicines and Medical Devices (NAMMD), www.anm.ro/en/_/HCS/SCD%20 no.%205_07.03.2012%20and%20Annexes.pdf

9. The European regulatory system for medicines, A consistent approach to medicines regulation across the European Union, European Medicines Agency, UK, 2016, EMA/716925/2016 www.ema.europa.eu/en/documents/leaflet/european-regulatory-system-medicines-european-medicines-agency-consistent-approach-medicines_en.pdf

10. EudraBook V1 – May 2015 / EudraLex V30 – January 2015, Compendium of EU pharmaceutical law, Publication year: 2015, ISBN: 978-92-79-44434-0, DOI: 10.2772/557259, Catalogue number: NB-06-15-186-EN-E, https://ec.europa.eu/health/documents/eudr alex_en

11. Burton, A., Smith, M. 2015. Torkel Falkenberg, Building WHO's global Strategy for Traditional Medicine. *European Journal of Integrative Medicine*, Volume 7(1) P. 13–15, ISSN 1876-3820, https://doi.org/10.1016/j.eujim.2014.12.007

12. Directive 2004/24/EC of the European Parliament and of the Council of 31 March 2004 amending, as regards traditional herbal medicinal products, Directive 2001/83/EC on the Community code relating to medicinal products for human use, *OJ L 136, 30.4.2004, p. 85– 90, Special edition in Romanian:* Chapter 13 Volume 044 P. 167–172.

13. European Medicines Agency, Committee on Herbal Medicinal Products(HMPC), Guideline on Good Agricultural and Collection Practice (GACP) for Starting Materials of Herbal Origin, 2006, UK, www.ema.europa.eu/en/documents/scientific-guideline/guideline-good-agricultural-collection-practice-gacp-starting-materials-herbal-origin_en.pdf

14. Ekor M. 2014. The growing use of herbal medicines: issues relating to adverse reactions and challenges in monitoring safety. *Frontiers in pharmacology*, Volume 4 P. 177. https://doi.org/10.3389/fphar.2013.00177

15. The National Standardization Body – ASRO, www.asro.ro/

16. Regulation (EU) No 1025/2012 of the European Parliament and of the Council of 25 October 2012 on European standardisation, amending Council Directives 89/686/EEC and 93/15/EEC and Directives 94/9/EC, 94/25/EC, 95/16/EC, 97/23/EC, 98/34/EC, 2004/22/EC, 2007/23/EC, 2009/23/EC and 2009/105/EC of the European Parliament and of the Council, *OJ L 316, 14.11.2012, p. 12–33.* https://eur-lex.europa.eu/legal-content/EN/ALL/?uri=celex:32012R1025

17. Ministry of Agriculture and Rural Development, Medicinal and aromatic plants guide, www.madr.ro/en/field-crops/medicinal-and-aromatic-plants.html

18. National Agency for Medicines and Medical Devices of Romania, Medicinal Products for Human Use, Bucharest, Romania, www.anm.ro/despre-institutie/despre-noi/

19. Law No. 491 Of 18 November 2003 On Medicinal And Aromatic Plants, www.global-regulation.com/translation/romania/3074133/law-no.-491-of-18-november-2003-on-medicinal-and-aromatic-plants.html

20. Law no. 95/2006 on healthcare reform, as republished Updated based on amending regulatory acts published in the Official Gazette of Romania, Part 1, before 30 September 2016, www.anm.ro/en/_/LEGI%20ORDONANTE/Titlul%20XVIII_Med_2016_EN%20.pdf

21. Maravela, G., Popescu, A. 2016. Life Sciences: product regulation and liability in Romania, A structured guide to product regulation and liability laws in Romania, www.lexology.com/library/detail.aspx?g=5ebfc32f-1def-42ed-81f1-26913ff4f80b

22. Order of the Minister of Health no. 131on approval of Rules on authorisation of human medicinal product wholesalers, Good Distribution Practice certification and registration of brokers of medicinal products for human use, Issued by: The Ministry of Healthpublished in:the Official Gazette of Romania, Part I no. 108 of11 February2016, www.anm.ro/en/_/ORDINE/Order%20of%20the%20Minister%20of%20Health%20no_131_2016EN.pdf

23. European Commission, Health and Food Safety Directorate-General Health systems and products, Medicinal products, VOLUME 2A, Procedures for marketing authorization, Chapter 1, Marketing Authorisation, July 2019, https://ec.europa.eu/health/sites/health/files/files/eudralex/vol-2/vol2a_chap1_en.pdf

Index

A

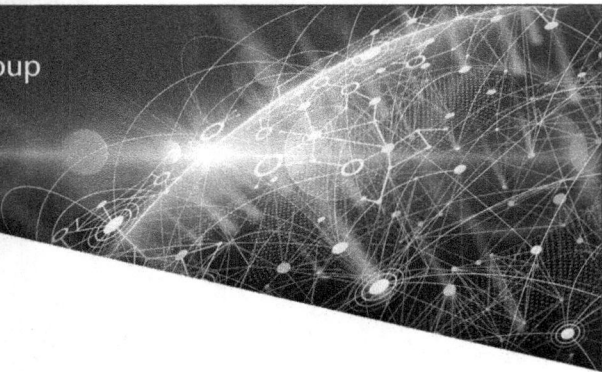

For Product Safety Concerns and Information please contact our EU
representative GPSR@taylorandfrancis.com
Taylor & Francis Verlag GmbH, Kaufingerstraße 24, 80331 München, Germany

www.ingramcontent.com/pod-product-compliance
Lightning Source LLC
Chambersburg PA
CBHW080121220326
41598CB00032B/4911

9 781032 219035